Energy Efficiency and Sustainable Development

Energy Efficiency and Sustainable Development

Editor: George Thomson

R CALLISTO REFERENCE

www.callistoreference.com

Callisto Reference,
118-35 Queens Blvd., Suite 400,
Forest Hills, NY 11375, USA

Visit us on the World Wide Web at:
www.callistoreference.com

ISBN: 978-1-63239-896-3 (Hardback)

The publisher's policy is to use permanent paper from mills that operate a sustainable forestry policy. Furthermore, the publisher ensures that the text paper and cover boards used have met acceptable environmental accreditation standards.

Trademark Notice: Registered trademark of products or corporate names are used only for explanation and identification without intent to infringe.

Printed in the United States of America.

Cataloging-in-Publication Data

Energy efficiency and sustainable development / edited by George Thomson.
   p. cm.
Includes bibliographical references and index.
ISBN 978-1-63239-896-3
1. Power resources. 2. Energy consumption--Environmental aspects. 3. Renewable energy sources. 4. Sustainable development. I. Thomson, George.
TJ163.2 .E54 2017
621.042--dc23

Table of Contents

Preface

The purpose of the book is to provide a glimpse into the dynamics and to present opinions and studies of some of the scientists engaged in the development of new ideas in the field from very different standpoints. This book will prove useful to students and researchers owing to its high content quality.

Energy efficiency is the process whereby the per unit consumption of electricity is greatly reduced. This book on energy efficiency and sustainable development discusses a variety of topics related to energy efficiency such as sustainable or green energy technology production, energy efficient building design and manufacture of efficient electric devices. The various studies that are constantly contributing towards advancing technologies and evolution of this field are examined in detail. This book includes some of the vital pieces of work being conducted across the world, on various topics related to energy efficiency. It will prove to be immensely beneficial to students and researchers in this field. For someone with an interest and eye for detail, this book covers the most significant topics in the field of energy efficiency and sustainable development.

At the end, I would like to appreciate all the efforts made by the authors in completing their chapters professionally. I express my deepest gratitude to all of them for contributing to this book by sharing their valuable works. A special thanks to my family and friends for their constant support in this journey.

Editor

Fe-Cu metastable material as a mesoporous layer for dye-sensitized solar cells

Abdul Hai Alami[1,2], Jehad Abed[2], Meera Almheiri[2], Afra Alketbi[2] & Camilia Aokal[2]

[1]Center for Advanced Materials Research, University of Sharjah, PO Box 27272 Sharjah, United Arab Emirates
[2]Sustainable and Renewable Energy Engineering Department, University of Sharjah, PO Box 27272 Sharjah, United Arab Emirates

Keywords
Dye-sensitized solar cells, mechanical alloying, mesoporous materials, metastable compounds

Correspondence
Abdul Hai Alami, Center for Advanced Materials Research, University of Sharjah, PO Box 27272 Sharjah, United Arab Emirates.
E-mail: aalalami@sharjah.ac.ae

Funding Information
No funding information provided.

Abstract

This study investigates the performance of dye-sensitized solar cells constructed with a Fe-Cu metastable material as the mesoporous layer on which a natural organic dye is applied. The synthesis of the Fe-Cu material is done via a high throughput process that produces nanosized particles from elemental metallic powders. Xanthophyll is singled out as the organic natural dye of choice among other dyes that were extracted, as it exhibited wider spectral absorptivity in terms of wavelength range and magnitude. Two compact solar cells were constructed and tested; one is a reference cell with a TiO_2 working electrode and the other with a Fe-Cu working electrode. The results show a better power conversion efficiency for the Fe-Cu-based solar cell 0.943% compared to 0.638% for the TiO_2, and the number of carriers in the former is found to be orders of magnitude higher than the latter (10^{19} vs. 10^{32}, respectively). A thorough optical, electrical, and thermal analysis of the Fe-Cu material is conducted and used to explain the obtained results.

Introduction

The dye-sensitized solar cells (DSSC) are formidable competitors to the industry-standard silicon-based solar cells due to their low cost, easy fabrication, and relatively high efficiencies, reaching up to 13% in porphyrin dye and cobalt (II/III) DSSC as reported by Grätzel group and others [1, 2]. Sensitizers are key cell components, critical to the widespread adoption of this technology. They possess excellent radiation absorbing properties in the visible wavelength range with their ability to mimic the light harvesting strategies found in nature to generate the required excitons. Many different sensitizers have been investigated for DSSCs applications, most of which use their carboxyl groups to enhance their attachment to the semiconductor [3]. The main types of sensitizers used nowadays are the metal complexes that use the metal to ligand (MTL) charge transfer phenomenon to increase photovoltaic performance, such as ruthenium complex sensitizers. It has been noted that DSSCs sensitized using pure organic sensitizers have higher absorption coefficients due to intramolecular π–π^* transitions. Their redox potentials, LUMO, and HUMO, are easily controlled for better performance and their ease of purification increases the chances of their acceptance in the market. Importantly, organic dyes have the availability aspect that metal complexes lack [4, 5]. Critical to achieving the reported efficiency levels, a proper choice of the nanostructured mesoporous substrate in order to achieve the desired dye molecules adsorption. This substrate is conventionally a metal oxide, usually a thin TiO_2 film, applied to the working electrode to maximize the surface area available for dye adsorption, optimize incident light harvesting, and enhance electrolyte diffusivity. The available molecular structures of TiO_2-containing mesoporous layers include: highly ordered nanorods with PCE of 2.9% [6, 7], spheres [8, 9], rice grain shapes and hollow fibers [10]; all pertaining to nanoscale dimensions. Some references in the literature have reported on the utilization of different materials to replace or be used in conjunction with the mesoporous semiconducting layer, for example, graphene has been used as the transparent conductive photoelectrode to great thermal and chemical stability [11, 12]. The use of organic perovskite electrodes is also a promising third generation solar cell technology employing lead or tin halide-based materials as the light harvesting electrodes, with reported efficiencies of around 20.1% in 2015, according to the National Renewable Energy Laboratory (NREL) [13].

The Fe-Cu intermetallic phase is attractive for many applications due to its high strength and traditionally attractive thermal and electrical properties [14–19]. But similar to other metastable intermetallic systems, for example, Ni-Ag, Cu-V, and Co-Cu, its synthesis suffers from the main drawback of limited immiscibility of its components as solid solutions due to their positive energy of mixing (around 13 kJ/mol for the Fe-Cu system) [20, 21]. Thus they will not form intermetallic compounds and will have negligible mutual solid solubility in equilibrium at temperatures below 700°C [22–26]. One effective and easy method to synthesize the Fe-Cu system with no conventional energy requirement in the form of applied heat or voltage is mechanical alloying (MA), which has the advantages of low-temperature processing, easy control of compositions, the production of relatively large amount of samples [20] and results in a significant extension of mutual solubility of the elements relative to the equilibrium values, which can be observed through X-ray diffraction patterns. This method involves ball-milling powders of the pure constituents to obtain the sought solid solutions. In this process, the coherent lattices of the pure metals undergo simultaneous shear induced deformation and thermal interdiffusion, with the resulting composition being determined by the equilibrium between the mechanically driven alloying and the diffusion-controlled decomposition [27–30]. It was found that low-energy ball milling of FCC and BCC metals leads to a refinement of the crystallite size to the nanometer scale [31–34]. With a work difference (ΔE_w) of around 0.43 eV, the Fe-Cu alloy system has a high intrinsic absorption coefficient that further optimizes its optical absorptance [35]. Also, any observed roughness of the microstructure allows the interreflection of incident irradiation in the UV–Vis range that reduces reflection and scattering losses [36], and decreases the impedance between space and the absorber, which also leads to better absorbance properties [37].

In this work an alternative to TiO_2 as a mesoporous substrate of the working electrode of a DSSC in the form of a Fe_{50}-Cu_{50} binary intermetallic phase is proposed and tested. The structural and optical properties of the obtained alloy are compared with those of TiO_2, and in addition, two DSSCs are constructed to evaluate the power, overall PCE, and fill factor (FF) for tested cells. The natural organic dye sensitizers (chlorophyll, anthocyanin, and xanthophyll) are extracted and processed locally and their absorptivities compared to choose the one with the widest spectral absorptivity in the UV–Vis wavelength. The Fe-Cu metastable alloy is produced via mechanical alloying, which is a facile and economic process that utilizes high-energy ball milling. It is an old but effective technique with high throughput, which will enhance the power-to-price ratio, increasing the appeal of DSSC in general as a third generation photovoltaic option. A comprehensive thermal, optical, and electrical investigation of the properties of the proposed Fe-Cu alloy is presented.

Experimental

Synthesis

The synthesis by MA takes place in a Retsch PM 100 planetary ball mill in a 25 mL stainless steel grinding bowl to mechanically alloy a starting amount of 9 g of high-purity copper (<425 μm, 99.5%) and iron (≥99%) powders, used as received from the supplier (Sigma-Aldrich, http://www.sigmaaldrich.com/united-states.html [United States of America]). A target composition 50:50 of Fe-Cu (% wt.) is used at a controlled milling speed of 600 rpm. Six 10-mm stainless-steel balls are used, making the filling ratio within the bowl 5:1. Milling is carried out for an hour at a time, pausing afterwards to cool the equipment and take a few milligrams of the powder for further characterization and testing. The run is to be terminated once microstructural changes become small, which in the present case took place after 6 h.

Microstructural analysis via SEM-EDX and XRD

The powder X-ray diffraction (XRD) patterns, plotted for five powder samples collected at a 2-h interval for 8 h, provide an insight into extent and progress of crystallization and the composition and grain structure of the developing solid solution. The X-ray patterns are recorded in the 2θ geometry between 40 and 90° at 0.02° 2θ/sec with a Bruker D8 Advance DaVinci multipurpose X-ray diffractometer with Cu Kα radiation operating at λ = 1.5406 Å, 40 kV tube voltage, and 40 mA current. The microstructural results are used to calculate important quantities such as the lattice parameter from Cohen's method, grain size by the Full-width half-max (FWHM) analysis with the Hall–Williamson method and Bragg's formula. Fused pieces of the material collected after 6 h milling time are examined under a scanning electron microscope (SEM) and the coupled energy dispersive X-ray spectrometer (EDX). The SEM is a VEGA3 XM by TESCAN, operating at 5 kV, whereas the EDX analysis is conducted with both map and point modes at the same operating voltage; the former was acquired during 3 min, whereas the latter was from four different spots of the sample during 30 sec live time.

Thermal analysis

Differential scanning calorimetry (DSC) is performed on the milled powders to provide insight on the energy of formation and mixing of the resulting compounds. The stability of these materials is due to the balance between the effectiveness of the MA process and thermal decomposition at high temperatures. For an endothermic reaction, heat flow indicates the phase shift within the solid solution (a peritectoid reaction). The calorimeter used is a Q20 from TA Instruments, running on 120 V_{ac}, 47–63 Hz, 500 W (4.5 amps) and equipped with a liquid nitrogen cooling system (LNCS) that allows automatic and continuous temperature control within a full range of −180°C to 550°C. A few milligrams of as-is and 2, 4, 6, and 8 h powders is encapsulated in an aluminum pan, and an empty reference pan sit on a thermoelectric disk surrounded by a furnace. As the temperature of the furnace is changed, heat is transferred to the sample and reference through the thermoelectric disk. The differential heat flow to the sample and reference is measured by area thermocouples. The phase formation, total enthalpy and heat flow through 0–500°C temperatures are examined for the current test.

Optical (spectroscopic) analysis

Spectral measurements of absorption in the ultraviolet, visible, and near-infrared (UV–Vis–NIR) regions were carried out on all the powders (as-is and 2, 4, 6, and 8 h milling time) with an Ocean Optics HR2000 high-resolution spectrometer. The HR2000 has a 300 lines per mm diffraction grating, 10 μm entrance slit, a Sony ILX511 2048-Pixel element linear CCD array detector, and is operating in the effective wavelengths range 300–1100 nm. The spectrometer is connected to a fiber optic reflection probe R200-7-SR, 2-m long, and of a 200-μm-core diameter. The reflection probe consists of a tight bundle of seven optical fibers in a stainless steel ferrule with six illuminating fibers around one axial read fiber, fixed at ~4 mm from the sample where losses due to scattering is assumed to be negligible, and all diffuse reflectance is collected at the probe. The source end of the reflection probe is connected to a tungsten halogen light source (Ocean Optics LS-1-LL). A reference surface in the form of a reflection standard (B0071519) is used to store baseline absorptance (0%) spectra to facilitate comparison between the various compositions. The integration time

was set to 30 msec to contain the intensity of the highest acquired peak. The recorded time-resolved spectra were averaged over 10 readings to increase the signal-to-noise ratio.

Electrical impedance and Mott–Schottky analysis

The impedance of the resulting Fe-Cu material is investigated and compared with values obtained for TiO_2 by virtue of the four-probe method impedance spectroscopy. The four probes, arranged as seen in Figure 1A, cover an area of 2 cm^2 and are connected to a Biologic

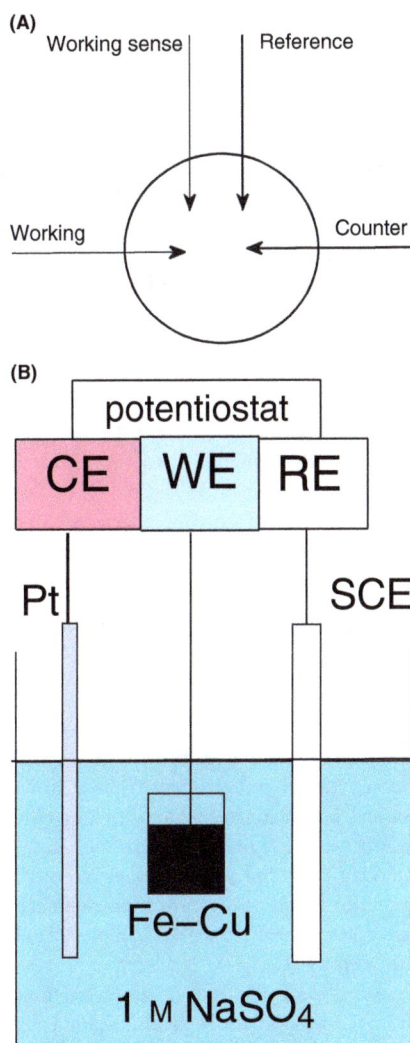

Figure 1. (A) Four-electrode arrangement for impedance spectroscopy and (B) Mott–Schottky cell setup.

VSP-300 potentiostat, providing impedance spectroscopy (10 μHz to 7 MHz) ± 12 V compliance, an automatic current range from 1 μA up to 500 mA (seven decades) and a potential resolution down to 750 nV. Impedance parameters were determined by fitting of impedance spectra using Z-view software and generating the pertinent Nyquist plot. The Fe-Cu material and the TiO_2 were deposited on respective glass substrates by spreading an emulsion of each in glycerol ($C_3H_8O_3$) and subsequently sintering the setup at 450°C for 30 min. It is noted that no phase change was observed afterwards in the postsintered XRD results. To investigate the interaction between the Fe-Cu layer and the electrolyte, a Mott–Schottky analysis was performed with the mesoporous FeCu deposited on ITO glass, which acts as the working electrode and immersed in an aqueous solution of 1 mol/L Na_2SO_4 (pH 8.5) with respect to a standard calomel (SCE) reference electrode and a Pt counter electrode [38]. The three-electrode arrangement for the electrochemical cell is shown in Figure 1B and is connected to a Biologic VSP-300 potentiostat providing an applied voltage range between −1 V and 1 V vs. SCE (20 potential steps recorded every 0.1 sec with waiting interval of 5 sec/step) and a potential resolution of 50 μV. The frequency range was from 0.1 Hz to 200 kHz (10 points per decade) with sinusoidal amplitude of 25 mV.

Surface area determination

The surface area of the resulting Fe-Cu material is determined by the BET method in an Autosorb iQ from Quantachrome USA machine, where 1 g of Fe-Cu and 1 g of TiO_2 (for comparison) are degassed for 16 h at 300°C then tested with adsorbent nitrogen at 77.4 K.

Natural organic dye extraction and absorptivity testing

The most commonly used natural organic dyes in literature are chlorophyll, anthocyanin, and xanthophyll. These dyes are extracted from plants rich in them, like spinach, cabbage, and red leaves, respectively. These plants are collected, washed with deionized water, dried by tissue paper and trimmed using scissors to remove the stalk, mid rib, and veins. A container full of sliced leaves is cooled downed by placement in liquid nitrogen to minimize degradation. Immediately after, leaves are soaked completely in a beaker filled with acetone (C_3H_6O). The acetone-leaves mixture is stirred using a stirring rod as the extraction process carries on overnight for 24 h in dim light. Later, liquid dye extracts are filtered through a funnel equipped with small pore filter paper and placed in a centrifuge at 10,000 rpm for 10 min. In the case of chlorophyll, the

beaker is covered with punched aluminum foil to allow natural vaporization of the acetone carrier for 48 h in dim light. On the other hand, each of the anthocyanin and xanthophyll are placed in a rotary evaporator to rotate and heat the filtrates up to the boiling point of extraction solution. The final products – chlorophyll, anthocyanin and xanthophyll extracts – are stored in a refrigerator to minimize any light-induced degradation [39–41].

In order to perform the absorptivity test, the dyes are diluted by distilled water, which is taken as the absorptivity baseline. The absorptivity measurement test is conducted using a PerkinElmer EZ301 spectrometer in ultraviolet–visible (UV-VIS) range (250 and 750 nm).

Solar cell fabrication and testing

To compare the effectiveness of utilizing the Fe-Cu material as a mesoporous material in DSSCs, two cells were constructed, one having TiO_2 as the photoelectrode, whereas the other had Fe-Cu. For depositing the former, the process starts with the preparation of TiO_2 paste by adding 1.5 mL of weak acetic acid (CH3COOH) to 1 g of titanium dioxide nanopowder (brookite nanopowder, <100 nm, 99.99%, used as received from Sigma-Aldrich) in a mortar while vigorously grinding by the pestle and then glycerol is added as a surfactant. The prepared homogenous paste is spread as a thin layer by employing the doctor blading technique on ITO-coated glass (15–25 Ω/sq., used as received from Sigma-Aldrich). The photoelectrode is then sintered by heating in a furnace at 450°C in static air for 30 min to remove solvents, surfactants, and any other organic materials.

As for the Fe-Cu photoelectrode, an identical process is applied, but instead of using acetic acid, glycerol is used to evenly spread the homogeneous layer on the ITO glass. The assembled cell is shown in Figure 2.

The electrolyte used is the classic I^-/I_3^- redox electrolyte prepared from 127 mg iodine crystals, 830 mg potassium iodide, and 10 mL ethylene glycol are mixed thoroughly until they are completely dissolved. This electrolyte will be the diffusion medium for ionic species between the working photoelectrode prepared above and the counter electrode made up of pure copper with carbon deposits on an area of 2.5 × 2.5 cm to enhance the surface area and consequently the conductivity. The operation mechanism (injection of electrons and diffusion of ionic species) of DSSCs is well explained and documented, especially for the I^-/I_3^- redox electrolyte [40].

The testing of the solar cell to measure its power conversion efficiency, fill factor, and IV characteristics are conducted using a solar simulator Xenon Arc Lamps setup, capable of providing irradiance between 0.1 and 1 suns (up to 1367 W/m²) is used to provide a constant

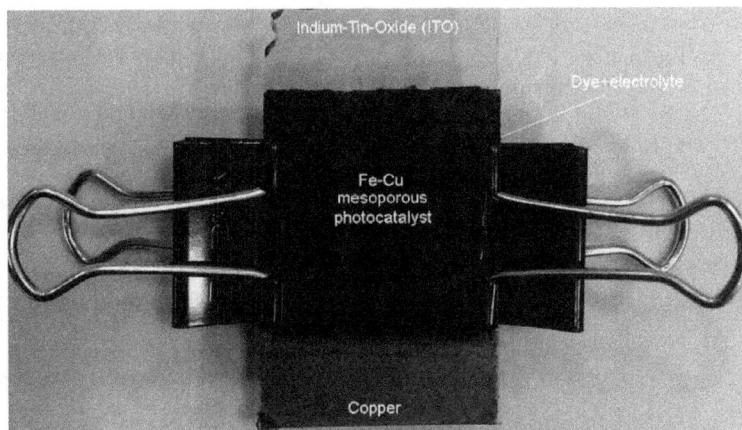

Figure 2. Assembled Fe-Cu sensitized solar cell.

irradiation on the cells, and was kept at a vertical distance of 30 cm at 1.5 AM at 25°C. The cell characterization is conducted using VSP-300 potentiostat from Biologic with a potential resolution of 1 μV and a control voltage of $\pm$10 V up to $\pm$48 V. The voltage was varied at a rate of 10 mV/sec and the current recorded at each point until the measured voltage reached the open circuit voltage, V_{OC}. A PV cell analysis software built into the potentiostat is also used to determine the power, efficiency, fill-factor (FF), and also the Nyquist plots of impedance for each cell configuration using the Z-fit postprocessor.

Results and Discussion

SEM and EDS

The SEM micrograph of the Fe-Cu system is shown in Figure 3A for a few milligrams removed after 6 h of milling. There is obvious presence of the FCC phase caused by the interdiffusion of Fe and Cu, resulting in different grain size distribution from a few nanometers to around 100 nm. Upon further magnification, Figure 3B shows a homogeneous phase with the apparent shearing effect from milling, which produces a lamellar structure [24]. An EDS analysis at two points (1) and (2) indicated on the figure shows a composition of 53% Fe and 47% Cu (%wt.) as evidence of successful interdiffusion and the formation of the new single phase. This is in contrast to other synthesis methods involving casting [21], deposition or electron-beam forming that reported severely segregated phases when the material is cooled below 700°C.

The lattice strain energy ΔU_{strain} can be calculated from the relation $\Delta U_{strain} = (\bar{E}\delta)(2\bar{r})^2\bar{d}$ where $\bar{E}$ is the average Young modulus for Cu and Fe, $\bar{d}$ the average

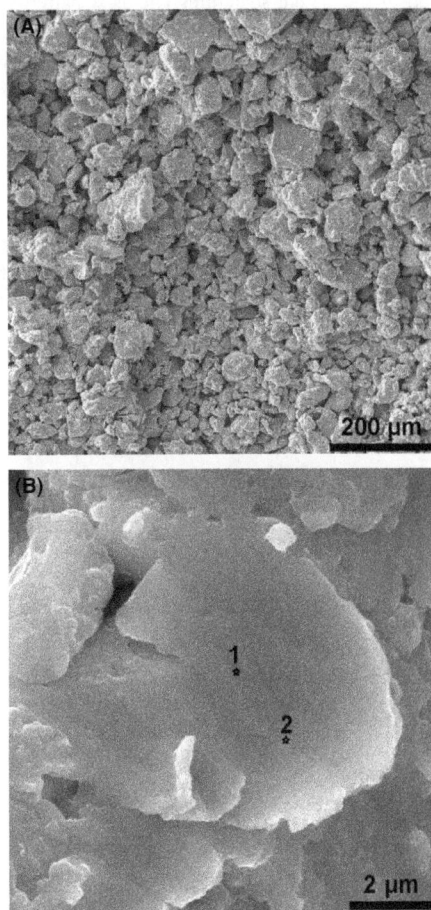

Figure 3. SEM photomicrographs of the resulting microstructure at (A) 250 × magnification, showing the granular microstructure, and (B) 20 k × magnification, showing homogeneous phases of Fe and Cu, with EDS test locations indicated at 1 and 2.

displacement of atoms ($\bar{d} = \delta\bar{r}$) and lattice distortion, δ, is calculated by using the equation $\delta = \sqrt{\sum_{i=1}^{n} X_i (1 - \frac{r_i}{\bar{r}})^2}$, where X_i is the fraction of the ith component (n here equals 2) and r_i and $\bar{r}$ are the ith and the average atom radii, respectively [42]. The lattice distortion energy is found to be 8.61 kJ/mol. Although the relations are non-linear and highly dependent on the Cu concentration in the Fe-Cu material, this value is still orders of magnitudes higher compared to the martensitic transformation in iron system (around 751 J/mol) that will form a homogenous single martensite phase with a high density of lattice defects [43].

XRD

The result of XRD analysis is shown in Figure 4 for the as-is and 2, 4, 6, and 8 h milling time, to follow the evolution of the crystal structure. The as-is patterns show intense sharp peaks of elemental iron and copper are slightly shifted towards lower angles, which are consistent with the similar atomic sizes of Fe and Cu present in the same mixture [24], and the recorded signals are strong with little noise. It is interesting to observe these peaks widening and shifting further as milling proceeds due to the interdiffusion of Fe and Cu atoms caused by the shearing action and the friction-induced temperature increase, promoting larger grain sizes. After 6 h, a single phase is observed in which Fe atoms are diffused in the Cu matrix, or an FCC Cu(Fe) solid solution, as corroborated by the results from SEM analysis in Figure 4B. These results are also consistent with previous reports [24, 44] and indicate the successful formation of the desired homogenous phase.

The grain size change with milling time is depicted in Figure 5 calculated from Bragg's law and Scherrer formula. It is noted from the figure that the grain size decreases rapidly as milling time increases, then it almost plateaus around 4–8 h, which is consistent with trends reported elsewhere [42, 44, 45].

Thermal analysis results

The results obtained from the DSC are shown in Figure 6 for all milling times and also for the as-is specimen. While the latter shows no phase reaction and its thermal response line remains flat, the rest of the samples show a trend that indicates more thermal stability as milling time increases. The results for 6 and 8 h show the least difference between the obtained peaks, indicating a limiting time for the milling process, which can be correlated with the XRD patterns of Figure 4. The results are inverted to show a positive scale and are in line with those found in previous work [18]. The peaks also occur at successively higher temperatures, indicating better phase stability.

Spectroscopy

The evolution of the microstructure with milling has impacted the optical absorptivity of the material, as can be

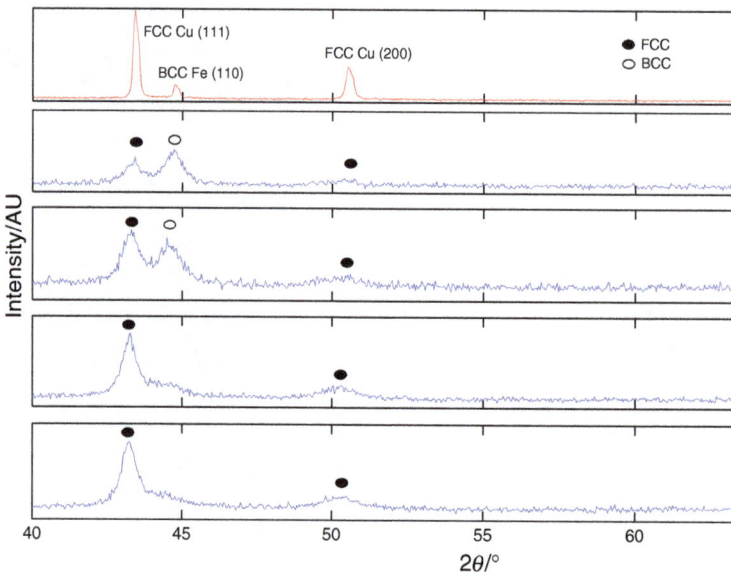

Figure 4. XRD plots for progression of milling times.

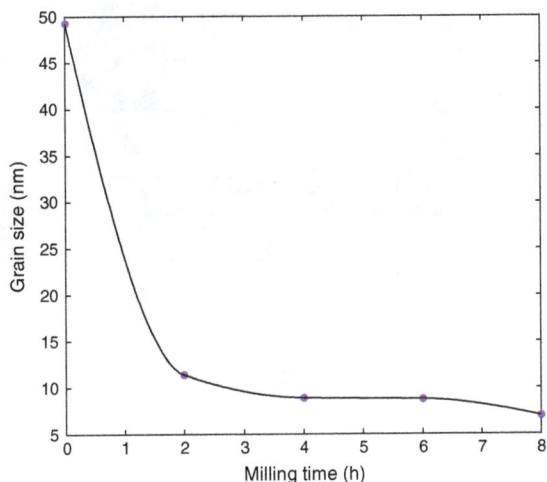

Figure 5. Grain size variation with milling time.

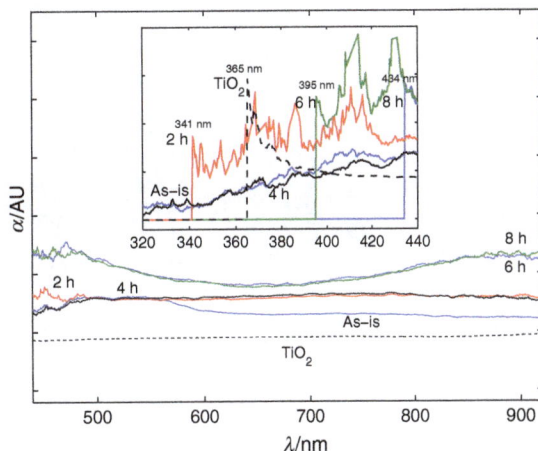

Figure 7. Absorptivity results at different milling times with TiO_2 superimposed in dotted line.

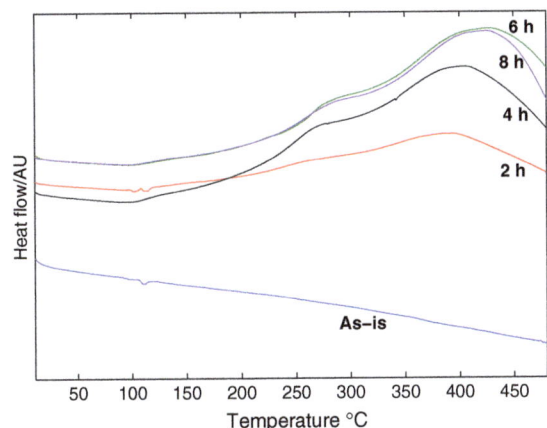

Figure 6. Differential scanning calorimetry heat flow plots versus milling times.

seen in Figure 7. The original powder shows a slight increase in absorptivity with respect to the mirror reference up to 500 nm, at which it plateaus until 550 nm, after which absorptivity decreases asymptotically towards 900 nm. As for milling times of 2, 6, and 8 h, the onset of absorptivity is seen to be an abrupt step increase at progressive wavelengths of 341, 396, and 434 nm, respectively. Another interesting observation from the curves can be made in conjunction with the XRD plots of Figure 3 for the 2–4 h on one hand and 6–8 h materials on the other. The former set shows a virtually constant absorptivity across the tested wavelength of Vis–NIR (450–900 nm) due to the existence of both FCC and BCC structures, whereas the latter start with a high onset value,

then assumes a flat parabolic shape with an apex at 650 nm, thought to be due to an all-FCC microstructure. Longer milling times of 6 and 8 h, exhibit a 63% and 81%, enhanced absorptivity, respectively, compared to as-is powder if the areas under the absorptivity curves are integrated from 450 to 900 nm. This enhancement is due in part to the roughness in microstructure that augments the probability of absorbing the diffracted and trapped light beams by adjacent grains. Roos et al. [46–48] have reported similar trends of optical properties of copper oxide thin films prepared by thermal and chemical oxidations and plotted against processing time, and the results are in general agreement of what is reported in this work.

It is also useful to compare the obtained optical properties of the prepared materials against TiO_2, a known photo catalyst that is used extensively in DSSCs. Thus, the optical absorptivity in the UV–Vis–NIR (280–900 nm) range was measured for an as-received TiO_2 powder (Sigma-Aldrich) and shown as a dashed line in Figure 6. The curve shows standard optical characteristics of the TiO_2 powder as reported in literature [49], where negligible absorptivity below 365 nm is replaced with a sharp increase in absorptivity followed by an exponential decay right before the 400 nm mark where it levels off as Figure 6 depicts. The advantage of using the suggested mesoporous Fe-Cu material is thus twofold, one is in the tuning of absorption response to higher wavelengths that can be well defended in situations where target solar cells are to be operated indoors or at different incident radiation intensities, and the other benefit is the obvious enhanced absorptivity magnitudes over the rest of the Vis–NIR spectrum which has many benefits for both solar thermal and solar photovoltaic applications.

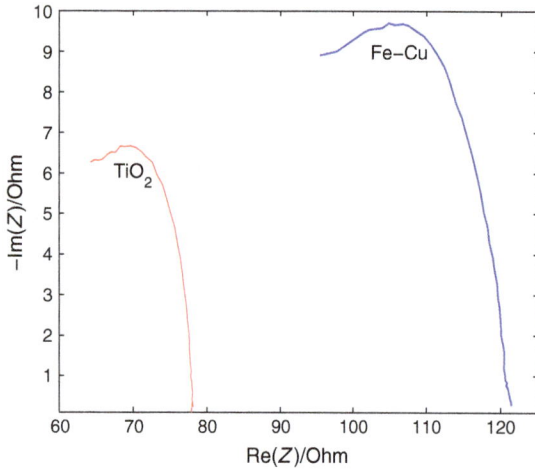

Figure 8. Impedance spectroscopy of 8 h milled Fe-Cu versus TiO_2.

Figure 9. (A) SEM of the Fe-Cu photoanode on ITO with the EDX map for (B) Fe, (C) Cu, and (D) oxygen; being only 11% wt. of the present components.

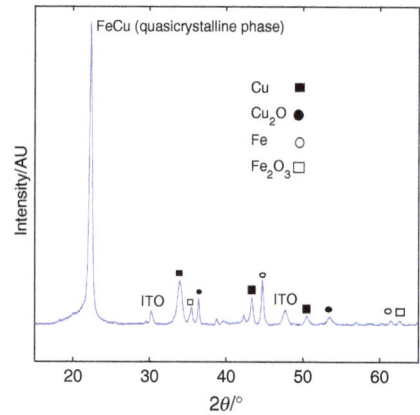

Electrical impedance

The electrical impedance of a thin film of an 8 h milled Fe-Cu versus TiO_2, each deposited on a glass substrate is shown in the Nyquist plot of Figure 8.

Compared with the near zero real impedance of the as-is starting Fe-Cu powder, the 8 h milling time displayed a large impedance value (around 120 Ω), larger than the one measured for titania (around 78 Ω) as seen in the figure, suggesting a fundamental change in the electric behavior of the new Fe-Cu composite compared to the as-is starting mix, akin to the thermal conductivity of an alloy being substantially less than that of its pure components. This result is important, especially that the traces of oxygen observed in the EDX analysis of the components indicates that no oxides were formed from either iron or copper as seen in the mapping of Figure 9, showing that oxygen as only 11% of the elements present, with Fe, Cu, and I as 42%, 43%, and 4%, respectively.

An XRD testing is also performed on the sintered Fe-Cu on ITO, and the results show a significant peak belonging to the FeCu FCC phase that is sharper than the one seen in the original XRD due to sintering on glass at 400°C and a shift towards $2\theta = 21°$, with minor peaks belonging mostly to elemental Fe and Cu that have been segregated from the unsintered mixture. This has resulted in using up the available oxygen in the form of Cu_2O and Fe_2O_3, with the former being more pronounced and is an intermediate phase of copper oxide that is less stable chemically and structurally than CuO [50]. From the XRD plot of Figure 10, the Fe-Cu materials shows good stability after the 50 cycles of solar cell testing while being in contact with the dye and electrolyte, exhibiting what is believed to be mild oxidation of the surface of the material occupied mostly by the Cu

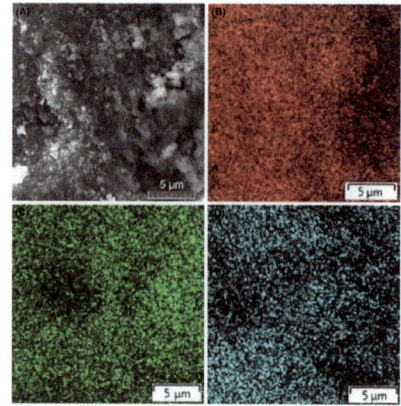

Figure 10. XRD of the Fe-Cu material deposited on ITO glass after solar testing.

molecules due to the presence of the hydroxyl (OH^-) species that caused the following pathway to take place [51]:

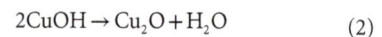

$$Cu + OH^- \rightarrow CuOH \tag{1}$$

$$2CuOH \rightarrow Cu_2O + H_2O \tag{2}$$

where the copper oxide film is gradually formed on the outer surface of the Fe-Cu material by the adsorption of OH^-, followed by the dehydration of copper hydroxide into Cu_2O. This is another indication of the formation of the FCC structure with Fe atoms being forced into the structure of the Cu grains, whereas the Fe_2O_3 oxidation takes place on the Fe atoms that diffused out and were segregated from the Fe-Cu material due to the annealing temperature [51–56].

The semiconductive behavior of the resulting material paves the way for using it in place of the more expensive titania that is apparently an inferior absorber in the useful solar spectrum as clearly seen in Figure 7. Since the Fe-Cu material exhibits this semiconductive behavior, an estimation of the resulting bandgap is shown in Figure 11. The band gap is determined from the traditional Tauc relationship; $\alpha.h\nu = A.(h.\nu-E_g)^n$, by plotting $(\alpha.h.\nu)^{1/n}$ versus the photoenergy h.v, where α is the absorption coefficient, A is the edge width parameter, E_g is the optical band gap value, and n is a constant dependent on the nature of the transition ($n = 1/2$ for a direct allowed transition and 2 for an indirect transition) [47, 48]. An extrapolation of the linear region of the plot for when the ordinate equals to zero, gives an absorption edge energy that corresponds to the value of the optical band gap E_g. With n being 1/2 for best fitting, the results in Figure 11 indicate direct transitions in Fe-Cu material with milling time. According to the figure, band gap results start to appear after 6 h of milling as elements diffusion becomes pronounced and the micrsostructure shifts to metastability, which is also inferred from the XRD plots of Figure 4. The obtained optical band gap values is 1.8 eV.

The semiconductive behavior of quasicrystalline materials has been observed by other researchers; some correlating the change in behavior with the increased role of disorder in the aperiodicity that has a profound effect on the metal-insulator transition [57], and others have reported a 2.2 eV bandgap value for the Fe-Cu solid solution and attributed it to a contribution similar to d transitions in the noble metal and follows the direct transitions from a virtual bound state to the Fermi level [58, 59]. In the particular case of quasicrystalline materials produced by mechanical alloying, the semiconductive behavior is generally attributed to the enhanced interdiffusion of elemental components that creates supersaturated structures that

leads to a metallic-covalent bonding conversion, and indeed a bandgap is reported by Takagiwa and Kimura [60] for an aluminum-based quasicrystalline material, and by our group [56].

Mesoporous media surface area

The results obtained from the BET testing show that the surface area of the Fe-Cu material is 0.78 m²/g, whereas that of TiO_2 equals 41.1 m²/g. While the fit for the volume adsorbed versus the differential pressure applied is strictly linear for TiO_2 ($R = 0.9995$), the variation in grains sizes for the Fe-Cu specimen as seen in the SEM of Figure 9 has caused inconsistencies in its adsorption as shown in Figure 12 and the fitted data shows a deviation from the linear behavior ($R = 0.867$).

Dye absorptivity results

The results obtained from testing chlorophyll, anthocyanin, and xanthophyll are shown in Figure 13. The spectral absorptivity in the UV–VIS range (250–750 nm) for each dye shows a high onset of absorption at the UV spectral range, compared to the distilled water reference. The three curves show a step fall after around 350, 400, and 450 nm for anthocyanin, chlorophyll, and xanthophyll, respectively. Xanthophyll exhibits superior absorptivity behavior, especially in the visible range (400–700 nm) where the drop in absorptivity is seen to be less abrupt than its two counterparts, never reaching zero even at the extreme end of the sampling spectrum (750 nm). This behavior, along with the broad spectral range of absorptivity of xanthophyll makes it an ideal choice for sensitizing the mesoporous materials of the experimental cells.

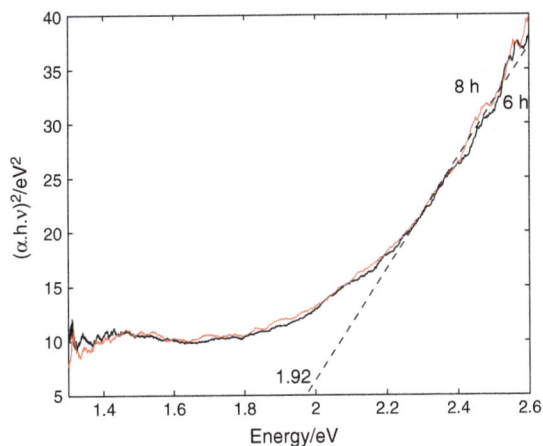

Figure 11. Energy bandgap of the resulting Fe-Cu material.

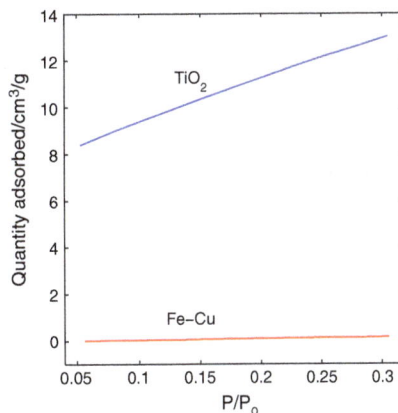

Figure 12. BET surface area for Fe-Cu versus TiO_2.

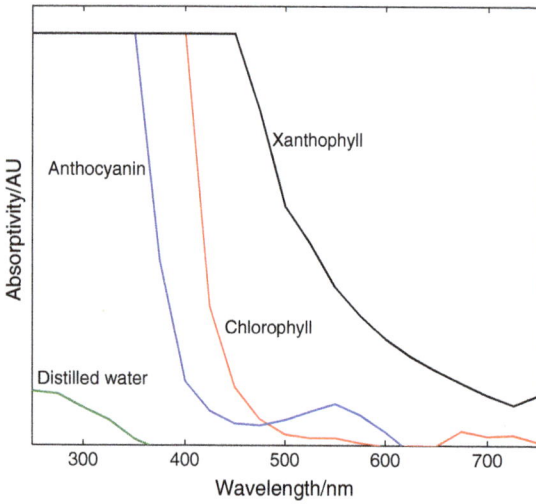

Figure 13. Absorptivity test of various natural organic dyes against distilled water.

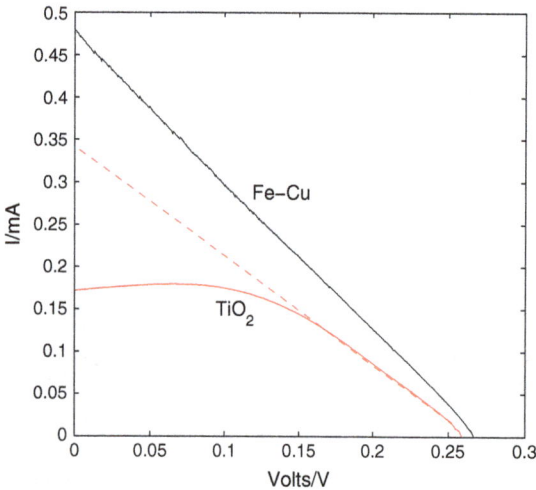

Figure 14. IV characteristic curves for TiO$_2$ and Fe-Cu xanthophyll-sensitized solar cells.

Solar cell test results

The results of the characteristics test for the manufactured DSSC are shown in Figure 14. It is noted that there is a linear decrease in the voltage of the Fe-Cu cell in proportion to the generated current, which is usually an indication of a large series resistance in multicrystalline silicon solar cells. The series resistance in solar cells is mostly due to contact resistance, and hence the recommendations to mitigate its effects are numerous, like changing the concentration of the electrolyte, the counter

Table 1. Comparison of parameters for the two DSSCs experimental cells.

Parameter	TiO$_2$-based cell	Fe-Cu-based cell
$I_{sc,}$ [mA]	0.173	0.53
$E_{oc,}$ [V]	0.248	0.269
FF (P_{max}/P_{Theo})[%]	48.4	25.0
η, [%]	0.638	0.943

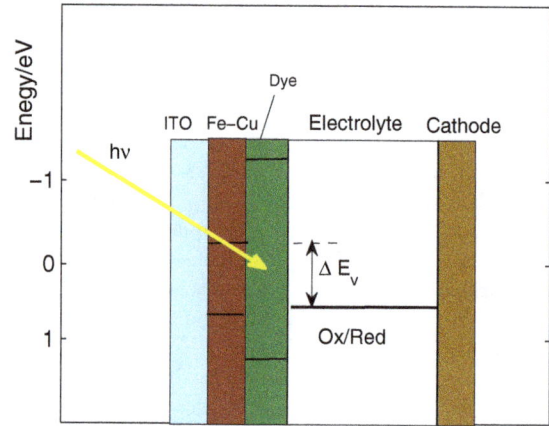

Figure 15. Illustration of energy levels diagram of a DSSC based on Fe-Cu mesoporous material.

electrode or the clips that connect the cell to the load. It is also noted from the curve that the series resistance in both have similar values (as shown from the line tangent to the TiO$_2$ IV curve in Figure 14).

In general, the Fe–Cu-based solar cell exhibits better performance when other characteristics are calculated, such as the fill factor, power, and efficiency. A summary of these parameters is shown in Table 1.

An explanation of the measured enhancement over the TiO$_2$-based cell is presented with reference to the energy levels diagram in Figure 10 that schematically shows the Fe-Cu bandgap to be around 1.8 eV (±0.9 eV at each side of the 0 eV mark), where the open circuit voltage, measured as ΔE_v is the potential difference between the dye/electrolyte interface and the upper end of the mesoporous material energy gap. The conduction band for the Fe-Cu has to be above 0 V versus normal hydrogen electrode (NHE) for the photovoltaic effect to take place (Fig. 15).

To determine the cell potential behavior at the Fe-Cu/electrolyte interface, a Mott–Schottky (MS) analysis is conducted using the potentiostat at frequencies 63, 79, and 100 Hz. The plots are shown in Figure 16. The MS analysis probes the depletion capacitance at a Schottky or p-n junction which is determined by the width of the

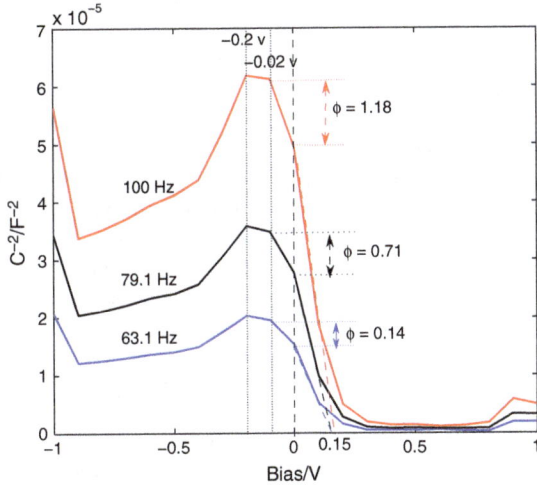

Figure 16. Mott–Schottky diagrams for Fe-Cu-based photoelectrode.

bias-dependent depletion region, hence the depletion capacitance, C, is also bias dependent and can be expressed as follows:

$$\frac{1}{C^2} = \frac{2(V_{bi} - V)}{A^2 q \epsilon \epsilon_0 N} \qquad (3)$$

where V is the applied bias, A is the device area, q is the elementary charge, ϵ is the compound dielectric constant, and ϵ_0 is the permittivity of free space. The built-in bias and doping density are then found by fitting equation 1 to the linear portion of the C^{-2} versus bias voltage plot, as shown by the converging dotted lines in Figure 16. The intersection of the line equals the flat band potential, whereas the slope of the line is used to estimate the carrier density, N, as the slope equals $2/(\epsilon \cdot \epsilon_0 \cdot q \cdot N)$. The Helmholtz capacitance on the electrolyte side of the interface is large enough that it is negligible since the observed capacitance, C, is given by $1/C = 1/C_H + 1/C_{SC}$, where C_H, and C_{SC}, are the capacitances of the Helmholtz layer and the semiconductor, respectively.

The MS plots reveal that the Fe-Cu behaves like a p-type semiconductor (SC), contrary to the n-type TiO$_2$. When the p-SC is in contact with an electrolyte solution, the electrons in the SC are transported to the vacant level of the electrolyte solution, which makes the SC positively charged and the electrolyte solution negatively charged. Upon irradiation, at the interface between the Fe-Cu layer and the electrolyte solution, the energy of which is larger than the bandgap (E_g) of the semiconductor (for Fe-Cu $E_g \sim 0.1.8$ eV), excitation of electrons from the valence band to the conduction takes place leaving holes in the valence band. The electron and the hole form the short-lived excitons. The space charge

layer, φ, which is measured as the difference between the C^{-2} value at the inflection point before crossing the zero-voltage bias (see Figure 16) for the three frequency values of 63 Hz, 79 Hz, and 100 Hz is measured as 0.14, 0.71, and 1.18 F^{-2}, respectively. The physical significance of the space charge layer is that a hole can migrate towards the SC interface, and the electron can migrate toward the inside of the SC bulk so that the chances of their recombination are smaller. When an electron donor is present in the contacted liquid phase, the holes can oxidize the donor in the electrolyte, and the electrons are transported first to the ITO glass conductive layer through Fe-Cu grain boundaries and then to the counter electrode, reducing electron acceptance there. From Figure 14, the flat band potential of the Fe-Cu layer is found to be 0.15 V compared with that of TiO$_2$ at −0.16 V [52]. To calculate the carrier density, whereas the relative permittivity of Fe-Cu is chosen to be in the order of magnitude of 10^3 [53], ϵ_0 the vacuum permittivity = 8.854×10^{-12} F/m, and q the elementary electric charge = 1.602×10^{-19} C; then the carrier density per meter cube is estimated to be 4.27×10^{32}/cm^3. This value is higher than the value reported for TiO$_2$ (6.96×10^{19}/cm^3) in literature [51], proving that the quasicrystalline Fe-Cu thin film can function as an excellent electron conducting material under irradiation conditions when a strong electron donor is present in the electrolyte. This explains the bigger current values in the IV curves of Figure 12 when Fe-Cu is used as a mesoporous layer instead of TiO$_2$, and with the facile and cheap synthesis procedure of Fe-Cu, the latter is considered a real competitor in the manufacture of dye-sensitized solar cells to drive the cost-per-watt even lower [54–56].

Conclusion

In this study, an economical and facile method for the production of a mesoporous material made from Fe-Cu metastable alloy system is investigated for DSSC photoelectrodes. High-energy ball milling, a mechanical alloying technique, was used to synthesize the alloy and the resulting microstructure was examined using SEM-EDS and XRD at 2 h intervals starting from the as-is powder up to the final duration of 8 h. The resulting alloy is a homogenous, FCC single phase of Fe diffused in Cu with no segregation of elements. The optical absorptivity in the Vis–NIR ranges for the alloy was measured for the same durations and the results show a strong correlation between the microstructural morphology and the optical behavior and enhancement of spectral absorptivity between 400 and 900 nm. The spectral absorptivity is enhanced up to 81% for the 8 h milled alloy compared to the original powder, and around 95% higher than TiO_2, which is the mesoporous material of choice for DSSCs. The increase in the impedance up to 120 Ω is an indication of the fundamental change of the electrical properties of the resulting phase after 8 h of milling, which has created the reported bandgap. Two DSSCs sensitized by the natural organic dye xanthophyll were constructed, one with Fe-Cu and the other with TiO_2 as the mesoporous layer, and the former had better performance in terms of power (0.078 vs. 0.022 mW, respectively) and efficiency (0.638% compared to 0.943%, respectively) for a 6.25 cm^2 cell. The carrier density for cells was also calculated using Mott–Schottky tests and Fe-Cu had orders of magnitude more carriers (4.27×1032 vs. $6.69 \times 1019/cm^3$) for the TiO_2. The Fe-Cu intermetallic phase is a promising mesoporous material due to its cost-effectiveness, mass-production capable synthesis technique, excellent spectral absorptivity in the UV–Vis range, recyclability, and good thermal and mechanical properties.

Conflict of Interest

None declared.

References

1. Green, M. A., K. Emery, Y. Hishikawa, W. Warta, and E. D. Dunlop. 2015. Solar cell efficiency tables (Version 45). Prog. Photovoltaics Res. Appl. 23:1–9.
2. Mathew, S., A. Yella, P. Gao, R. Humphry-Baker, B. F. E. Curchod, N. Ashari-Astani, et al. 2014. Dye-sensitized solar cells with 13% efficiency achieved through the molecular engineering of porphyrin sensitizers. Nat. Chem. 6:242–247.
3. Puxeu, L. 2014.Exploring novel dye concepts in dye sensitized solar cell, Thesis.
4. Chen, Z., F. Li, and C. Huang. 2007. Organic D-π-A dyes for dye-sensitized solar cell. Curr. Org. Chem. 11:1241–1258.
5. Hara, K., and N. Koumura. 2009. Organic dyes for efficient and stable dye-sensitized solar cells. Mater. Matters 92:4.4.
6. Mor, G. K., K. Shankar, M. Paulose, O. K. Varghese, and C. A. Grimes. 2006. Use of highly-ordered TiO_2 nanotube arrays in dye-sensitized solar cells. Nano Lett. 6(2):215–218.
7. Zhu, K., N. R. Neale, A. Miedaner, and A. J. Frank. 2007. Enhanced charge-collection efficiencies and light scattering in dye-sensitized solar cells using oriented TiO_2 nanotubes arrays. Nano Lett. 7(1):69–74.
8. Kim, Y. J., M. H. Lee, H. J. Kim, G. Lim, Y. S. Choi, N.-G. Park, et al. 2009. Formation of highly efficient dye-sensitized solar cells by hierarchical pore generation with nanoporous TiO_2 spheres. Adv. Mater. 21:3668–3673.
9. Park, Y.-C., Y.-J. Chang, B.-G. Kum, E.-H. Kong, J. Y. Son, Y. S. Kwon, et al. 2011. Size-tunable mesoporous spherical TiO_2 as a scattering overlayer in high-performance dye-sensitized solar cells. J. Mater. Chem. 21:9582–9586.
10. Chen, J., J. Wang, F. Bai, L. Hao, Q. Pan, and H. Zhang. 2013. Connection style and spectroscopic properties: theoretical understanding of the interface between N749 and TiO_2 in DSSCs. Dyes Pigm. 99:201–208.
11. Wang, X., L. Zhi, and K. Müllen. 2008. Transparent, conductive graphene electrodes for dye-sensitized solar cells. Nano Lett. 8:323–327.
12. Sun, S., L. Gao, and Y. Liu. 2010. Enhanced dye-sensitized solar cell using graphene-TiO_2 photoanode prepared by heterogeneous coagulation. Appl. Phys. Lett. 96:083113.
13. NREL. 2015. www.nrel.gov, last accessed 31 December 2015.
14. Liu, J. Z., A. van de Walle, G. Ghosh, and M. Asta. 2005. Structure, energetics, and mechanical stability of Fe-Cu bcc alloys from first-principles calculations. Phys. Rev. B 72:144109.
15. Hasebe, M. 1981. T. Nishizawa and Further study on phase diagram of the iron-copper system. Calphad 5:105–108.
16. Huang, X., and T. Mashimo. 1999. Metastable BCC and FCC alloy bulk bodies in Fe–Cu system prepared by mechanical alloying and shock compression. J. Alloy. Compd. 288:299–305.
17. Weeber, A. W. 1987. Application of the Miedema model to formation enthalpies and crystallisation temperatures of amorphous alloys. J. Phys. F: Met. Phys. 17:809.
18. He, J., J. Z. Zhao, and L. Ratke. 2006. Solidification microstructure and dynamics of metastable phase transformation in undercooled liquid Cu–Fe alloys. Acta Mater. 54:1749–1757.
19. Mazzone, G., and M. V. Antisari. 1996. Structural and thermodynamic factors of suppressed interdiffusion kinetics in multi-component high-entropy materials. Phys. Rev. B 54:441–446.
20. Ma, E., M. Atzmon, and F. E. Pinkerton. 1993. Thermodynamic and magnetic properties of metastable FexCu100– x solid solutions formed by mechanical alloying. J. Appl. Phys. 74:955–962.
21. Xu, J., U. Herr, T. Klassen, and R. S. Averback. 1996. Formation of supersaturated solid solutions in the immiscible Ni–Ag system by mechanical alloying. J. Appl. Phys. 79:3935–3945.
22. Hansen, M., and K. Anderko. 1965. Constitution of binary alloys, Issue 1, 2nd ed. McGraw-Hill, New York.

23. Lyasotsky, I., N. Dyakonova, D. Dyakonov, and E. Vlasova. 2008. Metastable phases and nanostructuring of Fe-Nb-Si-B base rapidly quenched alloys. Rev. Adv. Mater. Sci. 18:695–702.

24. Fu, G., Z. Hu, L. Xie, X. Jin, Y. Xie, Y. Wang, et al. 2009. Electrodeposition of nickel hydroxide films on nickel foil and its electrochemical performances for supercapacitor. Int. J. Electrochem. Sci. 4:1052–1062.

25. Gupta, R., N. Sukiman, M. Cavanaugh, B. Hinton, C. Hutchinson, and N. Birbilis. 2012. Metastable pitting characteristics of aluminium alloys measured using current transients during potentiostatic polarisation. Electrochim. Acta 66:245–254.

26. Li, Q. 2009. Formation of bulk ferromagnetic nanostructured $Fe_{40}Ni_{40}P_{14}B_6$ alloys by metastable liquid spinodal decomposition. SP Sci. China Press 1:1919–1922.

27. Das, N., J. Mittra, B. Murty, S. Pabi, U. Kulkarni, and G. Dey. 2013. Miedema model based methodology to predict amorphous-forming-composition range in binary and ternary systems. J. Alloy. Compd. 550:483–495.

28. Greenfield, M., and C. Pierce. 1973. Postweld aging of a metastable beta titanium alloy. Weld. J. 52:524–528.

29. Ravi, C., C. Wolverton, and V. Ozoliņš. 2006. Predicting metastable phase boundaries in Al–Cu alloys from first-principles calculations of free energies: the role of atomic vibrations. Europhys. Lett. 73:719.

30. Xu, J., G. S. Collins, L. S. J. Peng, and M. Atzmon. 1999. Deformation-assisted decomposition of unstable Fe50Cu50 solid solution during low-energy ball milling. Acta Mater. 47:1241–1253.

31. Oleszak, D., and P. Shingu. 1996. Nanocrystalline metals prepared by low energy ball milling. J. Appl. Phys. 79:2975–2980.

32. Jartych, E., J. K. Żurawicz, D. Oleszak, and M. Pękała. 2000. X-ray diffraction, magnetization and Mössbauer studies of nanocrystalline Fe-Ni alloys prepared by low- and high-energy ball milling. J. Magn. Magn. Mater. 208:221–230.

33. Roeder, J., J. Sculac, and M. Notis. 1984. The precipitation of iron in early smelted copper from Timna. Microbeam Anal. 243:243–246.

34. Ying-Yu, C., S. Rainer, and C. Y. Austin. 1984. Thermodynamic analysis of the iron-copper system I: the stable and metastable phase equilibria. Metall. Trans. A 15:1921–1930.

35. Kennedy, C. E. 2002. Review of mid- to high-temperature solar selective absorber materials. Technical Report NREL/TP-520-31267. National Renewable Energy Laboratory.

36. Alami, A. H., A. Allagui, and H. Alawadhi. 2015. Synthesis and optical properties of electrodeposited crystalline Cu2O in the Vis–NIR range for solar selective absorbers. Renewable Energy 82:21–25.

37. Rephaeli, E., and F. S. 2008. Tungsten black absorber for solar light with wide angular operation range. Appl. Phys. Lett. 92:211107-1–3.

38. Geldermon, K., L. Lee, and W. Donne. 2007. Falt-Band potential of a semiconductor: using the Mott-Schottky equation. J. Chem. Educ. 84:685–688.

39. Hoerner, L. 2013. Photosynthetic solar cells using chlorophyll and the applications towards energy sustainability. Thesis. University of South Florida, St. Petersburg.

40. Chien, C., and B. Hsu. 2013. Optimization of the dye-sensitized solar cell with anthocyanin as photosensitizer. Sol. Energy 98:203–211.

41. Patil, G., M. Madhusudhan, B. R. Babu, and K. Raghavarao. 2009. Extraction, dealcoholization and concentration of anthocyanin from red radish. Chem. Eng. Process. 48:364–369.

42. Zhang, Y., Y. J. Zhou, J. P. Lin, G. L. Chen, and P. K. Liaw. 2008. Solid-solution phase formation rules for multi-component alloys. Adv. Eng. Mater. 10:534–538.

43. Takaki, S., K. Fukunaga, J. Syarif, and T. Tsuchiyama. 2004. Effect of grain refinement on thermal stability of metastable austenitic steel. Mater. Trans. 45:2245–2251.

44. Gaffet, E., M. Harmelin, and F. Faudot. 1993. Far-from-equilibrium phase transition induced by mechanical alloying in the Cu-Fe system. J. Alloy. Compd. 194:23–30.

45. Lucas, F., B. Trindade, B. Costa, and G. Le Caër. 2003. The influence of pre-milling on the microstructural evolution during mechanical alloying of a $Fe_{50}Cu_{50}$ Alloy. J. Metastable Nanocrystalline Mater. 18:49–56.

46. Roos, A., T. Chibuye, and B. Karlsson. 1983. Properties of oxidized copper surfaces for solar applications I. Solar Energy Mater. 7:453–465.

47. Tauc, J., R. Grigorovici, and A. Vancu. 1966. Optical properties and electronic structure of amorphous germanium. Phys. Status Solidi B 15:627–637.

48. Erdogan, I. Y., and O. Gullu. 2010. Optical and structural properties of CuO nanofilm: its diode application. J. Alloy. Compd. 492:378–383.

49. Kormann, C., D. W. Bahnemann, and M. R. Hoffmann. 1988. Preparation and characterization of quantum-size titanium dioxide. J. Phys. Chem. 92:5196–5201.

50. Nakano, Y., S. Saeki, and T. Morikawa. 2009. Optical bandgap widening of p-type Cu_2O films by nitrogen doping. Appl. Phys. Lett. 94:022111.

51. Pike, J., S. W. Chan, F. Zhang, X. Wang, and J. Hanson. 2006. Formation of stable Cu_2O from reduction of CuO nanoparticles. Appl. Catal. A 303:273–277.

52. Kaneko, M., H. Ueno and J. Nemoto. 2011. Schottky junction/ohmic contact behavior of a nanoporous TiO_2

thin film photoanode in contact with redox electrolyte solutions. Beilstein J. Nanotechnol. 2:127–134.

53. Brian, Cantor (Book editor). 2005. Novel nanocrystalline alloys and magnetic nanomaterials, Brian Cantor. Institute of Physics Publishing, Sussex, UK.

54. Alami, A. H., A. Alketbi, M. Almheiri, and J. Abed. 2015. The Fe-Cu metastable nano-scale compound for enhanced absorption in the UV-Vis and NIR ranges. Metall. Mater. Trans. E Mater. Energy Syst. 2:229–235.

55. Alami, A. H., A. Alketbi, M. Almheiri, and J. Abed. 2016. Assessment of Al-Cu-Fe compound for enhanced solar absorption. Int. J. Energy Res. doi: 10.1002/er.3468 [Epub ahead of press].

56. Alami, A. H., A. Alketbi, and M. Almheiri. 2015. Synthesis and microstructural and optical characterization of Fe-Cu metastable alloys for enhanced solar thermal absorption. Energy Procedia 75 C:410–416.

57. Axel, F., F. Denoyer, and J. P. Gazeau. 2000. From quasicrystals to more complex systems, 13, 1, pp. 76. Springer-Verlag, Berlin Heidelberg.

58. Korn, D., H. Pfeifle, and G. Zibold. 1979. Optical properties of metastable CuFe solid solutions. J. Phys. F: Metal Phys. 9:1709–1–7.

59. Liu, B. X., C. H. Shang, L. J. Huang, and H.-D. Li. 1990. Fe-Cu icosahedral phase and its thermal and magnetic properties. J. Non-Cryst. Solids 117:785–788.

60. Takagiwa, Y., and K. Kimur. 2014. Metallic–covalent bonding conversion and thermoelectric properties of Al-based icosahedral quasicrystals and approximants, a review. Sci. Technol. Adv. Mater. 15:044802, (12 pp).

2

Field testing of thermoplastic encapsulants in high-temperature installations

Michael D. Kempe[1], David C. Miller[1], John H. Wohlgemuth[1], Sarah R. Kurtz[1], John M. Moseley[1], Qurat A. Shah[2], Govindasamy Tamizhmani[2], Keiichiro Sakurai[1,3], Masanao Inoue[4], Takuya Doi[4], Atsushi Masuda[4], Sam L. Samuels[5] & Crystal E. Vanderpan[6]

[1]National Renewable Energy Laboratory, 15013 Denver West Parkway, Golden, Colorado 80401
[2]Arizona State University Photovoltaic Reliability Laboratory, 7349 East Unity Avenue, Mesa, Arizona
[3]National Institute of Advanced Industrial Science and Technology, 1-1-1 Umezono, Tsukuba, Ibaraki 305-8568, Japan
[4]National Institute of Advanced Industrial Science and Technology, 807-1, Shuku-Machi, Tosu, Saga 841-0052, Japan
[5]DuPont Company, 200 Powder Mill Road, Wilmington, Delaware 19803
[6]Underwriters Laboratories, 455 East Trimble Road, San Jose, California

Keywords
Adhesives, creep, encapsulant, polymer, qualification standards, thermoplastic

Correspondence
Michael D. Kempe, National Renewable Energy Laboratory, 15013 Denver West Parkway, Golden, CO 80401.
E-mail: michael.kempe@nrel.gov

Funding Information
This work was supported by the U.S. Department of Energy under Contract No. DE-AC36-08-GO28308 with the National Renewable Energy Laboratory.

Abstract

Recently there has been increased interest in using thermoplastic encapsulant materials in photovoltaic modules, but concerns have been raised about whether these would be mechanically stable at high temperatures in the field. Recently, this has become a significant topic of discussion in the development of IEC 61730 and IEC 61215. We constructed eight pairs of crystalline-silicon modules and eight pairs of glass/encapsulation/glass thin-film mock modules using different encapsulant materials, of which only two were formulated to chemically crosslink. One module set was exposed outdoors with thermal insulation on the back side in Mesa, Arizona, in the summer (hot-dry), and an identical module set was exposed in environmental chambers. High-precision creep measurements ($\pm 20 \, \mu m$) and electrical performance measurements indicate that despite many of these polymeric materials operating in the melt or rubbery state during outdoor deployment, no significant creep was seen because of their high viscosity, lower operating temperature at the edges, and/or the formation of chemical crosslinks in many of the encapsulants with age despite the absence of a crosslinking agent. Only an ethylene-vinyl acetate (EVA) encapsulant formulated without a peroxide crosslinking agent crept significantly. In the case of the crystalline-silicon modules, the physical restraint of the backsheet reduced creep further and was not detectable even for the EVA without peroxide. Because of the propensity of some polymeric materials to crosslink as they age, typical thermoplastic encapsulants would be unlikely to result in creep in the vast majority of installations.

Introduction

There has been recent interest in the use of thermoplastic encapsulant materials in photovoltaic (PV) modules to replace chemically crosslinked materials, for example, ethylene-vinyl acetate (EVA). The related motivations include the desire to: reduce lamination time or temperature, reduce moisture permeation, use less corrosive materials, improve electrical resistance, or facilitate the reworking of a module during production. However, the use of any

thermoplastic material in a high-temperature outdoor environment raises safety and performance concerns. Therefore, there has been increased concern in the PV community regarding the possibility of viscoelastic creep prompting consideration for inclusion of a creep test into IEC 61730 and IEC 61215 [1–3]. Small areas of a module may reach much higher temperatures (>150°C) during the "hot-spot" test or during partial shading of a module without bypass-diode protection [4, 5]; but the localized nature of this occurrence is different from the situation of prolonged operation in the hottest module operating environments and mounting configurations. In very hot environments, modules are known to reach temperatures in excess of 100°C [6, 7]. One could envision an encapsulant with a melting point near 85°C with a highly thermally activated drop in viscosity, resulting in significant creep at 100°C.

Some early work with EVA encapsulation performed at Jet Propulsion Laboratories (JPL) did consider the issue of displacement during operation at high temperature [1, 8]. PV technology developers at that time speculated at the possibility of the displacement of components within a heated module operating in the field, but did not formally investigate to verify creep using a variety of modules deployed in a hot location. To specifically prevent creep, EVA that was crosslinked via a peroxide-initiated reaction was advocated at that time. The gel-content test was originally used at JPL as a means to quantify the content of insoluble crosslinked gel in EVA. The use of 65% gel was found to facilitate passing the sales qualification tests (which included the "melt/freeze" test at that time) and was therefore recommended by JPL. The use of EVA with at least 65% gel content was reaffirmed by Springborn Laboratories (later known as Specialized Technology Resources, Inc., or STR) [9], and presently, the use of EVA with 60–90% gel content is common in the industry.

The possibility of creep was more formally speculated in quantitative rheometry measurements of encapsulation materials [6, 7], motivating the study described in this paper. Characterization of the displacement from viscoelastic flow, identified here as "creep," should be distinguished from the "creep test" used to characterize the effects of densification associated with the process of physical aging [10]. "Creep" facilitated by prolonged exposure to high temperature is also different from the effects associated with rate-dependent loading [2].

This paper studies the potential hazard associated with creep in modules and how to test for such a problem [11, 12]. Modules fabricated with a variety of encapsulant materials are subjected to high temperatures and the resulting creep is documented as a function of temperature for each encapsulant type. The observed creep is compared with material-level tests to identify the best way to characterize phase transitions that could be predictive of creep in the field. The results are discussed to evaluate the hazard associated with creep for the materials studied and to propose both module- and material-level tests for evaluating the potential for creep. In the Experimental section, we describe the module construction, methods for measuring the phase transitions and rheological properties of the chosen encapsulant materials, modeling studies that were used to design the outdoor deployments, the

methodology for the outdoor deployments, the methodology for the step stress applied in environmental chambers, and the technique for measuring creep. In the Results section, we summarize the temperatures that were measured outdoors, the phase transition and rheological data collected for the encapsulant materials, the creep measured outdoors and in the chamber step-stress testing, and the high-pot test results. We also compare the creep measured outdoors and indoors with each other and with the phase transitions and rheological properties measured for the encapsulant materials. Using these observations, we propose a test designed to give confidence that creep rates will be negligible even in the hottest location. We also discuss evidence of material changes that are occurring during these tests. In the section Discussion: Creep and its Consequences, we summarize the conclusions, including the specific recommendation of module- and material-level tests to predict creep for products in the field.

Experimental

Encapsulant materials

Encapsulant materials chosen for this study were obtained from industrial manufacturers and are either being used, or under investigation for use, in PV modules (Table 1). However, without additional formulation-specific information, the product name is not of further use (from a scientific standpoint) in the context of this study. For the polydimethyl siloxane (PDMS) encapsulation, a different formulation was used for the thin-film mock modules than in the crystalline-silicon modules; however, both are sparsely crosslinked gels of similar composition, and the manufacturer of them indicated that they were extremely similar in composition. The PDMS values in Table 1 apply to the thin-film mock modules. Notably, the NC-EVA was formulated identically to a commercial EVA formulation but without the inclusion of a peroxide to promote curing during lamination.

Dynamic mechanical analysis (DMA) was performed on a TA Instruments ARES rheometer. Samples were tested in either a torsional configuration at temperatures below the melt or in a parallel-plate configuration near or above the melt transition. Data in Table 1 were measured at 0.1 rad/sec.

Differential scanning calorimetry (DSC) was performed using a TA Instruments DSC Q1000. Data were taken from the second of two consecutive cycles (from $-100 < T < 200°C$ or $-180 < T < 200°C$ for silicones) at the rate of 10°C/min in an N_2 environment.

Melt flow rate (MFR) measurements were performed on the Dynisco Melt Flow Indexer Model 4002. Testing weights included 0.225 kg, 0.95 kg, 2.06 kg, 4.12 kg (made

Table 1. Phase transitions determined by differential scanning calorimetry (DSC) and dynamic mechanical analysis (DMA). DMA glass transitions (T_g). DMA melting transitions (T_m) were determined when the phase angle was 45°, or at an inflection point in the modulus when a phase angle of 45° was absent. The crossover temperature (T_c) is where the phase angle of 45° occurs for materials with no melt transition.

Encapsulant material type and designation		Phase transitions					
		DSC			DMA at 0.1 rad/sec		
		T_g (°C)	T_m (°C)	T_f (°C)	T_g (°C)	T_m (°C)	T_c (°C)
Cured commercial PV EVA resin	EVA	−31	55	45	−30	47	
Commercial PV EVA Resin with all components but the peroxide	NC-EVA	−31	65	45	−28	69	
Polyvinyl butyral	PVB	15			17		121
Aliphatic thermoplastic polyurethane	TPU	2			3		84
Pt catalyzed, addition cure polydimethyl silioxane gel (mock modules)	PDMS-M	−158	−40	−80			
Pt catalyzed, addition cure polydimethyl silioxane gel (si modules)	PDMS-Si	−150	−40	−80			
Thermoplastic polyolefin #1	TPO-1	−43	93	81	−35	105	
Thermoplastic polyolefin #3	TPO-3	−44	61	55	−41	79	
Thermoplastic polyolefin #4	TPO-4	−34	106	99	−21	115	

by stacking two 2.06 kg weights), and 9.8 kg weights (made by stacking two 4.9 kg weights). Including the 0.11 kg weight of the piston, the net testing loads were 0.335, 1.06, 2.17, 4.23, and 9.91 kg, respectively. The tungsten carbide die had an orifice diameter of 2.095 ± 0.005 mm. The heating chamber allowed for steady-state temperature control to ±0.1°C.

Si module construction

The first module construction was a functional module with typical crystalline-silicon cells. It contained 42 156-mm multicrystalline, upgraded metallurgical-grade silicon cells with total dimension of 96.8 cm × 114.8 cm, providing an average module efficiency, at standard testing conditions (STC) [13], of 14.7 ± 0.3%. Because of a broken cell, one module had an efficiency of 12.7% and was not considered for this efficiency average. An Al frame was used with 3.18 mm tempered-glass superstrate, and a composite backsheet of Tedlar®-polyethylene terephthalate-Tedlar® (Dupont, Wilmington, Delaware) (TPT) construction. This backsheet composite construction was chosen because it is made of commonly used materials, and because it did not contain a low-vinylacetate, poly EVA seed layer. An EVA seed layer was also thought to be more likely to have adhesion compatibility concerns, whereas Tedlar® is surface treated to provide good adhesion to a wide variety of materials. During the course of this experiment, no delamination of the TPT from any interface was observed.

Because different methods are used for making PDMS encapsulated modules, the PDMS silicon modules were constructed at an outside vendor. Sixty 156-mm,

multicrystalline cells with a PET-based backsheet and Al frame were used for the PDMS modules, which had overall dimensions of 166.4 cm × 97.2 cm, and an average efficiency of 13.9 ± 0.7% at STC.

Thermal insulation was mounted to the back of the field-deployed modules to simulate mounting on an insulated roof. This is a realistic mounting configuration, representing the worst-case installation scenario on an insulated roof. Insulation was attached to the back of the module using a piece of 12.5-mm-thick plywood. Metal brackets, attached to the frame and the plywood, provided a gap of about 10 cm between the wood and backsheet. A 2.64 m^2·K/W [R15, 15 (h·ft^2·°F/Btu)] fiberglass mat insulation was inserted between the plywood and backsheet. The sides of the module were covered with duct tape to reduce air circulation around the perimeter and to prevent the insulation from being damaged (Fig. 1).

Thin-film mock module construction

The second module construction was designed to mimic a thin-film module with 61 cm × 122 cm dimensions, with a 3.18-mm-thick glass superstrate and substrate. These thin-film mock modules were not functional, containing only an F:SnO transparent conductive oxide (TCO) layer on the front piece of glass, and no PV cell present. The TCO was removed within 12.7 mm of the perimeter using laser ablation. The TCO was electrically connected to a conductive ribbon through a hole in the back glass to allow evaluation of safety compliance for the "Wet Leakage Current Test" [14, 15]. To mechanically fix and electrically isolate the ribbons during electrical testing, the ribbons were potted in a black silicone (Dow Corning 737).

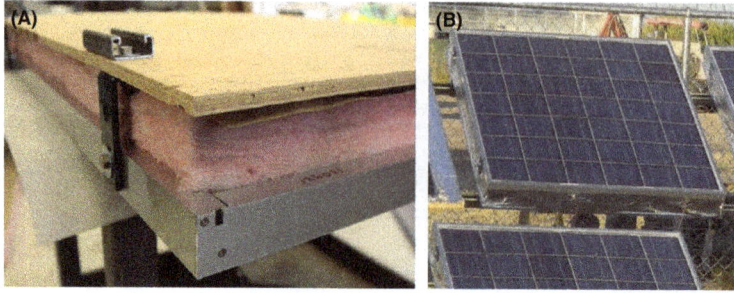

Figure 1. Photos of Si modules (A) during construction (B) as mounted in Mesa, Arizona.

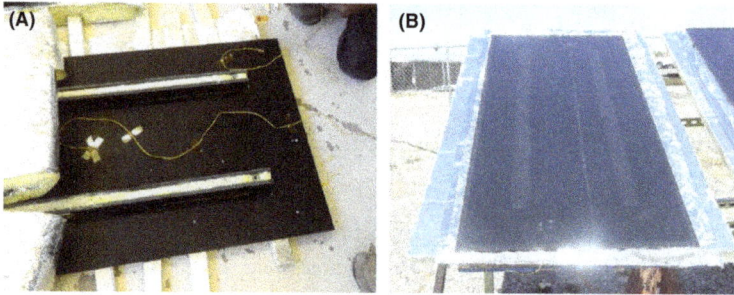

Figure 2. Photos of a thin-film mock module (A) during construction (B) after deployment in Mesa, Arizona.

To give the modules similar light absorptance—and hence, the thermal properties of a thin-film module—the backside of the glass substrate was painted black using Rust-Oleum® (Vernon Hills, Illinois) Universal flat-black enamel spray paint (P/N 245198). The absorptance of light by the paint, measured from the front glass side and weighted against the energy in the global solar spectrum [16], was 89.8%. This value compared favorably with the measured absorptance of representative Si, α-Si, and CdTe modules of 88.3, 87.8, and 89.4%, respectively.

Thin-film mock modules were mounted by adhesively attaching a 4 cm × 4 cm, 91-cm-long Unistrut fiberglass channel (McMaster-Carr, P/N 3261T34) on the back using Dow Corning 737 silicone, allowing the front piece of glass to move freely. No edge seal was present in the mock modules, so that the highest possible shear stress might be encountered. For the outdoor exposed modules, thermal insulation was applied to the back by filling the fiberglass channels with spray-in foam insulation, followed by a layer of 2.54-cm-thick fiberglass mat insulation (1.18 $m^2 \cdot K/W$ or R6.7), and a layer of 2.54-cm-thick polyisocyanurate sheeting (1.16 $m^2 \cdot K/W$ or R6.6). The two insulation layers were larger in width and height than the module to help reduce thermal gradients across the module, especially near the edges. The insulation had sections cut out to allow it to be placed around the

Unistrut channels. The insulation was held in place by a 1.25-cm-thick piece of plywood, attached to the Unistrut channels on the backside, and thus compressed the fiberglass insulation eliminating air flow on the back (Fig. 2).

Module temperature measurement

For these modules, and for the Si modules, K-type thermocouples (accuracy ±1°C) were placed in the center and 6 ± 2 cm from one corner on the backsides of the modules using Kapton tape. Module temperature was recorded at 6-min intervals. Additionally, one thin-film NC-EVA mock module was deployed in Golden, Colorado, and monitored at four locations (bottom-right corner, middle-left, middle-right, and the center) as indicated in Figure 7. Thermal images were also obtained using a FLIR Systems, Inc., ThermaCAM SC640 capable of resolving temperature variations to within 0.06°C.

Outdoor deployment

Modules were deployed at Arizona State University in Mesa, Arizona, from May to September 2011 on a rack inclined at the 33° latitude tilt and a 255° azimuth. This orientation was chosen to achieve the highest maximum module temperature possible. Because tracked systems do not have significantly restricted air flow on the back side, they operate at lower temperatures than roof mounted systems and were not considered for this study. This was determined using Typical Meteorological Year (TMY3) [14] data for Phoenix, AZ, along with the module temperature estimation equations from King et al. [17] for an insulated-back module accurate to ±5°C for Si cells. The maximum module temperature as a function of azimuth and tilt angle is shown in Figure 3. The maximum temperature is seen for all inclination angles at a southwest-facing azimuth of 255 ± 5°. For Phoenix in summer, this corresponds to a module facing the sun at 17:00 at the hottest part of the day (typically between 16:00 and 17:00). In Figure 3, there are two distinct peaks at the tilt angles of 28.3° and 34.5°. This can be interpreted as two different times (separated by several weeks) where some significantly hot days occurred and a different tilt angle helps to maximize the temperature. Arizona is at 33.4° latitude, so modules at the site were mounted at 33°, because we could not predict when the hottest days of the summer would occur.

Although the angles shown in Figure 3 demonstrate variability in maximum temperature from 95° to 106°C, the average temperature (including day and night) for a tilt of 33° was 49.03°C and 49.40°C for an azimuth of 180° and 255°, respectively. Changes in the array tilt only resulted in a modeled ±1°C average temperature, with azimuths around 240° generally producing the highest values. Thus, the change in azimuth was expected to increase the maximum temperature while not significantly affecting the average temperature.

The NC-EVA thin-film mock module exposed in Golden, Colorado, was mounted at a 180° azimuth (due south) and 40° latitude tilt [12] for aging in a moderately warm Steppe climate. For both the thin-film and silicon module types, this resulted in maximum measured temperatures between 102° and 104°C in Mesa, Arizona, and 93.5°C for the NC-EVA module in Golden, Colorado. The maximum temperatures measured in Arizona were very close to those predicted from the TMY3 data.

Mock module creep measurement

For the thin-film mock modules, the creep (displacement of the front glass relative to the back glass) was measured using a high-precision depth gauge, with 1-μm increments. The gauge was mounted to a flat plate to ensure that it was positioned perpendicular to the side of the module and in the plane of the glass. Creep measurement reproducibility was better than ±20 μm, Figure 4. For the Si modules, creep was monitored by comparison of optical and electro-luminescence (EL) images before and after exposure.

Indoor thermal stress

An identical set of thin-film mock modules and Si modules were exposed to heat in environmental chambers indoors. The highest-performing Si modules of each pair made were chosen for outdoor deployment. Only in the case of the TPO-4 modules was the issue of specimen selection significant. One of the cells was broken in the indoor exposed module during its manufacturing. Indoor aging was performed in an SPX Corp. (Thermal Product Solutions Division Charlotte, North Carolina.) T64RC-7.5 oven with active control of temperature, with the humidity unregulated after the laboratory ambient air was heated to chamber temperature. Indoor tests were therefore performed with the humidity less than ~15% relative humidity. The Si modules were placed vertically in the chamber resting on

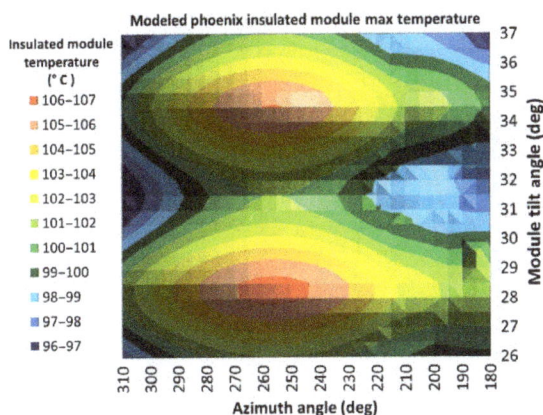

Figure 3. Maximum temperature predicted from TMY3 data and the King model [17] as a function of array tilt and azimuth angle (180° is due south).

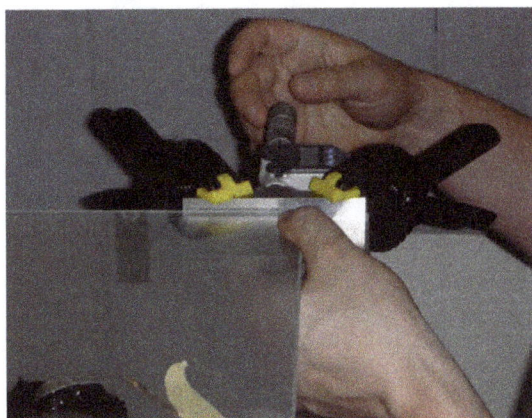

Figure 4. Photo of gauge used for measuring mock-module creep.

their frames, with the junction box (J-box) toward the top (the same orientation as was used outdoors).

Knowing that NC-EVA and TPO-3 had significantly lower melting temperature and were therefore more likely to creep, we began testing these materials at 65°C for 200 h in a step-stress test. After each thermal exposure, current-voltage (IV) curves, photographs, and EL images were obtained for the Si modules; and creep measurements were made for the mock modules. All test specimens were then placed back in the chamber with the test temperature increased by 5°C, for another 200 h. When a temperature of 85°C was reached, all the remaining samples constructed with the other six encapsulants were placed in the chamber for the step-stress test.

The Si modules were tested up to a temperature of 110°C, at which point testing was stopped to preserve the modules in working condition for a subsequent experiment. In contrast, the mock modules continued to be heated in 10°C increments after the 110°C step, up to a final temperature of 140°C.

Results

Outdoor exposure temperatures

The temperature at the center of the Arizona modules, which were mounted to give the maximum temperatures possible, achieved values in the upper 90°C range, with small excursions up to between 102° to 104°C (Fig. 5). However, for the thermocouples mounted near the corners of the modules, the maximum temperatures were 15° to 20°C cooler. A similar histogram for the Colorado-deployed

module produced temperatures with a maximum about 10°C cooler, but with a smaller, less pronounced peak at the highest temperatures. Both figures include the temperatures recorded during the night (typically the left of the figures) and day (right of the figures). The overall 10°C difference between the Arizona and Colorado data can be explained by the ~10°C higher ambient temperatures seen in Arizona. The peak in the high end of the Arizona module temperature data is attributed to the array orientation, where the highest irradiance from the sun occurs at the same time as the peak in ambient temperature. In Colorado, the module was oriented due south so that the peak solar irradiance occurred ~4 h before the greatest ambient temperature, flattening out the rightmost portion of module thermal profile.

Infrared (IR) images were taken to further qualitatively assess the temperature variation (Fig. 6A). For the crystalline-silicon module, a uniform temperature is observed within the module. The cooling around the perimeter is caused by better convective heat transfer and from efficient thermal conduction within the Al frame. Over the distance of 80 mm, that is, half a cell width, there is a 10°C change beyond which the temperature is uniform within ±2°C. For the thin-film mock module, there is less temperature variation between the sides of the module than between the top and bottom of the module (Fig. 6B). This is probably because of less insulation extending below and above the module than extends beyond the module at the sides (Fig. 2B). In all cases, the corner temperature was 15° to 20°C lower than the central temperature in both the IR images and the thermocouple histograms (Figs. 5 and 7); however, the majority

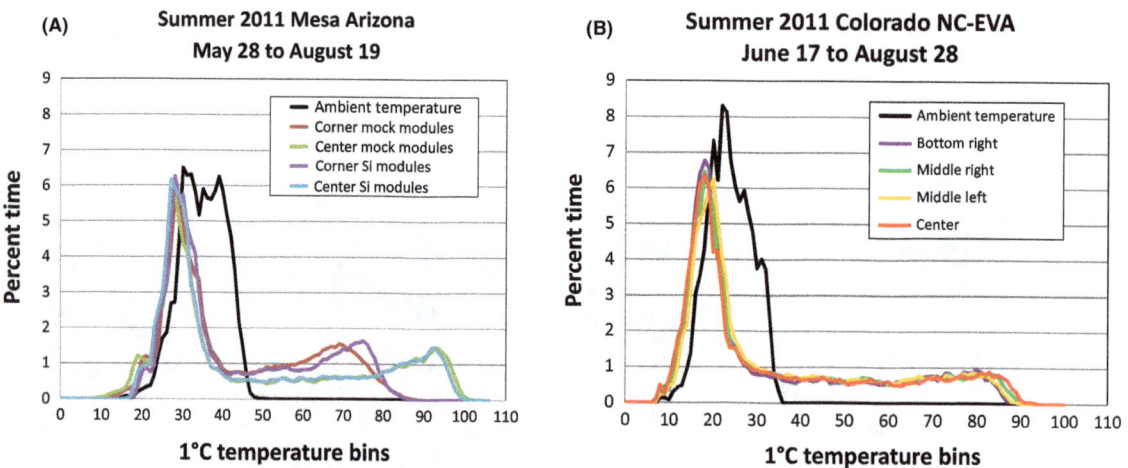

Figure 5. Summer temperature histograms for outdoor deployed modules. (A) Each profile represents the average of the eight modules for either the center or the corner thermocouple. Amongst each sample set, the individual histogram traces typically varied ±1°C or ±0.2%. (B) Histograms of four thermocouples placed on the Colorado-deployed NC-EVA mock module.

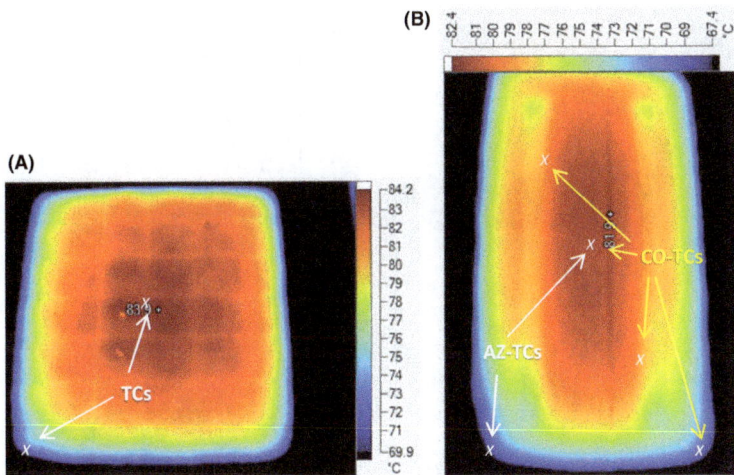

Figure 6. Infrared images of modules during deployment at the upper temperature ranges experienced. (A) Crystalline-silicon module temporarily set up in Colorado. (B) NC-EVA thin-film mock module deployed in Colorado. The locations of the thermocouples are indicated in the figure.

of the "active" region of modules is very close to the temperature indicated by the central thermocouple. Because the viscosity of a polymer can change rapidly near phase transitions and can have a strong variation in magnitude with temperature, the presence of a cooler perimeter could greatly affect the ability of the module to creep. A similar temperature differential appeared in both the thin-film mock and the crystalline-silicon modules, but the presence of an Al frame in the crystalline-silicon modules is expected to make it unlikely, but not impossible, to realize an installation configuration that would reduce this temperature heterogeneity significantly. It is this temperature differential that helps to significantly reduce the creep, despite most of the NC-EVA module being well above its melting point of around 65° to 69°C.

The modules were mounted individually, allowing air flow at the sides between modules, as opposed to a close-packed installation. Thus, it is possible to imagine an installation with modules mounted in a large array, on a flat rooftop, with higher temperatures or greater temperature uniformity than in Figures 5, 6, and 7. Furthermore, the record high temperature in Riyadh, Saudi Arabia, is 55°C—5°C hotter than the location of Phoenix—which would be expected to increase the module temperature by roughly 5°C in Riyadh.

However, the Si modules in Arizona were maintained open-circuited so that the 14.6% potential power output was converted to heat rather than electricity, as would be the case in an operating system drawing off power. To estimate the magnitude of this effect, we note that the maximum temperature of ~101°C for the NC-EVA module occurred when the ambient temperature was between 44.5° and 50°C. At elevated temperature, the

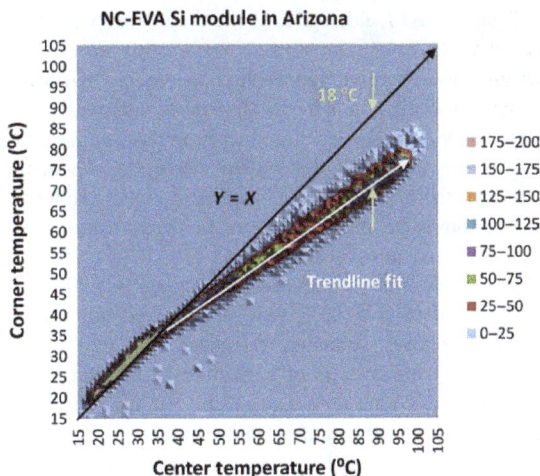

Figure 7. 2-D histogram of the center versus corner temperature of the NC-EVA Si module deployed in Arizona for the summer of 2011. The population of the data in 1 × 1°C bins is indicated in the figure legend.

14.6% module efficiency would be reduced by −0.5%/°C (relative), producing a module efficiency of ~9.4%. We estimate that this would reduce the ΔT of between ~56.5 and 51°C, to around 50°C above ambient for a maximum temperature of about 96°C if power was extracted from the module rather than converted to heat.

Modules are often mounted to encourage air flow. Even when building-integrated, a module is not likely to be mounted to a well-insulated roof surface (as chosen here to simulate a worst-case scenario) because such a configuration would run hot, compromising efficiency and PV durability [18–20]. We acknowledge that the high temperatures

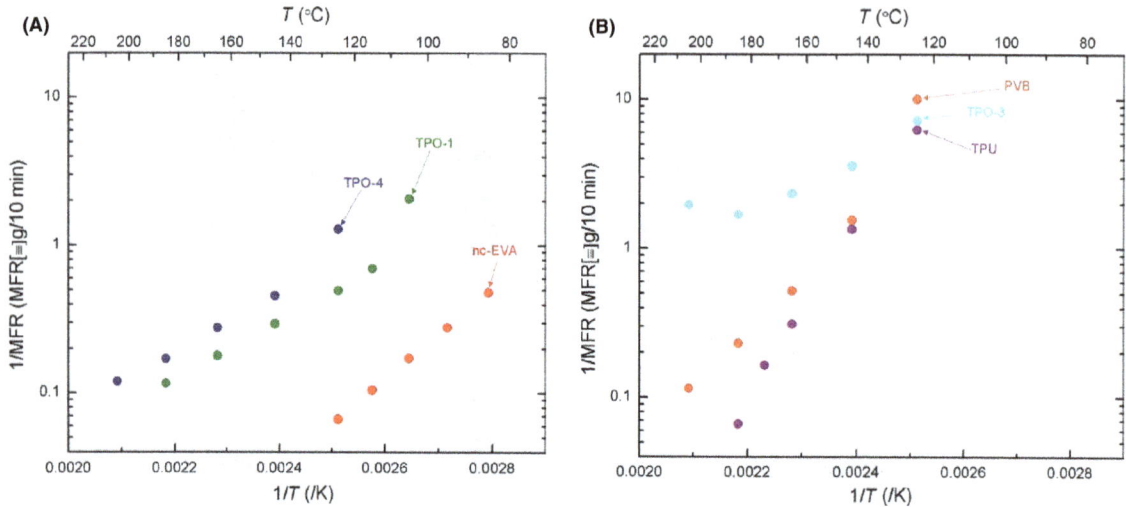

Figure 8. MFR results (with the linear Arrhenius fits shown) for (A) TPO-4, TPO-1, NC-EVA and (B) PVB, TPO-3, and TPU. Using the 4.23 kg test load.

in Figures 5 and 7, although rare, could exist for PV modules deployed at hot desert locations. Therefore, this experimental set-up is nearly the hottest possible configuration imaginable. Hotter installation would possibly run an additional 5° to 10°C hotter in a climate with a higher ambient temperature or for a significantly larger system where modules in the center of the arrays can get significantly hotter. However, to our knowledge, the hottest reported system temperatures do not exceed those of this study.

Inspection of the histograms (Figures 5 and 7) indicates that temperatures >100°C were experienced for a short amount of time (separately calculated at ~1 h), which over a 25 years lifetime of a module in Arizona would only sum to a total of around 25 h. The modules spent about 312 h at temperatures above 85°C over the course of the summer of 2011. The "damp-heat" qualification test runs at 85°C for 1000 h, that is, 3 years of exposure in Arizona. Similarly, in Colorado, the center of the module was above 85°C for only 57 h, that is, 1/18th of the duration of the damp-heat test. Thus, the current quali- fication test (1000 h at 85°C/85% RH) equates to a cumulative thermal exposure above 85°C that is less than observed in the most severe installations and potentially insufficient to evaluate the potential for creep.

Relative viscosity of materials by melt flow rate comparison

For the outdoor exposed modules, high-resolution optical photographs, EL, IV curves, and IR images were taken before and after exposure to look for creep and its effects. For all the modules, there were no overt signs of creep. However, it should be noted that the modules were

mounted with cell strings running vertically, where the ribbon provides greatest mechanical support. If the Si modules had been mounted in the more mechanically compliant horizontal orientation, there may have been a greater likelihood for discernible creep to have occurred. Figure 8 shows the melt flow rate (MFR) measurements for the polymers used in this experiment. The relative values of the quantity 1/MFR should correspond to the relative values of the zero shear viscosity (the viscosity limit as the strain rate approaches zero) and to the rela- tive expected amount of creep for the different materials. The NC-EVA crystalline-silicon module had the highest propensity to creep [12]. For the materials with melt transitions, the 1/MFR rose significantly beyond the range of measurement as the temperature is lowered near these transitions [12]. The viscous response of the NC-EVA near the upper limit of these experiments (~95°C) has a viscosity that would be expected to be several orders of magnitude lower than the other materials. Thus, if initial viscosity was the only consideration, one would expect the NC-EVA to creep orders of magnitude faster than the modules made using the other encapsulants [7].

Si module outdoor testing

As shown in Table 2, many of the modules, including TPO-1, TPO-3, TPO-4, and PDMS, experienced some per- formance losses after exposure in Arizona despite showing no visible signs of creep. Inspection of the EL images indicates this performance loss is coincident with the for- mation of cracks in the cells, Figure 9. The principal loss mode was through fill factor (FF), but because the modules have bypass diodes, the highest current-producing string

Table 2. Performance parameters for crystalline-silicon modules before and after aging. Uncertainty in these measurements is about 5% relative of the various reported values compared to actual, and ±0.5% for the values of the percent change. Yellow highlighted data are from modules where the EL images showed cracked cells. Module efficiency, open-circuit voltage, short-circuit current, fill factor, and maximum power are represented by the symbols η, V_{oc}, I_{sc}, FF, and P_{max}, respectively. Some numbers may be slightly inconsistent because of rounding.

Material and test location		Initial measurements					After Arizona exposure or 110°C environmental chamber exposure					%Change				
		η (%)	V_{oc} (V)	I_{sc} (A)	FF (%)	P_{max} (W)	η (%)	V_{oc} (V)	I_{sc} (A)	FF (%)	P_{max} (W)	η (%)	V_{oc} (V)	I_{sc} (A)	FF (%)	P_{max} (W)
EVA	Indoor	14.9	25.9	8.54	75.0	165	15.0	25.9	8.52	75.3	166	1.0	−0.1	−0.2	0.4	0.8
	ASU	15.1	26.2	8.54	75.0	168	15.1	26.2	8.41	76.1	168	−0.1	0.3	−1.5	1.4	0.1
NC-EVA	Indoor	14.9	26.0	8.54	75.0	166	15.0	25.9	8.51	75.7	167	0.5	−0.3	−0.4	0.9	0.8
	ASU	14.9	26.0	8.53	75.0	165	14.9	26.1	8.39	75.8	166	0.2	0.4	−1.7	1.1	0.3
TPU	Indoor	14.6	26.0	8.55	73.0	162	13.9	25.9	8.49	70.3	155	−4.8	−0.2	−0.7	−3.8	−4.6
	ASU	14.9	26.0	8.53	74.0	165	14.8	26.1	8.37	75.2	164	−0.4	0.3	−1.9	1.6	−0.5
PVB	Indoor	14.8	26.0	8.52	74.0	165	14.9	25.9	8.45	75.7	165	0.5	−0.4	−0.9	2.2	0.3
	ASU	14.9	26.0	8.53	75.0	166	14.8	26.2	8.39	75.1	165	−0.8	0.5	−1.7	0.1	−0.7
PDMS	Indoor	13.8	36.5	8.38	72.7	223	13.7	36.3	8.51	71.6	221	−0.7	−0.7	1.5	−1.5	−0.7
	ASU	13.9	36.7	8.43	72.9	225	13.6	36.6	8.39	71.4	220	−2.2	−0.1	−0.5	−2.1	−2.6
TPO-1	Indoor	14.8	26.0	8.45	75.0	164	14.9	26.0	8.47	75.5	166	0.7	−0.3	0.3	0.7	1.0
	ASU	14.2	25.9	8.51	72.0	158	13.8	26.0	8.46	69.7	153	−3.1	0.4	−0.6	−3.3	−3.1
TPO-3	Indoor	14.8	25.9	8.53	74.0	165	14.1	25.9	8.51	70.9	156	−5.0	0.0	−0.3	−4.4	−5.4
	ASU	14.9	26.0	8.50	75.0	165	12.3	26.2	7.97	65.5	137	−21	0.6	−6.7	−15	−21
TPO-4	Indoor	12.7	25.9	8.46	64.0	141	12.9	25.9	8.47	65.2	143	1.7	−0.2	0.1	1.8	1.4
	ASU	14.0	26.0	8.48	71.0	156	13.4	26.1	8.43	67.8	149	−4.5	0.5	−0.6	−4.7	−4.3

dominated the short-circuit current (I_{sc}) response, masking this degradation mechanism in most cases. Typically, *FF* loss was only a few percent stemming from only one or two cracked cells. Because cell cracking increases for high viscosity materials (compare Table 2 to Fig. 8), it is believed that the lamination of higher viscosity thermoplastic materials resulted in stressed cells that performed well initially but lost electrical connections later. This is evidenced by the much higher incidence of cell breakage during lamination of the TPO materials. However, it is also possible that higher modulus encapsulants could transfer more stress to cells, contributing to the increased breakage rate [2]. Although not statistically significant, it is noted that four of the six degraded modules were fielded in Arizona (the other two were stressed in the chamber), where they would have been exposed to prolonged thermal cycling and mechanical stresses, for example, wind load. Although encapsulant creep is not suspected to have significantly contributed to cell cracking, the results here indicate that, if lamination processes are not sufficiently optimized, higher viscosity encapsulant materials may mechanically stress cells during lamination (as was observed) or may facilitate crack formation during field deployment.

Mock module outdoor testing

Of the outdoor exposed modules, only the thin-film mock modules constructed using NC-EVA experienced significant

creepage as measured when deployed, Fig. 10A. All the other encapsulants (Fig. 10B) showed no creep within the uncertainty of the outdoor measurements. When measured during field exposure, only two of the corners were readily accessible, and the presence of the insulation created additional difficulties, limiting the accuracy of measurements during exposure. Therefore, a more detailed and accurate laboratory measurement was made both before and after outdoor exposure, Figure 11. Here, the insulation was removed, allowing creep measurements to be performed at all four corners, improving the accuracy of the measurements. These indoor creep measurements indicated that the TPO-3 module crept 0.090 ± 0.036 mm and the TPO-1 module crept 0.032 ± 0.024 mm (Fig. 10). No creep was detected for the other encapsulants in the comprehensive final measurement. Even though most of the thermoplastic encapsulants reached either the melt or rubbery state within the module during exposure in Arizona (Table 1), no movement was observed. The melting and freezing transitions determined by DSC were only about 10°C higher or 4°C lower, respectively, for TPO-3 as compared to NC-EVA, Table 1. This demonstrates that the presence of a melt or glass transition is not sufficient to predict the potential for creep.

Noting the absence of creep beyond day 110 in AZ, where the maximum temperature did not exceed 90°C, and that the Colorado-deployed module barely crept while rarely reaching temperatures above 90°C indicates that creep

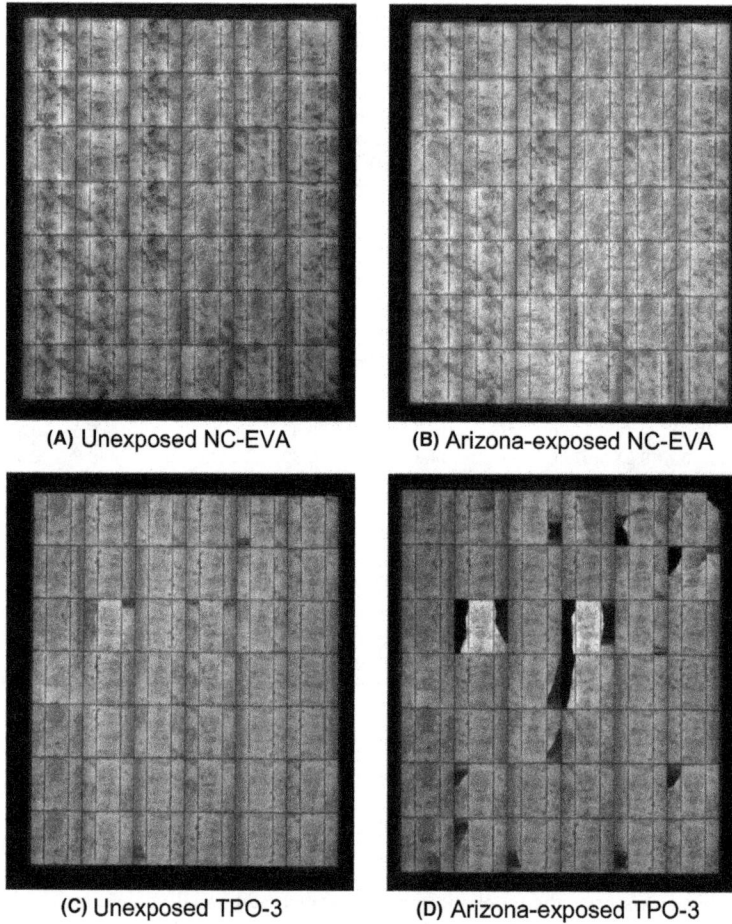

(A) Unexposed NC-EVA   **(B)** Arizona-exposed NC-EVA

(C) Unexposed TPO-3   **(D)** Arizona-exposed TPO-3

Figure 9. EL images of modules before and after exposure in Mesa, Arizona, at Arizona State University. (A) Unexposed module constructed with NC-EVA. (B) Arizona-exposed module constructed with NC-EVA. (C) Unexposed module constructed with TPO-3. (D) Arizona-exposed module constructed with TPO-3.

is possible for uncured EVA when the maximum module temperature approaches around 90°C, Figure 10. It is unlikely that one would expect to directly detect encapsulant creep in an outdoor-deployed module considering that: (1) module temperatures above 90°C will only occur in very hot environments when modules are mounted with very minimal air flow for heat transfer on the backside [21, 22], (2) significant temperature nonuniformities exist at the module periphery, restricting creep, (3) only the NC-EVA module, which has a very low viscosity, crept, and (4) the glass on these modules was unconstrained.

Over the first 30 days, creep was faster in the Arizona module (Fig. 10), but slowed down mid-summer despite the temperature being similar. Typical EVA formulations are known to crosslink as they age in the field [23]. This reduction in creep rate suggests that even without the peroxide additive, NC-EVA was crosslinking at these high temperatures in the field.

Mock module indoor testing

The MFR, Figure 8, which actually measures a rheological property, was not able to correctly rank order the materials in terms of creep. The value of 1/MFR is closely related to viscosity, but the presence of melt transitions produces large nonlinear changes in viscosity making extrapolation to operating conditions incorrect in these cases. Although TPO-1 and TPO-4 are much less viscous at temperatures above 120°C, TPO-3 is apparently much less viscous at module operating temperatures. Rheological properties used to assess the potential for creep must be measured at the module operating temperatures. The best rheological property to use for this purpose would be the zero shear viscosity.

Comparing Figures 11 and 12, more creep is seen in indoor experiments at lower temperatures relative to outdoor aging. Outdoors, the cool perimeter of the modules

(A)

(B)

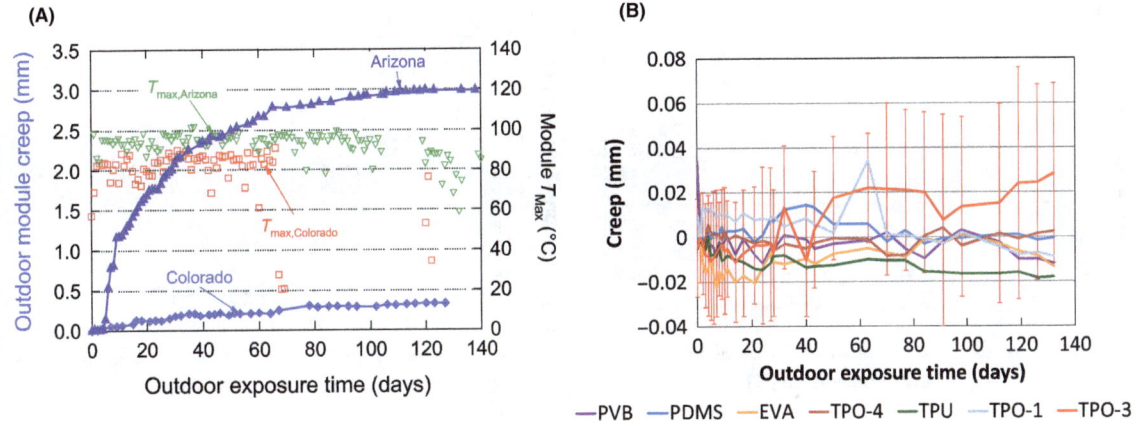

Figure 10. Measurement of thin-film mock module creep (relative displacement of top and bottom glass plates) during outdoor exposure. (A) Blue-creep distance, Green and Red-daily maximum module temperature in AZ and CO respectively. (B) Encapsulant types with relatively little to no creep. Error bars shown for TPO-3 only for clarity.

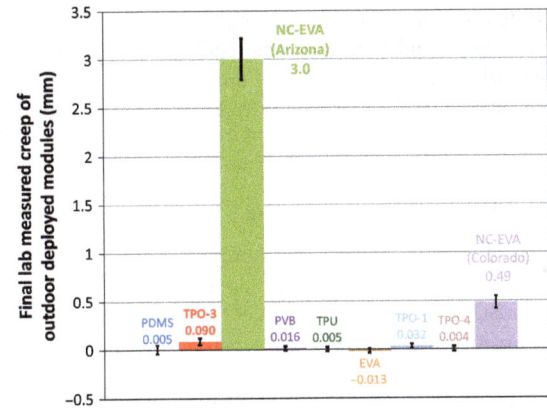

Figure 11. Measurement of thin-film mock module creep after field deployment in AZ or CO.

DSC is measuring the thermal effects of increased polymer mobility, but DMA is directly assessing the effect of phase transition on the bulk rheological properties.

The slope of the creep versus temperature curve for TPO-3 does not continue to increase rapidly beyond 95°C, suggesting that TPO-3 is chemically crosslinking at temperatures above 90°C. At temperatures above 110°C, the amount of creep in each cycle begins to decrease for TPO-3. Similar behavior is also seen for TPO-1, TPO-4, and PVB. The crosslinking and/or chain scission of these materials was characterized using gel-content measurements and multiangle laser light-scattering. This will be the topic of a subsequent paper.

In indoor testing, the TPU and the NC-EVA showed a continually increasing creep rate with increasing temperature. For NC-EVA, testing had to be stopped at 75°C because of excess creep, and we were not able to test at sufficiently high temperatures where the crosslinking chemistry was fast enough. The TPU similarly crept more than 1 cm at 125°C, contacting the chamber floor, preventing higher temperature exposure (see the dashed arrow in Fig. 12 for TPU). Bubbles began to form within the TPU module at 105°C and became located throughout the module after exposure at 110°C. The TPU is suspected to be unique in rapidly chemically decomposing at elevated temperatures. This decomposition may also explain the onset of creep is not coincident with the crossover temperature of 85°C.

Measuring the change in creep during a given testing step (Fig. 12B) is similar to measuring the reciprocal of the zero shear viscosity at the test temperature. Initially, as the temperature is increased, the creep rate increases; but at higher temperatures, the creep rate decreases, suggesting crosslinking of the materials, similar to the outdoor

limits motion. In indoor testing, the NC-EVA began creeping detectably at 75°C in the thin-film mock module construction, and at 80°C in the crystalline-silicon construction. Similarly, the TPO-3 and TPO-1 thin-film mock modules began to creep detectably at 90° and 105°C, respectively. PVB goes through a glass transition at around 16°C with no melt transition, but did not begin to creep until it was well above 100°C. Creep in PVB was mitigated because of its greater viscosity [12]. DMA measurements indicate that the phase angle of PVB reaches 45°, the crossover point, at 121°C, which corresponds with its onset of creep in indoor aging. However, the TPU had a crossover at a much lower temperature than the onset of creep. These temperatures for the onset of creep correspond better to the temperatures determined by DMA than those determined by DSC, Table 1. This is because

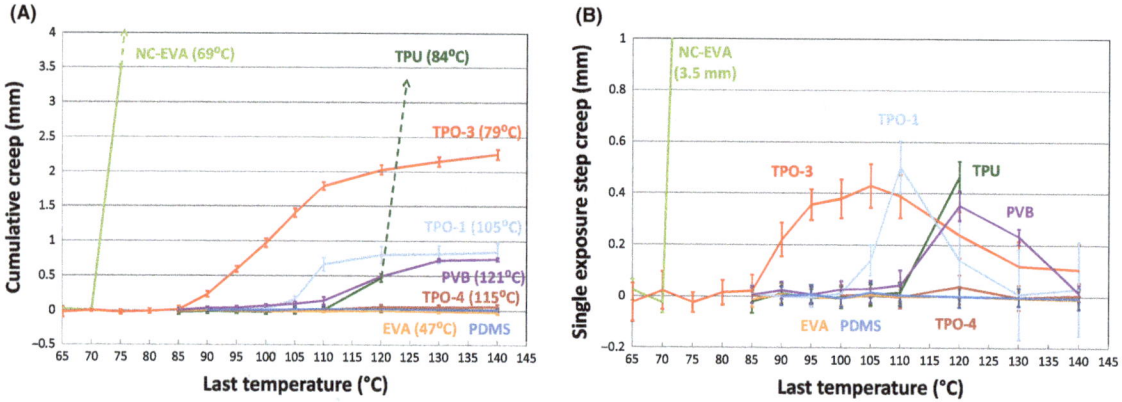

Figure 12. (A) Relative displacement of glass sheets in the thin-film mock modules after step-stress aging, of a duration of 200 h each. The indicated temperatures are the DMA-determined T_m or T_c from Table 1. (B) Creep experienced during a 200-h step stress cycle at the given temperature.

fielded mock modules. Noting that in the fielded modules, NC-EVA crept several mm, TPO-3 crept 0.9 mm, and TPO-1 only crept a barely detectable amount, the evaluation of zero shear viscosity at temperatures between 90° and 105°C would provide a prediction of initial creep during outdoor use. If such a test were extended for a long period of time, it could also highlight the tendency of a material to crosslink in response to heat, limiting the amount of creep.

Crystalline-silicon module indoor testing

In indoor studies, the onset of creep for the NC-EVA silicon module occurred at 75°C (Fig. 12A). Here, despite the fact that the sides of adjacent rows of cells appeared to touch, there was no discernible performance loss with

step stress tests up to 100°C, (Fig. 13A and B, Table 3). Even though this significant cell movement did not directly create a performance issue, it could cause a ground-fault safety issue, and it is likely that subsequent exposure to thermal cycling would increase the mechanical stress on the tabbing and solder bonds, increasing the long-term failure rates.

Wet insulation resistance

Despite creeping 3 mm, the outdoor-exposed NC-EVA mock module still passed the wet high-pot test. For all outdoor-exposed modules, the wet high-pot resistance was constant (within uncertainty) or increased upon exposure. However, some of the samples showed resistances >10,000 MΩ, the measurement limit for the instrument.

Table 3. Results of wet high-pot testing according to IEC 61730. Pass criterion is 40 MΩ/m².

	Arizona or Colorado outdoor deployment				Environmental chamber stress testing			
	Mock module		Crystalline Si module		Mock module		Crystalline Si module	
	Before exposure	After exposure	Before exposure	After exposure	Before exposure	After exposure	Before exposure	After exposure
Encapsulant type	(MΩ)	(MΩ)	(MΩ)	(MΩ)	(MΩ)	(MΩ)	(MΩ)	(MΩ)
Allowable Limit	57 (MΩ)		36.4 or 24.7 (MΩ)		57 (MΩ)		36.4 or 24.7 (MΩ)	
EVA	>10,000	>10,000	110	170	160	>10,000	100	190
NC-EVA	160	250	110	100	210	**<1**	100	140
PDMS	350	>10,000	1500	2500	290	350	1300	740
TPO-1	290	340	>10,000	>10,000	9100	390	>10,000	>10,000
TPO-3	250	360	1400	2800	300	380	1400	4100
TPO-4	360	>10,000	>10,000	>10,000	300	**Cracked**	>10,000	>10,000
PVB	160	690	**65**	**61**	110	**Cracked**	70	66
TPU	190	>10,000	150	130	270	**<1**	170	160
EVA (Colorado)	100	270						

Values in bold print are for modules that failed the wet high-pot test.

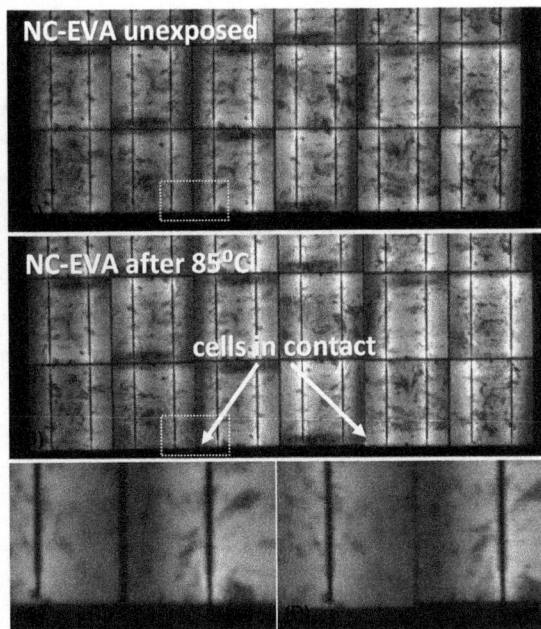

Figure 13. Electroluminescent images of the NC-EVA crystalline Si (A) before thermal exposure and (B) after the 85°C exposure step of the indoor step-stress test. (C, D) Detail from the bottom row of the module, insets as indicated by the boxes in A and B, respectively.

The wires for connecting to the TCO were not soldered, but were adhered to the TCO using a pressure-sensitive adhesive. The bond area is 0.635 cm × 107 cm (68 cm²), but poor electrical contact was possible. The TPO-1 and TPO-4 crystalline Si modules also occasionally demonstrated high, >10,000 MΩ, resistances, where the presence of a measurable IV curve indicates that electronic contact was obtained.

For the case of a mock module with a high-resistance polymer and a sufficient edge delete, the resistance through the glass was ~360 MΩ. For a 3.18-mm-thick glass, with a TCO area defined by the module size minus the edge-delete region, corresponding to a volume resistivity of glass of 7.9×10^{12} Ω·cm. The estimated volume resistivity is higher than typical literature values (~10^{11} Ω·cm) [24], probably due to the use of SiO_2 layers next to the TCO. Therefore, any values significantly higher than 360 MΩ indicate some loss of electrical contact. Values lower than 360 MΩ indicate significant current through the encapsulant, poor edge delete, or loss of adhesion somewhere within the package. Considering that the NC-EVA modules crept the most and were the only ones to show visible signs of flow, a large amount of creep (a few mm) can be tolerated before an immediate safety concern would appear in either module construction type. We only detected

failure of the wet high-pot test for samples with cracked glass or which experienced creep >1 cm.

Discussion: Creep and its Consequences

In addition to encapsulant properties, the module construction is an important factor to consider for creep. None of the crystalline Si modules demonstrated measureable creep when deployed outdoors. Despite approaching the maximum temperatures possible for a fielded module, none of the crystalline Si modules experienced a detectable safety or performance failure due to the use of a thermoplastic encapsulant. However, the modules were mounted with the cell strings arranged vertically, which may have reduced the propensity for the cells to be displaced.

Even though no movement was seen in the NC-EVA crystalline Si module, it is possible that movement on the order of 100 μm may have occurred. Such movement could put some additional mechanical stress on the interconnects (solder joints and ribbons) that might take years of aging (thermal and/or mechanical cycling) to yield an overt effect. If problems with undercured EVA exist, they are not likely to be a common occurrence, noting that (1) NC-EVA represents an extreme level of undercure, (2) the EVA will cure in time [23], (3) modules may have some level of tensile stress in them from the lamination process anyway, and (4) these modules were exposed to extreme conditions of heat. The study here predominantly applies to infant mortality—within the first few years of field exposure—and does not attempt to examine more prolonged aging.

The lesser creep in the Si modules relative to the mock modules can foremost be attributed to the fact that the polymeric backsheet is held in place by the frame, restricting component movement. Similarly, a typical thin-film module might be constructed using either a frame, or clips on the edges for mounting. Either of these package configurations would mitigate creep. If the glass in a thin-film module is fixed, only a few busbars within the encapsulant would be able to move, which would only happen in a very low-viscosity encapsulant. In this case, the more likely scenario is that flow would probably not directly affect the performance; but with sufficient time, polymer may flow out of the bottom of the module, creating a safety concern. Thus, an encapsulant would typically need to have a viscosity much lower than NC-EVA in Table 1 to pose an immediate safety or performance issue.

Within the PV industry, there is much attention to control of the degree of cure of the encapsulant during module processing. Some of this concern is because of potential problems due to module creep [1] and with poor adhesion during the qualification tests. This study has shown that even a nonperoxide-containing EVA would not be expected to pose a significant risk for prolonged creep. If fielded, an incompletely cured EVA module would be likely to cure in time with exposure to UV light and humidity. Fielded EVA modules typically have gel contents greater than 90%, despite not being initially cured to that level [1, 9, 23, 25]. However, for a module to pass the qualification tests, the amount of crosslinking is important. For example, in Figure 13, the cells were displaced enough that it is possible for a module to fail the visual inspection if this reduced a critical clearance distance or caused some components to contact each other even though the performance of the crystalline Si module was not decreased.

Many in the PV field have identified that an incompletely cured or improperly cured module might not have good adhesion retention [26]. This may occur, for example, if the peroxide crosslinking agent also functions to activate an adhesion-promoting compound. The current study indicates that concerns with adequate processing, as measured by gel content, are more likely to apply to adhesion retention than to component displacement.

The tests here were conducted to examine modules fielded at the highest temperatures expected for PV systems integrated into a well-insulated roof. Rack-mounted modules would clearly be expected to have a temperature ~15°C lower [17] than building-integrated or building-applied PV systems [21, 22]. The only ways to achieve higher temperatures would be if: modules were mounted in an even hotter environment than Phoenix, AZ, were mirror-augmented, mounted near a heat source such as a vent, or part of a much larger building-integrated array and mounted with no convective cooling on the backside on a well-insulated roof.

This study examined a few emerging candidate encapsulant materials, but the possibility still exists that a newer material may experience thermally induced failure modes (including creep) that might not be screened in the current qualification test methods. Consider that the maximum module temperature of 105°C is not examined within the present module qualification tests. Because the amount of time at temperatures above 85°C is very limited, a relatively short test at high temperature should be able to screen the majority of these failure modes. Therefore, it has been proposed to IEC to subject modules to a temperature of 105°C for 200 h as part of IEC 61215. Because it is likely that when deployed there may be cool spots restricting flow, this test represents a test with a significant safety factor. This proposal, in addition to hotspot testing, would screen for components and adhesives that might creep at temperatures above 85°C.

Conclusions

The use of 85°C and humidity freeze cycles (amongst other tests) in IEC standards necessitates crosslinking of typical EVA formulations to achieve gel contents in excess of around 65% to provide adequate adhesion properties and resistance to creep. However, this work indicates that even if an EVA encapsulant was formulated without peroxides to form chemical crosslinks, virtually no creep would be expected to be seen outdoors with typical module constructions and mounting. Only modules with an unrestrained front-glass were shown to have any propensity to creep outdoors. This absence of creep observations in fielded modules is due, in part, to the nonuniformity of temperature resulting in small lower-temperature areas that significantly resist creep, and to the restricted motion of front and backsheets by frames and mounting clips.

The materials tested produced creep profiles that indicate formation of crosslinks in response to heat and UV light or to heat only despite the presence of a polymer stabilizing formulation and in the absence of peroxides or other curing agents. Evidence was also presented that NC-EVA, TPO-1, and TPO-3 thermally crosslink despite the absence of peroxide above temperatures above ~100°C. This unintended crosslinking actually serves to further mitigate the potential for creep in some materials.

Often, researchers will consider the temperature of melting or glass transitions when estimating the likelihood of material creep. However, this work indicates that the onset of creep coincides reasonably well to the melting (or crossover) points determined by DMA using the phase angle of 45°, but not so well to the phase transitions indicated by DSC (compare Table 1 to Fig. 12). So when evaluating materials for the likelihood of creep, one must assess relevant rheological properties at the temperatures of interest. The zero shear viscosity would be the best predictor, but DMA-determined melting or crossover points are good indicators of potential problems.

Once the encapsulant has been chosen, quick screening tests could be performed to evaluate creep as a module test. To ensure adequate performance, it has been proposed to expose modules to 200 h at 105°C as part of IEC 61215. In general, the probability of module creep being significant is very low compared to other risks.

Acknowledgments

This work was part of a collaborative effort of a number of people contributing to test standard development at many institutions. The authors gratefully acknowledge the support of the following individuals: Adam Stokes, Alain Blosse, Ann Norris, Bernd Koll, Bret Adams, Casimir Kotarba (Chad), David Trudell, Dylan Nobles, Ed Gelak,

Greg Perrin, Hirofumi Zenkoh, James Galica, Jayesh Bokria, John Pern, Jose Cano, Kartheek Koka, Kate Stika, Keith Emery, Kent Terwilliger, Kolapo Olakonu, Masaaki Yamamichi, Mowafak Al-Jassim, Nick Powell, Niki Nickel, Pedro Gonzales, Peter Hacke, Ryan Smith, Ryan Tucker, Steve Glick, Steve Rummel, Tsuyoshi Shioda, and Yefim Brun. This work was supported by the U.S. Department of Energy under Contract No. DE-AC36-08-GO28308 with the National Renewable Energy Laboratory.

Conflict of Interest

None declared.

References

1. Cuddihy, E. F., A. Gupta, C. D. Coulbert, R. H. Liang, A. Gupta, P. Willis, et al. 1983. Applications of ethylene vinyl acetate as an encapsulation material for terrestrial photovoltaic modules. DOE/JPL/1012-87 (DE83013509). Pasadena, California.

2. Dietrich, S., M. Pander, M. Ebert, and J. Bagdahn. 2008. Mechanical assessment of large photovoltaic modules by test and finite element analysis. 23rd European Photovoltaic Solar Energy Conference. Valencia, Spain.

3. Wohlgemuth, J., M. Kempe, D. Miller, and S. Kurtz. 2011. Developing standards for PV packaging materials. SPIE, San Diego, California.

4. Wohlgemuth, J., and W. Herrmann. 2005. Hot spot tests for crystalline silicon modules. 31st IEEE Photovoltaic Specialists Conference. pp. 1062-1063, 3–7 January 2005. ISSN 0160-8371, Lake Buena Vista, Florida.

5. Wohlgemuth, J. 2013. Hot spot testing of PV modules. SPIE, San Diego, California.

6. Kurtz, S., K. Whitfield, D. Miller, J. Joyce, J. H. Wohlgemuth, M. D. Kempe, et al. 2009. Evaluation of high-temperature exposure of rack-mounted pohotovoltaic modules. 34th IEEE PVSC. Philadelphia, Pennsylvania.

7. Miller, D. C., M. D. Kempe, S. H. Glick, and S. R. Kurtz. 2010. Creep in photovoltaic modules: Examining the stability of polymeric materials and components. 35th IEEE Photovoltaic Specialists Conference (PVSC), 20–25 June 2010. Honolulu, HI.

8. Cuddihy, E. F., and P. B. Willis. 1984. Antisoiling technology: theories of surface soiling and performance of antisoiling surface coatings. DOE/JPL/1012-102 (DE85006658), Pasadena, California.

9. Holley, W. W., and S. C. Agro. 1998. Advanced EVA-based encapsulants, Final Report January 1993-June 1997. NREL/SR-520-25296, Golden, Colorado.

10. Ehrich, C., S.-H. Schulze, I. Hinz, M. Pander, and M. Ebert. 2011. Development of a novel technique for

mechanical characterization of polymer encapsulantes in laminated glass beams. 26th European Photovoltaic Solar Energy Conference and Exhibition. Pp. 3560–3565, Hamburg, Germany.

11. Kempe, M. D., D. C. Miller, J. H. Wohlgemuth, S. R. Kurtz, J. M. Moseley, Q.-U.-A. S. J. Shah, et al. 2012. A field evaluation of the potential for creep in thermoplastic encapsulant materials. Proceedings of the 38th IEEE Photovoltaic Specialists Conference. Austin, Texas.

12. Moseley, J. M., D. C. Miller, Q.-U.-A. S. J. Shah, K. Sakurai, M. D. Kempe, G. TamizhMani, et al. 2011. The melt flow rate test in a reliability study of thermoplastic encapsulation materials in photovoltaic modules. NREL/TP-5200-52586. Pp. 1–20, Golden, Colorado.

13. IEC, 61215, Crystalline silicon terrestrial photovoltaic (PV) modules - Design qualification and type approval. ed: Ed. 2. 2005.

14. UL 1703, Standard for Safety for Flat-Plate Photovoltaic Modules and Panels. 2002.

15. IEC, 61730,Photovoltaic (PV) module safety qualification - Part 1: Requirements for construction. ed: Ed. 1.0. 2004.

16. ASTM G173-03, Standard tables for reference solar spectral irradiances: direct normal and hemispherical on 37 tilted surface. 2012.

17. King, D. L., W. E. Boyson, and J. A. Kratochvil. 2004. Photovoltaic Array Performance Model. SAND2004-3535. Sandia National Laboratories, Albuquerque, NM.

18. Shrestha, B. L., E. G. Palomino, and G. TamizhMani. 2009. Temperature of rooftop photovoltaic modules: air gap effects. Proceedings of the SPIE - The International Society for Optical Engineering. San Diego, California.

19. Jaewon, O., and G. TamizhMani. 2011. BAPV modules: Installed-NOCT and temperature coefficients after 1-year exposure. 37th IEEE Photovoltaic Specialists Conference (PVSC), 000838-000841, 19–24 June 2011. Seattle, Washington.

20. Hrica, J., S. Chatterjee, and G. TamizhMani. 2011. BAPV array: thermal modeling and cooling effect of exhaust fan. Pp. 003144–003149 Photovoltaic Specialists Conference (PVSC). 37th IEEE. Seattle, Washington.

21. Chatterjee, S., and G. TamizhMani. 2012. BAPV arrays: Side-by-side comparison with and without fan cooling. 38th IEEE Photovoltaic Specialists Conference (PVSC), 537-542, 3–8 June 2012, Austin, Texas.

22. Hrica, J., S. Chatterjee, and G. TamizhMani. 2011. BAPV array: Thermal modeling and cooling effect of exhaust fan. 37th IEEE Photovoltaic Specialists Conference (PVSC), 3144-3149, 19–24 June 2011. Seattle, Washington.

23. Pern, F. J., and A. W. Czanderna. 1992. Characterization of ethylene vinyl acetate (EVA) encapsulant: effects of thermal processing and

weathering degradation on its discoloration. Sol. Energy Mater. Sol. Cells 25:3–23.

24. Seddon, E., E. J. Tippett, and W. E. S. Turner. 1932. The Electrical Conductivity of Sodium Metasilicate-Silica Glasses. J. Soc. Glass Technol. 16:450–477.

25. Klemchuk, P., M. Ezrin, G. Lavigne, W. Holley, J. Galica, and S. Agro. 1997. Investigation of the degradation and stabilization of EVA-based encapsulant

in field-aged solar energy modules. Polym. Degrad. Stab. 55:347–365.

26. Chapuis, V., S. Pélisset, M. Raeis-Barnéoud, H.-Y. Li, C. Ballif, and L.-E. Perret-Aebi. 2014. Compressive-shear adhesion characterization of polyvinyl-butyral and ethylene-vinyl acetate at different curing times before and after exposure to damp-heat conditions. Prog. Photovoltaics Res. Appl. 22:405–414.

3

Renewable energy auctions in South Africa outshine feed-in tariffs

Anton Eberhard[1] & Tomas Kåberger[2]

[1]Graduate School of Business, University of Cape Town, Private Bag X3, Rondebosch 7701, South Africa
[2]Energy and Environment, Chalmers University of Technology, SE-412 96 Gothenburg, Sweden

Keywords
Auctions, competitive tenders, prices, renewable energy, South Africa

Correspondence
Anton Eberhard, Graduate School of Business, University of Cape Town, Private Bag X3, Rondebosch 7701, South Africa.
E-mail: eberhard@gsb.uct.ac.za

Funding Information
No funding information provided.

Abstract

South Africa's Renewable Energy Independent Power Producer Procurement Program has run four competitive tenders/auctions since 2011, which have seen US$19 billion in private investment, and electricity prices of wind power falling by 46% and solar PV electricity prices by 71%, in nominal terms. Competitive tenders were introduced after an unsuccessful attempt to implement feed-in tariffs. The tenders incorporated standard, nonnegotiable contract documents, including 20-year Power Purchase Agreements and an Implementation Agreement whereby the Government of South Africa back-stops IPP payments by the national utility, Eskom. All of these projects have reached financial close to date and some are already delivering power to the grid. The financing success has been due in part to the requirements for commercial banks to undertake a thorough due diligence of projects prior to bids being offered. The details of the policy package described may be useful for other policy makers in countries developing policies for renewable energy deployment.

Introduction

Costs of renewable energy technologies have fallen during recent years. The cost reductions are the result of many different factors, some related to technologies, others to finance, national institutional development, and increased competition. These developments would not have been possible without generous supporting policies in some pioneering countries, as suggested by Wene [1].

Developments in different countries may vary significantly depending on policy, legislation and regulatory frameworks, procurement practices, and market conditions. There are significant opportunities for learning lessons from countries that are achieving price reductions [2]. South Africa offers an interesting example of a renewable energy auctioning system that is attracting significant investment at highly competitive prices. The data and analysis in this article draw, in part, from an earlier study by Eberhard et al. [3] and have been updated with the latest data extracted from the South African Department of Energy's Independent Power Producer Office.

The latest grid-connected renewable energy auctions in South Africa have seen prices fall to among the lowest in the world with solar PV prices as low as USc 6.4/kWh and the cheapest wind at USc 4.7/kWh.[1] Over four bid rounds, between 2012 and 2015, wind energy has fallen by 46% and solar PV by 71% (in nominal, local currency terms) (Fig. 1).

South Africa occupies a central position in the global debate regarding the most effective policy instruments to accelerate and sustain private investment in renewable energy.

To date, a total of 92 projects have been contracted and private sector investment totaling US$19 billion has been committed for projects totaling 6327 megawatt (MW). There have also been notable economic development commitments in the form of local manufacture, employment creation, black economic empowerment, and community development [4]. Important lessons can be learned for both South Africa and other emerging markets contemplating investments in renewables and other critical infrastructure investments [3].

South African Renewable Energy Programme		
Bid round	Wind ZARc/kWh	PV ZARc/kWh
Round 1	114	276
Round 2	90	165
Round 3	74	99
Round 4	62	79
Round 4		
Cheapest ZARc/kWh	56	77
Average USc/kWh	5.2	6.6
Cheapest USc/kWh	4.7	6.4

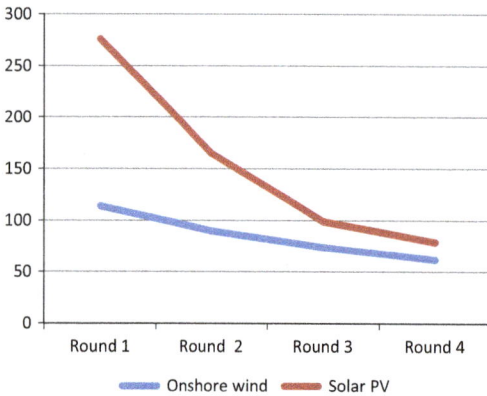

Figure 1. Average nominal bid prices in South Africa's renewable energy IPP program (ZARc/kWh). Source: Authors' compilation based on data provided by South Africa's DOE IPP Office.

From REFIT to REIPPPP

A REFIT policy was approved in 2009 by the national energy regulator of South Africa, NERSA. Tariffs were designed to cover generation costs plus a real, after tax return on equity of 17% that would be fully indexed for inflation. Initial published feed-in tariffs were generally regarded as generous by developers – 15.6 USc/kWh for wind, 26 USc/kWh for solar PV, and 49 USc/kWh for concentrated solar (troughs, with 6 h storage).[2] But considerable uncertainty about the nature of the procurement and licensing process remained. And the national utility, Eskom was less than enthusiastic in fully supporting the REFIT program by concluding power purchase agreements and interconnection agreements.

In March 2011, NERSA introduced a new level of uncertainty with a surprise release of a consultation paper calling for lower feed-in tariffs, arguing that a number of parameters, such as exchange rates and the cost of debt had changed. The new tariffs were 25% lower for wind, 13% lower for concentrated solar, and 41% lower for photovoltaic. Moreover, the capital component of the tariffs would no longer be fully indexed for inflation.

Importantly, in its revised financial assumptions, NERSA did not change the required real return for equity investors of 17% [3].

More policy and regulatory uncertainty was to come. Already concerned that NERSA's FITs were still too high, the Department of Energy and National Treasury commissioned the legal opinion that concluded that the feed-in tariffs amounted to noncompetitive procurement, and were therefore prohibited by the government's public finance and procurement regulations. The Department of Energy and National Treasury then took the lead on a reconsideration of the government's approach. The fundamental goal of achieving large-scale renewable energy projects with private developers and financiers remained the same. However, the structure of the transactions, including the feed-in tariffs, was to change significantly.

A series of informal consultations were held with developers, lawyers, and financial institutions throughout the first half of 2011. These meetings proved to be extremely important in terms of allaying market concerns resulting from the earlier REFIT process and providing informal feedback from the private sector on design, legal, and technology issues.

In August 2011, the Department of Energy (DOE) announced that a competitive bidding process for renewable energy would be launched, known as the Renewable Energy Independent Power Procurement Program. Subsequently, NERSA officially terminated the REFITs. Not a single megawatt of power had been signed in the 2 years since the launch of the REFIT program as a practical procurement process was never implemented, and the required contracts were never negotiated or signed. The abandonment of feed-in tariffs was met with dismay by a number of renewable energy project developers that had secured sites and initiated resource measurements and environmental impact assessments. But, it was these early developers who would later benefit from the first round of competitive bidding under REIPPPP.

Competitive tenders

In August 2011, a Request for Proposals was issued, and the next month a compulsory bidder's conference was held to address questions on bid requirements, documentation, power purchase agreements, etc. Some 300 organizations attended this conference. The REIPPPP program initially envisioned the procurement of 3625 MW of power over a maximum of five tender rounds. Another 100 MW was reserved for small projects below 5 MW that were procured in a separate small projects IPP program. Caps were set on the total capacity to be procured for individual technologies, the largest allocations were for wind and solar photovoltaics, with smaller amounts for concentrated solar, biomass, biogas, landfill gas, and hydro. The rationale for these caps was to limit the supply to be bid out, and therefore increased the level of competition among the different technologies and potential bidders.

The tenders for different technologies were held simultaneously. Interested parties could bid for more than one project and more than one technology. Projects had to be larger than 1 MW, and an upper limit was set on bids for different technologies, for example, 75 MW for a photovoltaic project, 100 MW for a concentrated solar project, and 140 MW for a wind project. Caps were also set on the price for each technology (at levels not dissimilar to NERSA's 2009 REFITs). Bids were due within 3 months of the release of the RFP, and financial close was to take place within 6 months after the announcement of preferred bidders.

The RPF was divided into three sections detailing: 1) general requirements, 2) qualification criteria, and 3) evaluation criteria. The documents also included a standard Power Purchase Agreement (PPA), an Implementation Agreement (IA), and a Direct Agreements (DA). The PPA was to be signed by the IPP and the Eskom, the off-taker. The PPAs specified that the transactions should be denominated in South Africa Rand and that contracts would have 20 year tenures from Commercial Operation Date (COD). The IAs were to be signed by the IPPs and the DOE and effectively provided a sovereign guarantee of payment to the IPPs, by requiring the DOE to make good on these payments in the event of an Eskom default. The IA also placed obligations on the IPP to deliver economic development targets. The DAs provided step-in rights for lenders in the event of default. The PPA, the IA and the DA were nonnegotiable contracts and were developed after an extensive review of global best practices and consultations with numerous public and private sector actors. Despite some bidder reservations regarding the lack of flexibility to negotiate the terms of the various agreements, the overall thoroughness and quality of the standard documents seemed to satisfy most of the bidders participating in the three rounds.

Bids were required to contain information on the project structure, legal qualifications, land, environmental, financial, technical, and economic development qualifications.

An important element of the design of the procurement process was to maximize the likelihood that winning bidders would able to execute the projects. Bidders had to submit bank letters indicating that the financing was locked-in highly unusual and basically a way to outsource due diligence to the banks. Effectively this meant that lenders took on a higher share of project development risk and this arrangement dealt with the biggest problem with auctions – the "low-balling" that results in deals not closing.

Further, the developers were expected to identify the sites and pay for early development costs at their own risk. A registration fee of US$1875 was due at the outset of the program. Bid bonds or guarantees had to be posted, equivalent to US$12,500 per megawatt of nameplate capacity of the proposed facilities, and the amount was doubled once preferred bidder status was announced. The guarantees are to be released once the projects come on line or if the bidder was unsuccessful after the RFP evaluation stage.

Project selection was based on a 70/30 split between price and economic development considerations. REIPPPP was able to adjust the normal government 90/10 split favoring price considerations in the procurement selection process. An exemption was obtained from the Public Preferential Procurement Framework Act in order to maximize economic development objectives.

The DOE IPP unit used a group of international and local experts to assess the bids. Many of these advisors had been involved in the initial design process. Given the scale of the investments, the competition anticipated, and the reputational risk identified, security, and confidentiality surrounding the evaluation process was extremely tight with 24-h voice and CCTV monitoring of the venue. Approximately 130–150 local and international advisors were used to develop the RFP and evaluate the bids in the first round, at a total cost of approximately US$10 million.

The bid evaluation involved a two-step process. First, bidders had to satisfy certain minimum threshold requirements in six areas: environment, land, commercial and legal, economic development, financial, and technical. For example, the environmental review examined approvals, while the land review looked at tenure, lease registration, and proof of land use applications. Commercial considerations included the project structure and the bidders' acceptance of the Power Purchase Agreement. The financial review included standard templates used for data collection that were linked to a financial model used by the evaluators. The technical specifications were set for each of the technologies. For example, wind developers were required to provide 12 months of wind data for the designated site and an independently verified generation forecast. The economic development requirements, in particular, were complex and generated some confusion among bidders.

Bids that satisfied the threshold requirements then proceeded to the second step of evaluation, where bid prices counted for 70% of the total score, with the remaining 30% of the score given to a composite score covering job creation, local content, ownership, management control, preferential procurement, enterprise development, and socioeconomic development. Bidders were asked to provide two prices: one fully indexed for inflation and the other partially indexed, with the bidders initially allowed to determine the proportion that would be indexed. In subsequent rounds, floors and caps were instituted for the proportion that could be indexed. The bids were evaluated using a standard financial model.

In the first round, 53 bids for 2128 MW of power-generating capacity were received. Ultimately 28 preferred bidders were selected offering 1416 MW for a total investment of nearly US$6 billion. Successful bidders realized that not enough projects were ready to meet the bid qualification criteria and that all qualifying bids were thus likely to be awarded contracts. Bid prices in the first round were thus close to the price caps set in the tender documents. Major contractual agreements were signed on November 5, 2012, with most projects reaching full financial close shortly thereafter. Construction on all of these projects has commenced with the first project coming on line in November 2013.

A second round of bidding was announced in November 2011. The total amount of power to be acquired was reduced, and other changes were made to tighten the procurement process and increase competition. Seventy-nine bids for 3233 MW were received in March 2012, and 19 bids were ultimately selected. Prices were more competitive, and bidders also offered better local content terms. Implementation, power purchase, and direct agreements were signed for all 19 projects in May 2013.

A third round of bidding commenced in May 2013, and again, the total capacity offered was restricted. In August 2013, 93 bids were received totaling 6023 MW. Seventeen preferred bidders were notified in October 2013 totaling 1456 MW. Prices fell further in round three. Local content again increased, and financial closure was expected in July 2014, but has been delayed a number of times because of uncertainties around Eskom transmission connections. A fourth round of bidding commenced in August 2014; 77 bids were received with 64 being compliant and 13 preferred bidders were announced in April 2015, totaling 1121 MW. Prices were so competitive that a further 13 projects were awarded totaling 1084.

Over the four bidding rounds, US$19 billion has been invested in 92 projects totaling 6327 MW.

Increased competition was no doubt the main driver for prices falling over the bidding rounds. But, there were other factors as well. International prices for renewable energy equipment have declined over the past few years due to a glut in manufacturing capacity, as well as ongoing innovation and economies of scale. REIPPPP was well positioned to capitalize on these global factors. Transaction costs were also lower in subsequent rounds, as many of the project sponsors and lenders became familiar with the REIPPPP tender specifications and process.

Now RE prices are reaching grid parity and there is the potential for other countries to explore how they can learn from the SA REIPPPP through lowering transaction costs and designing competitive tenders appropriate to local markets.

Conclusion

Over the past 4 years, South Africa's Renewable Energy Independent Power Producer Procurement Program has delivered remarkable investment and price outcomes which offer lessons for other countries on the potential benefits of competitive tenders or auctions.

Notes

[1] Prices fully indexed with inflation. ZAR/USD exchange deteriorated from 8 to 12 over period.
[2] These values are calculated at the exchange rate at the time of ZAR8/USD.

Conflict of interest

None declared.

References

1. Wene, C.-O., (2000). Experience Curves for Energy Technology Policy. OECD/IEA Paris. http://www.oecd-ilibrary.org/energy/experience-curves-for-energy-technology-policy_9789264182165-en
2. IRENA. 2015. Renewable energy auctions: a guide to design. International Renewable Energy Agency, Vienna.
3. Eberhard, A., J. Kolker, and J. Leigland. 2014. South Africa's renewable energy IPP procurement programme: success factors and lessons. Public Private Infrastructure Advisory Facility, World Bank, Washington, DC.
4. Baker, L., and H. L. Wlokas. 2015. South Africa's renewable energy procurement: a new frontier? Energy Research Centre. University of Cape Town, Cape Town, South Africa.

4

Influence of different SSF conditions on ethanol production from corn stover at high solids loadings

Arne Gladis*,†, Pia-Maria Bondesson, Mats Galbe & Guido Zacchi

Department of Chemical Engineering, Lund University, P.O. Box 124, SE-221 00 Lund, Sweden

Keywords
Fed-batch, prehydrolysis, solid loading, SSF, steam pretreatment

Correspondence
Pia-Maria Bondesson, Department of Chemical Engineering, Lund University, P.O. Box 124, SE-221 00, Lund, Sweden.

E-mail: Pia-Maria.Bondesson@chemeng.lth.se

Funding Information
The State Grid Corporation of China is most gratefully acknowledged for financial support of this project.

Abstract

In this study, three different kinds of simultaneous saccharification and fermentation (SSF) of washed pretreated corn stover with water-insoluble solids (WIS) content of 20% were investigated to find which one resulted in highest ethanol yield at high-solids loadings. The different methods were batch SSF, prehydrolysis followed by batch SSF and fed-batch SSF. Batch-SSF resulted in an ethanol yield of 75–76% and an ethanol concentration of 53 g/L. Prehydrolysis prior to batch SSF did not improve the ethanol yield compared with batch SSF. Fed-batch SSF, on the other hand, increased the yield, independent of the feeding conditions used (79–81%, 57–60 g/L). If the initial amount of solids during fed-batch SSF was lowered, the yield could be improved to some extent. When decreasing the enzyme dosage, the greatest decrease in yield was seen in the fed-batch mode (75%), while lower or the same yield was seen in batch mode with and without prehydrolysis (73%). This resulted in similar ethanol yields in all methods. However, the residence time to achieve the final ethanol yield was shorter using fed-batch. This shows that fed-batch can be a better alternative also at a lower enzyme loading.

Present address
†Department of Chemical and Biochemical Engineering, Center of Energy Resources Engineering, DTU Chemical Engineering, Technical University of Denmark, Søltofts Plads, DK-2800 Kgs. Lyngby, Denmark

Introduction

The demand for fuel is increasing alongside population growth due to requirements from an expanding transport and energy sector. However, it is well known that the combustion of fossil fuels has a negative impact on the environment [1]. One way of reducing environmental effects is to minimize the emission of greenhouse gases. This can be done by using biomass such as agricultural or forest residues to produce energy for heating or fuel. These materials bind atmospheric carbon dioxide during growth and can, therefore, reduce greenhouse gas emissions. However, progress in the introduction of biomass-based fuels is slow since their production is often more expensive than fossil fuels [2]. Therefore, efforts should be devoted to reducing the process cost, for example, increasing the overall yield from the substrate and increasing the ethanol concentration [2].

We have previously investigated the conversion of corn stover into ethanol, methane, and a lignin-rich solid residue [3–5]. The goal of those studies was to evaluate the combined production of ethanol and methane to determine which process or processes will result in the highest energy recovery from the raw material, while at the same time producing high amounts of ethanol. Two different process configurations were investigated, one based on whole slurry and the other on separate liquid and solid streams. The configuration that resulted in the highest energy recovery was the one in which the solid and the liquid phases were separated after pretreatment. The solid phase was used for ethanol production using simultaneous saccharification and fermentation (SSF), while the liquid phase was used for methane production by anaerobic digestion. However, although the energy recovery in ethanol, methane, and lignin was high (76–88%), the ethanol concentrations were only moderate (20–26 g/L).

The cost of producing ethanol from lignocellulosic materials is considerably higher than producing ethanol from starch or sugar, or fuels derived from oil. Less energy would be required in the distillation step if the ethanol concentration in the fermentation broth could be increased, leading to a reduction in the production cost [6]. One way of increasing the ethanol concentration is to use higher solids loading in SSF. However, it has been shown in many studies that higher solids loading decrease the ethanol yield [7–11]. The production cost is very sensitive to both the ethanol yield and the ethanol concentration. The latter is mainly due to the energy requirement for distillation, but also the capital cost, mainly for reactors (i.e. pretreatment, hydrolysis and fermentation tanks) as lower flow rates are processed.

One of the main problems associated with high solids loading is the increased resistance to mass transfer enhanced by the difficulty of proper mixing in the beginning [7, 12, 13]. However, other studies, for example, Kristensen et al. [14], conclude that insufficient mixing may not be the main cause for decreased yield, but suggest that inhibition may be the reason for decreased yield. Prehydrolysis of the substrate to reduce the viscosity before adding the yeast to the reactor improves mixing. This allows a higher hydrolysis temperature to be used, which is favourable for the enzymes. This commonly results in a shorter hydrolysis time and also a shorter running time at high viscosity, reducing stirring problems. Another option is to use the fed-batch mode in the SSF, where the amount of solids initially added to the reactor is smaller than in the batch mode, resulting in less power being required for stirring [15].

The aim of this study was to increase the ethanol concentration in SSF by increasing the water insoluble solids (WIS) content from 10% to 20%, while minimizing the effect on the ethanol yield. The same basic process configuration was used as in previous studies [3–5], that is, after pretreatment, the liquid and solid phase were separated, and the solid phase was used in SSF. In this study, different SSF conditions on the solid phase were investigated and compared with batch-wise SSF. The time and temperature during prehydrolysis prior to batch SSF were varied, and different fed-batch conditions were tested.

The effect of decreasing the enzyme dosage on the different SSF conditions was also investigated.

Materials and Methods

Raw material and steam pretreatment

Moist corn stover was provided by the State Grid Corporation of China (Handan City, Hebei Province in China). The corn stover was air dried at room temperature for some weeks, and turned over many times during drying to reduce the risk of molding during storage. The dry matter (DM) content after drying was 90%. The dried corn stover was soaked in an aqueous solution containing 0.4% H_3PO_4 by mass, at room temperature for 1 h. The liquid/solid ratio was 20 kg/kg dry corn stover. The material was then dewatered in a small laboratory high-pressure press (Tinkturenpressen HP5M; Fischer Maschinenfabrik GmbH, Burgkunstadt, Germany) to a DM content of 48–50%, and then steam pretreated. Steam pretreatment was carried out in a preheated 10 L reactor, as described previously [4], at 190°C for 10 min. Several runs were performed and then the material was thoroughly mixed to form a single large batch that was stored at 4°C.

Simultaneous saccharification and fermentation

Three kinds of SSF set-ups were performed: batch SSF, prehydrolysis followed by batch SSF and fed-batch SSF, as illustrated in Figure 1. For these set-ups different conditions were investigated. During prehydrolysis different times (4, 8, 24 and 48 h) and temperatures (45°C, 50°C, 55°C, and 60°C) of the prehydrolysis step were studied, and during fed-batch SSF, different substrate feeding strategies were investigated. The amount of material being added from the beginning (50% and 33% of total), the starting time of feeding (4, 8 and 12 h), the feeding interval (every second, fourth and 12th hour) and the total feeding time (12, 16 and 24 h) were investigated in six different feeding strategies (Table 1). The results from the different conditions in these cases were then compared with each other and with batch SSF. In addition, the effect of a decreasing enzyme dosage (7.5 FPU/g WIS instead of 10) was investigated in one case in the three kinds of SSF. All the experiments were performed on washed material except one that was run on unwashed material. In this case, the configuration was performed with fed-batch and feeding strategy F.

All SSF experiments were carried out in 2 L fermenters (Infors AG, Bottmingen, Switzerland) equipped with a pitched-blade impeller turbine and an anchor impeller. The final working weight in the fermenters was 1.0 kg.

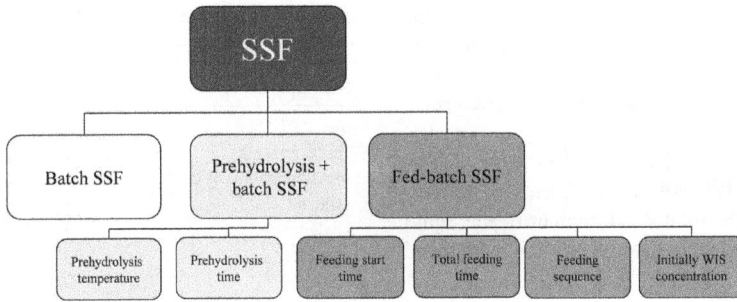

Figure 1. Overview of the modes of SSF and parameters studied. SSF, simultaneous saccharification and fermentation.

Table 1. Feeding strategies for fed-batch SSF.

Feeding strategy	WIS added at start as % of total amount used	Total feeding time (h)[1]	Time of first addition (h)	Feeding interval (h)
A	50	12	4	2
B	50	12	4	4
C	50	24	12	12
D	50	16	4	4
E	50	16	8	4
F	33	16	4	4

SSF, simultaneous saccharification and fermentation; WIS, water-insoluble solids.
[1]From the start of SSF until the last addition.

The material was washed before SSF, using the same procedure as in previous studies [3–5]. The pretreated material was dewatered in a filter press to a DM content between 40% and 50%. The same amount of water as had been pressed out was added, and the material was then dewatered to the same DM content as before water addition. In the case where the material was unwashed, the material was pressed to a DM content between 40% and 50% and the liquid was added back during SSF instead of sterile water.

The total duration of each of the SSF experiments was 144 h. The batch and batch with prehydrolysis experiments were carried out with WIS contents of 19.2% or 20%. The lower WIS content was due to a slightly incorrect measurement of the WIS content in the starting material. However, the difference in WIS is small, which will have a negligible effect on the final results. In the batch experiments all the solids, the enzymes, the nutrients and the yeast were added together at start. The SSF temperature was set to 35°C. In the batch experiments with prehydrolysis, only the enzymes, the nutrients and substrate were added initially. Four prehydrolysis temperatures were investigated: 45°C, 50°C, 55°C and 60°C, and

four residence times: 4, 8, 24 and 48 h. After prehydrolysis the reactor was cooled to 35°C and the yeast was added. The time required for the reactor to cool down was between 15 and 30 min.

The initial WIS content in fed-batch SSF was 13% or 9.5%. Yeast, enzymes, nutrients, and required water was initially added to the reactor together with 50% or 33% of the total amount of WIS, respectively. The remaining material was added in equal amounts after 4, 8 or 12 h at regular intervals (Table 1). The total amount of solid material corresponded to an overall WIS content of 20%. The washed pretreated material was diluted with sterile water to obtain the desired WIS concentration in all experiments. In the case of unwashed material pretreatment liquid, which was pressed out, was added instead. Nutrients were added to the fermenter in all experiments to final concentrations of 0.5 g/L $(NH_4)_2HPO_4$ and 0.025 g/L $MgSO_4 \cdot 7H_2O$. The enzyme mixture, Cellic CTec2 (Novozymes, Bagsvaerd, Denmark), was added at an amount corresponding to 7.5 or 10 FPU/g WIS (based on the overall amounts of pretreated material added). Dry baker's yeast, *Saccharomyces cerevisiae* (Jästbolaget AB, Rotebro, Sweden), was added to the fermenters to give a concentration of 3 g/L based on the volume when all material was added. The experiments started with the addition of the enzyme mixture and the yeast (except when performing prehydrolysis). The substrates with high WIS contents were difficult to mix, and stirring was thus started after some enzymatic hydrolysis had occurred (after ~1 h). The pH was set to 5 using a 10% NaOH solution (by mass). Samples were collected regularly throughout SSF for analysis using high performance liquid chromatography (HPLC).

Analysis

All DM contents were determined by drying the material in an oven at 105°C until constant weight. The WIS contents of the different slurries were determined from the

DM content of the slurry and the DM content of the hydrolysate, according to the method of Weiss et al. [16].

The compositions of raw corn stover and WIS after pretreatment were determined using the standard procedure "Determination of structural carbohydrates and lignin in biomass" from the National Renewable Energy Laboratory (NREL) [17]. All measurements were performed in triplicate. The total sugar content of the liquid fractions after pretreatment (hydrolysate) was analyzed according to the corresponding NREL procedure [18].

HPLC was used for the analysis of sugars, ethanol and by-products using a chromatographic system equipped with a differential refractive index detector (RID-10A) (both from Shimadzu, Kyoto, Japan). All samples were passed through a filter with a pore diameter of 0.20 μm prior to analysis to remove particles. The filtered samples were stored at $-20°C$ before analysis. The liquid from the slurry after steam pretreatment and samples from SSF were diluted if necessary, and analyzed using an Aminex HPX-87H column (Bio-Rad, Hercules, CA) at 50°C with 5 mmol/L H_2SO_4 as eluent, at a flow rate of 0.5 mL/min, to separate ethanol, lactic acid, acetic acid, formic acid, levulinic acid, HMF and furfural. The liquid from the NREL analysis, the liquid after steam pretreatment and that obtained after SSF were analyzed with an Aminex HPX-87P column (Bio-Rad) at 85°C with deionized water as eluent, at a flow rate of 0.5 mL/min, to separate monomeric sugars (glucose, xylose, galactose, arabinose and mannose). All acidic samples analyzed on this column had been previously neutralized using solid $CaCO_3$.

Results and Discussion

Raw material and pretreatment

The raw material consisted mainly of glucan (40.4%), xylan (24.5%) and lignin (22.5%, including both acid-soluble and acid-insoluble lignin), together with small amounts of galactan (3.0%) and arabinan (5.3%). The ash content was 2.3%. The corn stover was steam pretreated with 0.4% phosphoric acid in two different batches at 190°C for 10 min. The results of pretreatment of the two batches differed slightly, as can be seen in Table 2, where the glucan content and glucose concentrations were lower in the second batch, and the lignin content was somewhat higher. However, a mass balance on glucose results in the same glucan recovery (results not shown). Also the results from the batch experiments that were run with the two different batches gave similar yields (75% and 76%, respectively). Therefore, the results from SSF runs from the different pretreatment batches are comparable even though the lignin content differs.

Table 2. Compositions of the solid and liquid fractions in the two batches after pretreatment.

	Batch 1	Batch 2
Total solids (%)	13.5	13.9
WIS content (%)	10.2	10.6
Solid fraction (% of WIS)		
Glucan	59.6	55.7
Xylan	9.1	8.8
Galactan	nd	nd
Arabinan	0.5	0.7
Lignin	26.2	30.5
Ash	5.6	5.9
Liquid fraction (g/L)[1]		
Glucose	6.6	3.8
Xylose	21.4	19.1
Galactose	2.2	2.2
Arabinose	1.6	1.6

WIS, water-insoluble solids; nd, not detected.
[1]Oligomer and monomer sugars.

Effect on ethanol yield of increasing WIS

The ethanol yields obtained with batch SSF were similar to, or slightly higher than, those obtained in our previous study [5], where the same pretreatment conditions were used but only 10% WIS. In that study, the ethanol yield was found to be 74%, which can be compared with yields ranging from 75% to 76% in this study using 20% WIS in batch SSF. A similar pattern has also been observed in a previous study, where the ethanol yield did not decrease with increasing WIS concentration when washed material was used [19]. Others have reported a decrease in the ethanol yield from SSF with increasing solids loading [7, 9–11]. However, unwashed material was used in these studies, which contains all the inhibitors formed during the pretreatment. Therefore, one experiment was made in this study to verify that lower yields were obtained with unwashed material. The experiment was run in the same way as the SSF resulting in highest yield to have conditions that were suitable for the yeast (fed-batch, feeding strategy F). In this experiment, pretreatment liquid was added instead of water to get the correct WIS content. This resulted in only a small ethanol formation. However, when using the same pretreatment conditions and unwashed material in our previous study, ethanol was formed to the same extent as with washed material. It can thus be concluded that the ethanol yield does not decrease to the same extent with washed material as unwashed material. This was also confirmed in a study by Hodge et al. [20], showing that enzyme inhibition is due to soluble components rather than insoluble components. Mohagheghi et al. [21]. reported that washed material resulted in a decrease in ethanol yield with increasing solids loading, and concluded that the yeast cells were partially inhibited at ethanol concentrations above 55 g/L. This was not seen in the present study, where the ethanol yield did not decrease compared with that using 10% WIS, despite ethanol concentrations between 44 and 60 g/L.

Influence of prehydrolysis temperature on batch SSF

The yeast was added to batch SSF after 24 h of prehydrolysis at the temperatures 45°C, 50°C, 55°C and 60°C. The highest glucose concentration prior to yeast addition (111.5 g/L) was obtained with a prehydrolysis temperature of 50°C, while 55°C resulted in a slightly lower concentration of glucose (Fig. 2A). Prehydrolysis temperatures of 45°C and 60°C yielded lower glucose concentrations of 96.1 and 92.4 g/L, respectively. Fermentation was rapid, and started as soon as the yeast was added to the reactor in all cases. The highest ethanol concentration and yield were obtained with prehydrolysis temperatures of 45°C and 50°C (Fig. 2B and Table 3). The ethanol concentration decreased as the prehydrolysis temperature was increased above 50°C. The optimal temperature for the activity of the enzymes is 50°C, and they are partly deactivated at higher temperatures, resulting in lower yields. However, when investigating the ethanol yield, a lower temperature, for example, 45°C can also be used to obtain the same yield. A lower temperature can be benefi-

cial to decrease the cost of heating the reactor if the residence time is the same for both the processes. It has also been shown that the optimal hydrolysis temperature is depending on the residence time [22].

Influence of prehydrolysis time on batch SSF

The effect of different prehydrolysis times at a prehydrolysis temperature of 50°C on batch SSF was also evaluated. The glucose concentration, before yeast addition, was lower with shorter prehydrolysis time but the same after 24 and 48 h of prehydrolysis (Fig. 3A). The ethanol concentration and yield were slightly lower or approximately the same with prehydrolysis as without prehydrolysis (Fig. 3B and Table 3). The largest deviation was seen with prehydrolysis for 48 h. The glycerol concentration increased with longer prehydrolysis time (Table 3).

The ethanol yield was only slightly affected by varying the duration of prehydrolysis, but the glycerol concentration increased with increasing prehydrolysis time. This indicates that a longer prehydrolysis time favours cell growth. No effect, or a small negative effect, has been reported in some previous studies after prehydrolysis prior to batch SSF [7, 23], while a clear improvement in ethanol yield has been found in other studies when using prehydrolysis [7, 24]. However, in those studies was unwashed material used. The conclusion from Hoyer et al. [7, 24]. was that when batch SSF results in high yields, prehydrolysis does not improve the outcome. This was also confirmed in the present study. Although the effect on ethanol yield was small, prehydrolysis resulted in faster liquefaction. It has been shown that power consumption is highly dependent on the viscosity of the slurry [25]. Therefore, less stirring power will be required when using prehydrolysis than in batch SSF without prehydrolysis.

Fed-batch SSF

The fed-batch mode was employed to reduce stirring problems during SSF. No major differences were seen when varying the feeding interval or the total feeding time during fed-batch SSF (Fig. 4A). The ethanol yield was the same with 2-h feeding intervals and 4-h feeding intervals, during a total duration of 12 h. The same results were also obtained using a feeding interval of 12 h for a total feeding time of 24 h.

A total feeding time of 16 h was also investigated, where feeding was started after 4 or 8 h with a 4-h feeding interval. The second pretreatment batch was used in these experiments, which resulted in lower ethanol concentrations. The total ethanol yields were very similar in all experiments. These results is in agreement with results presented in previous studies where the feeding interval

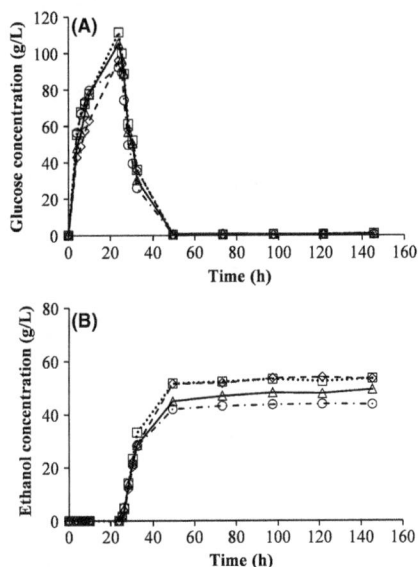

Figure 2. Measured concentrations of glucose (A) and ethanol (B) during batch simultaneous saccharification and fermentation with 24 h prehydrolysis at prehydrolysis temperatures of: 45°C (–◇–), 50°C (···□···), 55°C (—△—) and 60°C (– •O– •).

Table 3. Summary of SSF experiments. Ethanol yields and final concentrations of ethanol, glycerol and glucose.

Expt no.	SSF mode	Pretreatment batch	Experimental conditions[1]	WIS (%)[2]	Enzyme dosage (FPU/ g WIS)	Ethanol yield (%)[3]	Ethanol (g/L)	Glucose (g/L)	Glycerol (g/L)
1	Batch	1	0 h	19.2	10	75 (69)	53.8	0.5	4.0
2	Batch	2	0 h	20.0	10	76 (70)	53.4	0.3	3.8
3	Prehydrol.	1	24 h (45°C)	19.2	10	74 (68)	53.4	1.2	4.3
4	Prehydrol.	1	24 h (50°C)	19.2	10	75 (69)	53.5	1.0	5.1
5	Prehydrol.	1	24 h (55°C)	19.2	10	69 (63)	49.4	0.6	5.2
6	Prehydrol.	1	24 h (60°C)	19.2	10	60 (56)	44.8	0.4	5.3
7	Prehydrol.	1	4 h (50°C)	19.2	10	73 (67)	52.3	1.1	4.1
8	Prehydrol.	1	8 h (50°C)	19.2	10	72 (66)	51.5	0.5	4.6
9	Prehydrol.	1	48 h (50°C)	19.2	10	70 (65)	49.2	0.5	5.9
10	Fed-batch	1	A	20.0	10	79 (73)	60.2	0.5	3.9
11	Fed-batch	1	B	20.0	10	79 (73)	60.0	0.2	3.7
12	Fed-batch	1	C	20.0	10	79 (73)	59.5	0.2	4.1
13	Fed-batch	2	D	20.0	10	80 (73)	56.9	0.4	3.6
14	Fed-batch	2	E	20.0	10	80 (73)	56.8	0.2	3.6
15	Fed-batch	2	F	20.0	10	81 (75)	58.0	0.3	3.4
16	Fed-batch	2	F[4]	20.0	10	–	0.7	84.2	0.8
17	Batch	2	0 h	20.0	7.5	73 (67)	51.8	0.4	3.7
18	Prehydrol.	2	8 h (50°C)	20.0	7.5	73 (67)	52.6	0.3	3.4
19	Fed-batch	2	F	20.0	7.5	75 (69)	53.4	0.2	4.3

SSF, simultaneous saccharification and fermentation; WIS, water-insoluble solids.
[1]Duration of prehydrolysis and temperature (in brackets) during batch SSF, feeding strategy during fed-batch SSF.
[2]Final concentration when all material is added.
[3]Ethanol yield based on the glucan content in pretreated material and (raw material) as % of the theoretical.
[4]Unwashed material used.

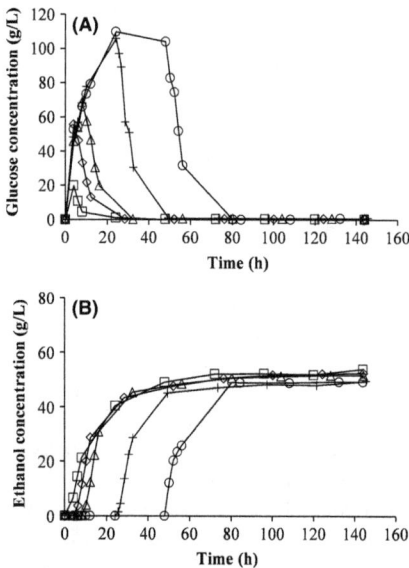

Figure 3. Measured concentration of glucose (A) and ethanol (B) during batch simultaneous saccharification and fermentation with prehydrolysis at 50°C for various times: 0 h (no prehydrolysis) (□), 4 h (◊), 8 h (Δ), 24 h (+), 48 h (O).

was varied [26, 27]. In those studies the total ethanol yield was also similar, independent on feeding interval. The initial solid loading was also varied in this series of experiments. One-third of the total WIS used in SSF was added at the beginning of one experiment (no. 15, Table 3) instead of 50% as in the other experiments. This might have resulted in a slightly higher ethanol concentration and yield than with an initial loading of 50% of the total WIS (Fig. 4B). More experiments are needed to verify if there is a significant difference when using a lower initial loading.

The ethanol yield was higher in fed-batch SSF than in batch SSF with and without prehydrolysis (Table 3). The glycerol concentration was similar to, or lower than, that obtained with batch SSF. This indicates that there was less cell growth during fed-batch SSF, resulting in more glucose being available for ethanol production. The findings of other studies are rather diverse. It has been reported that the ethanol yield was the same using fed-batch and batch SSF [21, 26], while others have reported fed-batch to be superior to batch SSF [28, 29]. In a study by Hoyer et al. [30], it was concluded that the enzyme-feeding strategy affected the ethanol yield. The strategy depends on a number of factors, for example, WIS content and

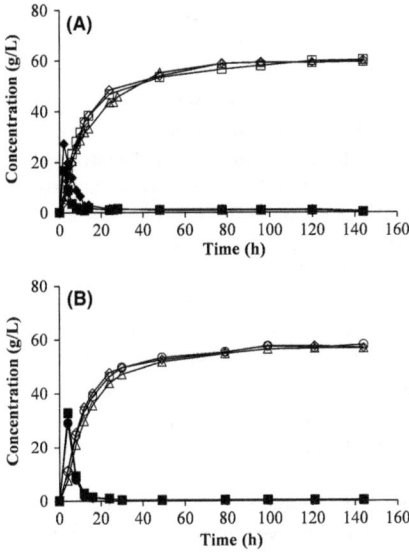

Figure 4. Concentrations of glucose (filled symbols) and ethanol (open symbols) during fed-batch SSF. (A) Fed-batch with feeding intervals of 2 h (□), 4 h (◊) and 12 h (Δ). (B) Fed-batch with a total feeding time of 16 h and feeding interval of 4 h, with different initial WIS loadings at different starting times: 50% of total WIS after 4 h (◊), 50% of total WIS after 8 h (Δ) and 33% of total WIS after 4 h (O). SSF, simultaneous saccharification and fermentation; WIS, water-insoluble solids.

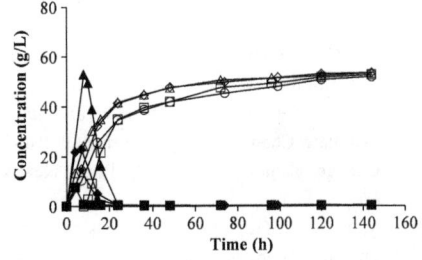

Figure 5. Concentrations of glucose (filled symbols) and ethanol (open symbols) during different SSF experiments. Batch SSF with 10 FPU/g WIS (◊); batch SSF with 7.5 FPU/g WIS (O); batch SSF with 8 h prehydrolysis at 50°C with 7.5 FPU/g WIS (□), fed-batch SSF with an initial WIS load of 33% of the total WIS and a total feeding time of 16 h with 7.5 FPU/g WIS (Δ). SSF, simultaneous saccharification and fermentation; WIS, water-insoluble solids.

inhibitor concentration. Different enzyme feeding strategies were not investigated in the present study.

Influence of enzyme loading

The enzyme loading can be reduced to minimize the overall production cost. However, this often results in a lower ethanol yield or a longer residence time. The enzyme loading was decreased from 10 to 7.5 FPU/g WIS to study the effects on SSF. A small or no decrease in ethanol yield was seen in batch SSF with and without prehydrolysis (comparing experiments 1 and 2 with 17 and experiment 8 with 18 in Table 3). But there is a difference in residence time as lower enzyme dosage results in longer residence time to reach the final ethanol yield. This is probably due to that more enzymes being partially or completely deactivated, but also because the amount of active enzyme is lower. The greatest decrease in ethanol yield was seen in fed-batch SSF (comparing experiment 15 with 19 in Table 3). The difference between the different configurations disappears with the lower enzyme dosage (Fig. 5). This could indicate that a lower enzyme dosage is more favourable using the batch mode than fed-batch. One reason is that all enzymes are added in the beginning, which can result in unproductive binding

of the enzyme on the initially added substrate. Using fed-batch, there is a risk that the enzymes will not efficiently hydrolyze the fed material.

Since the ethanol yield was similar in all configurations at the lower enzyme dosage, the residence time in the various configurations is an important factor to establish the most suitable SSF mode. The time to reach high ethanol concentration is lower with fed-batch than with batch with lower enzyme dosage. The residence time is also similar to the batch with original enzyme loading. As the residence time is shorter using the fed-batch mode, this indicates that fed-batch is still a better alternative, even when the enzyme loading is lowered by 25%.

Conclusions

Prehydrolysis did not improve the ethanol yield from batch SSF. A higher ethanol yield was seen when using fed-batch than when using batch, indicating that fed-batch is a better alternative at high solids loading. When decreasing the enzyme dosage, a small or no decrease in ethanol yield was seen in batch SSF. The decrease in ethanol yield was greater during fed-batch SSF, resulting in similar ethanol yield in both batch and fed-batch. However, the residence time to achieve the final ethanol yield is shorter with fed-batch. This shows that fed-batch is the better alternative also at a lower enzyme loading.

Acknowledgments

The State Grid Corporation of China is most gratefully acknowledged for financial support of this project.

Conflict of Interest

None declared.

References

1. IPCC. 2013. Climate change 2013: the physical science basis. Contribution of Working Group I to the Fifth Assessment Report of the Intergovernmental Panel on Climate Change. Intergovernmental Panel on Climate Change, Cambrige University Press, New York, NY.

2. Carriquiry, M. A., X. Du, and G. R. Timilsina. 2011. Second generation biofuels: economics and policies. Energy Policy 39:4222–4234.

3. Bondesson, P.-M., M. Galbe, and G. Zacchi. 2013. Ethanol and biogas production after steam pretreatment of corn stover with or without the addition of sulphuric acid. Biotechnol. Biofuels 6:11.

4. Bondesson, P.-M., M. Galbe, and G. Zacchi. 2014. Comparison of energy potentials from combined ethanol and methane production using steam-pretreated corn stover impregnated with acetic acid. Biomass Bioenergy 67:413–424.

5. Bondesson, P.-M., A. Dupuy, M. Galbe, and G. Zacchi. 2015. Optimizing ethanol and methane production from steam-pretreated, phosphoric acid-impregnated corn stover. Appl. Biochem. Biotechnol. 175:1371–1388. doi: 10.1007/s12010-014-1358-4

6. Wingren, A., M. Galbe, and G. Zacchi. 2003. Techno-economic evaluation of producing ethanol from softwood: comparison of SSF and SHF and identification of bottlenecks. Biotechnol. Prog. 19:1109–1117.

7. Hoyer, K., M. Galbe, and G. Zacchi. 2009. Production of fuel ethanol from softwood by simultaneous saccharification and fermentation at high dry matter content. J. Chem. Technol. Biotechnol. 84:570–577.

8. Humbird, D., A. Mohagheghi, N. Dowe, and D. J. Schell. 2010. Economic impact of total solids loading on enzymatic hydrolysis of dilute acid pretreated corn stover. Biotechnol. Prog. 26:1245–1251.

9. Jorgensen, H., J. Vibe-Pedersen, J. Larsen, and C. Felby. 2007. Liquefaction of lignocellulose at high-solids concentrations. Biotechnol. Bioeng. 96:862–870.

10. Öhgren, K., A. Rudolf, M. Galbe, and G. Zacchi. 2006. Fuel ethanol production from steam-pretreated corn stover using SSF at higher dry matter content. Biomass Bioenergy 30:863–869.

11. Varga, E., H. B. Klinke, K. Reczey, and A. B. Thomsen. 2004. High solid simultaneous saccharification and fermentation of wet oxidized corn stover to ethanol. Biotechnol. Bioeng. 88:567–574.

12. Kadić, A., B. Palmqvist, and G. Lidén. 2014. Effects of agitation on particle-size distribution and enzymatic hydrolysis of pretreated spruce and giant reed. Biotechnol. Biofuels 7:77.

13. Rosgaard, L., P. Andric, K. Dam-Johansen, S. Pedersen, and A. S. Meyer. 2007. Effects of substrate loading on enzymatic hydrolysis and viscosity of pretreated barley straw. Appl. Biochem. Biotechnol. 143:27–40.

14. Kristensen, J. B., C. Felby, and H. Jorgensen. 2009. Yield-determining factors in high-solids enzymatic hydrolysis of lignocellulose. Biotechnol. Biofuels 2:11.

15. Palmqvist, B., and G. Liden. 2012. Torque measurements reveal large process differences between materials during high solid enzymatic hydrolysis of pretreated lignocellulose. Biotechnol. Biofuels 5:57.

16. Weiss, N. D., J. J. Stickel, J. L. Wolfe, and Q. A. Nguyen. 2010. A simplified method for the measurement of insoluble solids in pretreated biomass slurries. Appl. Biochem. Biotechnol. 162:975–987.

17. Sluiter, A., B. Hames, R. Ruiz, C. Scarlata, J. Sluiter, D. Templeton, et al. 2008. Determination of structural carbohydrates and lignin in biomass. Laboratory analytical procedure, NREL, Golden, CO.

18. Sluiter, A., B. Hames, R. Ruiz, C. Scarlata, J. Sluiter, and D. Templeton. 2006. Determination of sugars, byproducts, and degradation products in liquid fraction process samples. Laboratory analytical procedure, NREL, Golden, CO.

19. Lu, Y., Y. Wang, G. Xu, J. Chu, Y. Zhuang, and S. Zhang. 2010. Influence of high solid concentration on enzymatic hydrolysis and fermentation of steam-exploded corn stover biomass. Appl. Biochem. Biotechnol. 160:360–369.

20. Hodge, D. B., M. N. Karim, D. J. Schell, and J. D. McMillan. 2008. Soluble and insoluble solids contributions to high-solids enzymatic hydrolysis of lignocellulose. Bioresour. Technol. 99:8940–8948.

21. Mohagheghi, A., M. Tucker, K. Grohmann, and C. Wyman. 1992. High solids simultaneous saccharification and fermentation of pretreated wheat straw to ethanol. Appl. Biochem. Biotechnol. 33:67–81.

22. Tengborg, C., M. Galbe, and G. Zacchi. 2001. Influence of enzyme loading and physical parameters on the enzymatic hydrolysis of steam-pretreated softwood. Biotechnol. Prog. 17:110–117.

23. Öhgren, K., J. Vehmaanperä, M. Siika-Aho, M. Galbe, L. Viikari, and G. Zacchi. 2007. High temperature enzymatic prehydrolysis prior to simultaneous saccharification and fermentation of steam pretreated corn stover for ethanol production. Enzyme Microb. Technol. 40:607–613.

24. Hoyer, K., M. Galbe, and G. Zacchi. 2013. The effect of prehydrolysis and improved mixing on high-solids batch simultaneous saccharification and fermentation of spruce to ethanol. Process Biochem. 48:289–293.

25. Palmqvist, B., M. Wiman, and G. Lidén. 2011. Effect of mixing on enzymatic hydrolysis of steam-pretreated spruce: a quantitative analysis of conversion and power consumption. Biotechnol. Biofuels 4:10.

26. Rudolf, A., M. Alkasrawi, G. Zacchi, and G. Lidén. 2005. A comparison between batch and fed-batch simultaneous saccharification and fermentation of steam pretreated spruce. Enzyme Microb. Technol. 37:195–204.

27. Zhang, M., F. Wang, R. Su, W. Qi, and Z. He. 2010. Ethanol production from high dry matter corncob using fed-batch simultaneous saccharification and fermentation after combined pretreatment. Bioresour. Technol. 101:4959–4964.

28. Nilsson, A., M. J. Taherzadeh, and G. Lidén. 2001. Use of dynamic step response for control of fed-batch conversion of lignocellulosic hydrolyzates to ethanol. J. Biotechnol. 89:41–53.

29. Rudolf, A., M. Galbe, and G. Liden. 2004. Controlled fed-batch fermentations of dilute-acid hydrolysate in pilot development unit scale. Appl. Biochem. Biotechnol. 113:601–617.

30. Hoyer, K., M. Galbe, and G. Zacchi. 2010. Effects of enzyme feeding strategy on ethanol yield in fed-batch simultaneous saccharification and fermentation of spruce at high dry matter. Biotechnol. Biofuels 3:14.

5

Greenhouse gas emissions from domestic hot water: heat pumps compared to most commonly used systems

Bongghi Hong[1] & Robert W. Howarth[2]

[1]Department of Ecology and Evolutionary Biology, Cornell University, 103 Little Rice Hall, Ithaca, New York 14853
[2]Department of Ecology and Evolutionary Biology, Cornell University, E309 Corson Hall, Ithaca, New York 14853

Keywords
Heat pump water heater, lifecycle emission, methane, natural gas, shale gas, technology warming potential

Correspondence
Bongghi Hong, Department of Ecology and Evolutionary Biology, Cornell University, 103 Little Rice Hall, Ithaca, NY 14853.

E-mail: bh43@cornell.edu

Funding Information
This study was funded by the Wallace Global Fund, the Park Foundation, and Cornell University.

Abstract

We estimate the emissions of the two most important greenhouse gasses (GHG), carbon dioxide (CO_2) and methane (CH_4), from the use of modern high-efficiency heat pump water heaters compared to the most commonly used domestic hot water systems: natural gas storage tanks, tankless natural gas demand heaters, electric resistance storage tanks, and tankless electric resistance heaters. We considered both natural gas-powered electric plants and coal-powered plants as the source of the electricity for the heat pumps, the thermal electric storage tanks, and the tankless electric demand heaters. The time-integrated radiative forcing associated with using a heat pump water heater was always smaller than any other means of heating water considered in this study across all time frames including at 20 and 100 years. The estimated amount of CH_4 lost during its lifecycle was the most critical factor determining the relative magnitude of the climatic impact. The greatest net climatic benefit within the 20-year time frame was predicted to be achieved when a storage natural gas water heater (the most common system for domestic hot water in the United States) fueled by shale gas was replaced with a high efficiency heat pump water heater powered by coal-generated electricity; the heat pump system powered by renewable electricity would have had an even greater climatic benefit, but was not explicitly modeled in this study. Our analysis provides the first assessment of the GHG footprint associated with using a heat pump water heater, which we demonstrate to be an effective and economically viable way of reducing emissions of GHGs.

Introduction

Each of the last three decades has been consecutively the warmest on record since the start of the industrial revolution, and the global temperature will continue to rise to potentially dangerous levels within 10–30 years without an immediate reduction in greenhouse gas (GHG) emissions [1, 2]. The two most important GHGs responsible for accelerating climate change are carbon dioxide (CO_2) and methane (CH_4) that are released when carbon-based fossil fuels are burned for heat and energy. CH_4 has received relatively less attention than CO_2, yet is quite important in climate change. Including indirect effects, the Intergovernmental Panel on Climate Change (IPCC) recognizes in its fifth assessment report (AR5) that CH_4 has 120 times greater radiative forcing than CO_2 on a

mass basis during the time both gasses are in the atmosphere [2], updated from its previous fourth assessment report (AR4) of 100 [3] and the third report (AR3) of 85 [4]. The IPCC AR5 concludes that the current radiative forcing by CH_4 is almost 1 W m^{-2}, compared to 1.66 W m^{-2} for CO_2. The model of Shindell et al. [1] indicates it is even more critical to control CH_4 emissions than CO_2 emissions if we are to slow the rate of global warming over the coming few decades: reducing CO_2 emissions has little effect on warming over this time period due to lags in the climate system, whereas reductions in CH_4 emissions have an immediate influence [1, 5].

Heat and energy for human use can be generated from different sources (e.g., electricity from coal or natural gas), and it is useful to have a tool or methodology that allows us to assess potential climate impacts of alternative choices.

Although CH_4 is a more powerful GHG than CO_2, it has a shorter lifetime of about 12 years [6, 7], making it challenging to compare two technologies that result in differing amounts of CO_2 and CH_4 emissions during their lifecycles. The technology warming potential (TWP) approach introduced by Alvarez et al. [8] provides a framework for evaluating a technology for heat and energy generation against a reference technology and has been applied for energy policy assessment and design [9]. The TWP approach can be considered as a two-step process: first, emissions of GHGs, most importantly CO_2 and CH_4, resulting from the application of each technology is estimated. For ease of comparison, the emissions are estimated on a unit energy basis, such as kilograms CO_2 emitted per megawatt-hour (MWh) electricity generated. Next, for each GHG emitted its cumulative radiative forcing over time is computed, and the total radiative forcing over all the GHGs emitted from the alternative technology is compared with that from the reference technology.

Alvarez et al. [8] used the TWP approach and evaluated climatic impacts of generating electricity from a combined cycle natural gas power plant (with an efficiency of 50%) relative to that from a supercritical pulverized coal power plant (39%). They concluded that, at the level of CH_4 emissions they estimated to be associated with the production and use of the natural gas (2.1%) [10] and based on the radiative forcing from the IPCC AR4, generating electricity from natural gas is relatively less damaging to the climate than from coal across all time frames considered. This analysis provides a useful context for discussions and policy development over the future use of natural gas, although there is room for reconsideration. For example, published literature estimating CH_4 emission rates has been rare, but since Howarth et al. [11] reported their first estimates many more measurements and estimates have become available, many of which are higher than the EPA estimate [10] that Alvarez et al. [8] used: an inverse modeling study by Miller et al. (3.6% or above) [12], a satellite data analysis by Schneising et al. (9.5%, shale gas, upstream emission only) [13], a review paper by Brandt et al. (3.6–7.1%) [14], and estimates from aircraft campaigns [15, 16]. Howarth [5] reviewed these studies and suggested the best available data for CH_4 emissions indicates rates of 3.8% for conventional natural gas and 12% for shale gas over the full lifecycle, well to final consumer. In light of the new emission estimates and radiative forcing from the IPCC AR5 (which are higher than those from the IPCC AR4 report) that became available after the analysis by Alvarez et al. [8], we provide a reevaluation of the electricity generation scenario in Data S1, which shows that generating electricity from natural gas can be more damaging to the climate than from coal

at the current level of CH_4 emission, especially within the coming few decades.

Electricity generation makes up a large portion of the use of natural gas in the United States (33% in 2009, the year on which the analysis by Alvarez et al. [8] is based), but many other uses are also very important, and the GHG emissions associated with most other uses of natural gas have remained largely unexplored. Alvarez et al. [8] also analyzed the use of natural gas as a long-distance transportation fuel, but that is a very minor use. Residential (23%) and commercial (15%) uses make up other large portions of natural gas use, a substantial portion of which is for space and water heating [17]. According to the U.S. Energy Information Administration (EIA)'s Residential Energy Consumption Survey (RECS) report, in 2009 ~1.9×10^{12} MJ of energy was delivered to the U.S. households for water heating (18% of total energy delivered to households) mainly in the form of natural gas (1.3×10^{12} MJ) and electricity (4.5×10^{11} MJ) (http://www.eia.gov/consumption/residential/). Most of the water heaters in the U.S. households have storage tanks (98% in 2009), although tankless water heaters have shown higher energy use efficiencies and may potentially help reduce household energy expenditure as well as GHG emissions associated with water heating.

Another promising new technology for heating water is heat pump water heaters. Powered by electricity, heat pump water heaters operate by transferring heat from the air into the tank [18]. Although currently making up only about 1% of new water heater sales in the U.S., they are becoming more popular (34,000 and 43,000 units sold in 2012 and 2013, respectively; http://www.energystar.gov/) because of their high efficiencies and their potential to improve overall home energy use efficiencies [19]. Their potential for reducing GHG emissions has not been evaluated up to now.

In light of these considerations, here we apply the TWP approach of Alvarez et al. [8], revised with updated CH_4 emissions and radiative forcing values, to evaluate the climatic impacts of using heat pump water heaters relative to other ways of heating water (natural gas and electric resistance water heaters, with and without storage tanks). To test its sensitivity, TWP was calculated under differing assumptions as to the source of electricity and efficiencies of water heaters and power plants. We also introduce a new web-based tool developed for this study, allowing anyone to perform the analysis described in this paper and evaluate his or her own scenarios.

Methods

In section "Emission factors," we describe the estimation of the lifecycle emissions of GHGs (CO_2 and CH_4)

associated with heating water with different water heaters considered in this study: heat pump water heaters as well as storage/tankless natural gas and electric resistance water heaters. Section "Technology warming potential" describes how the time-integrated radiative forcing resulting from these emissions is calculated and compared using the TWP approach. Finally, section "Web-based tool" introduces a web-based tool for making the TWP calculation. This study builds on the previous assessment by Alvarez et al. [8] on the use of natural gas for electricity generation, reconsidered with a number of updates as described in detail in Data S1.

Emission factors

Table 1 summarizes the CH_4 and CO_2 emission factors, expressed in kg emitted per GJ of water heated, resulting from the use of five different types of water heaters: heat pump water heater, storage natural gas water heater, tankless natural gas water heater, storage electric resistance water heater, and tankless electric resistance water heater. Energy factors for these water heaters were obtained from the Air-conditioning, Heating, and Refrigeration Institute (AHRI, http://www.ahrinet.org/) that provided

information on 5478 AHRI-certified water heaters as of December 2015. Excluding inactive or discontinued products yielded energy factors for 120 heat pump water heaters (2.21–3.39 with an average of 2.79), 1093 storage natural gas water heaters (0.57–0.82 with an average of 0.64), 378 tankless natural gas water heaters (0.82–0.99 with an average of 0.87), 587 storage electric resistance water heaters (0.9–0.95 with an average of 0.94), and 97 tankless electric resistance water heaters (0.96–1.0 with an average of 0.99).

CH_4 emission factors for the natural gas water heaters are highly sensitive to the estimated emissions of CH_4 during its lifecycle (production, processing, distribution, and use). CH_4 emission factors for the heat pump and electric resistance water heaters are also sensitive to the CH_4 emission rate if the electricity powering these water heaters comes from natural gas power plants instead of coal power plants. In their TWP analysis of electricity generation, Alvarez et al. [8] used emission estimates from U.S. EPA [10] reporting that 2.4% of CH_4 in the natural gas withdrawn in 2009 was lost to the atmosphere (broken down to 1.5%, 0.2%, 0.5%, and 0.3% lost during the field production, processing, transmission/storage, and distribution, respectively). Alvarez et al. [8]

Table 1. CH_4 and CO_2 emission factors used in this study associated with heating water using heat pump, natural gas, and electric resistance water heaters (kg GJ^{-1}).

Efficiency	Source of electricity	CH_4 emission reference	GHG	Heat pump water heater	Natural gas water heater		Electric resistance water heater	
					Storage	Tankless	Storage	Tankless
Base	Coal	Alvarez et al. [8]	CH_4	0.069	0.82	0.60	0.20	0.19
			CO_2	86.2	86.5	63.6	256	243
		Howarth et al. [11], conventional	CH_4	0.069	1.38	1.02	0.20	0.19
			CO_2	86.2	86.5	63.6	256	243
		Howarth [5], shale	CH_4	0.069	4.02	2.96	0.20	0.19
			CO_2	86.2	86.5	63.6	256	243
	Natural gas	Alvarez et al. [8]	CH_4	0.33	0.82	0.60	0.97	0.93
			CO_2	42.0	86.5	63.6	125	119
		Howarth et al. [11], conventional	CH_4	0.44	1.38	1.02	1.30	1.23
			CO_2	42.0	86.5	63.6	125	119
		Howarth [5], shale	CH_4	1.72	4.02	2.96	5.10	4.85
			CO_2	42.0	86.5	63.6	125	119
Best case	Coal	Alvarez et al. [8]	CH_4	0.046	0.64	0.53	0.17	0.16
			CO_2	58.1	67.5	55.9	207	197
		Howarth et al. [11], conventional	CH_4	0.046	1.08	0.89	0.17	0.16
			CO_2	58.1	67.5	55.9	207	197
		Howarth [5], shale	CH_4	0.046	3.14	2.60	0.17	0.16
			CO_2	58.1	67.5	55.9	207	197
	Natural gas	Alvarez et al. [8]	CH_4	0.23	0.64	0.53	0.81	0.77
			CO_2	29.0	67.5	55.9	103	98.2
		Howarth et al. [11], conventional	CH_4	0.30	1.08	0.89	1.08	1.02
			CO_2	29.0	67.5	55.9	103	98.2
		Howarth [5], shale	CH_4	1.18	3.14	2.60	4.23	4.01
			CO_2	29.0	67.5	55.9	103	98.2

assumed that natural gas power plants receive their fuel directly from the transmission system (large, long-distance pipelines) and excluded emissions of CH_4 associated with the distribution loss from smaller, local pipelines (0.3%), resulting in an overall CH_4 emission rate of 2.1% (instead of 2.4%) associated with generating electricity using natural gas. This implies that all the 0.3% distribution loss, which corresponds to 1.4 Tg of CH_4 in 2009 according to the EPA estimates [10], must have occurred while natural gas was used for purposes other than electricity generation. Since electric power plants consumed 33 percent of natural gas produced in 2009 [20], the absence of 0.3% loss in electricity generation translates into additional 0.15% loss in other uses of natural gas (including for water heating), resulting in the overall loss of 2.55%.

After an extensive literature review, Howarth [5] concluded that the best available information suggests the CH_4 emission rate of 3.8% for conventional natural gas (1.3% from upstream emissions and 2.5% from downstream emissions) and 12% for shale gas (9.5% from upstream emissions and 2.5% from downstream emissions). Note that as of 2013, 60% of natural gas in the United States was from conventional sources and 40% from shale gas and other unconventional sources [20]. A new study also reports ~2.5% CH_4 emissions for the distribution system in Boston, MA [21]. Following the distribution loss assumption by Alvarez et al. [8] described above, we excluded emissions of CH_4 associated with the distribution loss (1%, which is 40% of the total downstream loss as assumed by Alvarez et al. [8]) when natural gas is used for electricity generation, resulting in the overall emission rate of 2.8% for conventional gas and 11% for shale gas. The absence of 1% loss in electricity generation translates into additional 0.5% loss in other uses of natural gas, resulting in the overall loss of 4.3% for conventional gas and 12.5% for shale gas.

In consideration of electricity generation scenario, Alvarez et al. [8] assumed an efficiency of 50% for a combined cycle natural gas power plant and 39% for a supercritical pulverized coal power plant, both of which are somewhat higher than the efficiencies reported by the U.S. EIA as the average operating heat rates of natural gas and coal power plants in 2009 (42% and 33%, respectively; http://www.eia.gov/electricity/annual/html/epa_08_01.html). When performing the base efficiency calculation (Table 1), we applied the same power plant efficiencies as used by Alvarez et al. [8] and assumed average efficiency factors obtained from the AHRI (http://www.ahrinet.org/) for the water heaters considered in this study. When performing the "best case" efficiency calculation, natural gas power plants were assumed to be 60% efficient as reported for the GE Flex 60 units (http://www.ge-energy.com/) and coal

power plants to be 48% efficient for the ultrasupercritical units [22]. Maximum water heater efficiency factors obtained from the Energy Star (http://www.energystar.gov/) and the AHRI were used under the best case scenario.

The emission factors used by Alvarez et al. [8] for the electricity generation scenario were 3.1 kg-CH_4 MWh^{-1} and 397 kg-CO_2 MWh^{-1} for a combined cycle natural gas power plant, and 0.65 kg-CH_4 MWh^{-1} and 814 kg-CO_2 MWh^{-1} for a supercritical pulverized coal power plant. The following is an example of how the CH_4 emission factor for a heat pump water heater, powered by electricity from shale natural gas, was determined under the best case scenario: 3.1 kg-CH_4 MWh^{-1} (=0.86 kg-CH_4 GJ^{-1}) adjusted for maximum power plant efficiency (60%) and methane emission for shale gas associated with electricity generation (11%) yields 3.77 kg-CH_4 GJ^{-1}. Approximately six percent of electricity production may be lost in transmission and distribution when delivered to homes (EIA; http://www.eia.gov/tools/faqs/faq.cfm?id=105&t=3). The emission factor becomes 4.01 kg-CH_4 GJ^{-1} after accounting for this loss, and 1.18 kg-CH_4 GJ^{-1} after applying the efficiency factor of a heat pump water heater (3.39). To estimate how much CH_4 is emitted for a tankless natural gas water heater to heat the same amount of water, we again start with the same emission factor from Alvarez et al. [8] (3.1 kg-CH_4 MWh^{-1}), use the heat rate (efficiency) of a natural gas power plant applied by Alvarez et al. [8] to calculate the emission factor associated with delivering natural gas to power plants (0.432 kg-CH_4 GJ^{-1}), and in turn use it to estimate the emission factor associated with delivering shale natural gas to homes (2.57 kg-CH_4 GJ^{-1}). Finally, by applying the efficiency factor of a tankless natural gas water heater (0.99), we obtain the CH_4 emission factor of 2.60 kg-CH_4 GJ^{-1}. All the emission factors in Table 1 were computed by adjusting these emission rates and efficiencies as they apply to each scenario.

Technology warming potential

In this study, TWP is calculated as the sum of time-integrated radiative forcing from CH_4 and CO_2 emitted by applying one of the four alternative technologies (heating water using a natural gas or electric resistance water heater, with or without a storage tank) divided by that from the reference technology (heating water using a heat pump water heater). TWP can be calculated under three different implementation scenarios: pulse, service life, and fleet conversion [8]. In the case of a permanent fleet conversion, the technology is applied indefinitely by replacing it with an identical unit at the end of its service life. Fleet conversion TWP at year t is calculated as:

$$\text{TWP}(t) = \frac{E_{a,CH_4} \times A_{CH_4} \times \left(\tau_{CH_4} t - \tau_{CH_4}^2 \left(1 - \exp(-\frac{t}{\tau_{CH_4}}) \right) \right) + E_{a,CO_2} \times A_{CO_2} \times \left(a_0 \frac{t^2}{2} + \sum_{i=1}^n a_i \left(\tau_i t - \tau_i^2 \left(1 - \exp(-\frac{t}{\tau_i}) \right) \right) \right)}{E_{r,CH_4} \times A_{CH_4} \times \left(\tau_{CH_4} t - \tau_{CH_4}^2 \left(1 - \exp(-\frac{t}{\tau_{CH_4}}) \right) \right) + E_{r,CO_2} \times A_{CO_2} \times \left(a_0 \frac{t^2}{2} + \sum_{i=1}^n a_i \left(\tau_i t - \tau_i^2 \left(1 - \exp(-\frac{t}{\tau_i}) \right) \right) \right)} \qquad (1)$$

where: E_{a,CH_4} = CH_4 emission factor for the alternative technology (kg GJ^{-1}), E_{a,CO_2} = CO_2 emission factor for the alternative technology (kg GJ^{-1}), E_{r,CH_4} = CH_4 emission factor for the reference technology (kg GJ^{-1}), E_{r,CO_2} = CO_2 emission factor for the reference technology (kg GJ^{-1}), A_{CH_4} = radiative efficiency of CH_4 (W m^{-2} kg^{-1}), A_{CO_2} = radiative efficiency of CO_2 (W m^{-2} kg^{-1}), τ_{CH_4} = lifetime coefficient of CH_4 (years), τ_i = lifetime coefficients of CO_2 (years).

TWP greater than one indicates that the cumulative radiative forcing from choosing the alternative technology at year t is higher than the reference technology. Emission factors (E_{a,CH_4}, E_{a,CO_2}, E_{r,CH_4} and E_{r,CO_2}) used in this study are given in Table 1 and described in section "Emission factors". Radiative efficiency (radiative forcing per unit mass increase in atmospheric abundance) of CH_4 was updated from 1.82×10^{-13} W m^{-2} kg^{-1} in AR4 [3] to 2.11×10^{-13} W m^{-2} kg^{-1} in AR5 [2], reflecting updated knowledge of the magnitude of the indirect radiative effects of CH_4 on tropospheric ozone and stratospheric water vapor [23]. The remaining part of equation (1) describes the time-integrated change in GHGs after they are released into the atmosphere. This time-integrated portion can be replaced with one that considers single pulse emissions of CH_4 and CO_2 in the same way as done in the global warming potential (GWP) calculation, or that considers service lives of different technologies: CH_4 and CO_2 are emitted continuously throughout the year at a constant rate during the time periods from $t = 0$ to $t = t_{max}$ over which the technology is applied (for example, a single power plant generating electricity over its full service life of 50 years). Equations for pulse and service life TWP are given in Alvarez et al. [8] and also in Data S1.

Web-based tool

We developed a web-based tool evaluating the GHG footprint of using a heat pump water heater relative to those of alternative technologies (Fig. 1). Written in JavaScript [24], the first version of the tool along with the source code is currently available at http://www.eeb.cornell.edu/howarth/methane/tool.htm and runs on all major web browsers. It allows the user to select the scenario to evaluate (e.g., heating water with a heat pump water heater vs. a storage natural gas water heater), method of radiative

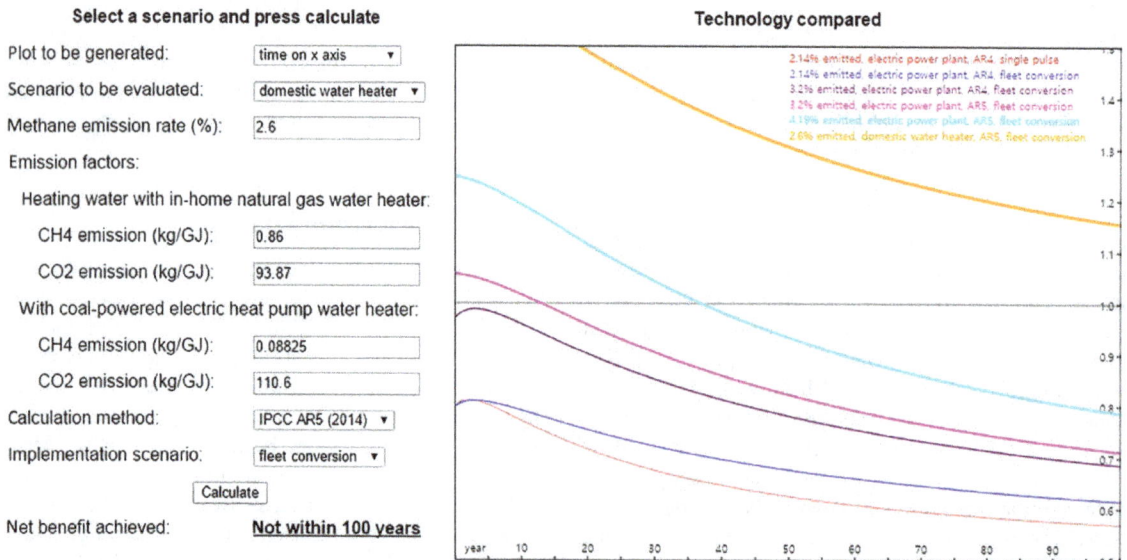

Figure 1. User interface of the web-based tool for evaluating the greenhouse gas footprint from using a heat pump water heater, available at www.eeb.cornell.edu/howarth/methane/tool.htm.

forcing calculation (AR3, AR4, or AR5) and implementation scenario (pulse, service life, or fleet conversion). The user can also choose the type of plot to be generated (time or percent emission on the x-axis) and change the CH_4 emission rate or time frame considered. The CH_4 emission factor for the natural gas-based technology is directly proportional to the assumed CH_4 emission rate (section "Emission factors"), and the user can try different emission rates and test their impacts on the TWP calculation. The user can also build his or her own scenarios by directly entering CH_4 and CO_2 emission factors for the reference and alternative technologies (Fig. 1). One way of making use of this tool, although not developed for this purpose, would be to calculate GWP by setting E_{r,CH_4} and E_{a,CO_2} (eq. 1) to zero and E_{a,CH_4} and E_{r,CO_2} to one, and making the "pulse" calculation.

Results and Discussion

Under the base efficiency condition, the time-integrated radiative forcing due to CH_4 and CO_2 emissions associated with heating water with a storage natural gas water heater (which is the most common way of heating water in the U.S.) was greater than that with a heat pump water heater powered by electricity from coal across all time scales considered in this study (up to 100 years), regardless of the method of radiative forcing calculation (AR3, AR4, or AR5) or implementation scenario (pulse, service life, or fleet conversion) (Fig. 2). TWPs were higher when calculated using the AR5 method and under the fleet conversion scenario, although the magnitude of their variation was relatively small compared to the differences due to the emission estimates chosen (from Alvarez et al. [8] for building on the previous TWP analysis, Howarth et al. [11] for conventional gas and Howarth [5] for shale gas). Our analysis indicates that if the natural gas from shale is used for heating water in homes, the accumulated radiative forcing from using a natural gas water heater can be as much as six times higher than from using a heat pump water heater within the first year of their installations, and still more than five times higher after 20 years.

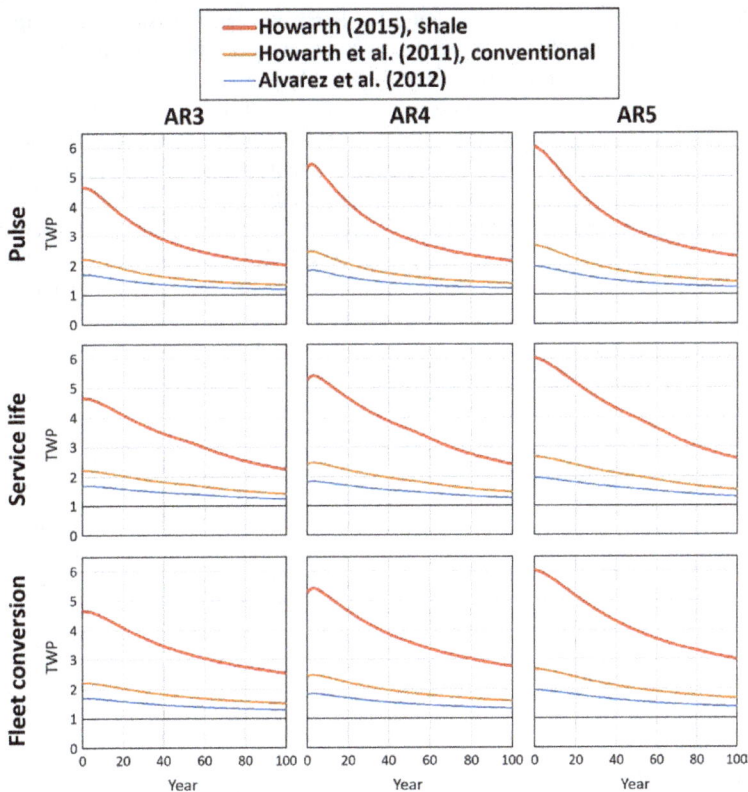

Figure 2. Time-integrated radiative forcing resulting from heating water with a storage natural gas water heater, relative to that from a heat pump water heater powered by electricity from coal. Technology warming potentials are calculated from three emission estimates (Alvarez et al. [8], Howarth et al. [11] for conventional gas and Howarth [5] for shale gas), three calculation methods (AR3, AR4, and AR5), and three implementation scenarios (pulse, service life, and fleet conversion).

The maximum CH_4 emission rate below which the time-integrated radiative forcing associated with using natural gas is smaller than that with an alternative technology for all time horizons considered has been referred to as a "break-even point" [8, 25, 26]. The break-even point for generating electricity from natural gas versus coal was estimated to be 3.2% by Alvarez et al. [8], who based their estimation on the IPCC AR4 parameters [3]. When updated with AR5 parameters that reflect the most recent scientific information on the methane's indirect effects on tropospheric ozone and stratospheric water vapor [2], this break-even point was revised to be 2.7% (Data S1). The break-even point associated with heating water in homes with a storage natural gas water heater relative to a heat pump water heater was estimated to be a CH_4 emission rate of 0.2%. All published CH_4 emission rates are well above this threshold.

TWPs of using a storage tank natural gas water heater relative to a coal-electric-powered heat pump water heater increased linearly with increasing CH_4 emission rate (Fig. 3). The rate of increase (slope) of the 20-year scale TWP (0.34 per %) was about twice that of the 100-year scale (0.16 per %). The CH_4 emission rate at which both technologies yield the same cumulative forcing (i.e., the storage tank natural gas- and coal-electric-based water heaters have the same climatic impact) was close to 0.2% at both the 20- and 100-year scales. Shale gas was only

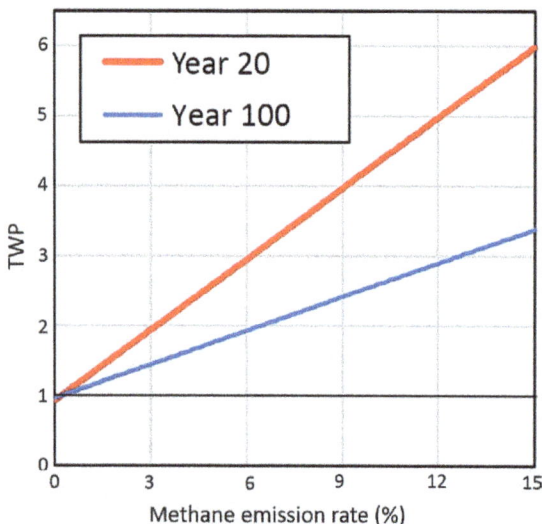

Figure 3. Time-integrated radiative forcing resulting from heating water with a storage natural gas water heater, relative to that from a heat pump water heater powered by electricity from coal. Technology warming potentials are calculated as a function of methane emission rate (%) at two time horizons (20 and 100 years) applying the AR5 calculation method and fleet implementation scenario.

15% of the natural gas production in 2009, but had risen to 40% in 2013 [20]. Applying the methane emission estimates from Howarth et al. [11] and Howarth [5] for conventional (3.8%) and shale (12%) gas, the overall emission rate is likely to have increased from ~5% to 7% during this time period, increasing the 20-year TWP for this scenario by about 0.7.

TWPs at 20 and 100 years calculated with the AR5 parameters and under the fleet conversion scenario were always greater than 1, regardless of the type of the water heater compared with the heat pump water heater (natural gas or electric resistance water heater, storage or tankless), efficiencies of power plants and water heaters (base or best case efficiencies) and the source of electricity powering the heat pump and electric resistance water heaters (natural gas or coal) (Table 2). Tankless water heaters yielded lower TWPs than storage water heaters because of their higher efficiencies, and the difference was greater in the natural gas water heaters than in the electric resistance water heaters. TWP values of the electric water heaters were constant over time because the relative magnitude of CO_2 and CH_4 emissions associated with using the electric resistance versus heat pump water heaters stayed the same, only varied by the difference in their efficiencies. The best case scenarios (in which the maximum available efficiencies were assumed for all power plants and all water heaters) always resulted in higher TWPs than those from corresponding base case scenarios, suggesting that the benefit from using heat pump water heaters would be greater when best available technologies are applied. In this study, we have not explicitly modeled the use of heat pump water heaters powered by electricity from renewable sources, but these obviously would have far lower GHG emissions yet.

The highest TWP (5.96) was observed at year 20 in the comparison between the storage natural gas water heater versus heat pump water heater powered by electricity from coal, under the best case efficiency and with emissions from shale gas (Table 2). This TWP was decreased to 2.59 when the electricity powering the heat pump water heater was assumed to come from natural gas, also from shale formations. It appears that, under the low methane emission rate, the net gain of replacing the existing natural gas water heater with a new heat pump water heater would be higher if the electricity powering the heat pump water heater comes from natural gas instead of coal. However, if producing natural gas involves high emissions of CH_4 as is the case for the shale gas, higher net climatic benefits would be gained if electricity comes from coal.

Although it is important to note that replacing any of the water heaters considered in this study with a heat pump water heater will give net climatic benefits over all

Table 2. Time-integrated radiative forcing from heating water with storage/tankless natural gas and electric resistance water heaters relative to that with a heat pump water heater.

Efficiency	Source of electricity	CH$_4$ emission reference	Time scale (year)	Natural gas water heater		Electric resistance water heater	
				Storage	Tankless	Storage	Tankless
Base	Coal	Alvarez et al. [8]	20	1.79	1.32	2.97	2.82
			100	1.38	1.01	2.97	2.82
		Howarth et al. [11], conventional	20	2.38	1.75	2.97	2.82
			100	1.66	1.22	2.97	2.82
		Howarth [5], shale	20	5.15	3.79	2.97	2.82
			100	2.98	2.19	2.97	2.82
	Natural gas	Alvarez et al. [8]	20	2.25	1.65	2.97	2.82
			100	2.17	1.60	2.97	2.82
		Howarth et al. [11], conventional	20	2.61	1.92	2.97	2.82
			100	2.41	1.77	2.97	2.82
		Howarth [5], shale	20	2.28	1.68	2.97	2.82
			100	2.24	1.65	2.97	2.82
Best case	Coal	Alvarez et al. [8]	20	2.07	1.72	3.57	3.39
			100	1.60	1.32	3.57	3.39
		Howarth et al. [11], conventional	20	2.76	2.28	3.57	3.39
			100	1.92	1.59	3.57	3.39
		Howarth [5], shale	20	5.96	4.94	3.57	3.39
			100	3.46	2.86	3.57	3.39
	Natural gas	Alvarez et al. [8]	20	2.55	2.11	3.57	3.39
			100	2.46	2.04	3.57	3.39
		Howarth et al. [11], conventional	20	2.96	2.45	3.57	3.39
			100	2.73	2.26	3.57	3.39
		Howarth [5], shale	20	2.59	2.14	3.57	3.39
			100	2.54	2.10	3.57	3.39

Technology warming potentials are calculated applying the AR5 calculation method and fleet implementation scenario.

time frames (up to 100 years), it is equally important to recognize that avoiding the use of natural gas-based technology may make it possible to prevent the acute adverse impacts on the climate that may take place in a short foreseeable future (e.g., within the next 20 years), especially when the natural gas is produced from high emission sources [26, 27]. It has been suggested that an increase in global mean temperature by 1.5–2°C above the 1900 baseline, that could happen within the next 15–35 years, may push the earth past a critical threshold into an alternate state for the climate system [1]. Finding currently available alternatives to the use of natural gas-based technologies and thereby reducing the CH$_4$ emissions immediately would be essential for slowing the climate change over the coming decades.

In 2009, total energy delivered to households for water heating in the U.S. was 1.9×10^{12} MJ (18% of energy delivered to households; http://www.eia.gov/consumption/residential/), 68% of which (1.3×10^{12} MJ) was in the form of natural gas and 24% (4.5×10^{11} MJ) was electricity. Only 2.3% of households had tankless water heaters in 2009. Applying the weighted average efficiency factor for the natural gas water heaters (0.64) and the emission

factors derived from Howarth et al. [11] and Howarth [5], we estimate that ~8.3×10^{11} MJ of water was heated by the natural gas water heaters in 2009, and the associated GHGs released into the atmosphere were 1.5×10^9 kg of CH$_4$ and 7.2×10^{10} kg of CO$_2$. Under the best case efficiencies considered in this study, only 2.5×10^{11} MJ of energy would have been needed to heat this much water with the heat pump water heaters, releasing 1.5×10^8 kg of CH$_4$ and 4.0×10^{10} kg of CO$_2$ if we assume that about 66% of the electricity used by the heat pump water heaters comes from the coal and the rest from the natural gas (http://www.eia.gov/electricity/). Likewise, we estimate that ~3.3×10^8 kg of CH$_4$ and 8.9×10^{10} kg of CO$_2$ were released in 2009 while heating 4.3×10^{11} MJ of water in homes with electric resistance water heaters. Again under the best case scenario, 1.3×10^{11} MJ of electricity would have been needed to heat the same amount of water with the heat pump water heaters, releasing 7.7×10^7 kg of CH$_4$ and 2.0×10^{10} kg of CO$_2$. Combining these two sources, the 2009 emissions associated with heating water in homes in the U.S. potentially could have been 2.3×10^8 kg of CH$_4$ and 6.0×10^{10} kg of CO$_2$ (summing up to be a total of

0.08 Pg-CO_2 equivalents) instead of 1.8×10^9 kg of CH_4 and 1.6×10^{11} kg of CO_2 (0.32 Pg-CO_2 equivalents). The net saving would be 0.24 Pg-CO_2 equivalents, which is about 2.8% of the total GHG emissions from fossil fuel use in the U.S. in 2009 (8.4 Pg-CO_2 equivalents [5]) and about 26% of those from the residential and commercial uses, assuming that they comprise ~11% of the total GHG emissions from fossil fuels. In terms of household expenditure, the U.S. households in 2009 spent about 1.1 cents per MJ natural gas delivered and 3.1 cents per MJ electricity delivered. Multiplying these prices to the energy consumption in 2009 results in 29 billion dollars for heating water, consisting of 15 billion dollars for using natural gas and 14 billion dollars for electricity. Replacing the natural gas and electric resistance water heaters with heat pump water heaters and using the electricity generated by high-efficiency power plants could have reduced this spending to 12 billion dollars (7 and 10 billion dollars saved by replacing natural gas and electric resistance water heaters, respectively). Since there were 58 and 47 million households with natural gas and electricity as the main water heater in 2009, the savings in water heating expenses would have been 120 and 214 dollars per household, respectively.

The above consideration provides a strong argument that replacing existing water heaters with high-efficiency heat pumps is an effective and economically viable way of reducing emissions of GHGs. It may be refined further as new relevant information on the water heater efficiencies and CH_4 emission rates become available. For example, typical heat pump water heaters switch to electric resistance mode as ambient temperatures approach freezing point [28]. Thus, the results from the electric resistance water heaters serve as theoretical lower bounds (worst case) of the application of the heat pump water heaters, subject to season and geographic region. Using natural gas water heaters is more damaging to the climate than using electric resistance water heaters powered by electricity from coal, if the natural gas is produced from shale (Table 2).

After Howarth et al. [11, 29, 30], more literature has been published reporting estimates of lifecycle emissions of CH_4 [12–16, 31] that are thoroughly reviewed by Howarth [5, 26]. In our study, we chose to use three estimates: Alvarez et al. [8] for building on the previous TWP analysis, Howarth et al. [11] for conventional gas and Howarth [5] for shale gas. We believe that the CH_4 emission from conventional gas (3.8%), first proposed in Howarth et al. [11] and rigorously reexamined in Howarth [5], is a well-supported, robust estimate that serves as a lower bound of the methane emission associated with the natural gas production in the U.S. today. Several recent studies have estimated upstream methane emissions from

shale gas and other unconventional natural gas development using integrated measurement techniques, including airplane flyovers that produced highly variable results [15, 16, 31]. Among the estimation methods available at the moment, the satellite data (on which our shale estimate from Howarth [5] is based) may produce the most robust estimates as they integrate in space and over a longer time period, whereas other methods like the aircraft campaigns are one of several estimates based on relatively short-term observations. This satellite-based estimate is about 20-fold greater than the estimate by Allen et al. [32], a study that worked closely with industry to measure emissions from various component processes of shale gas development. Two papers published in 2015 [33, 34] including one in Energy Science & Engineering cast serious doubts on the Allen et al. [32] estimate that could have been biased downward by the sensor failure of the Bacharach Hi-Flow® Sampler used in their study. It is conceivable that the shale emission estimate that we used in our study (12%) represents the emission levels resulting from the lack of rigorous controls and it may decline with time following improved methane regulations; however, at most, we believe it would come down to the estimate for the conventional natural gas, the current best estimate of which is 3.8% [5].

Concluding Thoughts

Natural gas is often portrayed as a "bridge fuel," with the implication that it is a preferable energy source over other carbon-based fossil fuels resulting in a less adverse impact on the climate, desirable to be used until carbon-free technologies are mature and in place [25, 35]. The analysis presented here provides counterarguments to this idea in two different aspects: that (1) using natural gas-based technologies can result in even higher emissions of GHGs than using coal-based technologies, and that (2) technologies that can support carbon-free sources of energy already exist in an economically viable way. The electricity powering the heat pump water heaters can be generated from carbon-free energy sources such as wind, solar, hydropower, and geothermal sources that have little or no emissions directly related to electricity production; yet, even when powered by electricity from coal and natural gas, total GHG emissions from generating domestic hot water with heat pumps is less than directly using natural gas. Therefore, currently available modern technologies like the heat pump water heater are in fact a true bridge to the clean energy environment.

Our analysis provides the first assessment of the GHG footprint involved with using heat pump water heaters compared to other common ways of heating water at U.S. homes. In the future, we hope to expand this type of analysis to evaluate the use of alternative clean energy sources for energy needs where natural gas currently provides a large amount of the energy. Recent studies [36, 37] show that a transition to a society that is driven only by renewable energy sources can be accomplished in a cost-effective way using the commercially available technologies such as the heat pumps. This transition may also be expedited with the help of web-based tools such as the one introduced here, that allow anyone to incorporate, evaluate, and share new information on the alternative energy sources as it becomes available.

Acknowledgments

This study was funded by the Wallace Global Fund, the Park Foundation, and Cornell University. We thank Dennis P. Swaney, Roxanne M. Marino and Anthony R. Ingraffea for reviewing an early draft of this paper and providing helpful suggestions.

Conflict of Interest

None declared.

References

1. Shindell, D., J. C. I. Kuylenstierna, E. Vignati, R. van Dingenen, M. Amann, Z. Klimont, et al. 2012. Simultaneously mitigating near-term climate change and improving human health and food security. Science 335:183–189.

2. IPCC. 2013. Climate change 2013: the physical science basis. Intergovernmental Panel on Climate Change. Available at: https://www.ipcc.ch/report/ar5/wg1/ (accessed 1 December 2015).

3. IPCC. 2007. IPCC Fourth Assessment Report (AR4), Working Group 1, the physical science basis. Intergovernmental Panel on Climate Change. Available at: http://www.ipcc.ch/publications_and_data/ar4/wg1/en/contents.html (accessed 1 December 2015).

4. IPCC. 2001. Climate change 2001: the scientific basis. Contribution of Working Group I to the third assessment report of the Intergovernmental Panel on Climate Change. Available at: https://www.ipcc.ch/ipccreports/tar/wg1/ (accessed 1 December 2015).

5. Howarth, R. W. 2015. Methane emissions and climatic warming risk from hydraulic fracturing and shale gas development: implications for policy. Energy Emission Control Technol. 3:45–54.

6. Joos, F., M. Bruno, R. Fink, T. F. Stocker, U. Siegenthaler, C. Le Quéré, et al. 1996. An efficient and accurate representation of complex oceanic and biospheric models of anthropogenic carbon uptake. Tellus 48B:397–417.

7. Siegenthaler, U., and F. Joos. 1992. Use of a simple model for studying oceanic tracer distributions and the global carbon cycle. Tellus 44B:186–207.

8. Alvarez, R. A., S. W. Pacala, J. J. Winebrake, W. L. Chameides, and S. P. Hamburg. 2012. Greater focus needed on methane leakage from natural gas infrastructure. Proc. Natl. Acad. Sci. USA 109:6435–6440. doi: 10.1073/pnas.1202407109

9. Thomson, H., J. J. Corbett, and J. J. Winebrake. 2015. Natural gas as a marine fuel. Energy Policy 87:153–167.

10. EPA. 2011. Inventory of U.S. greenhouse gas emissions and sinks: 1990–2009. 14 April 2011. U.S. Environmental Protection Agency, Washington, DC. Available at: http://epa.gov/climatechange/emissions/usinventoryreport.html (accessed 1 December 2015).

11. Howarth, R. W., R. Santoro, and A. Ingraffea. 2011. Methane and the greenhouse gas footprint of natural gas from shale formations. Clim. Change Lett. 106:679–690. doi: 10.1007/s10584-011-0061-5

12. Miller, S. M., S. C. Wofsy, A. M. Michalak, E. A. Kort, A. E. Andrews, S. C. Biraud, et al. 2013. Anthropogenic emissions of methane in the United States. Proc. Natl. Acad. Sci. USA 110:20018–20022.

13. Schneising, O., J. P. Burrows, R. R. Dickerson, M. Buchwitz, M. Reuters, and H. Bovensmann. 2014. Remote sensing of fugitive emissions from oil and gas production in North American tight geological formations. Earths Future 2:548–558. doi: 10.1002/2014EF000265

14. Brandt, A. F., G. A. Heath, E. A. Kort, F. O. O'Sullivan, G. Pétron, S. M. Jordaan, et al. 2014. Methane leaks from North American natural gas systems. Science 343:733–735.

15. Karion, A., C. Sweeney, G. Pétron, G. Frost, R. M. Hardesty, J. Kofler, et al. 2013. Methane emissions estimate from airborne measurements over a western United States natural gas field. Geophys. Res. Lett. 40:4393–4397.

16. Peischl, J., T. B. Ryerson, K. C. Aikin, J. A. de Gouw, J. B. Gilman, J. S. Holloway, et al. 2015. Quantifying atmospheric methane emissions from the Haynesville, Fayetteville, and northeastern Marcellus shale gas production regions. J. Geophys. Res. Atmos. 120:2119–2139.

17. EIA. 2014. Natural gas consumption by end use. Energy Information Agency, U.S. Department of Energy. Available at: http://www.eia.gov/dnav/ng/ng_cons_sum_dcu_nus_a.htm (accessed 1 December 2015).

18. Shapiro, C., S. Puttagunta, and D. Owens.2012. Measure guideline: heat pump water heaters in new and existing homes. U.S. Department of Energy, Washington, DC. Available at: http://www.nrel.gov/docs/fy12osti/53184.pdf (accessed 1 December 2015).

19. Widder, S. H., J. M. Petersen, G. B. Parker, and M. C. Baechler 2014. Impact of ducting on heat pump water heater space conditioning energy use and comfort. PNNL-23526, Pacific Northwest National Laboratory, Richland, WA. Available at: http://labhomes.pnnl.gov/experiments/hpwh.stm (accessed 1 December 2015).

20. EIA. 2014. P. 260 in Natural gas annual 2014. Office of Oil, Gas, and Coal Supply Statistics, U.S. Department of Energy, Washington, DC. Available at: http://www.eia.gov/naturalgas/annual/pdf/nga14.pdf (accessed 1 December 2015).

21. McKain, K., A. Down, S. M. Raciti, J. Budney, L. R. Hutyra, C. Floerchinger, et al. 2015. Methane emissions from natural gas infrastructure and use in the urban region of Boston, Massachusetts. Proc. Natl. Acad. Sci. USA 112:1941–1946.

22. Zhang, D. 2013. Ultra-supercritical coal power plants: materials, technologies and optimisation. Woodhead Publishing, Philadelphia, PA. ISBN 978-0857091161.

23. Shindell, D. T., G. Faluvegi, D. M. Koch, G. A. Schmidt, N. Unger, and S. E. Bauer. 2009. Improved attribution of climate forcing to emissions. Science 326:716–718. doi: 10.1126/science.1174760

24. Flanagan, D. 2010. P. 1100 in JavaScript: the definitive guide. 6th ed. O'Reilly Media, Sebastopol, CA. ISBN 0-596-80552-7.

25. Banks, D., and G. Taraska. 2013. US natural gas use must peak by 2030. Center for American Progress. Available at: http://www.americanprogress.org (accessed 1 December 2015).

26. Howarth, R. W. 2014. A bridge to nowhere: methane emissions and the greenhouse gas footprint of natural gas. Energy Sci. Eng. 2:47–60.

27. Howarth, R. W., and A. Ingraffea. 2011. Should fracking stop? Yes, it is too high risk. Nature 477:271–273.

28. Northwest Energy Efficiency Alliance (NEEA). 2015. Combination ductless heat pump & heat pump water heater lab and field tests. Energy 350, Portland, OR. Available at: https://neea.org/docs/default-source/reports/ (accessed 1 December 2015).

29. Howarth, R. W., R. Santoro, A. Ingraffea. 2012. Venting and leakage of methane from shale gas development: reply to Cathles et al. Clim. Change 113:537–549. doi: 10.1007/s10584-012-0401-0

30. Howarth, R. W., D. Shindell, R. Santoro, A. Ingraffea, N. Phillips, and A. Townsend-Small. 2012. Methane emissions from natural gas systems. Background paper prepared for the National Climate Assessment, Reference # 2011-003. Office of Science & Technology Policy Assessment, Washington, DC. Available at: http://www.eeb.cornell.edu/howarth/publications/Howarth_et_al_2012_National_Climate_Assessment.pdf (accessed 1 December 2015).

31. Pétron, G., G. Frost, B. T. Miller, A. I. Hirsch, S. A. Montzka, A. Karion, et al. 2012. Hydrocarbon emissions characterization in the Colorado Front Range – a pilot study. J. Geophys. Res. 117:D04304. doi: 10.1029/2011JD016360

32. Allen, D. T., V. M. Torres, K. Thomas, D. W. Sullivan, M. Harrison, A. Hendler, et al. 2013. Measurements of methane emissions at natural gas production sites in the United States. Proc. Natl. Acad. Sci. USA 110:17768–17773.

33. Howard, T., T. W. Ferrarab, and A. Townsend-Small. 2015. Sensor transition failure in the high flow sampler: implications for methane emission inventories of natural gas infrastructure. J. Air Waste Manag. Assoc. 65:856–862.

34. Howard, T. 2015. University of Texas study underestimates national methane emissions at natural gas production sites due to instrument sensor failure. Energy Sci. Eng. 3:443–455. doi: 10.1002/ese3.81

35. Pacala, S., and R. Socolow. 2004. Stablization wedges: solving the climate problem for the next 50 years with current technologies. Science 305:968–972.

36. Jacobson, M. Z., R. W. Howarth, M. A. Delucchi, S. R. Scobies, J. M. Barth, M. J. Dvorak, et al. 2013. Examining the feasibility of converting New York State's all-purpose energy infrastructure to one using wind, water, and sunlight. Energy Policy 57:585–601.

37. Jacobson, M. Z., M. A. Delucchi, A. R. Ingraffea, R. W. Howarth, G. Bazouin, B. Bridgeland, et al. 2014. A roadmap for repowering California for all purposes with wind, water, and sunlight. Energy 73:875–889.

6

Comparison of the indoor performance of 12 commercial PV products by a simple model

Georgia Apostolou[1], Angèle Reinders[1,2] & Martin Verwaal[1]

[1]Design for Sustainability, Faculty of Industrial Design Engineering, Delft University of Technology, Landbergstraat 15, 2628CE Delft, The Netherlands
[2]Department of Design, Production and Management, Faculty of Engineering Technology, University of Twente, P.O. Box 217, 7500AE Enschede, The Netherlands

Keywords

Indoor, irradiance, modeling, performance, PV cells, PV products

Correspondence

Georgia Apostolou, Design for Sustainability, Faculty of Industrial Design Engineering, Delft University of Technology, Landbergstraat 15, 2628CE Delft, The Netherlands. g.apostolou@tudelft.nl

Funding Information

The authors acknowledge the Faculty of Industrial Design Engineering (IO) of TU Delft for supporting this research.

Abstract

This article presents a simple comparative model which has been developed for the estimation of the performance of photovoltaic (PV) products' cells in indoor environments. The model predicts the performance of PV solar cells, as a function of the distance from a spectrum of artificial (fluorescent light, halogen light, and light-emitting diodes) and natural light. It intends to support designers, while creating PV-integrated products for indoor use. For the model's validation, PV cells of 12 commercially available PV-powered products with power ranging from 0.8 to 4 mWp were tested indoors under artificial illumination and natural light. The model is based on the physical measurements of natural and artificial irradiance indoors, along with literature data of PV technologies under low irradiance conditions. The input data of the model are the surface of the solar cell (in m^2), the wavelength-dependent spectral response (SR) of the PV cell, the spectral irradiance indoors, and solar cell's distance from light sources. The model calculates solar cells' efficiency and power produced under the specific indoor conditions. If using the measured SR of a PV cell and the irradiance as measured indoors, the model can predict the performance of a PV product under mixed indoor light with a typical inaccuracy of around 25%, which is sufficient for a design process. Measurements revealed that under mixed indoor lighting of around 20 W/m^2, the efficiency of solar cells in 12 commercially available PV products ranges between 5% and 6% for amorphous silicon (a-Si) cells, 4–6% for multicrystalline silicon (mc-Si) cells, and 5–7% for the monocrystalline silicon (c-Si) cells.

Introduction

The term product-integrated photovoltaic (PIPV) [1, 2] is used for all types of products that contain solar cells in one or more of their surfaces, aiming at providing power during the use of a product. The application of photovoltaics (PV) as a power source for consumer products is already common for more than 30 years, since the first solar calculators. Thereafter, integrated solar cells in consumer products became more popular and at present PV cells are applied in both indoor and outdoor applications, for example, lanterns, chargers, speakers, bike lights, solar watches, etc. A product can be defined as a PIPV product mainly by the existence of integrated PV cell(s) on at least one of its surfaces and the internal use of energy, which is generated by the PV cell(s) [1] (see Fig. 1, example of PIPV).

Figure 2A depicts the basic power system of a PV-powered product. Basic elements are solar cell, an energy storage device (i.e., a capacitor or battery), and a diode to prevent discharging of the battery through the solar panel. Matching of the battery voltage with the solar cell is done by creating a small solar panel with the right number of solar cells in series for an appropriate voltage.

Figure 1. Solar-powered keyboard by Logitech [3].

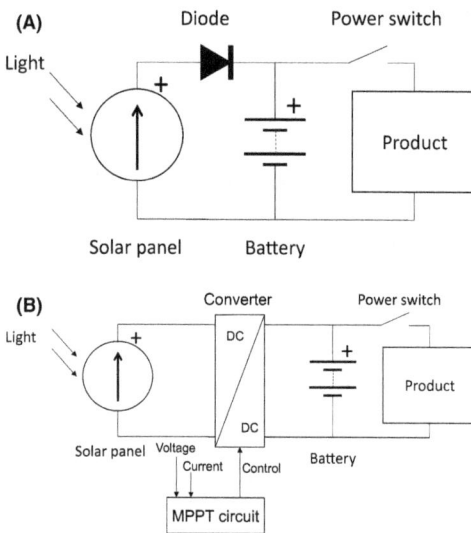

Figure 2. (A) Simple and (B) advanced circuit scheme of a PV product.

In more advanced systems (Fig. 2B), a DC/DC converter matches the solar panel voltage and the battery voltage.

After systematic investigation of 90 PV products [1], we conclude that some combinations of PV cells and batteries seem more common than others; a-Si cells combined with Li-ion batteries are common for PV products that are mainly used indoors (around 16%) [1], while a-Si cells combined with alkaline or nickel-based batteries are used in PV products for both indoors and outdoors (around 20%) [1]. Furthermore, c-Si or m-Si cells with Li-ion or nickel-based batteries (NiMH, NiZn) are used for PV products that are mostly used in outdoor environments (around 28%) [1].

Three categories of PV products can be distinguished: PV products for indoor use, for outdoor use and for both indoor and outdoor use, called "mixed." Another distinction made concerns the PV product size. Two

categories are distinguished, the small or thin PV products and the large or thick, covering product sizes in the range of 2.7×10^{-4} m^2 (the product with the smallest PV cell area among the tested products) to 87×10^{-4} m^2 (the product with the largest PV cell area among the tested), respectively.

The integration of PV cells into consumer products creates several advantages. The environmental benefits [5] could be significant, by entirely avoiding the use of primary batteries, or by enhancing the integration of rechargeable batteries. Thus, energy efficiency would be increased and battery waste would be reduced [6, 7]. Second, the use of artificial light as an irradiance source for in-house PV products does not require electricity [8]. Consumer electronics typically include rechargeable batteries that need to be connected to the grid for charging. Solar-powered products are operated autonomously and can provide independence and convenience to the user, as they can be charged when no grid connection is available, provided that sufficient light is available.

Although PIPV market is rapidly growing [2], there are still many issues that have not been extensively analyzed which mainly concern the use of PIPV indoors. The prominent issue is that while most of the PV products perform well under direct sunlight, they have a remarkable drop in their performance indoors [9]. The efficiency of solar cells is usually measured under STC conditions (AM1.5 spectrum, 1000 W/m^2, 25°C). However, the indoor spectrum is often a combination of natural and artificial light, and the irradiance levels range between 0 and 100 W/m^2. At low irradiance conditions, below 100 W/m^2, solar cells perform differently, which is something that should be taken along in a design of a product. This is therefore the core scope of our study; the effect of indoor irradiance conditions on the design of PV-powered product or PIPV. In this study, we focused on PIPV containing PV technologies that occur most often [1, 10], that is, crystalline silicon (c-Si), multicrystalline silicon (mc-Si), and amorphous silicon (a-Si), under artificial irradiance of compact fluorescent lamps (CFL), light-emitting diodes (LED), incandescent light, and indoor irradiance originating from solar light.

At the moment literature is limited regarding research done on solar cells' performance in indoor environments. Several researchers studied the PV cells' performance under low irradiance conditions [11–13], indoor light conditions, and light spectra [14–16], the spectral irradiance of various PV technologies under different irradiance conditions, methods for optimal design of PV-powered products [1, 2, 11, 17–21], and the development of simulation tools for irradiance conditions and energy calculations of PV-powered devices [22–24]. Some studied methods are the use of CAD software for the simulation of indoor irradiance [23, 24], the use of ray tracing programs, such as the radiance or the DAYSIM [22], or spectral irradiance measurements of low intensities in indoor environments [14].

Based on literature, it seems that there are no models available, which estimate the performance of PV cells under low indoor irradiance, which has been measured, and which comprises both natural and artificial irradiance. Most models that are available calculate the efficiency of PV cells under high irradiance, such as under standard test conditions or under very specific simulated weak irradiance (e.g., 10 or 100 W/m^2), which can differ from measured indoor irradiance.

Our modeling approach is based on the performance of the abovementioned PV technologies under indoor irradiance conditions. For the validation of the model, we used a sample of 12 commercially available PV products for which we compare simulated results with real measurements indoors, under various irradiance conditions. The analyzed products' sample consists of small PV products for either indoor or outdoor use: four PV-powered lighting products, two solar toys, three PV-powered chargers, a solar keyboard, a solar computer mouse, and one PV kitchen weight scale.

The structure of our article is as follows: in the "Indoor Light" section, we shortly discuss indoor irradiance. Next we explain the modeling approach in the "Model Description" section. In the next section (Experiments), the experimental setup is presented which was used to validate our model using measurements of 12 PV products, and then results derived from simulations with the model are described. Finally, the discussion and conclusions of this study are presented.

Indoor Light

In this section, we will introduce indoor irradiance as a mixture of artificial and natural light. With natural light we mean light originating from the sun, also called irradiance. With artificial light we mean light originating from artificial light sources, also called lamps. Below we will shortly address both types of light in the context of indoor environments with windows and interiors containing objects and light sources.

Indoor natural light

During daytime, light indoors is usually a mixture of sunlight and artificial light, depending on the time of the day. The share of sunlight entering a room depends on the surface area of its windows, their orientation, and the degree of overcast [25], as well as the geographic location, the season (the date), and the time of the day.

Solar irradiance that enters a room is dependent on the distance between an open aperture – a window – and the point of observation, and the obscuration by the open aperture. In cases of large windows over the whole width of a room, the attenuation – the gradual loss in the intensity of solar irradiance – is mainly caused by the distance to and the reflectance of the window. The overall irradiance level therefore depends on the architecture of the building, and interior characteristics like the surface reflectance of the walls, ceiling, and floors.

Generally, the irradiance levels outdoors in northern Europe at mid-summer range between 1000 W/m^2 irradiance at a clear day and around 325 W/m^2 at a diffuse day/overcast [26]. Irradiance levels indoors are significantly lower, because the amount of transmitted light through a windowpane broadly depends on the type of glass, cover materials, size, and type of frame. Literature shows that at a distance of 1 m from a single glazing, the radiant power has reduced below 40% of the outdoor measured value, leading to values of 400 and 130 W/m^2 for a clear and an overcast day, respectively. At a greater distance, for example, 5 m from the window, the radiant power decreases even more, reaching 93% of the value outdoors. In case of a double-glass insulated window, the decrease in the radiant power at 1 and 5 m from the window will be around 70% and 97%, respectively [9, 27].

Indoor artificial light

When natural light indoors is inadequate, artificial lights can provide additional light. The irradiance then depends on the amount, sort, and location of the lights that are turned on. Typical artificial light sources are CFL, LED, and incandescent lamps. Figure 3 presents the light spectra of three types of artificial lighting: CFL lamp (Megaman compact reflector GU10, BR0709i, 9 W, 78 mA, 220–240 V, 50/60 Hz, 3000 K Warmwhite), LED (Gamma 230 V–50 Hz, 4.2 W), and halogen lamp (Twistalu, Philips B9, 35 W, 230 V, 40D).

Each light source (halogen, LED, and CFL) was mounted inside a specially designed box with dimensions 67 × 30 × 30 cm. The lamps were placed at a distance of 55 cm from the base of the box. The spectroradiometer was placed inside the boxes, so that ambient light from the room could not affect the measurements. The sensor of the spectroradiometer was placed just under the lamp at a distance of 35 cm from it. The spectral irradiance of each lamp was measured using a spectroradiometer (StellarNet Fiber Optic Spectrometer SCal-C10122012, of type Black C-SR-50, BW-16), which was connected to a computer.

It is noticeable that each light radiates at a specific range of wavelength, which is characteristic for the

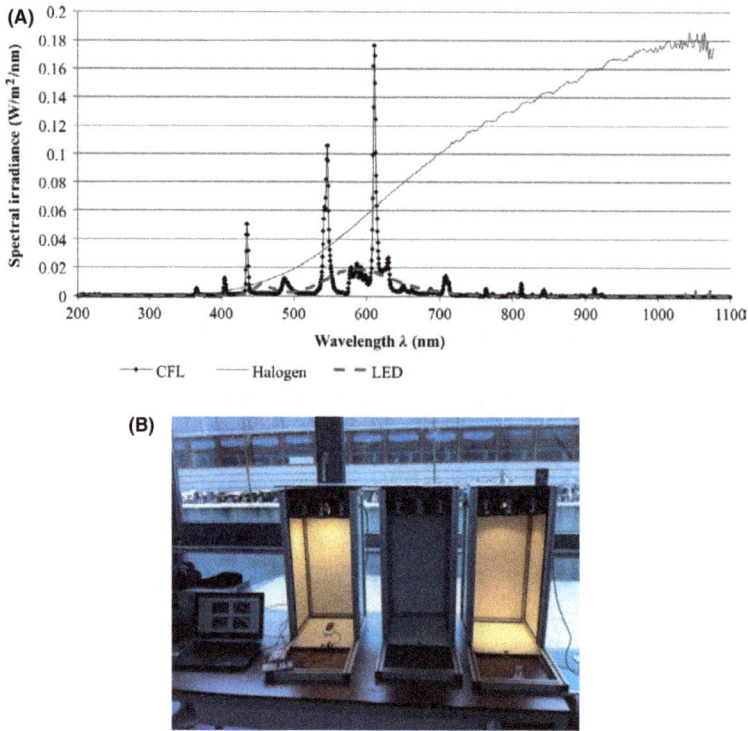

Figure 3. (A) Spectral irradiance of CFL, LED, and halogen lamp in W/m²/nm. Measurements taken at the Applied Labs of TU Delft on 21 January 2014, and (B) the experimental setup that was used for the measurement of the lamps' spectrum.

physical functioning of these different lamp types. Since different solar cell technologies have different band gaps and different spectral responses (SRs), they use only a dedicated part of the light spectrum. For example, artificial light emitted by incandescent lamps has a spectral range between 350 and 2500 nm, while an LED has a range of only 400–800 nm². Lamps like LEDs and CFLs contain most of their power in certain peaks in the visible spectrum between 390 and 700 nm, whereas halogen lamps radiate a considerable amount of their power in the infrared region of the spectrum. This means that depending on the technology of the solar cells, exposure to different light technologies results in different efficiencies and power output by these solar cells.

Model Description

With the objective to estimate the performance of PV products' cells indoors, an analytical model was created in Microsoft Excel. The model combines measurements of natural and artificial irradiance in several rooms, with literature data and measured data regarding the performance

and the SR of PV technologies under low irradiance levels, that is, below 1000 W/m² [11–13, 28–30]. The input data of the model are the surface of the solar cell (in m²), the measured SR (in A/W under STC) of the PV technology that is used by a certain PV product, the measured mixed indoor irradiance (in W/m²/nm), and solar cell's distance from light sources (in m). The model executes spectrally distributed calculations and delivers solar cell's efficiency and power produced (in W) under specific indoor conditions (Fig. 4). Finally, the simulated results of the efficiency and maximum power of the PV products' cells are compared with the measured values of efficiency and power of the PV cells using their measured I-V curves. This later step is executed to assess the model's accuracy.

Mathematical equations

Equation (1) gives the general formula for the calculation of the efficiency η of a solar cell:

$$\eta\,(\%) = \frac{P_{\text{mpp}}}{P_{\text{in}}} \times 100\% = \frac{I_{\text{SC}} V_{\text{OC}} FF}{P_{\text{in}}} \times 100\% \tag{1}$$

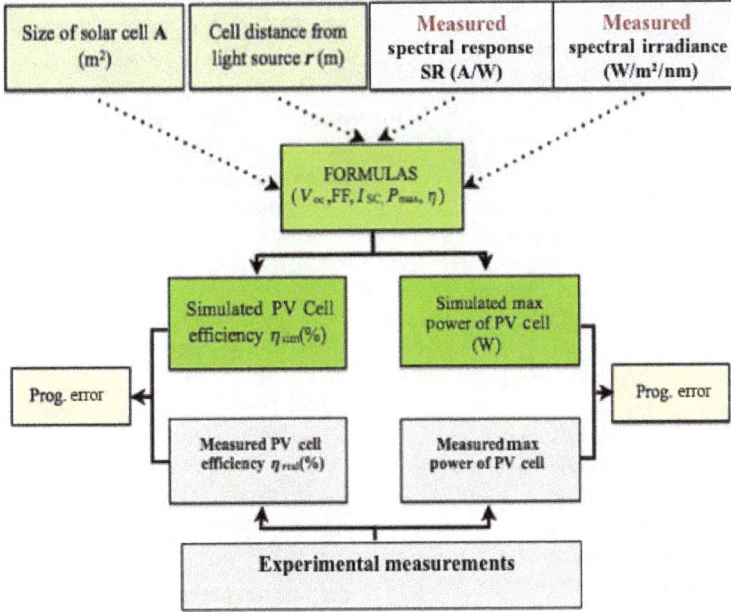

Figure 4. Schematic depiction of the analytical model created for the estimation of the performance of PIPV cells indoors [4, 31].

where I_{sc} is the short circuit current (in A), V_{oc} the open circuit voltage (in V), FF the fill factor ($-$), P_{mpp} the measured power in the maximum power point (mpp) (in W), and P_{in} the power (in W) of the irradiance hitting a solar cell. The FF was calculated by equation (2):

$$FF = \frac{I_{mpp} \cdot V_{mpp}}{I_{sc} \cdot V_{oc}} \qquad (2)$$

where the "mpp" values are the maximum power point values (I_{mpp} [in A]: current at the maximum power point, and V_{mpp} [in V]: voltage at the maximum power point), which is also the maximum output of the solar cell. The I_{sc} is the short circuit current and V_{oc} the open circuit voltage.

Using the Shockley equation in one-diode model (3), it can be assumed that ideally, the short circuit current I_{SC} is equal to the photocurrent I_{Ph} [32–34].

$$J = J_{ph} - J_0 \left[\exp\left(\frac{eV}{k_B T}\right) - 1 \right] \qquad (3)$$

where J is the current density produced by the solar cell (A/m^2), J_0 the saturation current density (A/m^2), J_{ph} the generated photocurrent density (A/m^2), e the elementary charge ($1.60217662 \times 10^{-19}$ Coulomb), V the applied voltage across the terminals of the diode (V), k_B the Boltzmann's constant ($1.3806488(13) \times 10^{-23}$ J/K), and T the temperature (K).

The short circuit current I_{SC} is calculated using equation (4).

$$I_{SC}(\lambda) \cong I_{Ph}(\lambda) = \int E(\lambda)\, SR(\lambda) d\lambda \qquad (4)$$

where $E(\lambda)$ is the spectral irradiance (W/m^2 nm) and $SR(\lambda)$ is the SR of the solar cell (A/W).

$$I_{SC}(\lambda) = \int \left(SR(\lambda) \times \left[E_{natural}(\lambda) + E_{artificial}(\lambda) \right] \right) d\lambda \qquad (5)$$

where $E_{natural}$ is the spectral irradiance indoors originating from the sun (in W/m^2nm), and $E_{artificial}$ the spectral irradiance indoors originating from artificial lights (in W/m^2nm). The integral of the short circuit current from 191 to 1076 nm wavelength is the total short circuit current of the cell. The range in which the measurements were conducted is defined by the wavelength range of the measurement equipment – the spectroradiometer. The specific spectroradiometer can measure irradiance (in W/m^2) in a range of 185 and 1078 nm. For our measurements, we have chosen a wavelength range from 191 to 1076 nm due to irregularities of the measurements around the edges (e.g., above 1000 nm there was high infrared radiation). A presentation of results in this range does not affect the accuracy of the measurements and the calculation of the short circuit current of the PV cell.

Modeling of the indoor irradiance

Due to the small size of artificial lights compared to their distance to a PV product's cells, we assume in our model that lamps are point sources. Therefore, the irradiance at each point can be calculated using equation (6), which follows the inverse square law.

$$I_{rps} = \frac{I_{ps}}{r^2} \qquad (6)$$

where I_{rps} is the irradiance at a specific distance from the lighting source (W/m²), I_{ps} is the intensity (luminous or radiant power per unit solid angle) of the lighting source (W/sr), and r is the distance of the object from it (m). However, equation (6) cannot be used in case of linear light sources, such as fluorescent light tubes.

For the investigation of indoor natural irradiance, associated with the distance from windows, multiple measurements were performed in an office and a workshop at TU Delft. Figure 5 depicts the results after 3 days of measurements at a north-oriented office. The specific orientation was chosen as the worst case scenario (less irradiance during the day) and also due to its availability to conduct tests there.

On 3 December 2013, the measured irradiance outside of the window of the office was 37 W/m², on 2 April 2014 it was 91 W/m², and on 4 June 2014 it was 146 W/m². The orientation of the measurement device (in this case a spectroradiometer) during the measurements was horizontal (placed flat on a table). In Figure 5, the distance-to-window rule is described as indicated by equation (7).

For the measurements that are presented in Figure 5, no solar cell was used. More specifically, a spectroradiometer was used, which was placed at different distance from the window. Artificial lighting is not used during this test. At each position of the spectroradiometer's sensor, we got one measurement. Figure 5 presents the indoor irradiance as measured, including possible reflections,

transmissions, etc. In the model we do not account the reflections and transmissions because it is not possible. The transmissions and reflections of irradiance differentiate broadly in each case. Thus, it is not possible to be calculated or even predicted. The only reflection that can be assumed is the one coming from the window's glass.

The measurements exposed that the irradiance changes approximately with the reciprocal of the distance:

$$I_{rs} = \frac{I_s}{r} \qquad (7)$$

where I_{rs} is the irradiance at a specific distance from the lighting source (W/m²), I_s is the intensity (luminous or radiant power per unit solid angle) of the lighting source (W/sr), and r the distance (m).

Equations (6) and (7) are used for the calculation of the irradiance at a specific distance from the artificial light sources and windows.

Experiments

Measuring I-V curves of PV cells

The equipment used for the measurements of the PV products' cells' performance (I-V curves) include a data acquisition module and an electronic circuit (Fig. 6). The I-V measurement setup is in-house designed by Martin Verwaal at the Applied Labs of the Industrial Design Engineering department of Delft Technical University. This circuit consists of a charging capacitor (C1) with low series resistance, ranging from 10 to 1000 μF, a current measuring resistor (R1) ranging from 1 to 100 Ω, a discharging MOSFET BUK9535 (T1), and a start switch (SW1). The MOSFET has a low ON resistance (<35 mΩ at Vgs = 5V). Resistor R2 is 1 kΩ and it is added to keep the gate of the MOSFET in the normal state at a low level, so the MOSFET is not conducting.

First, the PV cell of the product has to be disconnected from the products' electronic circuit and connected to

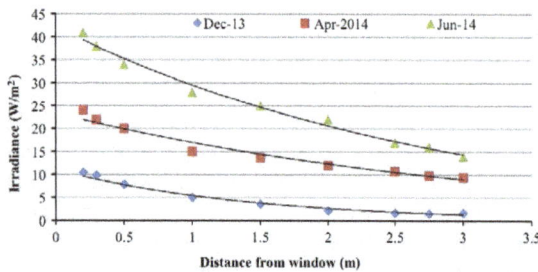

Figure 5. Irradiance measurements conducted on three different days: 3 December 2013, 2 April 2014, and 4 June 2014 at a north-oriented office, TU Delft, the Netherlands [4, 31].

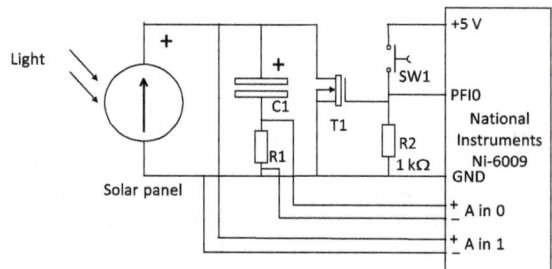

Figure 6. Circuit diagram used for the measurement of I-V curves.

the measuring circuit. The capacitor is discharged through the MOSFET by pressing switch SW1. The capacitor is initially emptied (at 0 V). When the switch is released, the PV cell recharges the capacitor. Voltage in the capacitor does not sweep from zero to V_{oc}, but the capacitor is charged until the V_{oc}. At the same time the data acquisition module records the solar panel voltage and charging current. A Labview program controls the measurement and presents the I-V curves of the tested PV cells.

Measurements were taking place under mixed indoor lighting conditions and with solar cells of various types. Therefore, the size of the capacitor and the value of the current measuring resistor had to be adjusted each time. For indoor measurements, it proved to be very important to integrate measurements over 1 or more power line cycles (multiples of 20 msec) due to the flickering of the artificial light sources. The value of the current measuring resistor was kept as low as possible, but high enough to keep the accuracy below 0.1% of the total range. In our case, the measuring range was 100 mV, which gives an accuracy of 100 μV. For example, in order to get a measuring range of 10 mA, a resistance of 10 Ω was used. The accuracy in the current measurement was 1% of the measured value, as well as the accuracy of the measuring resistor (1%). The absolute voltage accuracy was 0.1% of the range (10 mV). The rate of the measurement depends on the PV cell and the capacitor and it is estimated to be in a range of 100 msec to 10 sec.

Measuring indoor irradiance

Spectrally distributed measurements of indoor irradiance were conducted in offices and laboratories of the Department of Design Engineering of TU Delft in the Netherlands. Natural light and three types of artificial light sources, as described in "Indoor artificial light" section, compact fluorescent lamp (CFL), LED, and halogen lamp were measured. A StellarNet Fiber Optic Spectroradiometer, type Black-Comet-SR, model C-SR was used for the measurements of spectrally distributed irradiance of both natural and artificial lighting. The accuracy of the instrument is 5%, with bandwidth 0.5 nm and wavelength range of 190–1080 nm. It contains a probe for a CR2 miniature cosine receptor for UV-VIS-NIR.

Figure 7 depicts the irradiance spectrum indoors, at an office environment, under mixed light (natural light indoors and artificial light by CFL lamps 58 W, light color 830 Warmwhite, Philips Master TL-D, 58W/830), and the spectrum of light outdoors, just outside the window of the specific office. Both measurements were conducted at the same time. The sensor was placed horizontally during both measurements, indoors and outdoors, at a distance of 50 cm inside and outside the window, respectively. For the outdoor measurements of irradiance, we used the spectroradiometer and we placed a cover on the top of the sensor (CR2-AP ~4.8%, cosine receptor 2 – aperture ~4.8%), which lets only a very small percentage of solar irradiance through (namely ~4.8% of the total irradiance). The cosine receptor CR2 has a wavelength of 200–1100 nm, diameter 1/4 inches, and field of view 180°. Total irradiance outside the office was measured around 33 W/m², while indoor irradiance was around 10.5 W/m². It is interesting to notice the difference between these two spectra, regarding their values. The window glass (double glazing) cuts almost two thirds of the measured outdoor irradiance and permits only one third of it to pass indoors. The glass manufacturer is Glaverbel,

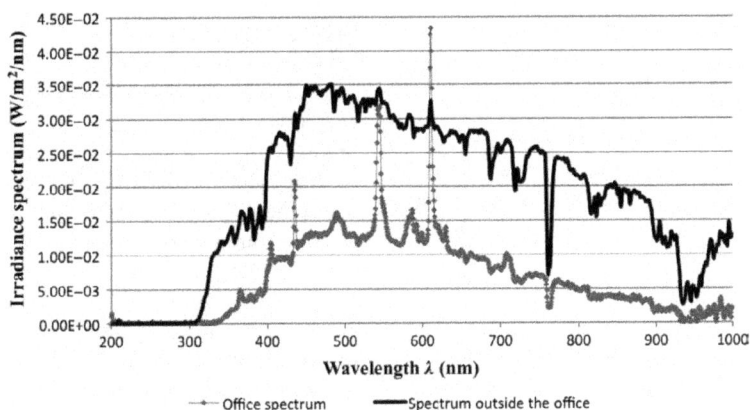

Figure 7. Irradiance measurements indoors under mixed indoor lighting and outside the window at the Applied Labs of TU Delft, the Netherlands, on 21 January 2014.

and the type is Thermobel 0.5 Stopray. Stopray means that there is a triple silver coating that stops direct IR sunlight with around 80% and lets the visible sunlight through.

Observing the curve that depicts the outdoor spectral irradiance, we can see only the spectrum of natural light, while indoors the curve seems to contain both natural light that enters the room, as well as artificial light originating from fluorescent lamps. The existence of artificial fluorescent light is clear, due to the peaks of the curve at specific wavelengths, some of which are at the 437, 547, and 612 nm, typical characteristic of the specific light.

External quantum efficiency and SR measurements

Measurements were conducted for the calculation of the PV cells' SR under standard test conditions (STC). For that purpose, we used a built-in house setup of the Photovoltaic Materials and Devices (PVMD) group at TU Delft. This setup consisted of a Newport illuminator/monochromator, a probes' holder, a chopper, and a lock-in amplifier. The PV cell of each product was placed at a stable position and at a distance of around 2 m from the monochromatic light source to measure the external quantum efficiency (EQE). The measured EQE at this stage of the procedure was not at STC. In order to calculate the EQE and consequently the SR of the tested PV cells at STC, we also used a solar simulator calibrated at AM1.5 (Super Solar Simulator WACOM, Model WXS-90S-L2, AM1.5GMM, Serial No. 07061501, 1ϕ, 230V, 18A) for the calculation of the short circuit current I_{SC} of the solar cells of each product. Using the correlation of the short circuit current under STC and under the monochromatic light, we calculate the EQE at STC and from there the SR at STC, using equations (8) and (9).

$$EQE_{STC} = EQE_{mon.light} \cdot \frac{J_{sc_{stc}}}{J_{sc_{mon.light}}} \qquad (8)$$

where EQE is the external quantum efficiency and J_{SC} is the short circuit current density (mA/cm^2) at STC and under the monochromatic light measurements (mon.light).

The SR of the PV products' cells is calculated by equation (9):

$$SR = \frac{q}{hc} \lambda \cdot EQE \qquad (9)$$

where SR is the spectral response of the solar cell (A/W), EQE is the external quantum efficiency, λ is the wavelength (nm), q the elementary electric charge (~1.6 × 10^{-19} C), h is Planck's constant (~6.626 × 10^{-34} Js), and c the speed

of light in vacuum (m/sec). As equations (8) and (9) show, the EQE and SR are wavelength dependent. EQE is dependent on the J_{SC} (see eq. 8), which also depends on the wavelength.

The PV products that were tested using the above-described equipment are illustrated in Figure 8.

The PV cells are tested including encapsulant material and contacts. Figure 9 illustrates a sample of the tested products' PV cells, as used during the measurements. Dissimilarities in measured and simulated values are a result of the damages in PV cell's surface, the PV cells' connection, the type of the coating material, or lesions of the PV cell.

Figures 10, 11, and 12 present the results from the measurements of the SR data for c-Si, mc-Si, and a-Si cells at STC, respectively. The literature-reported SR of each technology, at irradiance levels between 1 and 1000 W/m^2 [13], is also included in Figures 10, 11, and 12, as well as the measured SR of each one of the products' PV cells.

Figure 10 presents the SR of the c-Si cells, as measured at STC, for 4 PV products: the Little Sun light, the Voltaic bag, the frog toy, and the WakaWaka light. There is also one extra line in the graph, which depicts literature data for the SR of c-Si. It is noticeable that the literature-reported SR of c-Si is far higher than the measured values. In reality, the SR of the cells that consumer PV products use could hardly compare with the SR of the laboratory fabricated PV cells intended for larger applications. Furthermore, even PV cells of the same technology, which are fabricated by other manufacturers, could have deviations in their SR, as it is broadly influenced by the transmissions and reflections of the cell's surface.

The SR of the frog toy's PV cell is the highest among the other cells and reaches around 57% of the literature data of SR for c-Si. The lowest SR is noticed for the Little Sun light and the voltaic bag, which seem to be less than 3% of the literature values.

Figure 11 presents the SR of mc-Si cells and includes seven lines: one for the literature-reported SR of mc-Si and six lines, which depict the SR of six commercial PV products that use mc-Si cells. These products are the Sunnan light, the Solio charger, the car toy, the Ranex lights, the Logitech keyboard, and the kitchen weight bar. The literature-reported SR of mc-Si is the highest compared to the products' values. The SR of the car toy reaches around 46% of the literature data for the mc-Si's SR, while the lowest value among the tested products is the one of the Sunnan lamp, which is around 8% of literature value.

Figure 12 presents the SR of a-Si cells. It includes three lines: one for the literature-reported SR of a-Si cells and two lines, which depict the SR of two commercial PV

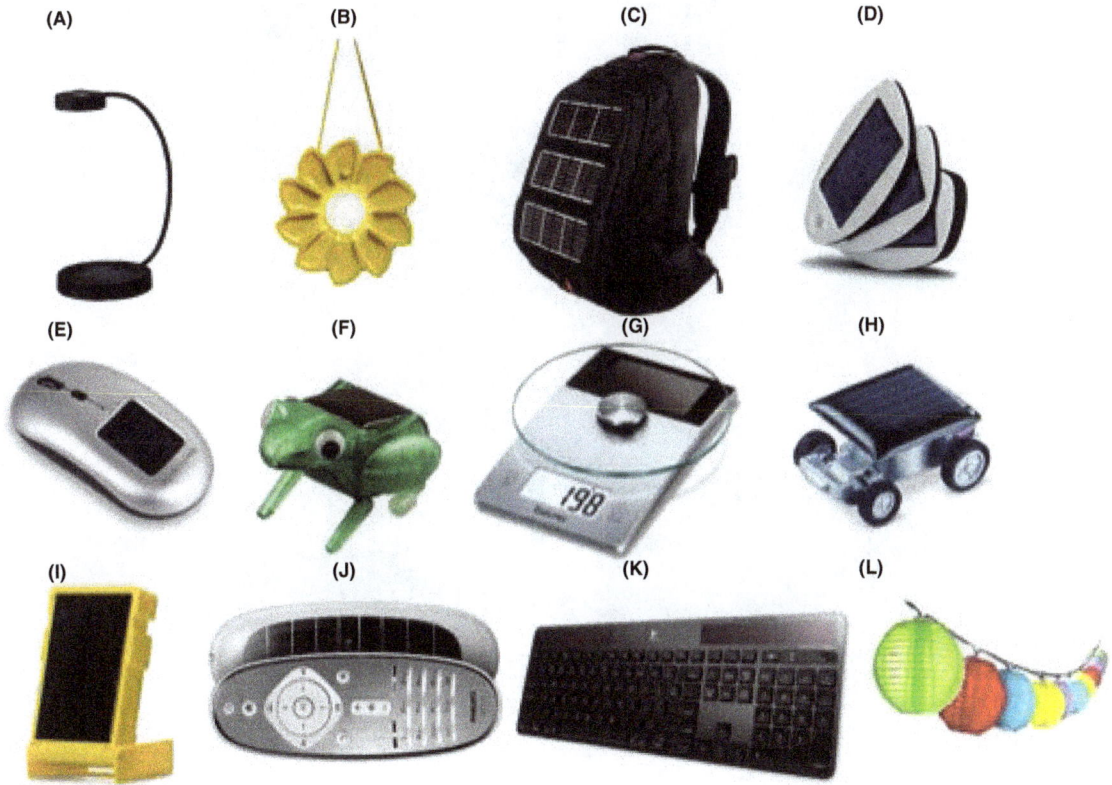

Figure 8. Tested PV products: (A) IKEA Sunnan lamp, (B) Little Sun solar-powered lamp, (C) Voltaic solar bag, (D) Solio charger, (E) solar mouse by Bondidea, (F) frog toy, (G) PV kitchen weight scale, (H) car toy, (I) WakaWaka light, (J) Philips remote control, (K) solar wireless keyboard by Logitech, (L) Ranex lights (figures attached from Google).

products – the Philips remote control and the PV-powered mouse. Between the two products, the Philips remote control has the lowest SR, which is around 7% of the literature data for the SR of a-Si cells.

Looking at the SR curves of the Figures 10, 11, and 12, it arises that at wavelengths below 400 nm, the PV cell absorbs most of the incident light and therefore the SR of the cell is low. At wavelengths 400–1000 nm the cell approaches the ideal, while at wavelengths above 1100 nm the SR significantly decreases, finally falling down to zero.

Results

The model described in "Model Description" section was applied twice (first and second simulations). In order to clarify the differences between the two simulations, in Table 1 we present the inputs, variables, and measurements that took place during the first and the second simulations. As Table 1 shows, for the first simulation we measured the spectral irradiance indoors with the

equipment described in "Measuring indoor irradiance" section, as well as the current–voltage (I-V) curves of the products' PV cells under mixed indoor light, using the equipment described in "Measuring I-V curves of PV cells" section. The inputs of the model for the first simulation were the literature data of the SR of c-Si, mc-Si, and a-Si cells [13], the measured indoor irradiance, the surface of the PV cell, and its distance from the artificial and natural light sources. The outcomes were the calculation of the maximum power and the efficiency of the PV cells. Finally, the model's outcomes are compared with the results of the maximum power and efficiency of the cells, as derived from the measurements, and the model error is calculated.

For the second simulation, extra measurements took place (see Table 1). At this stage except from the indoor irradiance and the I-V curves of the tested PV products' cells, we also measure the EQE and the SR of each PV cell at STC, using the equipment described in "External quantum efficiency and SR measurements" section. The inputs of the model are the same as in the first

Figure 9. Some of the tested products' PV cells, as used during the experiments.

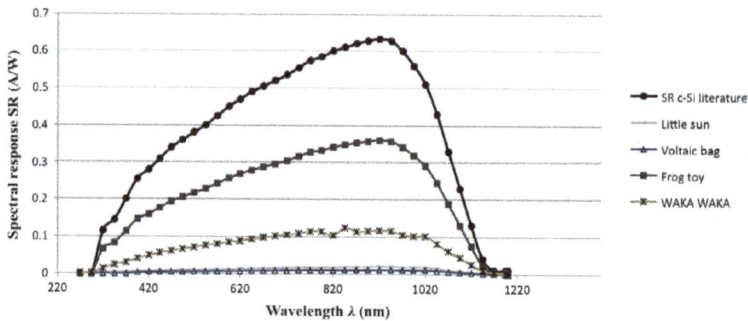

Figure 10. Spectral response of c-Si cells as reported in literature [13], and measured for the tested PV products [4, 31] under STC.

simulation with one important difference; for the second simulation, we use as input the measured SR of the PV cells at STC and not the literature data. The model's outcomes are the maximum power and efficiency of the PV cells, which are again compared with the measured values.

The purpose of these two simulations is to understand the role of the SR in the performance of a solar cell and the big deviations of the measured values compared to the literature data.

For the simulations, it is important to clarify that the active solar cell surface is standardized for all the PV products at 10 cm². All PV products' cells were tested at a of distance 50 cm from the window and the cells were horizontally placed during the measurements (flat on a table). During the first simulation, the estimation of the fill factor and the open circuit voltage was done according to the irradiance levels on the cell. The necessary data for this calculation were obtained from measurements performed by Randall (2001). By

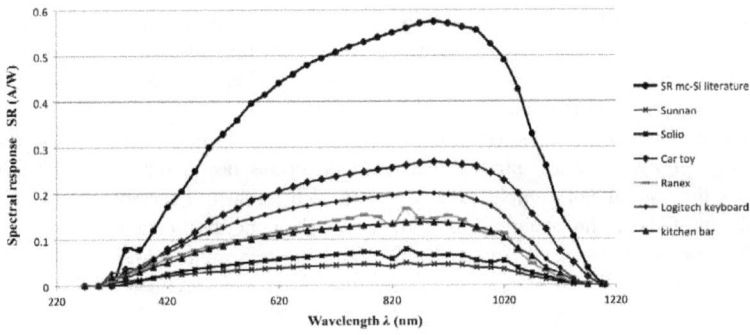

Figure 11. Spectral response of mc-Si cells as reported in literature [13], and measured for the tested PV products [4, 31] under STC.

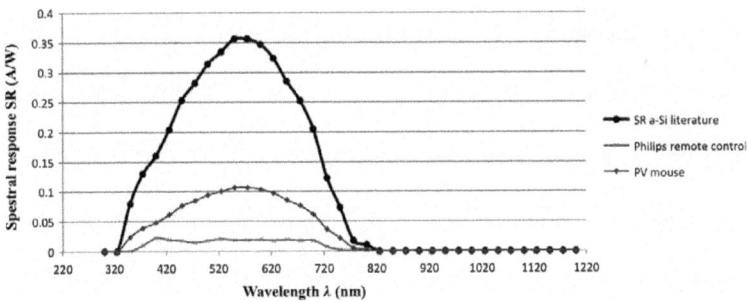

Figure 12. Spectral response of a-Si cells as reported in literature [13], and measured for the tested PV products [4, 31] under STC.

Table 1. Model's inputs and variables for the first and the second simulations.

Simulations	Measurements	Inputs	Variables
First simulation	1. Spectral irradiance indoors, 2. I-V curves of products' cells under mixed indoor light	a. Literature data of SR, b. Spectral irradiance indoors, c. Solar cell's surface, d. Cell's distance from the light sources	1. P_{max} (mW), 2. η (%)
Second simulation	1. Spectral irradiance indoors, 2. EQE and SR under STC, 3. I-V curves of products' cells under mixed indoor light	a. Measured SR under STC, b. Spectral irradiance indoors, c. Solar cell's surface, d. Cell's distance from the light sources	1. P_{max} (mW), 2. η (%)

processing the data in Excel, logarithmic equations were found that best fit the experimental results. It was calculated that according to Randall (2001), the voltage depends on the irradiance based on the following relation: $V_{OC} = 0.045 \ln(E) + 0.2931$, with $R^2 = 0.98572$. This is a linear variation on a logarithmic scale. Using this equation, the program computes the V_{oc} that corresponds to the irradiance reaching the surface of the cell. The V_{oc}, for the second simulation, was measured for each PV product's solar cell using the I-V curve

measurement setup that was described in "Measuring I-V curves of PV cells" section.

There exist several reasons to present results from simulations using real solar cells adding to simulations based on theoretical values of solar cells from literature, namely:

1. We would like to show the performance of real solar cells in commercially available PIPV to show experts in the field of energy technologies, in particular PV researchers, that a lot of improvement can be made in

increasing the performance of PIPV by using solar cells with a better performance, better said with a higher efficiency.

2. Second, to show product designers that the use of literature data of variables such as SR, cells' efficiencies, etc., during the design process of a PV product may result in assumptions that would not comply with the reality. If using literature data, the assumptions and calculations regarding the performance of the designed product will be too positive.

By showing the difference between simulations based on data from literature and data from existing cells, we provide valuable information to researchers and product designers.

Table 2 presents the cells of the tested PV products, as used in the measurements. The table illustrates the PV cells of each product, it presents the PV technology of each cell, as well as the short circuit current, the open circuit voltage, and the maximum power of each cell, as measured using the equipment described in "Measuring I-V curves of PV

Table 2. The tested products' PV cells, as used during the experiments.

PV products' cells	Product name	PV cell type	Full cell area (cm^2)	I_{SC} (mA) (under CFL light)	V_{OC} (V) (under CFL light)	P_{mpp} (mW) (under CFL light)
	Sunnan	mc-Si	57	0.57	2.56	0.62
	Little Sun	c-Si	36	0.35	3.22	0.62
	Voltaic bag	c-Si	87	0.49	6.54	1.65
	Solio	mc-Si	17	0.31	3.7	0.58
	Frog toy	c-Si	4.8	0.24	0.52	0.076
	Philips control	a-Si	30	0.27	5.37	0.77
	Car toy	mc-Si	2.7	0.31	0.24	0.041
	WakaWaka	c-Si	65.7	1.4	1.45	1.1
	Ranex	mc-Si	8.1	0.63	1.62	0.6
	Logitech keyboard	mc-Si	35.24	0.32	4.7	1.64
	PV mouse	a-Si	9.53	0.09	6.9	0.41
	Kitchen bar	mc-Si	44	0.32	6.1	0.96

cells" section. The values of I_{SC}, V_{oc}, and P_{mpp} that are presented in this table are measured under fluorescent light. It can be seen from equation (5) that I_{SC} depends on the available light spectrum. Besides, the short circuit current differs for each type of PV technology, as each material has different SR and solar cells have different surface area. These factors greatly explain the variation of a PV cell's performance under different types of illumination.

First simulation

For the first simulation, the model has been applied by using the SR as given in the literature [13], as it is also discussed in the introduction of "Results" section. The measurements and results presented in Table 3 were conducted on 21 January, 26 September, and 3 October 2014. The measured spectral irradiance on 21 January 2014 outside the window of the office, where the PV products were tested, was around 33.4 W/m^2. Indoors, at the laboratory the irradiance was ranging between 8 and 12 W/m^2, and composed of a mixture of artificial fluorescent light and natural light. On 26 September and 3 October 2014, the measured irradiance at the laboratory was ranging between 10 and 13 W/m^2 (see Table 3).

Table 3 shows the measured and simulated results for the maximum power P_{max} (mW) and efficiency η (%) of the tested PV products' cells under mixed indoor irradiance. It indicates that the simulated efficiency of the PV products' cells during the first simulation was exceeding the measured value. From Table 3, it seems that a well-performing PV product under low indoor irradiance is the solar keyboard, for which the measured and simulated efficiency is 10.7% ($\pm$0.1) and 10.9% ($\pm$1.1),

respectively. Other sufficiently performing products indoors are the frog toy with measured and simulated efficiency 6.4% ($\pm$0.2) and 7.3% ($\pm$2.2), respectively, and the PV-powered mouse with 6.2% ($\pm$0.1) and 7.3% ($\pm$2.3), respectively. On the other hand, bad-performing PV products for use at low indoor irradiance are the Philips remote control with measured and simulated efficiency 2.5% ($\pm$0.1) and 7.0% ($\pm$2.1), the Sunnan lamp with 2.1% ($\pm$0.2) and 3.8% ($\pm$1.1), and the WakaWaka light with 1.2% ($\pm$0.1) and 4.2% ($\pm$1.3), respectively. The measured results that are presented in Table 3 were conducted using the equipment described in "Measuring I-V curves of PV cells" section.

The results show that the model can predict the efficiency of the PV products' cells under mixed indoor irradiance with a typical inaccuracy of around +30% (see Table 3). Besides, except from the irradiance, it seems that there are other factors that influence PV cells' performance under mixed irradiance conditions. In order to define these factors and succeed a more accurate result, we continued with the second round of simulations of our model, where we measured the specific SR of the products' cells.

Second simulation

During the second simulation, we used the SR as measured under STC, for the 12 different PV product's cells, instead of literature data. Other inputs of the model for the second simulation were the measured indoor irradiance, the surface of the PV cell, and its distance from the artificial and natural light sources, as discussed in the introduction of "Results" section. The measured values

Table 3. Measured and simulated maximum power P_{max} (mW) and efficiency η (%) of various PV products' cells under mixed indoor irradiance.[1]

PV product	Product function	PV cell type	Total irradiance[2] (W/m^2)	Measured P_{max} (mW)	Simulated P_{max} (mW)	Measured efficiency η (%)	Simulated efficiency η (%)
Sunnan	Lighting	mc-Si	9.9 ± 0.5	1.2 ± 0.5	1.0 ± 0.3	2.1 ± 0.2	3.8 ± 1.1
Little Sun	Lighting	c-Si	10.2 ± 0.5	1.2 ± 0.6	1.5 ± 0.5	4.0 ± 0.2	6.2 ± 1.9
Voltaic bag	Charger	c-Si	8.6 ± 0.4	3.7 ± 0.5	2.3 ± 0.7	5.4 ± 0.5	6.0 ± 1.8
Solio	Charger	mc-Si	10.2 ± 0.5	0.8 ± 0.2	0.9 ± 0.3	4.5 ± 0.1	3.8 ± 1.1
Frog toy	Moving	c-Si	10.2 ± 0.5	1.3 ± 0.7	1.8 ± 0.5	6.4 ± 0.2	7.3 ± 2.2
Philips control	Charger	a-Si	10.5 ± 0.5	0.8 ± 0.4	1.7 ± 0.5	2.5 ± 0.1	7.0 ± 2.1
Car toy	Moving	mc-Si	10.5 ± 0.5	1.1 ± 0.3	1.4 ± 0.4	5.8 ± 0.1	6.0 ± 1.8
WakaWaka	Lighting	c-Si	10.9 ± 0.5	0.8 ± 0.3	1.5 ± 0.5	1.2 ± 0.1	4.2 ± 1.3
Ranex	Lighting	mc-Si	9.5 ± 0.5	0.3 ± 0.1	0.5 ± 0.1	3.6 ± 0.04	3.6 ± 1.1
Logitech keyboard	Entertainment	mc-Si	11.5 ± 0.5	3.3 ± 0.1	4.1 ± 0.3	10.7 ± 0.1	10.9 ± 1.1
PV mouse	Entertainment	a-Si	10.5 ± 0.5	0.7 ± 0.2	1.2 ± 0.5	6.2 ± 0.1	7.3 ± 2.3
Kitchen weight bar	Cooking	mc-Si	10.5 ± 0.5	1.2 ± 0.4	1.7 ± 0.6	3.6 ± 0.6	5 ± 1.6

[1]Mixed indoor irradiance: natural and artificial irradiance measured indoors.
[2]Irradiance measured with spectroradiometer.

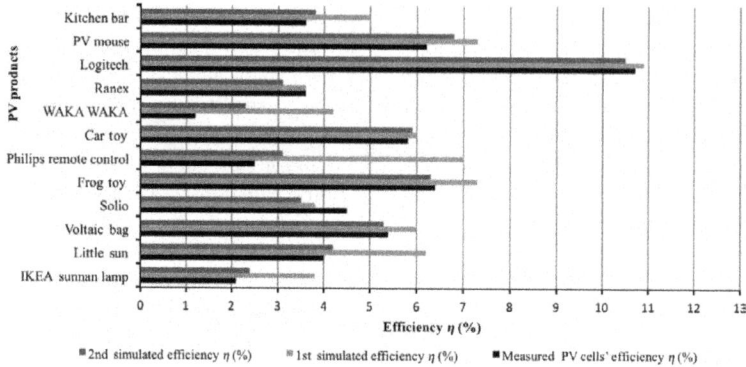

Figure 13. Measured and simulated PV product cell efficiency η (%) under mixed indoor lighting, for the first and second simulation [4, 31].

of the SR of the c-Si, mc-Si, and a-Si cells were significantly lower than that reported in literature, as it is also presented in Figures 10, 11, and 12.

As Figure 13 demonstrates the PV cells' efficiency resulting from the second simulation was lower than the values of the first simulation, and they slightly deviated from the measured efficiency. The lesser the SR of a PV cell, the lower the short circuit current and the lower the cell efficiency. In Figure 13, the measured and the simulated PV cells' efficiency for the tested PV products are presented for the first and the second simulation round. The Philips remote control seems to have the biggest deviation of PV cells' efficiency between the first and the second simulation, which ranges from 7% to 3%, respectively. This deviation is a result of the SR of the product's PV cells (a-Si), whose measured SR is significantly lower (<10%) than the literature data [13] (see Fig. 12). Other products with big deviations between the first and the second simulated efficiency are the Little Sun light, with 6.2% efficiency during the first round and 4.2% for the second simulation, and the WakaWaka light, with 4.2% simulated efficiency in the first round and 2.3% for the second round, respectively.

As Figure 13 indicates the best-performing PV product is the solar-powered keyboard, with simulated efficiency for the second simulation round around 10.5%, while the measured value is estimated at 10.7%. The frog toy also performs sufficiently with simulated and measured efficiency around 6.3% and 6.4%, respectively. On the other hand, the Sunnan lamp seems to be one of the bad-performing products, with simulated efficiency around 2.4% in the second round of simulations, while the measured efficiency is 2.1%. For the second simulation, results revealed that the model can predict PV cells' efficiency under mixed indoor irradiance with higher accuracy than the first simulation, reaching a typical inaccuracy of approximately +18% (see Fig. 13).

The simulated results show that the model can sufficiently predict the performance of the PV products' cells at an indoor environment. More specifically, the simulated results of the PV cells' efficiency η for mixed indoor light were slightly different from the measured values, with a typical inaccuracy of around 30%. However, the simulated results of the PV cells' maximum power P_{max} were even closer to the measured values, with a typical error around 25% (see Table 3). Trying to increase model's accuracy, primary conditions (e.g., amount of indoor irradiance, spectrum of artificial light, distance of the PV cells from light sources, and test room) were very specific.

In general, model's inaccuracy is a result of some unspecified conditions and features of the products. In detail, an important factor, which is responsible for the low PV cells' performance is the unknown cover material of the PV cell (coating). Each PV cell was surrounded by a plastic surface (cover), whose material is not known. This also results to an undefined transmittance of the cover material. Other factors, which limit down the efficiency of the PV cells, are the orientation of the cell, possible damages on the product's surface (e.g., scratches, dust, fingerprints, lesions on the front cover), as well as indoor reflections of irradiance, or shadows in the interior of the room from the surroundings (e.g., furniture, curtains, outdoor shadows, etc.). However, the most important factor, which is responsible for the deviations in results, is the unknown SR of the PV technology that each product uses. The measured SR of the tested PV cells at STC was significantly lower than the literature data of SR for c-Si, mc-Si and a-Si cells (see Fig. 10–12).

Discussion and Conclusions

This article describes a model, which estimates the performance of PV cells at an indoor environment and under mixed indoor light that partially contains outdoor light. The most significant variables in this model are the SR of the PV product's cell and the indoor irradiance. The model has been validated by two different simulations: (1) using the SR as given in the literature (under STC) and (2) using the SR as measured (under STC) for 12 different PV products with either x-Si or a-Si solar cells. It is due to the limited research in this field and the related lack of data from other studies regarding modeling of product-integrated PV, the SR of PV cells under mixed indoor lighting, as well as cells' performance under low lighting conditions, that the results of this study could not be compared in full extent with existing findings. However, we assume that now that this basic model exists, students, researchers, and designers can use it to design or evaluate indoor PV products with the purpose to improve their performance. The results of the model are precise enough for product design, using measured SR curves the accuracy is typically in the order of 30%. This is due to the low irradiance conditions, deviations between measured SR at STC and the actual SR at low irradiance conditions, and the bad quality of commercially applied PV cells in PV products.

In Figure 14, literature-reported efficiencies are presented for c-Si, mc-Si, and a-Si cells [1, 15, 31, 35]. Figure 14 shows that for mc-Si cells the efficiency at 100 W/m^2, which is around 12%, drops significantly at 10 W/m^2 down to 5%, whereas the efficiency of a-Si cells seems to be relatively constant around 5–6% at different levels of irradiance. Finally, c-Si cells' efficiency decreased from 13% at 100 W/m^2, to 5–6% at 10 W/m^2, and 5% at 1 W/m^2. The PV cell efficiencies presented in Figure 14 are based on literature (Kan, 2006; Martin A. Green K. E., 2013) and are not outcomes of the authors' research. Based on the literature,

the spectrum and temperature of the cells are AM1.5 and 25°C. However, the Rsh seems to be low in the low intensities and therefore the efficiency of the cells is low, too.

From literature [22–24] it follows that the simulation or modeling of the PV products cells' performance indoors has still not sufficiently been investigated. There are several factors that influence the performance of PV products in an indoor environment, such as the level of indoor irradiance, the performance of the PV cells under low irradiance conditions, the type of illumination [36] the interaction of the user with the product, and the system's energy losses. Therefore, we propose a new simple model capable of predicting the PV performance under various illumination conditions, which would provide basic support during the design of a PV product's energy system.

The results of the second set of simulations show that under mixed indoor lighting conditions, the simulated PV cells' efficiency slightly deviates from the measured values, with a typical inaccuracy of around +18%. Additionally, the model practically forecasts a PV product's cells performance under artificial illumination, with a typical inaccuracy of around +29% for CFL and LED lighting. Measurements with higher accuracy are quite difficult to obtain, since indoor irradiance reaches just a few tenths of Watts/m^2, which is close to the measurement limits of irradiance sensors. Besides, the efficiency of PV cells under these conditions is rather low. The model's results hence expose the abovementioned fact and are considered satisfactorily accurate. We have found that under mixed indoor lighting of around 20 W/m^2, the efficiency of solar cells in 12 commercially available PV products ranges between 5% and 6% for a-Si cells, 4% and 6% for multicrystalline silicon (mc-Si) cells, and 5% and 7% for the monocrystalline silicon (c-Si).

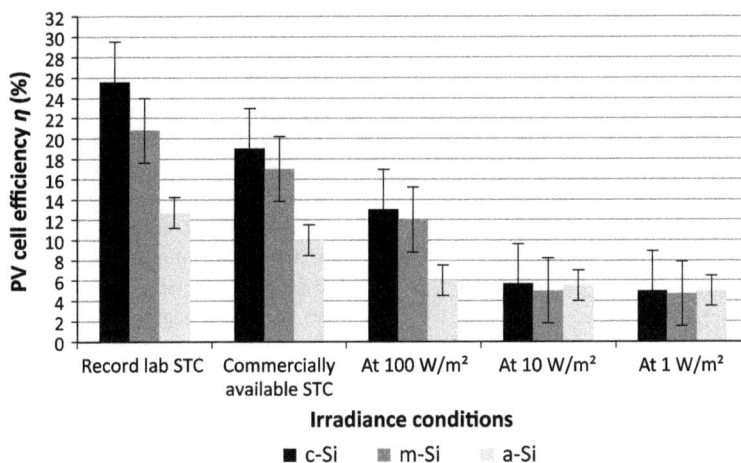

Figure 14. PV cell efficiencies of c-Si, mc-Si, and a-Si at different irradiance conditions, respectively, at STC (AM1.5, 25°C, 1000 W/m^2), 100, 10, and 1 W/m^2 as reported in literature [1, 15, 31, 35], with absolute error ±1%.

Figures 10–12 have shown that the SRs of tested PV cells at AM1.5 deviate considerably from literature data. They are typically around 70–80% lower and in some cases even more than 90% less, as presented in "Experiments" section. The significantly low SR of commercial PV products' cells happens due to low quality of the cells applied. The cutting of PV cells in small pieces – to be applied in PV products' surfaces – and their condition, for example, soiling of cell's surface, possible scratches, cracks, and other damage play a crucial role on the measured SR. Consequently, the use of low quality PV cells leads to PV products with low performance. Furthermore, it is essential to stress here that another reason for the dissimilarities in the SRs is that in this study PV products were not tested as single PV cells, but as assembled devices with several interconnected PV cells (see Fig. 9 and Table 2).

It is also important to be aware of the fact that the SR of the PV cells as measured at STC (1000 W/m²) has been used for modeling at 10 W/m². This is due to the measurement range of solar simulators, which usually does not cover the very low irradiance range used in our model and due to the unavailability of PV cells' SR data under low irradiance conditions as provided by manufacturers.

Finally, because of our purpose to support designers in their design processes to realize indoor PV products with higher performance than the existing ones, we consider the accuracy of this model as being rather acceptable.

Acknowledgments

The authors acknowledge the Faculty of Industrial Design Engineering (IO) of TU Delft for supporting this research. Besides, we thank Dimitris Deligiannis and Remko Koornneef for their assistance and consultation during the measurements, and the Photovoltaic Materials and Devices group, Faculty of Electrical Engineering, Mathematics and Computer Science (EWI) of TU Delft.

Conflict of Interest

None declared.

References

1. Apostolou, G., and A. H. M. E. Reinders. 2014. Overview of design issues in product-integrated photovoltaics. Energy Technol. 2:229–242.
2. Reinders, A. H. M. E., and van Sark W. G. J. H. M.. 2012. Product-integrated photovoltaics. Pp. 709–732 in A. Sayigh, ed. Comprehensive renewable energy, vol. 1. Elsevier, Oxford.
3. NOTEBOOK CHECK, Logitech: Wireless Solar Keyboard K760 für Mac, iPad und iPhone. Available at http://www.notebookcheck.com/ Logitech-Wireless-Solar-Keyboard-K760-fuer-Mac-iPad-und-iPhone.75344.0.html, von Ronald Tiefenthäler (accessed 30 October 2015).
4. Apostolou, G., M. Verwaal, and A. H. M. E. Reinders. 2014. Modeling the performance of product integrated photovoltaic (PIPV) cells indoors. Proceedings of 26th European Photovoltaic Solar Energy Conference (EU PVSEC), Oral Presentation, Amsterdam 2014, The Netherlands, Pg. 3535-3540, DOI: 10.4229/EUPVSEC20142014-6BO.9.5.
5. Durlinger, B., A. Reinders, and M. Toxopeus. 2010. Environmental benefits of PV powered lighting products for rural areas in South East Asia: A life cycle analysis with geographic allocation. Proceedings of 35th IEEE PVSC, Hawaii, USA, 2010, Pp. 2353–2357.
6. Reinders, A. H. M. E. 2002. Options for photovoltaic solar energy systems in portable products. in I. Horvath, P. Li, J. Vergeest. Proceedings of TMCE 2002, Fourth International Symposium April 22-26, 2002. Wuhan, P.R. China ISBN 7-5609-2682-7.
7. Veefkind, M. J. 2003. Industrial design and pv-power, challenges and barriers. in s. n. ed. ISES solar world congress 2003: solar energy for a sustainable future. s.l. ISES.
8. Durlinger, B., A. Reinders, and M. Toxopeus. 2012. Life cycle assessment of solar powered lighting products for rural areas in South East Asia, Components for PV Systems, PV System Engineering and Standards, Socio-economic Aspects, and Sustainability. Proceedings of 25th EU PVSEC, Valencia, Spain, 2012, Pp. 3917–3925.
9. Kan, S.Y. 2006. SYN-Energy in solar cell use for consumer products and indoor applications. PhD dissertation, Final Report 014 –28– 213, NWO/NOVEM, Technical University of Delft, Delft, The Netherlands.
10. Apostolou, G., and A. H. M. E. Reinders. 2012. A comparison of design features of 80 PV-powered products. Proceedings of the 27th EUPVSEC, Poster Presentation, Frankfurt 2012, Germany, Pp. 4227–4232, doi: 10.4229/27thEUPVSEC2012-5BV.1.25
11. Randall, J. F., C. Droz, M. Goetz, A. Shah, and J. Jacot. Comparison of 6 photovoltaic materials across 4 orders of magnitude of intensity. Proceedings of 17th EUPVSEC, October 2001, Munich, Germany, Pp. 603–606.
12. Randall, J. F., and J. Jacot. 2003. Is AM 1.5 applicable in practice? Modeling eight photovoltaic materials with respect to light intensity and two spectra. Renew. Energy 28:1851–1864.
13. Reich, N. H., van Sark W. G. J. H. M., E. A. Alsema, S. Y. Kan, S. Silvester, van der Heide A. S. H., et al. Weak light performance and spectral response of different solar cell types. Proceedings of 20th EUPVSEC, June 2005, Barcelona, Spain, Pp. 2120–2123.

14. Apostolou, G., A. H. M. E. Reinders, and M. Verwaal. 2012. Spectral irradiance measurements in a room fit for indoor PV products. Proceedings of the 27th EUPVSEC, Poster Presentation, Frankfurt 2012, Germany, Pp. 4240–4244, doi: 10.4229/27thEUPVSEC2012-5BV.1.30

15. Freunek (Müller), M., M. Freunek, and L. M. Reindl. 2013. Ideal and empirical maximum efficiencies for indoor photovoltaic devices. IEEE J. Photovol. 3:59–64.

16. Müller, M., J. Wienold, W. D. Walker, and L. M. Reindl. Characterization of indoor photovoltaic devices and light. In: Proc. of 34th IEEE Photovoltaic Specialist Conference. Philadelphia: IEEE, 2009, 738–743, poster award.

17. Alsema, E. A., B. Elzen, N. H. Reich, W. G. H. J. M. Sark, van Kan S. Y., S. Silvester, et al. 2005. Towards an optimized design method for PV-powered consumer and professional applications - The syn-energy project. Pp. 1981–1984 in s. n. ed. Proceedings of the 20th European Photovoltaic Solar Energy Conference. WIP Renewable Energies, München.

18. Reinders, A. H. M. E., J. C. Diehl, and H. Brezet. 2012. The power of design: product innovation in sustainable energy technologies. Wiley, Delft, Netherlands.

19. Veefkind, M. J., N. H. Reich, B. Elzen, S. Y. Kan, S. Silvester, M. Verwaal, et al. 2006. The design of a solar powered consumer product, a case study. Pp. 2306–2311 in S. n. ed. Proceedings of going green care innovation 2006. International Care Electronics, Vienna.

20. Timmerman, M. B. 2008. Review of existing knowledge of product integrated PV for industrial design engineers, Internal Research Paper, Master of Sustainable Energy Technology, University of Twente, Enschede, The Netherlands.

21. Reich, N. H., M. J. Veefkind, E. A. Alsema, B. Elzen, and vanSark W. G. H. J. M.. 2006. Industrial design of a PV powered consumer application: case study of a solar powered wireless computer mouse. pp. 2306–2311 in S. n, ed. 21st European Photovoltaic Solar Energy Conference, Proceedings of the International Conference. Fraunhofer IWM, München.

22. Müller, M., H. Hildebrandt, W. D. Walker, and L. M. Reindl. Simulations and measurements for indoor photovoltaic devices. In: Proc. 24th European Photovoltaic Specialist Conference, Hamburg, Germany: WIP, 2009, Pp. 4363–4367.

23. Reich, N. H., van Sark W. G. J. H. M., deWit H., and A. H. M. E. Reinders. Using CAD software to simulate complex irradiation conditions: predicting the charge yield of solar cells incorporated into PV powered consumer products. Oral presentation, Proceedings of 34th IEEE Photovoltaic Specialists Conference, Philadelphia 2009.

24. Reich, N. H., van Sark W. G. J. H. M., E. A. Alsema, de Wit H., and A. H. M. E. Reinders. A CAD based simulation tool to estimate energy balances of device integrated PV systems under indoor irradiation conditions. Proceedings of 23rd European Photovoltaic Solar Energy Conference, PV Systems, Off-grid applications, Valencia, Spain, 2008, pp. 3338–3343.

25. Randall, J. F. 2005. Designing indoor solar products: photovoltaic technologies for AES. PhD disseration, John Wiley and Sons, Ltd, ETHZ Switzerland.

26. European Commission JRC- Photovoltaic Geographical Information System - Interactive Maps. Available at http://re.jrc.ec.europa.eu/pvgis/apps4/pvest.php (accessed October 30, 2015).

27. Wen, J., and T. F. Smith. 2002. Absorption of solar energy in a room. Sol. Energy 72:283–297.

28. Reich, N. H., van Sark W. G. J. H. M., E. Alsema, R. Lof, R. Schropp, W. Sinke, et al. 2009. Crystalline silicon cell performance at low light intensities. Sol. Energy Mater. Sol. Cells 93: 1471–1481.

29. Reich, N. H., vanSark W. G. J. H. M., and W. C. Turkenburg. 2011. Charge yield potential of indoor-operated solar cells incorporated into product integrated photovoltaic (PIPV). Renew. Energy 36:642–647.

30. Randall, J. F., and J. Jacot. 2002. The performance and modelling of 8 photovoltaic materials under variable light intensity and spectra. Published in World Renewable Energy Congress VII & Expo.

31. Apostolou, G., M. Verwaal, and A. H. M. E. Reinders. 2014. Estimating the performance of product integrated photovoltaic (PIPV) cells under indoor conditions for the support of design processes. Proceedings of 40th IEEE Photovoltaic Specialists Conference, Poster Presentation, Denver 2014, Colorado, Pp. 0742–0747. doi: 10.1109/PVSC.2014.6925027

32. Besheerab, A. H., A. M. Kassemc, and A. Y. Abdelazizd. 2014. Single-diode model based photovoltaic module: analysis and comparison approach. Electr. Pow. Compon. Syst. 42:1289–1300.

33. Rekioua, D., and E. Matagne. Optimization of photovoltaic power systems – modelization, simulation and control. Springer London Dordrecht Heidelberg New York.

34. Quaschning, V. Understanding renewable energy systems, Volume 2, First published by Earthscan in the UK and USA in 2005.

35. Green, M. A., K. Emery, Y. Hishikawa, W. Warta, and E. D. Dunlop. 2015. Solar cell efficiency tables (Version 45). Prog. Photovolt: Res. Appl. 23:1–9.

36. Li, Y., N. J. Grabham, S. P. Beeby, and M. J. Tudor. 2015. The effect of the type of illumination on the energy harvesting performance of solar cells. Sol. Energy 111:21–29.

7

Multi angle laser light scattering evaluation of field exposed thermoplastic photovoltaic encapsulant materials

Michael D. Kempe[1], David C. Miller[1], John H. Wohlgemuth[1], Sarah R. Kurtz[1], John M. Moseley[1], Dylan L. Nobles[1], Katherine M. Stika[2], Yefim Brun[2], Sam L. Samuels[2], Qurat (Annie) Shah[3], Govindasamy Tamizhmani[3], Keiichiro Sakurai[4], Masanao Inoue[4], Takuya Doi[4], Atsushi Masuda[4] & Crystal E. Vanderpan[5]

[1]National Renewable Energy Laboratory, 1617 Cole Boulevard, Golden, Colorado 80401
[2]DuPont Company, 200 Powder Mill Road, Wilmington, Delaware 19803
[3]Polytechnic Campus, Arizona State University, 7349 East Unity Avenue, Mesa, Arizona
[4]National Institute of Advanced Industrial Science and Technology, 1-1-1 Umezono, Tsukuba, Ibaraki 305-8568, Japan
[5]Underwriters Laboratories, 455 East Trimble Road, San Jose, California

Keywords
Adhesives, creep, EVA encapsulant, polymer, qualification standards, thermoplastic

Correspondence
Michael D. Kempe, National Renewable Energy Laboratory, 1617 Cole Boulevard, Golden, CO 80401.

E-mail: michael.kempe@nrel.gov

Funding Information
This work was supported by the U.S. Department of Energy under Contract No. DE-AC36-08-GO28308 with the National Renewable Energy Laboratory.

Abstract

As creep of polymeric materials is potentially a safety concern for photovoltaic modules, the potential for module creep has become a significant topic of discussion in the development of IEC 61730 and IEC 61215. To investigate the possibility of creep, modules were constructed, using several thermoplastic encapsulant materials, into thin-film mock modules and deployed in Mesa, Arizona. The materials examined included poly(ethylene)-co-vinyl acetate (EVA, including formulations both cross-linked and with no curing agent), polyethylene/polyoctene copolymer (PO), poly(dimethylsiloxane) (PDMS), polyvinyl butyral (PVB), and thermoplastic polyurethane (TPU). The absence of creep in this experiment is attributable to several factors of which the most notable one was the unexpected cross-linking of an EVA formulation without a cross-linking agent. It was also found that some materials experienced both chain scission and cross-linking reactions, sometimes with a significant dependence on location within a module. The TPU and EVA samples were found to degrade with cross-linking reactions dominating over chain scission. In contrast, the PO materials degraded with chain scission dominating over cross-linking reactions. Although we found no significant indications that viscous creep is likely to occur in fielded modules capable of passing the qualification tests, we note that one should consider how a polymer degrades, chain scission or cross-linking, in assessing the suitability of a thermoplastic polymer in terrestrial photovoltaic applications.

Introduction

In the manufacturing of the photovoltaic modules there is a desire to use new thermoplastic encapsulant materials. This is motivated by the desire to reduce lamination time or temperature, reduce moisture permeation with new materials [1], use less corrosive materials [2, 3], improve electrical resistance [4, 5], or facilitate the reworking of a module after lamination. However, the use of any thermoplastic material in a high-temperature outdoor environment, where creep may occur, could raise safety and performance concerns. Therefore, there has been an increased concern in the photovoltaic (PV) community regarding the possibility of viscoelastic creep and has brought about concerns for the testing of modules according to IEC 61730 and IEC 61215 [6–8]. Small areas of a module may reach much higher temperatures (>150°C) during the "hot-spot" test or during partial shading of a module without bypass-diode protection [9, 10]; but the localized nature of this occurrence is different from the situation of prolonged operation in the hottest environments and mounting configurations. In very hot environments, modules are known to reach temperatures in excess of 100°C [11, 12]. One could envision an encapsulant with a melting point near 85°C with a highly thermally activated drop in viscosity, resulting in significant creep at 100°C. Creep is distinguished here from delamination, which may also occur when the degree of cure is less than intended, where a primer may not activate as intended [13].

Some early work with poly(ethylene)-co-vinyl acetate EVA encapsulation performed at the Jet Propulsion Laboratory (JPL) did consider the issue of creep during operation at high temperature [6, 14]. PV technology developers at that time were concerned that there was the possibility of displacement of the components within a heated module operating in the field, but did not formally investigate to verify creep using a variety of modules deployed in a hot location. To specifically prevent creep, EVA, that was cross-linked via a peroxide-initiated reaction, was advocated at that time. The use of 65% gel was found to facilitate passing the sales qualification tests (which included the "melt/freeze" test at that time) and was therefore suggested by JPL [6]. The use of EVA with at least 65% gel content was reaffirmed by Springborn Laboratories (later known as Specialized Technology Resources, Inc., or STR) [15], and currently, the use of EVA with 60–90% gel content after lamination is commonly used in the industry. However, if one wants to switch to a thermoplastic material, gel content is not a useful metric to evaluate creep. Therefore, one must find other ways to verify proper processing of a new encapsulant material.

Examination of EVA in fielded modules reveals gel contents above 90% [16, 17]. Often when EVA degrades, it leaves a clear border around the perimeter of the cell and a yellow/brown inner area. It has been observed that in the clear portion, the amount of oxygen incorporation and UV absorber (Cyasorb 531) was higher, but the gel content was slightly lower. Here, oxygen ingress is slow enough such that it is consumed on the cell perimeter and the oxygen serves to increase the rate of chain scission.

At elevated temperatures, EVA degrades by thermal deacetylation with the formation of acetic acid [16, 17]. However, at ambient or use conditions, this reaction pathway is unimportant and long-term fielded EVA samples have been shown to have <1% change in the amount of residual acetate groups [2, 18]. However, even this small amount of acid formation can lower the pH sufficiently to increase degradation rates of PV cells and interconnections.

Prior work studied the possible hazards associated with creep in modules, including how to screen for creep and how to test for the results of creep [19, 20]. Modules were fabricated with a variety of encapsulant materials, including polyethylene vinyl acetate (EVA, both cross-linked and with no curing agent), polyethylene/polyoctene copolymer (PO), poly(dimethylsiloxane) (PDMS), polyvinyl butyral (PVB), and thermoplastic polyurethane (TPU). These modules were subjected to high temperatures using outdoor aging (in Mesa, AZ or Golden, CO) or indoor aging (in a temperature step-stress test performed in a chamber), and the resulting creep was documented. The greatest creep was observed for EVA with no curing system present (3 mm in Mesa, AZ and 0.3 mm in Golden, CO), while the next greatest displacement was on the order of 30 μm. Curiously, the rate of creep for the fielded EVA specimens was seen to decrease, even though the enabling ambient temperature remained warm through the summer months of the exposure period. The observed creep was correlated with material-level tests to identify the best way to characterize phase transitions that could be predictive of creep in the field. The creep measured outdoors and indoors was found to correlate with each other, where the nonuniform temperature distribution within a module as well as the packaging components (frame or mounting clips) was found to greatly limit creep. Creep was also particularly correlated with the melting point, the phase transitions and rheological properties measured for the encapsulant materials. These data suggested that some noncuring thermoplastics were cross-linking, but did not present conclusive evidence to that effect.

The effort described here adds to the aforementioned study [19–21]. We demonstrate that, in some cases, most notably noncuring EVA (NC-EVA), a significant factor in the reduction of encapsulant creep is the formation of cross-links in the absence of cross-linking agents. The ability to cross-link polymers in the absence of specific chemistry to do so very significantly reduces the likelihood of damage resulting from creep in a field-deployed module. We explore the importance of proximity to the edge of the module to assess the role of diffusion of reactants and/or reaction products in the dominance of chain scission over cross-linking reactions.

Experimental Procedure

The specifics of the materials and construction are more thoroughly covered in M. D. Kempe et al. [21], but we will present a brief description of the sample preparation here.

Module construction

Encapsulant materials being used, or under investigation for use, in PV modules were obtained from industrial manufacturers. The nomenclature for as well as the phase

Table 1. Differential scanning calorimetry (DSC) and dynamic mechanical analysis (DMA) determined phase transitions.

Encapsulant material type and designation		Phase transitions					
		DSC			DMA at 0.1 rad/sec		
		T_g (°C)	T_m (°C)	T_f (°C)	T_g (°C)	T_m (°C)	T_c (°C)
Cured commercial PV EVA resin	EVA	−31	55	45	−30	47	
Commercial PV EVA resin with all components but the peroxide	NC-EVA	−31	65	45	−28	69	
Polyvinyl butyral	PVB	15			17		121
Aliphatic thermoplastic polyurethane	TPU	2			3		84
Pt catalyzed, addition cure polydimethyl siloxane gel (mock modules)	PDMS-M	−158	−40	−80			
Pt catalyzed, addition cure polydimethyl siloxane gel (Si modules)	PDMS-Si	−150	−40	−80			
Thermoplastic polyolefin #1	TPO-1	−43	93	81	−35	105	
Thermoplastic polyolefin #3	TPO-3	−44	61	55	−41	79	
Thermoplastic polyolefin #4	TPO-4	−34	106	99	−21	115	

DMA glass transitions (T_g) were determined as the peak in the phase angle, and the DMA melting and freezing transitions (T_m and T_f, respectively) were determined when the phase angle was 45° except for the cross-linked PDMS [poly(dimethylsiloxane)] and poly.EVA where an inflection point in the modulus was used. Because PVB and thermoplastic polyurethane (TPU) did not have melt transitions, the critical temperature for significant flow (T_c) was determined as the point where the phase angle was 45°. Data are taken from M. D. Kempe, et al. [21].

characteristics of these materials are summarized in Table 1. A noncuring poly(ethylene-vinyl acetate) (NC-EVA) was formulated identically to a standard EVA formulation but without the inclusion of a peroxide to promote curing during lamination.

In previous experiments [19, 21], both crystalline silicon and thin-film mock samples were tested. Because we wanted to redeploy the silicon modules, only the thin-film mock modules were destructively evaluated in this work. The thin-film mock modules were constructed using two pieces of 3.18-mm thick, 61 × 122 cm glass obtained from a thin-film PV manufacturer. The rear surface of the back plate was painted black to simulate the optical absorption of a thin-film module. Thin-film mock modules were mounted by adhesively attaching fiberglass channel on the back, allowing the front piece of glass to move freely.

For the thin-film mock modules, the creep (displacement of the front glass relative to the back glass) was measured using a high-precision electronic depth gauge that was incremented ±1 μm. This gauge was mounted to a flat plate to ensure that it was positioned perpendicular to the side of the module, and in the plane of one of the glass plates. With this setup, the repeatability of creep measurement was better than ±20 μm.

Modules were deployed in Mesa, Arizona at the Arizona State University from May until September 2011 on a rack tilted at 33°latitude tilt and a 255 azimuth to more directly face the sun at the hottest part of the day. Additionally, a single NC-EVA thin-film mock module was exposed in Golden, Colorado at a 180 azimuth and 40°latitude tilt [8]. Insulation was placed on the backside

to simulate a close-roof installation (resulting in maximum measured temperatures between 102°C and 104°C in Mesa, Arizona, AZ or Golden, CO)).

The modules were also examined indoors in environmental chambers in a step-stress experiment using a replicate set of modules. The temperatures applied during the test ranged from 65 to 110°C, with a 5°C increment, then from 110 to 140°C in a 10°C increment using a 200-h exposure time at each step.

Size exclusion chromatography and multiangle laser light scattering

Following field deployment, the formation of polymer chain cross-links on the mock modules was evaluated using size exclusion chromatography (SEC) in conjunction with multiangle laser light-scattering (MALLS, Waters Corporation GPCV 2000 instrument 34 Maple StreetMilford, MA 01757 USA) and viscometric detection (using a capillary viscometer detector CV from Waters). For the tetrahydrofuran (THF) solvent, four SEC styrene-divinyl benzene 8 × 300 mm columns from Shodex (4-1, Shiba Kohen 2-chome, Minato-ku, Tokyo, 105-8432, Japan) were used for separation. These were two linear GPC KF806M, one GPC KF802, and one GPC KF-801 columns. Columns were run at 40°C, with a flow rate of 1.0 mL/min, an injection volume of 0.219 mL, and a run time of 90 min.

Samples were cut out of the mock modules using a ceramic saw blade (Wale Apparatus Co., Inc.400 Front StreetHellertown, PA 18055) enabling samples to be taken at different distances from the edge. For the TPU and EVA, polymer was removed from the glass by extraction for 72 h at 40°C, utilizing mild agitation, in a solution of THF; however, this typically left behind some EVA as an insoluble fraction. For EVA, the residual glass-polymer specimens were then additionally soaked in trichlorobenzene (TCB) overnight at 140°C, with slight agitation, to solubilize some of the remaining polymer. The TCB soluble measurements of EVA represent material not soluble in THF, but soluble in TCB. For the thermoplastic polyolefins (TPO), the SEC measurements were done using only TCB solvent.

The TCB-based system utilized two Styragel HT 6E and one Styragel HT 2 styrene-divinyl benzene columns from Waters for SEC separation. The TCB columns used a temperature of 140°C, a flow rate of 1.01 mL/min, an injection volume of 0.2095 mL, and a run time of 80 min.

Samples of "virgin" (unaged) polymer film samples were dissolved in THF and TCB, as appropriate, at room temperature and 150°C, respectively, to serve as controls. All solutions were made with approximately 1 mg/mL polymer concentration.

Gel content measurement

Gel content measurement was accomplished using a Soxhlet extractor with toluene, THF, mixed isomeric xylenes, or methanol as the solvent. Test material was obtained in approximately 1 g samples by cutting out small samples using a ceramic wet saw, and subsequently separating the polymer from the glass manually using a blade. Polymer samples were taken at various distances from the edge of the glass.

Results and Discussion

NC-EVA

Three NC-EVA samples were removed from the mock modules deployed in Golden, Colorado and Mesa, Arizona. One cut from the edge of the module, one about 2 cm from the edge, and another about 4 cm. The unexposed sheets were completely soluble in both THF and TCB. For the exposed samples, a qualitative assessment during the preparation for SEC experimentation indicated that only ~40% of the NC-EVA was soluble in THF. Then, of this THF-insoluble fraction another ~40% (relative to initial amount) was soluble in TCB, leaving ~20%, which was not soluble in either THF or TCB. Qualitatively, this is in agreement with rigorous gel content measurements using toluene in a Soxhlet extractor (Fig. 1), where 5.5–35.4% gel was measured at various locations. This NC-EVA was formulated with EVA pellets containing trace amounts of butylated hydroxyl toluene (BHT) to which was added an hindered amine light stabilizer (HALS), methacryloxypropyl trimethoxysilane (Z6030), a phenylphosphonite, a benzophenone-based UV absorber, but no peroxide [5, 18].

The cross-linking of the NC-EVA in the fielded module was further investigated using SEC in conjunction with MALLS and viscometry detection (Fig. 2). For the THF soluble fraction, very little change in the molecular weight distribution was seen for either location within the specimens. The sample taken 4 cm from the edge for the Colorado module has a high-molecular-weight fraction, but otherwise, the THF soluble fractions are similar to the unexposed material. The EVA specimens contain

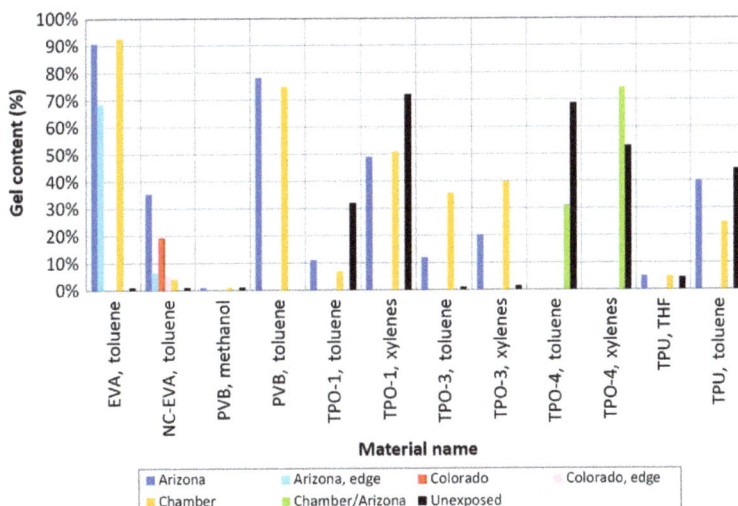

Figure 1. Results of gel content measurements using a Soxhlet extractor. "Edge" samples were taken <1 cm from the side of the module, all other samples were taken about 4 cm from the side.

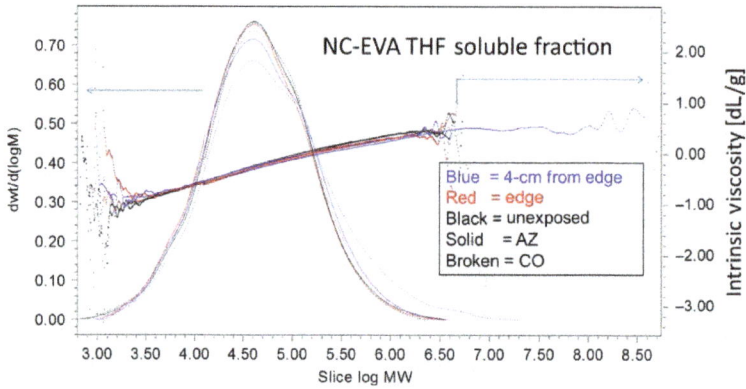

Figure 2. Weight fraction determined by SEC-MALLS (SEC, size exclusion chromatography) and viscometry for the NC-EVA (poly[ethylene]-co-vinyl acetate) thin-film mock module after exposure in Golden, Colorado (dotted lines) and Mesa, Arizona (solid lines). Samples dissolved in tetrahydrofuran (THF) by extraction at 40°C overnight. Average molecular weight (MW) of each slice of polymer distribution is in units of g/mol as determined by SEC-MALLS. Right axis, intrinsic viscosity (dL/g) of each slice. Lines are normalized to the area under the curve.

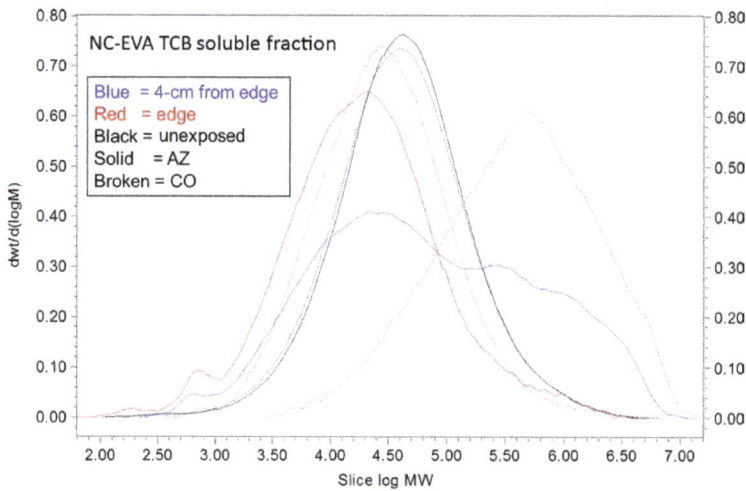

Figure 3. Weight fraction determined by SEC-MALLS for the NC-EVA poly(ethylene-co-vinyl acetate) thin-film mock module after exposure in Golden, Colorado (dotted lines) and Mesa, Arizona (solid lines). Samples obtained from tetrahydrofuran (THF) insoluble fraction by extraction in TCB at 150°C overnight. Molecular weight (MW) is in units of g/mol determined by (SEC, size exclusion chromatography) SEC-MALLS. Lines are normalized to the area under the curve.

33 wt% vinyl acetate. Qualitative experiments indicate that above about 30 wt% the solubility of EVA in THF is good, but below 30 wt% the solubility in THF is poor [22]. This indicates that even very slight modification/ degradation of EVA molecules is sufficient to render them significantly less soluble in THF consistent with our data.

In contrast, the SEC profiles of the TCB-solubilized samples varied greatly depending on the exposure site, and on the distance from the module edge (Fig. 3). Samples taken at distances of 2 and 4 cm from the edge have a higher molecular weight fraction than the control indicating that significant cross-linking is occurring (Fig. 4).

The molecular weight distributions for Colorado and Arizona are very different. Of note in Table 2 are the apparently lower values for M_n in TCB versus THF. It is believed that this is an experimental artifact resulting from the silane component being isorefractive in THF but not in TCB. Because of the significantly lower temperature experienced in Colorado [19, 21], one would expect to see less degradation, that is cross-linking, than for the Arizona sample. Also of significance is the low-molecular-weight fractions, which if unmodified should have been previously extracted in THF. The absence of low-molecular-weight fractions at a distance of 4 cm from

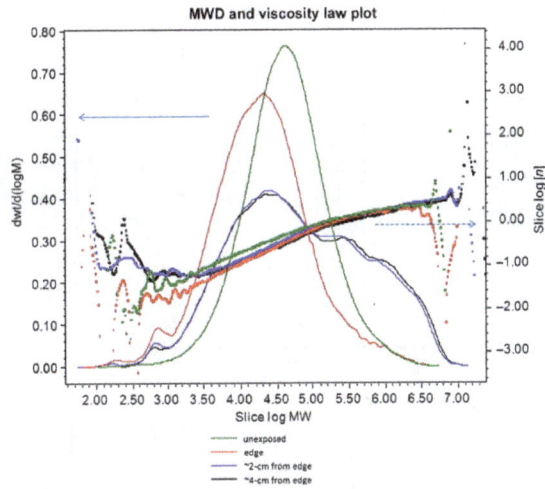

MWD and viscosity law plot

Figure 4. Weight fraction determined by (SEC, size exclusion chromatography) SEC-MALLS and viscometry for the NC-EVA thin-film mock module after exposure in Arizona. Samples dissolved in TCB by extraction at 150°C overnight. Molecular weight (MW) is in units of g/mol as determined by SEC-MALLS. Right axis, intrinsic viscosity (dL/g) of each slice. Lines are normalized to the area under the curve.

Table 2. NC-EVA poly (ethylene-co-vinyl acetate) size exclusion chromatography, multiangle laser light scattering

Sample	M_n (g/mol)	M_w (g/mol)	M_z (g/mol)	PDI
Tetrahydrofuran (THF)	20,000	98,000	450,000	4.8
Unexposed	20,000	98,000	450,000	4.8
Trichlorobenzene (TCB)	15,000	100,000	620,000	6.5
Unexposed	16,000	98,000	510,000	6.3
Mesa, Arizona				
TCB soluble fraction				
Edge	20,000	89,000	380,000	4.5
~2 cm from the edge	20,000	85,000	330,000	4.2
~4 cm from the edge	21,000	100,000	440,000	4.9
THF soluble fraction				
Edge	5,500	67,000	630,000	12.1
~2 cm from the edge	8,700	390,000	2,200,000	45.2
~4 cm from the edge	11,000	380,000	2,100,000	35.9
Golden, Colorado				
TCB soluble fraction				
Edge	20,000	88,000	340,000	4.4
~4 cm from the edge	21,000	190,000	2,100,000	8.9
THF soluble fraction				
Edge	13,000	66,000	390,000	5.2
~4 cm from the edge	100,000	980,000	3,100,000	8.7

M_n is the number average molecular weight, M_w is the weight average molecular weight, M_z is the Z average molar mass, and PDI is the polydispersity index which is equal to M_w/M_n.

the edge in the Colorado sample indicates that the process of dissolution in TCB at 140°C is not significant in forming these fractions. Noting that low-molecular-weight

fractions are more prevalent for the edge samples suggest that O_2 may play an important role in the formation of low molecular weight THF-insoluble fractions.

Gel content measurements of both the Colorado and the Arizona test modules (Fig. 1) demonstrate a significantly lower gel content near the edge and a higher gel content away from the edge for NC-EVA. The same trend was also seen in the EVA with peroxide where the gel content of the edge sample was 68.5%, and 4 cm from the edge it was 90.7%, consistent with other observations [16, 17]. This is presumably because of an interaction with O_2 at the periphery.

Numerous observations of fielded modules have observed discoloration of EVA in the center area of a cell; however, greater oxygen incorporation is observed around the cell perimeter [18]. Similarly, it has been observed that the gel content in the more discolored EVA at the center of the cell is higher than EVA at the perimeter in aged samples [16]. Thus, yellowing and cross-linking are correlated with lower oxygen incorporation, and lower cross-link densities are correlated with higher oxygen incorporation and oxidative bleaching. We hypothesize that oxygen can act to increase the rate of chain scission, but that its absence allows cross-linking chemistries to dominate more. Of particular note, is the observation that samples at a distance of 2 and 4 cm from the edge have nearly identical molecular weight profiles (Fig. 4). This implies that oxygen availability is the same at 2 and 4 cm, therefore the polymer environment is very likely anaerobic at distances >2 cm from the edge of the cells, consistent with the observed size of brown/yellow areas in some field deployed EVAs.

Marais and Hirata [23] measured the solubility of O_2 in 33 wt% EVA at 25°C at 0.0023 cm³ (STC)/cm³/cm Hg (or 5.6×10^{-5} g/cm³) and a diffusivity (D) of about 4.0×10^{-7} cm²/sec. For the 140-day exposure, the characteristic distance for oxygen ingress ($\sqrt{DT}$) at a temperature of 25°C is 4.8 cm. For comparison, at 25°C the diffusivity of water in EVA is comparable at 4.8×10^{-7} cm²/sec, but its solubility at 100% RH is 100 times higher at 0.0021 g/cm³ [1, 24–26]. Considering that the perimeter of 2 cm is where all of this diffusant is consumed, a linear gradient at 25°C would result in 0.07 g/cm³ of water consumed in the perimeter of 2 cm over a 10-year period. Because EVA is not observed to be this highly degraded when fielded for long periods, water is not completely consumed so quickly upon diffusion into EVA. At the higher module temperatures in field deployment, oxygen would diffuse further than this estimate at 25°C. Thus, it is highly likely that oxygen ingress is enhancing chain scission in EVA, and is limited by its low solubility in EVA.

If the degree of branching is constant as a function of molecular weight, then the lines for intrinsic viscosity would be straight lines when plotted against Log(*MW*). The slight concave down curvature in Figures 2 and 4 indicates increased amounts of branching at higher molecular weights. For the THF soluble fractions, the difference in intrinsic viscosity between the aged samples is undetectable. But for the TCB soluble fraction there is a relative reduction in the intrinsic viscosity for low-molecular-weight fractions, indicating that they are becoming more compact and branched.

For the NC-EVA samples deployed in both Mesa, Arizona and Golden, Colorado, cross-linking was observed resulting from combined thermal/UV aging. Both SEC and gel content measurements indicate an increase in the frequency with which degradation results in chain scission for locations where, presumably, oxygen is able to diffuse. Very small amounts of degradation are able to significantly reduce the solubility of EVA in THF. Regardless of the initial degree of cure, or even the total absence of cure agents, typical PV EVA formulations are therefore expected to cure as they age.

Thermoplastic polyolefins

In this work three different thermoplastic polyolefins were evaluated. Thermoplastic polyolefin #1 (TPO-1) was designed to be a material that might be susceptible to creep, but still pass the qualification test because its melting transition was above 85°C. However, when fielded, the movement of the glass was barely detectable (~30 μm) [19, 20]. Thermoplastic polyolefin #4 (TPO-4) is similar to TPO-1, but with a higher melting point and viscosity, which resulted in no detectable movement. Finally, thermoplastic polyolefin #3 (TPO-3) was made by a different manufacturer and has been used in the construction of commercial modules. TPO-3 has a melting point between 61°C and 79°C (depending on the measurement method) which is just a bit higher than EVA. Because it had a much higher viscosity than NC-EVA [20], the 0.090 mm of creep was not sufficient to see the effects of chain scission/cross-linking in the creep response curve for the outdoor exposed modules. However, a significant increase in viscosity was seen in the indoor stressed samples at temperatures >90°C [21]. All three of these materials were analyzed in SEC to evaluate cross-linking and chain scission reactions.

All three TPO samples were completely soluble in TCB at 150°C and were analyzed in this solvent using the multidetector SEC system (Fig. 5). Similar to NC-EVA, for all three materials the sample taken from the edge showed the greatest reduction in molecular weight indicating that a similar O_2-induced chain scission mechanism may be active at the periphery. By far, the greatest shift was for TPO-4, followed by TPO-1, then TPO-3. Similarly,

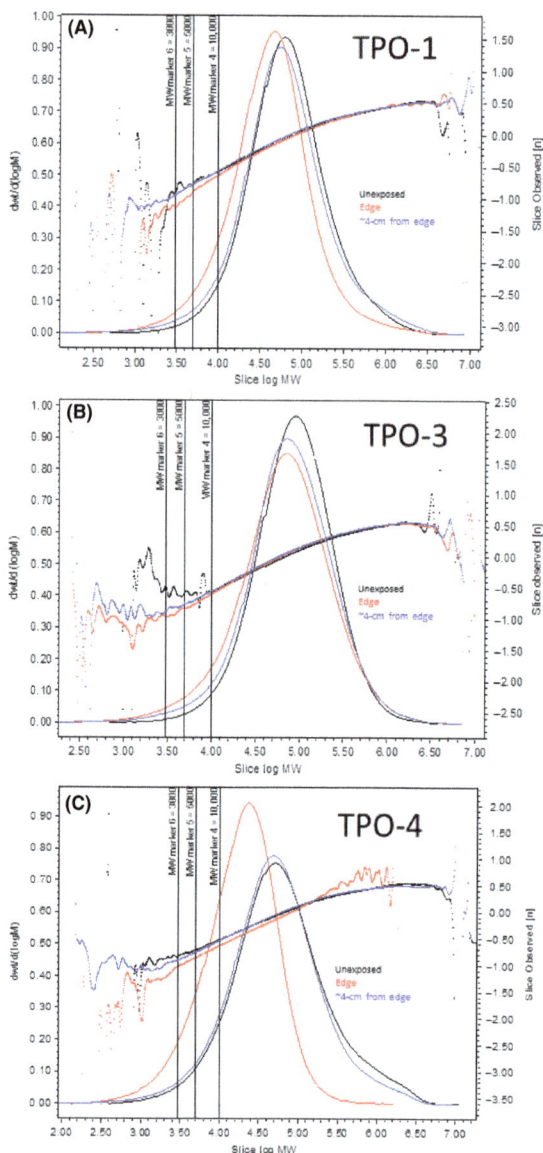

Figure 5. Weight fraction determined by SEC-MALLS (SEC, size exclusion chromatography) and viscometry for the TPO-1 thin-film mock module after exposure in Arizona. Samples dissolved in TCB by extraction at 150°C overnight. Molecular weight (MW) is in units of g/mol. Right axis, intrinsic viscosity (dL/g) in slice, horizontal lines. Blue: 4 cm from the edge; red: edge sample; black: unexposed. (A) TPO-1, (B) TPO-3, and (C) TPO-4. Lines are normalized to the area under the curve.

the samples taken at a distance of 4 cm from the edge showed a small drop in molecular weight. This is in stark contrast to the NC-EVA samples which experienced significant cross-linking in the interior anaerobic regions. This difference in cross-linking behavior between EVA and these TPOs could be a result of the stabilization

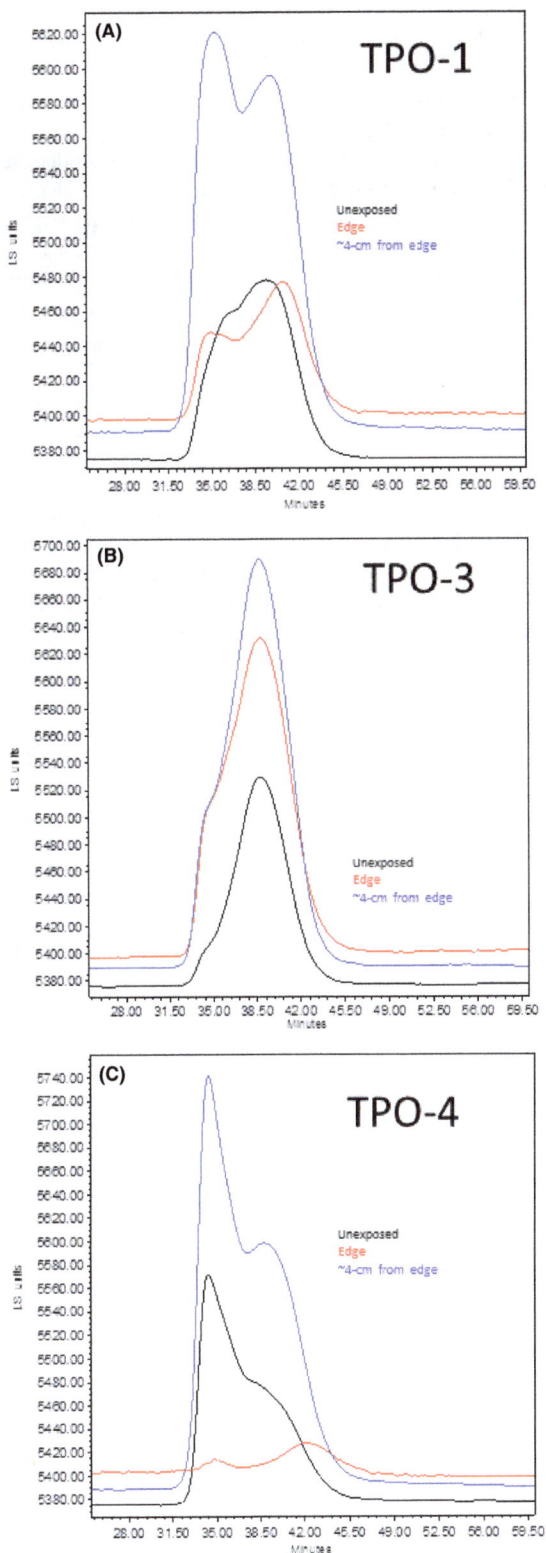

chemistry used, or more likely there could be a greater ability of EVA to form cross-links through a mechanism involving the acetate monomers as seen elsewhere [27]. Tertiary carbons in the backbone are prone to experiencing chain scission. Therefore, it is likely that the chain scission reactions dominate throughout the polyolefin copolymers. In combination, these effects can explain the difference in the cross-linking/chain scission behavior of EVA as compared to TPOs.

Similar to EVA, the slight concave down curvature in Figure 5 indicates the presence of increased amounts of branching at higher molecular weights. For TPO-1 and TPO-4, and to a lesser extent TPO-3, the aged samples showed a relative reduction in intrinsic viscosity near the edge but not at 4 cm distance from the perimeter. Thus, O_2 is degrading the polymers, making them more compact and branched.

The light-scattering measurements at an angle of 90° also provide insight into the cross-linking behavior of the materials (Fig. 6). Here, the strength of the signal is related to the product of concentration and molecular weight. Thus, when two elution peaks are seen in light-scattering detectors, the first peak consists of a small amount of very large molecular weight material, and the second peak is of a much lower molecular weight material where the signal strength is attributable to significantly larger quantities of polymer. Looking at the changes in the relative ratios of these peaks gives a sensitive indication of changes in the large molecular weight fraction.

For TPO-1, the unaged material has a small amount of large molecular weight material. Upon exposure, the position of this large molecular weight fraction moves to larger molecular weights and becomes more pronounced (Fig. 6). Comparison with Figure 5 indicates that this larger molecular weight fraction is only a very small part of the total polymer. Also of note is the fact that the sample taken from a distance of 4 cm from the edge demonstrates more cross-linking than the sample on the edge of the module. Although chain scission dominates the degradation for TPO-1, there is still some evidence cross-linking reactions have occurred. Overall, these changes produced decreases in the mass average molecular weight (M_w) near the edge, and slight increases further in the module, while increasing the polydispersity of the sample as it ages (Table 3).

For TPO-4, there is initially a significant amount of large molecular weight material (Fig. 5) such that the light-scattering chromatograph is dominated by the large molecular weight fraction (Fig. 6). Upon exposure, the

Figure 6. Light-scattering chromatographs of TPOs measured at an angle of 90°. Blue: 4 cm from the edge; red: edge sample; black: unexposed. (A) TPO-1, (B) TPO-3, and (C) TPO-4.

Table 3. TPO size exclusion chromatography, multiangle laser light scattering in TCB.

Sample	M_n (g/mol)	M_w (g/mol)	M_z (g/mol)	PDI
TPO-1				
Unexposed	36,200	124,000	472,000	3.4
Edge	20,200	81,200	454,000	4.0
~4 cm from the edge	26,100	134,000	738,000	5.1
TPO-3				
Unexposed	50,400	135,000	337,000	2.7
Edge	27,300	130,000	461,000	4.8
~4 cm from the edge	35,000	137,000	439,000	3.9
TPO-4				
Unexposed	24,400	172,000	1,040,000	7.0
Edge	8,980	30,100	96,200	3.4
~4 cm from the edge	18,300	135,000	868,000	7.4

M_n is the number average molecular weight, M_w is the weight average molecular weight, M_z is the Z average molar mass, and *PDI* is the polydispersity index which is equal to M_w/M_n.

small molecular weight fraction becomes more distinct. In the edge sample, M_w is reduced much more than the number average molecular weight (M_n), indicating a higher probability that large chains are reduced, resulting in a reduction in the PDI (Table 3). Neither the light-scattering data nor the molecular weight distribution can clearly demonstrate the formation of cross-links. However, considering that TPO-4 and TPO-1 were made by the same manufacturer and intended to differ primarily in their melting temperatures, it is likely that cross-linking is occurring in TPO-4 but that it is masked by a much stronger chain scission reaction.

For TPO-3, there is initially a weak signal for a large molecular weight fraction in the light-scattering data. After exposure, the large molecular weight fraction becomes more pronounced. Here, M_w remains relatively unchanged after exposure, but M_n is reduced (Table 3). For M_w to be constant, chain scission and cross-linking reactions must be occur at a similar rate. In Figure 5B, both the edge and 4-cm distance from the edge samples demonstrate the formation of a large molecular weight tail which is more pronounced for the edge sample. Thus, oxygen may be enhancing both cross-linking and chain scission reactions slightly in this material.

While TCB at 150°C was a good solvent for all the TPOs, the xylenes and toluene solvent did not produce complete solubility in the Soxhlet extractor (Fig. 1). For TPO-1, toluene was a better solvent than xylenes; but for TPO-4, xylenes were a marginally better solvent than toluene. Similarly, toluene was a poor solvent for PVB and TPU producing gel contents between 28.8% and 78.2%. But when PVB and TPU were tested using methanol and THF in the Soxhlet extractor, gel contents of 0% and ~4.9% were obtained, respectively. This illustrates

how solvent quality, if it is marginal or poor, can give misleading results. Additionally, one must be cautious with TCB at 150°C because the thermal history of the extraction could alter the results, though we did not see evidence for this in our experiments. In the Soxhlet extractor, even with xylenes boiling around 140°C, the liquid is condensed and significantly cooled when in contact with the polymer. One would therefore not expect to see significant degradation of the polymers, even when xylenes are used.

For TPO-1 and TPO-4, a higher gel content was seen for the unexposed sample (Fig. 1). A reduction in the molecular weight is consistent with the SEC experiments, but the poor quality of the solvent leaves open the question of the amount of gel in the materials both before and after the experiment.

Toluene was a better solvent than xylenes for TPO-3 (Fig. 2), but both solvents indicate nearly zero gel content initially. Even though the aged samples indicate significant increases in gel content, the SEC data indicated there was only a small change in the molecular weight distribution (Table 3). Toluene is probably just barely able to dissolve this material such that the small amount of cross-linking occurring is able to significantly affect the measurement of the gel content without really changing the molecular weight distribution significantly.

For all three TPOs, chain scission seems to dominate over cross-linking reactions and is enhanced by the presence of oxygen in the perimeter regions. For TPO-1 and TPO-3 (and possibly TPO-4) there are also some cross-linking chemistry occurring but with a less noticeable dependence on the presence of oxygen. Dominance of chain scission over cross-linking was unexpected because indoor experiments [19, 21] with TPO-1 and TPO-3 mock modules clearly demonstrate a reduction in creep rates above 110°C and 90°C, respectively. This could be because UV light is largely responsible for chain scission, or that the small amount of cross-linked fractions here has a large effect on the rheology of the polymers.

Thermoplastic polyurethane

Even after exposure in Arizona region, the thermoplastic polyurethane (TPU) was highly soluble in THF and could be run in the SEC system using THF. Gel content measurements in THF (Fig. 1) indicate that 5.1, 4.9, and 4.5% of the polymer were not soluble for samples at a distance of 4 cm from the edge in the Arizona region, environmental chamber, and unexposed conditions, respectively. Upon exposure, a large molecular weight tail was seen on the GPC curve due to cross-linking of polymer chains, and there is significant indication of cross-linking in the light-scattering chromatographs (Fig. 7B). However, no signs of chain scission were seen in the mass

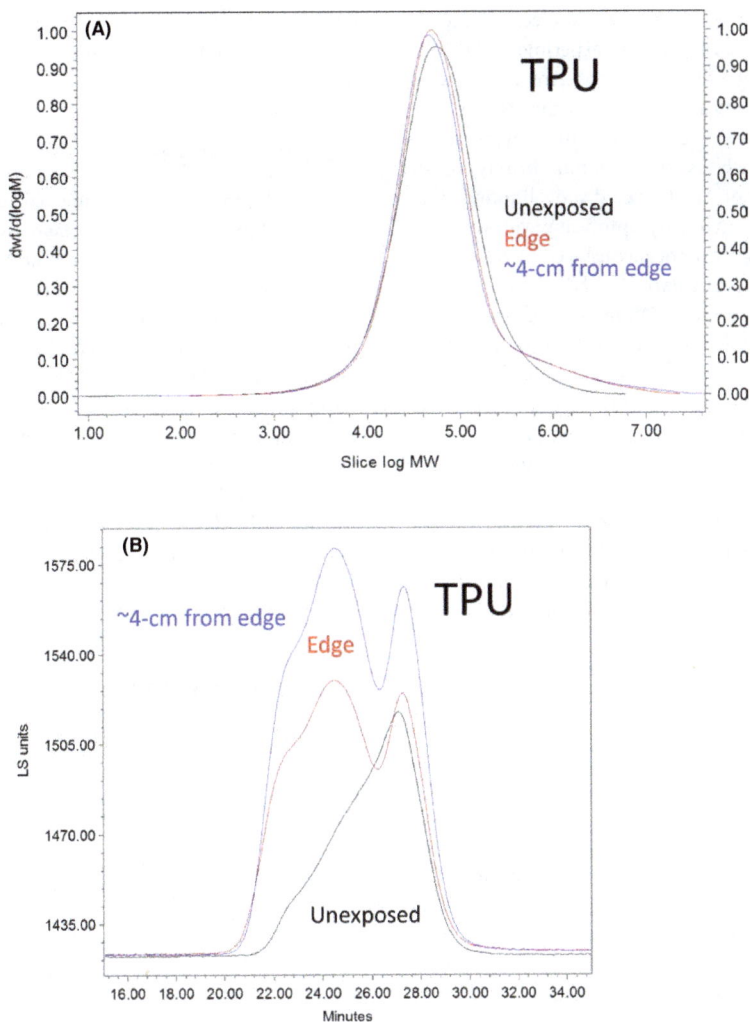

Figure 7. Weight fraction determined by SEC-MALLS (SEC, size exclusion chromatography) and viscometry for the thermoplastic polyurethane thin-film mock module after exposure in Arizona. Samples dissolved in tetrahydrofuran (THF) by extraction at 40°C overnight. Molecular weight (MW) is in units of g/mol. Blue: 4 cm from the edge; red: edge sample; black: unexposed. Lines are normalized to the area under the curve.

chromatograms (Fig. 7A). The absence of small scale for cross-linking is further supported by the fact that M_n actually increased upon exposure (Table 4). Thus, a significant amount of polymer is becoming highly cross-linked.

Initially, the TPU seems to have three bumps in the light-scattering plot (Fig. 7). After exposure, the two larger molecular weight fractions become more pronounced and shifted to larger molecular weights. The shift is slightly larger for the sample near the edge indicating that oxygen has a small additional effect, increasing the rate of cross-link formation. This could be indicative of the use of a block copolymer or a polymer blend where each different polymer types have differing ability to form cross-links.

Table 4. Thermoplastic polyurethane (TPU) size exclusion chromatography, multiangle laser light scattering in a THF (tetrahydrofuran) solvent.

Sample	M_n (g/mol)	M_w (g/mol)	M_z (g/mol)	PDI
TPU				
Unexposed	20,700	89,900	370,000	4.3
Edge	24,700	176,000	2,540,000	7.1
~4 cm from the edge	23,900	229,000	5,160,000	9.6

Conclusions

Modules were deployed in a very hot environment with insulation on their backside in an effort to reproduce one of the worst case situations where creep of the encapsulant might occur. In previous experiments [19, 21], even when noncuring polymer encapsulant formulations were used, only insignificant creeping of the cells or other module components was observed. This was attributed to several factors such as some nonuniformity in the temperature distribution creating cold spots limiting creep. Measurable creep, occurring asymptotically with time, was observed for the more homogeneously constructed thin-film mock modules, particularly for NC-EVA.

For the NC-EVA and the TPOs, the rate of cross-linking and chain scission reactions vary with position in the sample indicating that a chemical species must be entering or leaving the module package affecting the kinetics. It is plausible that oxygen ingress is increasing the rate of chain scission relative to cross-linking and increasing the overall degradation rate.

The EVAs and the TPU tested here increased in their polymer molecular weight attributable to the formation of cross-links. For TPU, no evidence for chain scission was observed, but for EVA chain scission was observed and was enhanced presumably by the presence of oxygen in the perimeter areas. In TPU, these changes did not result in an increase in gel content, but in EVA an increase in gel content was seen even in the chain scission-prone perimeter regions with the NC-EVA sample.

In contrast, the TPOs experienced an overall decrease in molecular weight with aging. SEC data for TPO-1 and TPO-3 indicate some cross-linking reactions are occurring, but this is much less prominent than chain scission reactions. The overall reduction in molecular weight is a small effect and not likely to result in module creep as demonstrated elsewhere [19–21]. When using any thermoplastic material in a PV module, one should determine if cross-linking or chain scission dominates degradation reactions sufficiently to cause a concern for a reduction in molecular weight and subsequent creep.

The materials and "module" designs examined here help to identify the requirements for a module that could be subject to creep during a 25-year field lifetime. It would be difficult to construct an encapsulant that would experience creep when deployed. The encapsulation would have to have a low melting transition (≤70°C as in NC-EVA) providing a low viscosity at operating conditions, the module would have to use a mounting scheme that did not fix the relative positions of the front and back sheets, the module would have to be installed in a very hot climate in a hot mounting configuration, and the encapsulation would have to age predominantly by chain scission rather than cross-linking. Such a module construction might still pass 1000 h of 85°C exposure without creep but fail in the field. Examination at 90°C or 105°C for 200 h (MST 56) has

been recently added to IEC 61730–2 to more rigorously screen for modules that might be prone to creep over prolonged time.

Acknowledgments

This work was part of a large collaborative effort of a number of people working on standards developed at many institutions. The authors gratefully acknowledge the support of the following individuals: Adam Stokes, Alain Blosse, Ann Norris, Bernd Koll, Bret Adams, Casimir Kotarba (Chad), David Trudell, Ed Gelak, Greg Perrin, Hirofumi Zenkoh, James Galica, Jayesh Bokria, John Pern, Jose Cano, Kartheek Koka, Keith Emery, Kent Terwilliger, Kolapo Olakonu, Masaaki Yamamichi, Mowafak Al-Jassim, Nick Powell, Niki Nickel, Pedro Gonzales, Peter Hacke, Ryan Smith, Ryan Tucker, Steve Glick, Steve Rummel, and Tsuyoshi Shioda. This work was supported by the U.S. Department of Energy under Contract No. DE-AC36-08-GO28308 with the National Renewable Energy Laboratory.

Conflict of Interest

None declared.

References

1. Kapur, J., K. Proost, and C. A. Smith. 2009. Determination of moisture ingress through various encapsulants in glass/glass laminates. Pp. 001210–001214 in Photovoltaic Specialists Conference (PVSC), 2009 34th IEEE.
2. Kempe, M. D., G. J. Jorgensen, K. M. Terwilliger, T. J. McMahon, C. E. Kennedy, and T. T. Borek. 2007. Acetic acid production and glass transition concerns with ethylene-vinyl acetate used in photovoltaic devices. Sol. Energy Mater. Sol. Cells 91:315–329.
3. Meyer, S., S. Timmel, S. Richter, M. Werner, M. Gläser, S. Swatek, et al. 2014. Silver nanoparticles cause snail trails in photovoltaic modules. Sol. Energy Mater. Sol. Cells 121:171–175.
4. Reid, C. G., S. ferrigan, I. Fidalgo, and J. T. Woods. 2013. Contribution of PV encapsulant composition to reduction of potential induced degradation (PID) of crystalline silicon PV cells. 28th European Photovoltaic Solar Energy Conference and Exhibition.
5. Reid, C. G., J. G. Bokria, and J. T. Woods. 2013. UV Aging and Outdoor exposure correlation for EVA PV encapsulants. Pp. 882508-882508-11 in 2013 SPIE, San Diego, California.
6. Cuddihy, E. F., A. Gupta, C. D. Coulbert, R. H. Liang, A. Gupta, and P. Willis, et al. 1983. Applications of ethylene vinyl acetate as an encapsulation material for terrestrial photovoltaic modules. DOE/JPL/1012-87 (DE83013509).

7. Dietrich, S., M. Pander, M. Ebert, and J. Bagdahn. 2008. Mechanical assessment of large photovoltaic modules by test and finite element analysis.

8. Wohlgemuth, J., M. Kempe, D. Miller, and S. Kurtz. 2011. Developing standards for PV packaging materials. *SPIE*. Proceedings of SPIE Conference

9. Wohlgemuth, J., and W. Herrmann. 2005. Hot spot tests for crystalline silicon modules. Pp. 1062–1063*31st IEEE Photovoltaic Specialists Conference, 3-7 Jan. 2005.*

10. Wohlgemuth, J.. 2013. Hot Spot Testing of PV Modules. *SPIE, San Diego, California.*

11. Kurtz, S., K. Whitfield, D. Miller, J. Joyce, J. H. Wohlgemuth, and M. D. Kempe, et al. 2009. Evaluation of high-temperature exposure of rack-mounted pohotovoltaic modules. *34th IEEE PVSC.*

12. D. C. Miller, M. D. Kempe, S. H. Glick, and S. R. Kurtz. 2010. Creep in photovoltaic modules: Examining the stability of polymeric materials and components. *35th IEEE Photovoltaic Specialists Conference (PVSC),* 20–25 June 2010.

13. Miller, D. C. 2013. Examination of a standardized test for evaluating the degree of cure of EVA encapsulation (#1066/0731). National Renewable Energy Laboratory, Golden, Colorado.

14. Cuddihy, E. F., and P. B. Willis. 1984. Antisoiling technology: theories of surface soiling and performance of antisoiling surface coatings. *DOE/JPL/1012-102 (DE85006658).*

15. Holley, W. W., and S. C. Agro. 1998. Advanced EVA-based encapsulants. Final Report January 1993-June 1997 *NREL/SR-520-25296.*

16. Pern, F. J., and A. W. Czanderna. 1992. Characterization of ethylene vinyl acetate (EVA) encapsulant: effects of thermal processing and weathering degradation on its discoloration. Sol. Energy Mater. Sol. Cells 25:3–23.

17. Patel, M., S. Pitts, P. Beavis, M. Robinson, P. Morrell, N. Khan, et al. 2013. Thermal stability of poly(ethylene-co-vinyl acetate) based materials. Polym. Testing 32:785–793.

18. Klemchuk, P., M. Ezrin, G. Lavigne, W. Holley, J. Galica, and S. Agro. 1997. Investigation of the degradation and stabilization of EVA-based encapsulant in field-aged solar energy modules. Polym. Degrad. Stab. 55:347–365.

19. Kempe, M. D., D. C. Miller, J. H. Wohlgemuth, S. R. Kurtz, J. M. Moseley, and Q.-U.-A. S. J. Shah, et al. 2012. A field evaluation of the potential for creep in thermoplastic encapsulant materials. *Proceedings of the 38th IEEE Photovoltaic Specialists Conferences.*

20. Moseley, J. M., D. C. Miller, Q.-U.-A. S. J. Shah, K. Sakurai, M. D. Kempe, and G. TamizhMani, et al.,2011. The melt flow rate test in a reliability study of thermoplastic encapsulation materials in photovoltaic modules. Pp. 1–20 *NREL/TP-5200-52586.*

21. Kempe, M. D., D. C. Miller, J. H. Wohlgemuth, S. R. Kurtz, J. M. Moseley, and Q.-U.-A. S. J. Shah, et al., 2015. Field Testing of Thermoplastic Encapsulants in High-Temperature Installations. *Submitted to Energy Science & Engineering.*

22. Lux, C., U. Blieske, E. Malguth, and N. Bogdanski. 2013. "Variations in Cross-Link Properties of EVA of Un-Aged and Aged PV-Modules. *29th European Photovoltaic Solar Energy Conference and Exhibition.*

23. Marais, S., Y. Hirata, D. Langevin, C. Chappey, T. Nguyen, and M. Metayer. 2002. Permeation and Sorption of Water and Gases Through EVA Copolymers Films. Mater. Res. Innovations 6:79–88.

24. Kempe, M. D. 2006. Modeling of rates of moisture ingress into photovoltaic modules. Sol. Energy Mater. Sol. Cells 90:2720–2738.

25. Hulsmann, P., D. Philipp, and M. Kohl. 2009. Measuring temperature-dependent water vapor and gas permeation through high barrier films. Rev. Sci. Instrum. 80:113901.

26. McIntosh, K. R., N. E. Powell, A. W. Norris, J. N. Cotsell, and B. M. Ketola. 2011. The effect of damp-heat and UV aging tests on the optical properties of silicone and EVA encapsulants. Prog. Photovoltaics Res. Appl. 19:294–300.

27. Rätzsch, M., and U. Hofmann. 1991. The Crosslinking of Ethylene-Vinyl Acetate Copolymers with Sodium Alcoholates. J. Macromolecular Sci.: Part A - Chemistry 28:145–157.

8

Parametric analysis and systems design of dynamic photovoltaic shading modules

Johannes Hofer[1], Abel Groenewolt[2], Prageeth Jayathissa[1], Zoltan Nagy[1] & Arno Schlueter[1]

[1]Architecture and Building Systems, Institute of Technology in Architecture, ETH Zurich, John-von-Neumann Weg 9, 8093 Zürich, Switzerland
[2]Institute for Computational Design, University of Stuttgart, Keplerstrasse 11, 70174 Stuttgart, Germany

Keywords
Building integrated photovoltaics, energy efficiency, partial shading, solar tracking, system design, thin film PV

Correspondence
Johannes Hofer, Architecture and Building Systems, Institute of Technology in Architecture, ETH Zurich, John-von-Neumann Weg 9, 8093 Zürich, Switzerland. E-mail: hofer@arch.ethz.ch

Funding Information
This research has been financially supported by CTI within the SCCER FEEB&D (CTI.2014.0119) and by the Building Technologies Accelerator program of Climate-KIC.

Abstract

Shading systems improve building energy performance and occupant comfort by controlling glare, natural lighting, and solar gain. Integrating PV (photovoltaics) in shading systems opens new opportunities for BIPV (building integrated photovoltaics) on façades. A key problem of such systems is mutual shading among PV modules as it can lead to electrical mismatch losses and overheating effects. In this work, we present a new modeling framework, which couples parametric 3D with high-resolution electrical modeling of thin-film PV modules to simulate electric energy yield of geometrically complex PV applications. The developed method is able to predict the shading pattern for individual PV modules with high spatio-temporal resolution, which is of great importance for electrical system design. The methodology is applied to evaluate the performance of different dynamic BIPV shading system configurations, as well as its sensitivity to façade orientation and module arrangement. The analysis shows, that there is a trade-off between tracking performance and mutual shading of modules. Distance between modules is a critical parameter influencing the amount of mutual shading and hence limiting solar irradiation and electricity generation of PV shading systems using solar tracking. Planning of module string configuration, PV cell orientation, and location of bypass diodes according to partial shading conditions, reduces electrical mismatch losses and results in significantly higher electricity generation. The integration of parametric 3D and electrical modeling opens new possibilities for PV system design and dynamic control optimization. Though the analysis focuses on BIPV, the method is useful for the planning and operation of solar tracking systems in general.

Introduction

In order to reduce fossil primary energy demand and its effects on the global climate, energy strategies aim at increasing efficiency along with the use of renewable sources. As the building sector contributes significantly to energy use and greenhouse gas emissions, the EU aims for all newly constructed buildings to have close to zero net energy consumption by 2020. The realization of this goal requires local energy harvesting, and the integration of PV (photovoltaic) systems in buildings is an obvious choice for many climatic regions. Due to strong cost reductions and technical developments of PV systems in recent years, there is a growing opportunity for integrating PV elements into buildings. Implementations of BIPV (building integrated photovoltaics) include rooftop installations, external building walls, or semitransparent façades. Even though façade PV systems receive less irradiation than rooftop and ground installations, they open new application possibilities for BIPV in the urban context and can substantially contribute to electricity generation, with higher production in winter and potentially lower diurnal and seasonal variations. Moreover, façade BIPV systems can replace conventional building materials, such as the shading system, thus avoiding related costs and environmental impacts [1].

Building shading systems are an important architectural element, simultaneously influencing building appearance, its energy performance, and occupant comfort. Shading systems can control glare, natural lighting and solar gain, thereby offering reductions in energy demand for heating and cooling. Architectural implementations range from static elements and traditional blinds to automated shading systems. Dynamic shading systems that are capable of reacting to their external environment can result in high-energy savings when compared to static shading systems [2]. The mechanics required to actuate such systems couples seamlessly with the mechanics required for façade integrated PV solar tracking. With the increased use in large window openings and curtain walls in today's architecture, there is a growing opportunity to integrate PV elements into such shading systems [3–7].

A key problem with many PV installations, in particular for tracking systems and tightly integrated building applications, is shading among PV modules and by objects in the environment. Module shading can lead to electrical mismatch losses and overheating effects, thereby affecting power output and system reliability [8, 9]. Possible proposed solutions to reduce these adverse effects include optimized PV module distribution and tracker control to circumvent shading [10–13] or electrical configurations of a PV array to minimize power losses, such as the inclusion of bypass diodes or series/parallel circuit designs [8, 14–16].

Several approaches to model the effect of shading on PV systems power output have been proposed in the literature [8, 17–28]. Only few of the developed models have a resolution high enough to capture the electrical effects of small area shadows, such as those from antennas, chimneys, or other PV modules [28]. In addition, state-of-the-art modeling tools for PV systems simulation consider shading only for simplified conditions, that is, flat, rectangular PV modules, uniform module positioning, and regular building geometries and environments. Some recent studies use architectural 3D modeling software such as AutoCAD or SketchUp for the calculation of PV module shading and combine it with PV electrical modeling at cell level resolution [28–30]. Such a 3D modeling framework provides much higher flexibility than conventional tools in defining arbitrarily shaped objects, realistic building environments, and elevation profiles. The approach is very promising for integrated architectural and technical planning of BIPV systems and provides an opportunity for the development of innovative BIPV solutions.

In this work, we combine for the first time a high-resolution shading analysis performed with the Rhinoceros 3D software [31] and the parametric Grasshopper plugin [32] with electrical modeling of thin-film PV modules at subcell level resolution. We employ the model to analyze mutual shading, solar insolation, and electric energy yield of dynamically actuated PV modules in a building environment. The method we have developed can accurately predict the shading pattern for individual modules with high spatial resolution. The geometrical simulation is coupled with irradiation data and applied to assess solar insolation for various dynamic shading system configurations, including one-axis actuated horizontal and vertical louver blinds, as well as more complex geometries, such as two-axis actuated rectangular modules in a diamond pattern. The irradiation and shading simulation is coupled with a high-resolution electrical model of monolithic thin-film PV modules to evaluate the effect of different electrical configurations on array efficiency.

We begin in Methodology by describing the framework and methodology used for the parametric 3D and electrical simulations. In Solar Insolation Analysis, we analyze module shading and resulting solar insolation as a function of tracking system type, façade orientation, and module distance. In Electrical Performance Analysis, we discuss the electrical system design and performance as a function of relevant parameters for the different shading system configurations analyzed. In Experimental Implementations, we show experimental measurement results of thin-film PV module performance in partial shading conditions and progress leading to building integration. In Discussion and Conclusions, we summarize the results and discuss limitations and possible advancements of this work.

Methodology

General approach

In order to study both the appearance and solar insolation on an array of dynamic PV modules attached to the building envelope, we have developed a framework for the parametric 3D design and calculation of module shading as well as solar irradiance at high resolution. The results are analyzed as a function of different input parameters, such as module arrangement and tracking control. For further analysis of the electrical performance, the shading pattern and solar irradiance are coupled with an electrical model which calculates the characteristic current–voltage (I-V) curves of PV modules as a function of time. On the basis of this, we evaluate electric energy yield for different module string interconnections and other electrical design parameters. The general workflow is illustrated in Figure 1.

The developed modeling framework is versatile and principally any configuration in terms of module shape, arrangement, and motion control can be simulated. For practical relevance we focus in this study on a rectangular PV module design, three types of façade-attached shading system types (Fig. 2A), and one or two-axis solar tracking. Parameters varied include façade orientation, distance between modules,

Input parameters

Site & Building   **PV system geometry**
Weather data   **Motion control**
                        Electrical design

Parametric 3D model

Module positioning and shading calculation
using Rhinoceros / Grasshopper

Solar radiation analysis

Electric model

Simulation and design of thin-film PV modules
and array configuration in Matlab

Electrical performance
analysis

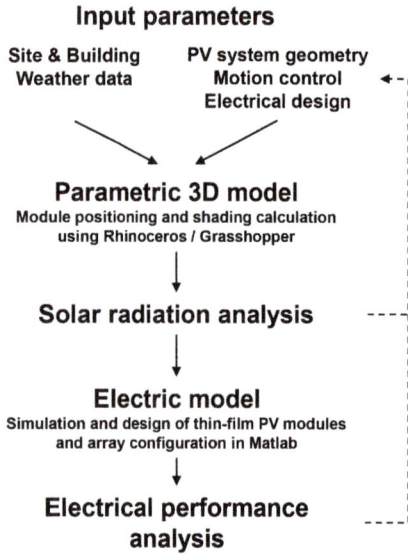

Figure 1. Modeling framework for illustration of the workflow. The dashed line represents ways to optimize geometry, motion control, and electrical design parameters based on system performance.

as well as electrical design parameters. The goal of the simulation is to analyze solar insolation for complex dynamic shading situations and to find the best electrical layout for a given shading system configuration and control strategy. Generally, the integration of dynamic 3D and electrical modeling allows to further optimize module shape, arrangement, and control, minimizing mutual shading and maximizing electric energy yield. In the following, the methods of the parametric 3D and electrical model are explained.

Parametric 3D model

Module positioning

The workflow for the parametric 3D design of shading systems has been implemented in the Rhinoceros 3D software [31] using the Grasshopper plugin [32]. The resulting designs consists of multiple rectangular surfaces that represent PV modules, each of which can be rotated in the solar azimuth and/or the altitude direction, within a specified range of angles. The number of modules and their positions are controlled by indicating coordinates

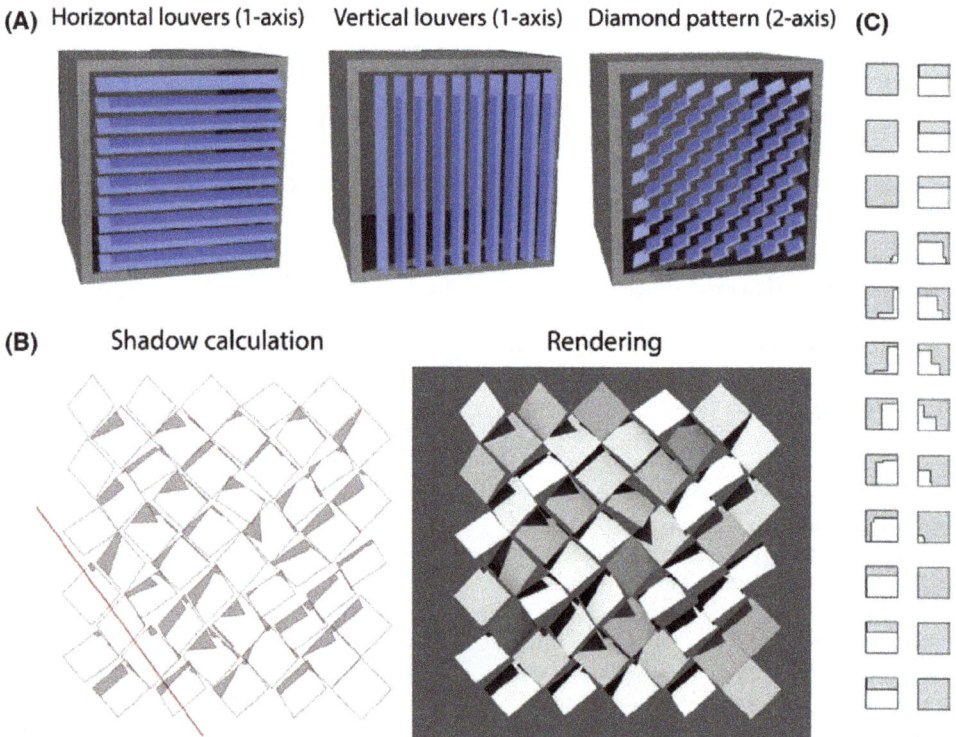

Figure 2. (A) Renderings of analyzed module configurations and mutual shading for a specific sun position. Depending on the system, the modules orient toward the sun using one or two-axis tracking. (B) Screenshot of the shading calculation performed in Rhino-Grasshopper for arbitrary module positions. The red line indicates the direction of the sun. For comparison a rendering of the 3D model is shown on the right. (C) Projection of shadows for a single module over the course of a day; gray represents the shaded area.

in 3D space at which the modules should be placed. This is done by specifying parameters that generate a rectangular or diamond pattern grid.

To exemplify, we generate and simulate different horizontal and vertical louver shading systems with single-axis solar tracking and a diamond configuration with dual-axis tracking in front of a room with a dimension of 3.6 m width and 2.8 m height (see Fig. 2A). The size of individual PV modules is fixed at 40 by 40 cm. In the reference case, the distance between louver blinds and modules in the diamond pattern is fixed at 10% of the module dimension, resulting in seven blinds with nine modules in series for the horizontal and vice versa for the vertical louver system. The diamond configuration is composed of 50 individual modules.

In our framework, one-axis tracking is simulated by rotating the modules in the direction of the sun along a horizontal or vertical axis. The center of rotation lies at an adjustable distance behind the module plane, so that actual mechanical rotation systems can be simulated. The orientation of the modules is defined by aligning the module's normal vector to the projection of the sun's direction on the plane perpendicular to the module's axis of rotation. For two-axis solar tracking, the modules orient exactly perpendicularly to the sun direction. Horizontal (azimuth) and vertical (altitude) tracking angles can be constrained to any angle range, so that tracking systems with physical limits on the range of rotation can be simulated.

Calculation of shading and solar irradiation

For each module, the calculation of solar irradiation is simulated at a resolution of 50 by 50 grid points, corresponding to an irradiance matrix with 2500 elements per module and instance. The dimension of these grid elements corresponds to the dimension of subcells used for the assessment of shading effects on the electrical PV module performance.

Solar irradiation is calculated separately for each module m and grid point p ($G_{tot,m,p}$) as a function of time t based on the three-component model [33, 34], for which the radiation on an inclined surface is calculated as the sum of direct beam ($G_{dir,m,p}$), diffuse sky ($G_{dif,m,p}$), and diffuse reflected radiation ($G_{ref,m,p}$) as

$$G_{tot,m,p}(t) = G_{dir,m,p}(t) + G_{dif,m,p}(t) + G_{ref,m,p}(t). \quad (1)$$

To assess these components, the module shading fraction or detailed shape (depending on the desired resolution) as well as the amount of radiation coming from the sky versus radiation from the ground needs to be known.

Within the parametric 3D model, the solar position is generated using the DIVA plugin [35], which provides vectors in the direction of the sun for any location on earth at any moment in time, based on the method described in [36]. The location for our analysis is Zurich, Switzerland (latitude 47.37°N, longitude 8.55°E). The sun direction vectors are used for shadow calculations and solar tracking. The shape of shadows on the modules is calculated based on vector algebra, using the sun light direction, panel geometry, and optionally external geometry as input. First, the module's normal vector is compared to the sun direction to check if sunlight could be received by the module. If so, the input geometry is intersected with the module's plane; the geometry that is located behind the module's plane logically does not generate shadows on the module's front side. The remaining geometry is then projected on the module plane using the method described in [37]. To verify this method, the resulting shadows have been compared with those generated by a physically based rendering engine [38], which is based on [39] (Fig. 2B). By generating sequences of sun positions at fixed time intervals, the geometric behavior of a tracking system can be studied. For any moment in time, shading on the panels (including mutual shading) can be studied both numerically and graphically (Fig. 2C). As the shadow shapes are generated as vector graphics, shading grids at any resolution can be quickly exported.

The sky view factor (or diffuse radiation factor) is approximated separately for each module by creating a hemispherical mesh, connecting the midpoints of the faces of this mesh to the midpoint of the module and checking these connections for intersections with panel geometry and external geometry. The area of the faces belonging to lines that are not intersected with any geometry are then added. Dividing the sum of this addition by the total surface area of the hemisphere results in the approximated view factor.

Direct radiation is calculated as the product of the shading vector ($SV_{m,p}$), which is zero for shaded and one for not shaded areas, the projection of the module area on its normal plane relative to incident solar radiation ($NP_{m,p}$), and direct normal irradiance (G_{dni})

$$G_{dir,m,p}(t) = SV_{m,p}(t) \cdot NP_{m,p}(t) \cdot G_{dni}(t). \quad (2)$$

Diffuse sky radiation is the product of the sky view factor ($SF_{m,p}$) and diffuse irradiance on the horizontal plane ($G_{dif,hor}$)

$$G_{dif,m,p}(t) = SF_{m,p}(t) \cdot G_{dif,hor}(t). \quad (3)$$

Diffuse reflected radiation is the product of the fraction of ground light irradiance reflected toward the module, the albedo of reflecting objects (a) and global irradiance on the horizontal plane ($G_{glo,hor}$)

$$G_{ref,m,p}(t) = (1 - SF_{m,p}(t)) \cdot a \cdot G_{glo,hor}(t). \quad (4)$$

In Eq. 4 it is assumed, that the sum of the fraction of light reflected from the ground to the module and the sky view factor for this module equals one. Note that currently only a single albedo value is used, not distinguishing between different reflecting objects.

The irradiation data used in this work was exported from the Meteonorm software [40] and is based on long-term measurements of a MeteoSwiss weather station in Zurich, Switzerland (Zurich SMA, Fluntern). The total annual global insolation on a horizontal plane at this location is 1120 kWh/m^2. For the analysis of solar insolation presented in Section 3, hourly solar irradiance is averaged per month and only one average day per month is simulated. PV electricity generation is strongly dependent on the specific shading and insolation condition. Therefore, the full hourly resolution was used as input data for the analysis of electric performance in Section 4. The weather data is shown in Figure 3.

While the time resolution of shadow calculations is determined by the time step set in the sun positioning algorithm (which in the DIVA plugin can be set to any value), the time resolution of irradiation on the modules is determined by solar irradiance input (one hour for the used weather data).

Electrical model

One main reason for energy generation losses of BIPV systems is partial shading of PV modules due to surrounding buildings, trees, or other objects [16]. Another reason, particularly for PV integrated in dynamic shading systems, is mutual shading between PV modules. Varying insolation due to partial shading of PV modules leads to different I-V characteristics of PV cells and consequently to electrical mismatch within the module. Furthermore, modules exposed to different shading conditions lead to mismatch and power losses in the whole PV array [8, 41], which can cause hot-spot heating and damage of cell encapsulation materials affecting system reliability [8, 18, 19].

Modeling the effect of shading on PV systems power output has been extensively investigated in the literature [8, 17–28]. The impact of different cell shapes and orientations on the I-V characteristics of monolithic thin-film PV modules can be assessed with a two-dimensional SPICE circuit simulation [18, 41, 42]. We employ a similar model implemented in MATLAB. Figure 4A–D illustrate different levels of this model from array to module to subcell. In this two-dimensional simulation, each module consists of 50 cells in series and each cell is subdivided into 50 parallel connected subcells. The subcell I-V curve is modeled based on the standard equivalent circuit model with a single diode, one series and one shunt resistance [43]. For the parameters of the model we use values of a commercial CIGS (copper indium gallium selenide) module [44, 45] with an exponential increase in the shunt resistance at low irradiance [43, 45]. In case of partial shading, some of the cells are in reverse bias and act as a load. In order to account for this effect, the reverse characteristic of cells is modeled according to [46, 47], similar to the measured characteristic for CIGS solar cells [48]. The module I-V curve is calculated by summing the current of parallel connected subcells and subsequently the voltage of series connected cells. Since the module shading matrix and the network of subcells within a module have the same spatial resolution, both models can be directly coupled to simulate module electric

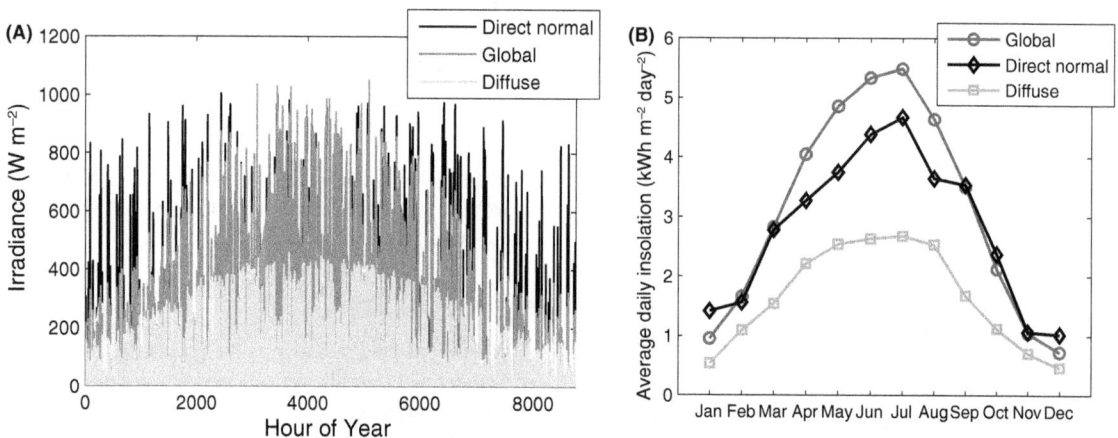

Figure 3. (A) Direct normal, diffuse horizontal, and global horizontal irradiance over 1 year at the reference location with hourly resolution. (B) Average daily insolation per month.

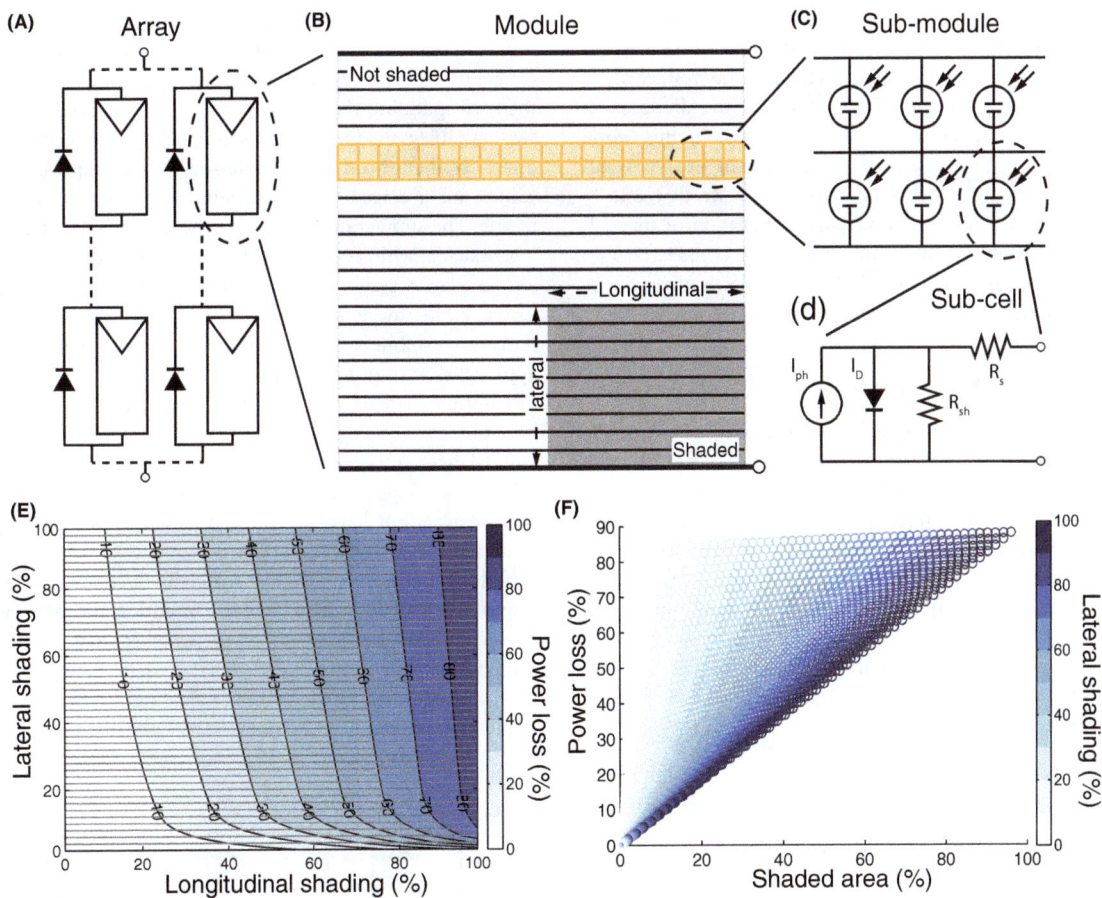

Figure 4. (A) Array of several PV modules with external bypass diodes in series-parallel connection. (B) Typical thin-film PV module with long rectangular cells connected in series. The module is partially shaded. (C) The module is modeled as a 2D grid of parallel-series connected subcells. (D) Each subcell is modeled based on the standard equivalent circuit. (E) Simulation of power loss as a function of lateral and longitudinal shading. In this case, the irradiance in the shaded region is reduced by ca. 80% relative to the full irradiance in the unshaded region. Horizontal lines indicate the orientation of PV cells. (F) Power loss as a function of shaded area. Color scale indicates the amount of lateral shading. Power loss is linear at 100% lateral shading, but deviates significantly at small lateral shading.

performance for a specific shading pattern and solar irradiance. Figure 4E and F show the simulated module power loss at the MPP (maximum power point) as a function of shading parallel (longitudinal) and perpendicular (lateral) to cell elongation.

Note that different from [18, 41, 42] contact sheet resistances between individual subcells have been neglected. Another limitation of this work includes the assumption of homogenous temperature distribution of PV cells in shaded and unshaded regions. Previous studies analyzed the feedback of power dissipation on module temperature distribution and showed that due to electrical mismatch from partial shading, temperature differences between cells in shaded and unshaded regions occur [9, 18]. Considering such electro-thermal effects is important for the prediction

of hot-spots, potentially leading to PV module degradation and permanent damage [9]. Infrared thermal imaging of partially shaded PV modules analyzed in this study has shown, that module temperature distribution is dominated by direct light absorption. The model we developed offers subcell temperature resolution, however, further analysis and experimental validation will be required to accurately account for module temperature distribution based on both effects, direct light absorption and electrical dissipation due to cell mismatch. With the current model setup we observe good agreement between measured and simulated data and consider both simplifications to be of minor influence on the general conclusions of this study.

The array I-V curve is simulated by summing the voltage of modules connected in series strings and then the current of strings connected in parallel. Usually modules have integrated bypass diodes that conduct in case of reverse bias due to shading and limit the voltage drop over a module to the voltage at which the diode starts to conduct (ca. 0.6 V for silicon diodes). In this work, we consider three cases for the placement of bypass diodes, which are detailed in Section 4. Note that due to manufacturing tolerances, varying degradation or dirt, PV modules exhibit slightly different current–voltage characteristics. Mismatch losses induced by such phenomena are neglected in the analysis.

Solar Insolation Analysis

Sun-tracking ability, module shading, and consequently solar insolation on PV modules varies significantly as a function of system configuration parameters, such as motional degree of freedom (one/two-axis tracking), façade orientation, and module spacing. In this section, we first analyze solar tracking performance and mutual shading for different shading system configurations using one and two-axis solar tracking and then present results of the average daily and monthly distribution of solar insolation for various façade orientations and module distances.

Figure 5A depicts the projection of module area on its normal plane relative to incident solar radiation over the course of a day for each month of the year and Figure 5B the corresponding visible (not shaded) module area. These two variables are important as they are proportional to direct irradiation on the module surface (see Eq. 2). The values are averaged over all modules of a south facing façade. The simulation is done for one and two-axis tracking systems with 10% module spacing. For comparison also a planar system without tracking ability is shown. As expected, for a fixed plane the area normal to sun radiation exhibits strong seasonal and daily variations. It approaches zero in the morning and evening and is lower in summer than in winter. Due to solar altitude tracking, horizontal louvers have an improved daily performance and balance out seasonal fluctuations. Vertical louvers with azimuth tracking orient toward the sun in the morning and evening, but have a minimum at solar noon as they cannot track solar altitude. The two-axis tracking system completely cancels out seasonal and daily fluctuations. The improvements of tracking systems orienting the module toward the sun are countered by increased mutual shading (Fig. 5B). While for a static planar system there is no mutual shading at all, the effects of shading increase using single-axis tracking and even further with dual-axis

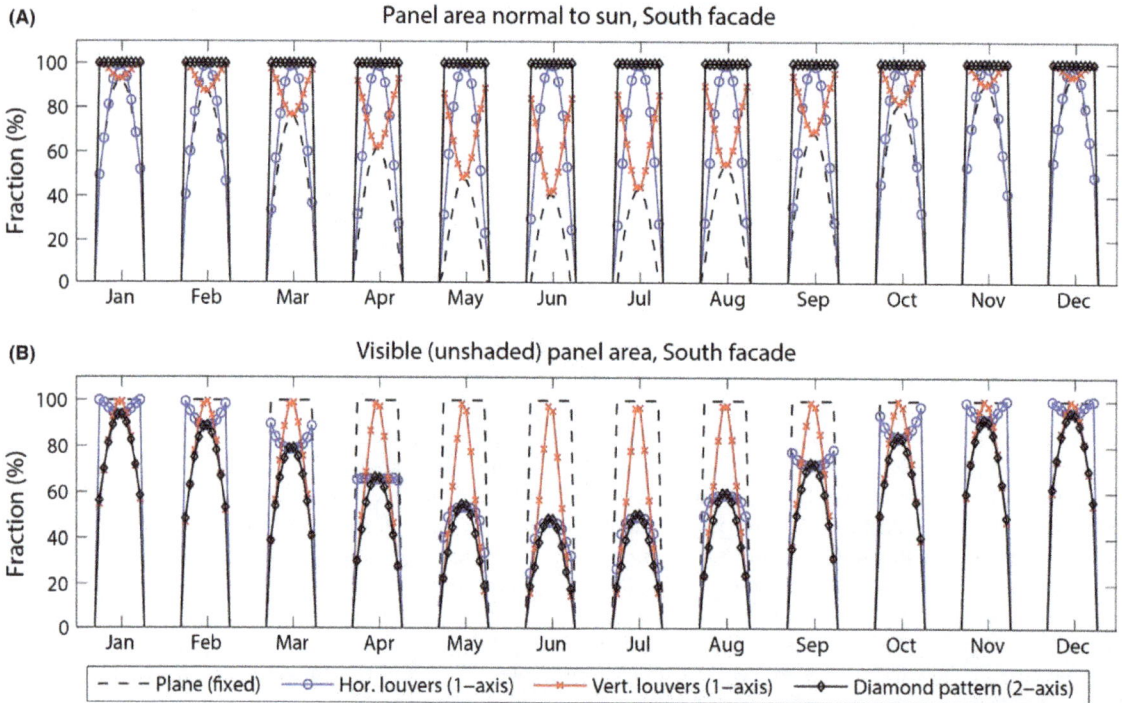

Figure 5. (A) Fraction of the module area oriented normally to the sun over the course of a day for four different solar tracking systems on a south facing façade. (B) Visible, not shaded module area.

tracking. For all systems we observe a trade-off between module mutual shading and solar tracking performance in terms of projected normal area, in particular for narrow spaced module arrangements.

The results of module tracking performance and mutual shading serve as inputs to the calculation of daily and monthly solar insolation shown for a horizontal louver system using solar altitude tracking in Figure 6. The irradiance has a different daily profile depending on façade orientation and season. East, west, or south oriented facades receive most sunlight in the morning, evening or at noon, respectively. As such, daily variations in solar insolation can be partly balanced by using PV modules for different façade orientations. Interestingly for this tracking system on a south facing façade, direct insolation in winter is relatively high but drops during summer months due to mutual shading. This leads to relatively low seasonal variation in solar insolation. The effect is less influential for east or west oriented facades, which exhibit less mutual shading in summer and have strong seasonal variations.

Mutual shading and solar insolation have been evaluated for different façade orientations as a function of the distance between modules, that is, the distance between rows or columns of single-axis or individual modules of dual-axis tracking systems in percent of the module dimension. Figure 7A shows the decrease in mutual shading with increasing module distance for a south and east facing façade. Module shading fraction is calculated as the average amount of mutual shading of all modules within the façade assembly at times, when there is direct light irradiation on the respective façade. For a south facing façade, the mutual shading fraction is high for all systems in summer and low in winter. This trend can be also compared to the results shown Figure 5B. The variation with module distance is smallest for the vertical

and horizontal louver system in summer and winter, respectively. For the east oriented façade, shading is high in winter, particularly for the vertical louver and diamond configuration tracking system. This is because an east facing façade receives direct light in winter only for a short amount of time, during which these two systems generate high mutual shading. The horizontal louver tracking system is highly shaded in summer and less in winter. The dependence is opposite for the vertical louver system.

The amount of shading and its reduction with increasing module distance lead to a variation in insolation per module area as shown in Figure 7B. At small module distance, the absolute amount of insolation is similar for all systems, but at a higher level for the east façade in summer and for the south façade in winter, respectively. In turn, the variation in insolation is higher for the south façade in summer and for the east façade in winter. Overall the increase in insolation with module distance is highest for the two-axis tracking system.

Further analysis not shown in Figure 7 reveals, that even though the insolation per module area increases with increasing module distance, the absolute amount of insolation on all modules of the façade decreases. This is because absolute module area decreases more rapidly than relative insolation increases. Note that an increase in the module spacing corresponds to less reduction in module area for the one-axis than for the two-axis tracking system, due to a variation in the module distance in only one instead of two dimensions. For conventional ground-mounted PV tracking systems [49–51], the ratio of module relative to ground area is known as the ground cover ratio and the trade-off between land occupation and energy yield is discussed in [10, 52]. Comparing the performance of the different tracking systems analyzed

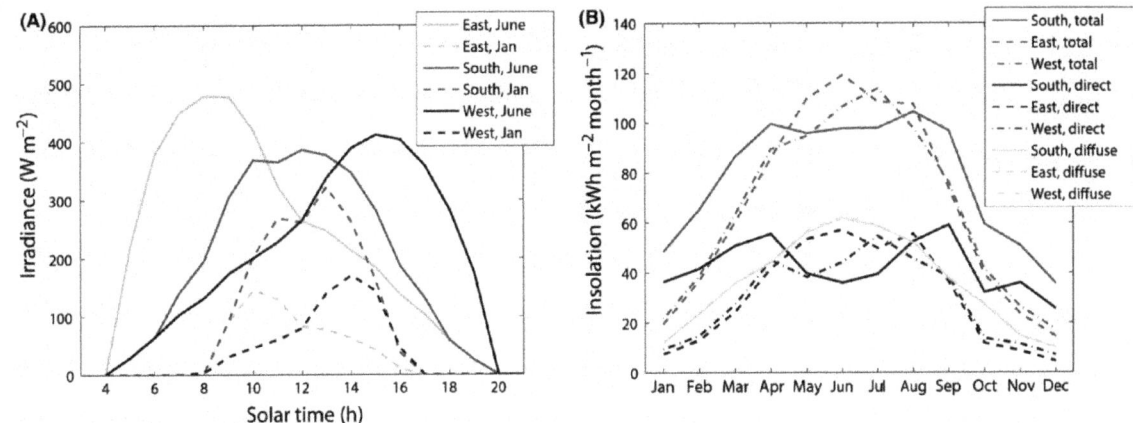

Figure 6. (A) Average daily solar irradiance per module area for a horizontal louver solar tracking system in June and January at the location of Zurich. (B) Average monthly insolation per module area by façade orientation.

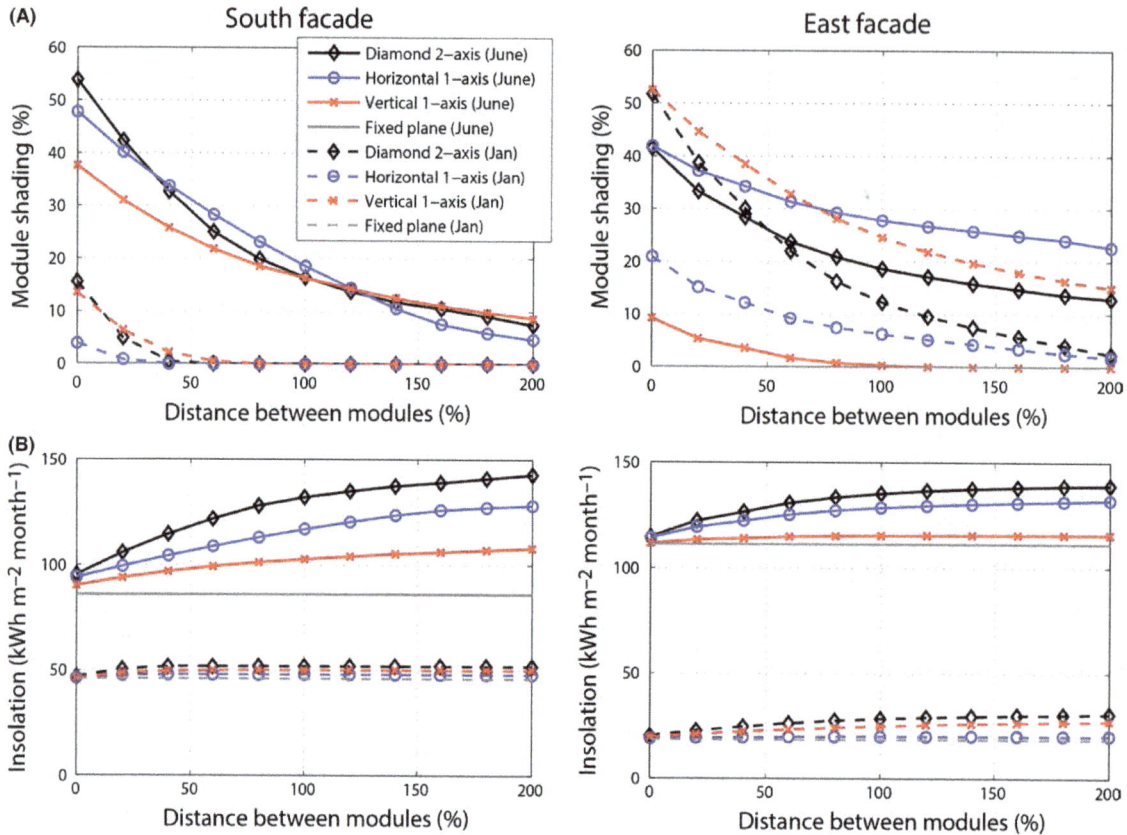

Figure 7. (A) Average module mutual shading and (B) insolation per module area as a function of the distance between modules for south (left) and east (right) oriented façade shading systems in June and January.

in this work relative to the ground cover ratio instead of the module distance, shows that the horizontal louver one-axis and diamond two-axis tracking systems achieve the highest insolation per façade area in summer. In winter, the best performance is reached by vertical louver and two-axis tracking systems. On an annual basis, the two axis-tracking system reaches highest insolation per façade area.

Electrical Performance Analysis

Knowledge of the irradiance per module can be used to design electrical configurations that reduce power losses. Due to mutual shading, the irradiance per module can vary significantly depending on the shading system type and the location of the module within the façade assembly. In this section, we first evaluate the shading patterns by configuration for a south oriented facade shading system. On the basis of this, we analyze suitable electrical designs and present simulation results of power production.

Shading analysis and system design

In shaded conditions the photocurrent produced by a PV module decreases significantly. Because series connected PV modules are forced to operate at the same current, shading of a single module within a string of modules connected in series limits the current of other not shaded modules. On the hand, parallel-connected modules are forced to operate at the same voltage. Since the voltage decrease during partial shading is comparably low, parallel connection of modules leads to lower mismatch and power losses in partially shaded conditions than series connection [8, 16, 19]. There is, however, a limit to the degree of parallelism, because currents in parallel-connected modules, and hence resistive (ohmic) losses, are higher. In addition, wiring is more complex for parallel connected modules and usually for the inverter a certain minimum input voltage is required. For this reason, modules are often connected in series-parallel arrays by first forming series connection of modules in strings and then connecting those strings in parallel. In

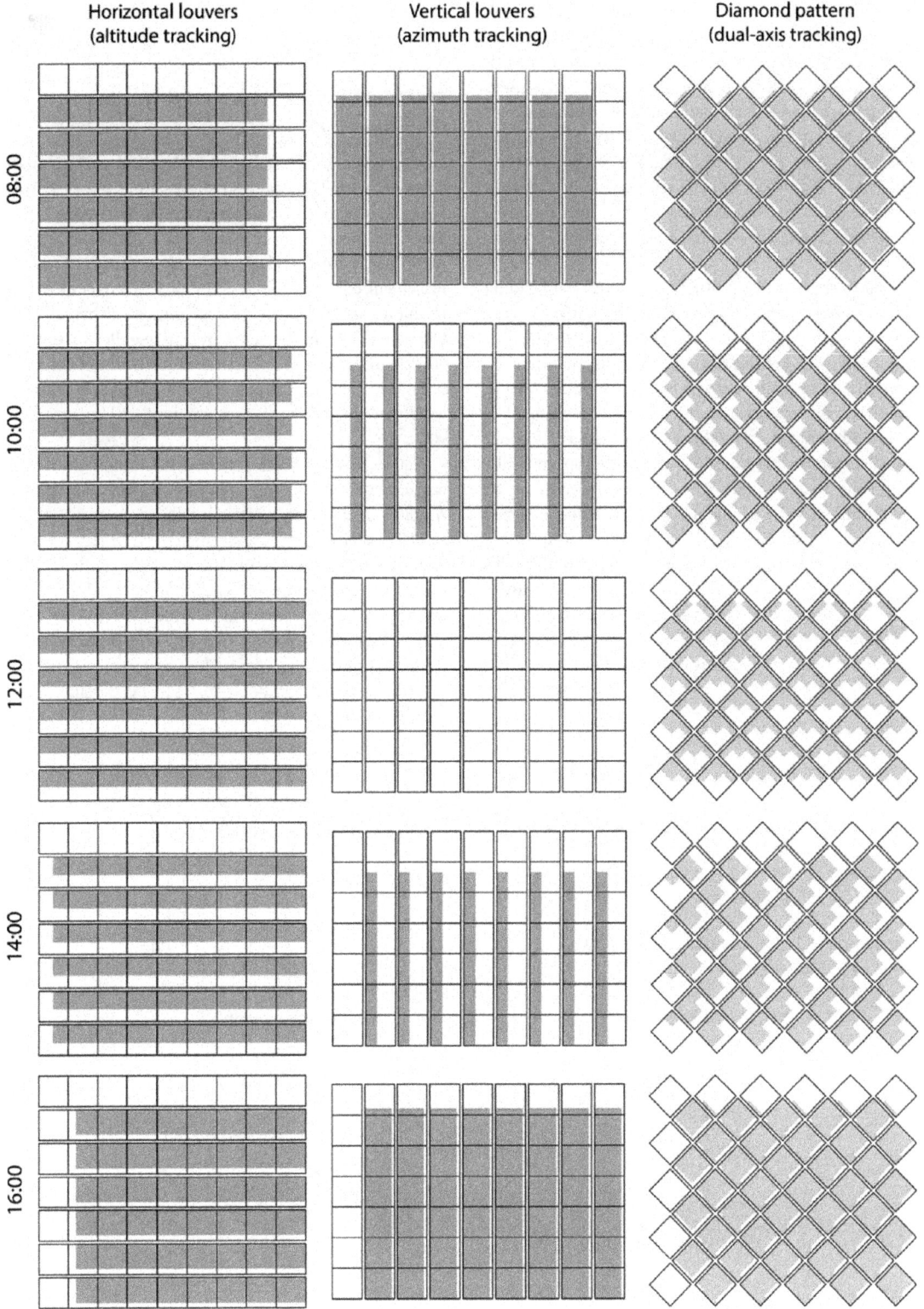

Figure 8. Projected mutual shading patterns over one clear day in June for a south oriented façade.

order for these strings to work most efficiently, the current at the maximum power point should be similar for PV modules connected in a string. For the analyzed PV shading systems, we therefore connect modules with similar shading pattern and solar irradiance over the course of a day in series strings. In particular, we try to avoid the situation that only one or few modules within a string are shaded, while others are fully illuminated. Figure 8 depicts calculated mutual shading among modules of a south oriented façade for the three shading systems analyzed.

The figure shows the 2D projection of mutual shading for five instances from morning to afternoon of a clear day in June. For horizontal and vertical louver systems, shadows follow the direction of the axis of rotation. Therefore, it is straightforward to connect modules within louver blinds in series, which also facilitates wiring. For the diamond configuration, the shading pattern is more complex. Generally, modules located in the top and subsequent row, the left and right column, as well as all central modules receive similar irradiance. For the analyzed arrangement of 50 modules, it is, however, impossible to find string connections of more than two modules exactly matching this shading pattern. For this reason, we constrain the analysis to two cases of either five or ten modules in series as shown in Figure 9. Note that configuration B is unsymmetrical and favors the shading situation in the morning. The string connections shown

in Figure 9 will be analyzed further in the next section.

Electrical performance of different configurations

In this section, we analyze the electrical performance in terms of array I-V characteristic and maximum power as a function of relevant electrical design parameters. In addition to possible string connections shown in Figure 9, these parameters include PV cell orientation and the placement of bypass diodes. For each string configuration, we model three cases for the placement of bypass diodes: (1) none; (2) one bypass diode in parallel to the entire 50 cells of a module; (3) two submodules of 25 cells in series, each with one bypass diode in parallel. Furthermore, we study the effect of a rotation of cell orientation by 90°.

Figure 10 shows the resulting maximum array power per module area for one clear day in June. For the horizontal louver system, vertical cell orientation (indicated as 90°) leads to significantly higher electricity generation than horizontal cell orientation. This can be explained by shading of the upper part of most modules (Fig. 8) and the stronger decrease in module MPP for lateral than longitudinal shading (Fig. 4E and F). In contrast, electricity generation of the vertical louver system, is higher for horizontal cell orientation. This can be explained by the

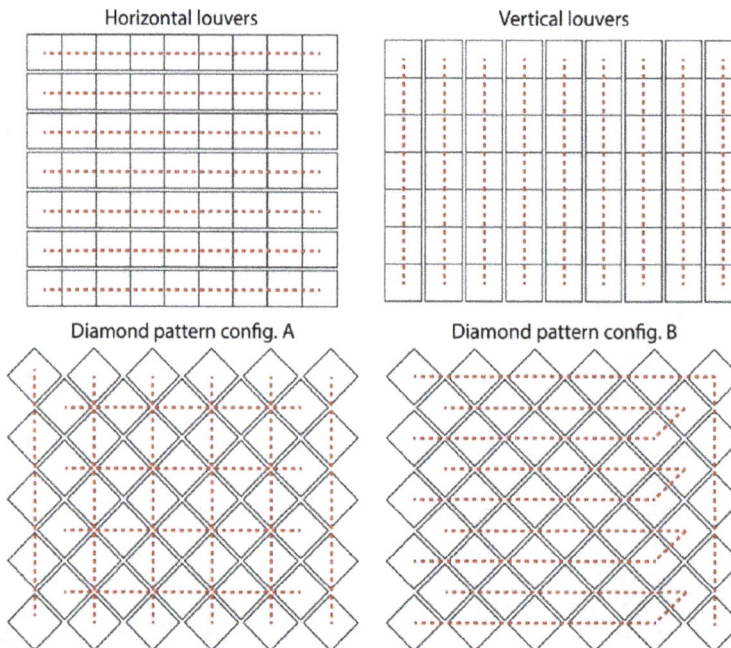

Figure 9. Possible string connection of modules based on shading analysis.

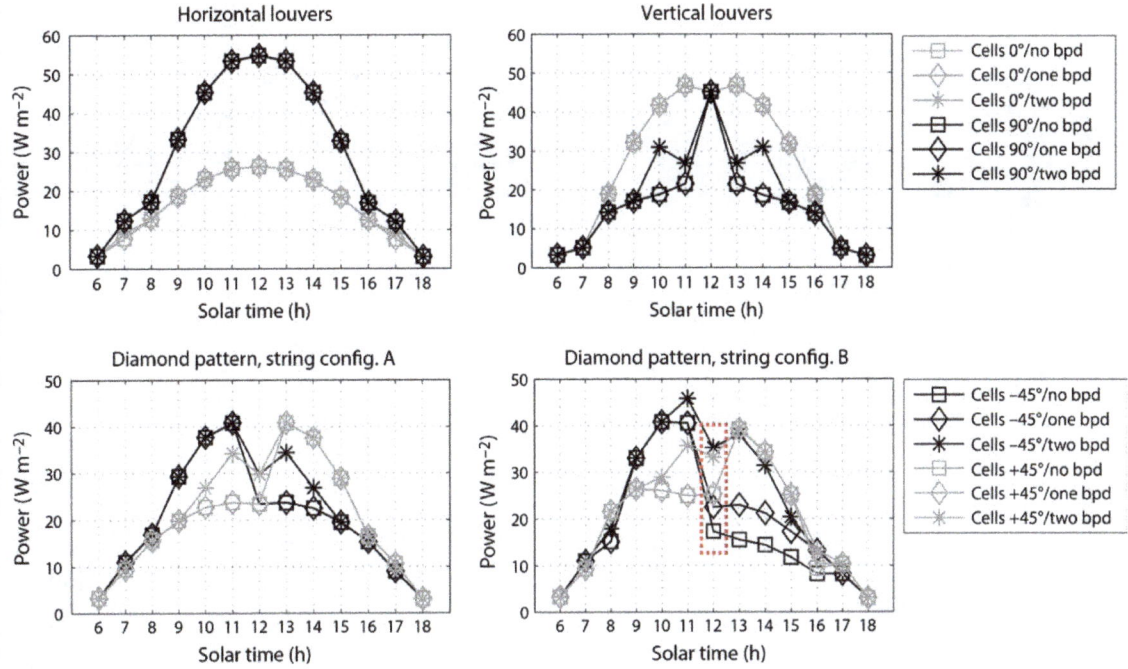

Figure 10. Maximum power per module area during a clear day in June. For each configuration, two different cell orientations and three options for the placement of bypass diodes (bpd) are simulated. The highlighted area is analyzed further in Figure 11.

inverted shading pattern. At solar noon, cell orientation is irrelevant for the vertical louver system, because modules are not shaded (Fig. 8). For the diamond pattern system in both string configurations, −45° cell orientation leads to higher electricity generation in the morning and +45° in the afternoon. Our analysis shows, that power losses caused by differing morning and evening shading conditions, can be reduced by integration of two bypass diodes per module. While string configuration A leads to a symmetrical electricity generation profile for different cell orientations, it is asymmetric for configuration B due to the nonuniform string design (Fig. 9).

The results shown in Figure 10 are on a highly aggregate level. For many applications it is of interest to know the exact I-V characteristic of modules, strings, and array, e.g. for the design of optimal electrical interconnections [14–16] and MPPT (maximum power point tracking) methods [30, 53–55]. Figure 11A shows the I-V curves of all 50 modules for one specific instance in Figure 10 (diamond configuration, string layout B, at 12o'clock). With one integrated bypass diode, reverse bias losses are limited, because the diode conducts below certain a threshold voltage. With two bypass diodes, module I-V curves exhibit a stepwise profile for some shading conditions. The integration of module bypass diodes leads to multiple steps for the I-V curves of strings (Fig. 11B)

and arrays (Fig. 11C), as well as multiple peaks in the P-V (power-voltage) characteristic of arrays.

Varying shading conditions and solar irradiation influence the relative performance of the configurations analyzed in this section. In an analysis of monthly electricity generation of those different configurations, we find that the order from lowest to highest electric yield remains the same as the comparison for a single clear day in Figure 10 suggests. However, the absolute difference varies significantly throughout the year. As an example, Figure 12 shows monthly electricity production of the horizontal louver tracking system for horizontal and vertical PV cell orientation. Results are based on hourly weather data and shading simulations. The difference in electric energy yield between horizontal and vertical PV cell orientation is low from November to January and high from March to October. The main reason for this trend is a decrease in module shading during winter (Fig. 7A).

System comparison

In this section, we compare the performance of the three PV shading system configurations analyzed (cf. Fig. 2A) with regard to solar insolation, system efficiency, and electric energy yield. As can be seen from Figure 7, raising

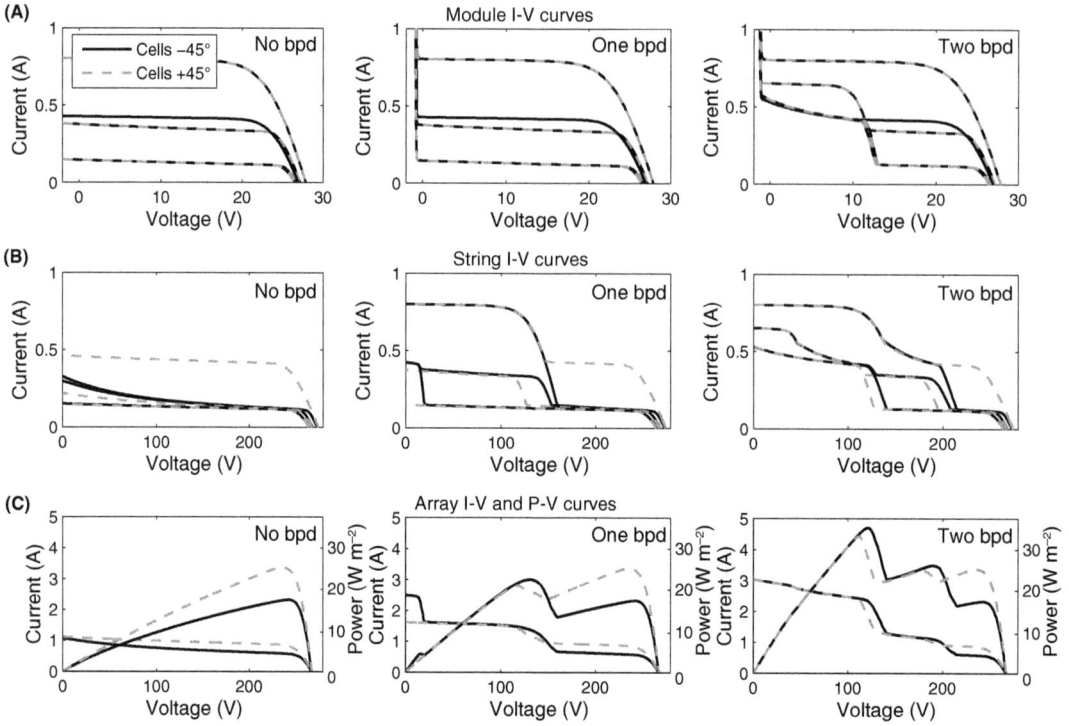

Figure 11. (A) I-V curves of individual modules for one specific instance shown in Figure 10 (diamond configuration, string layout B, at 12 o'clock). The diagram shows the I-V curves of all 50 modules (mostly overlapping) for either −45° or +45° cell orientation. From left to right, the number of bypass diodes per module increases from zero to two. (B) I-V curves of different strings with 10 modules in series. (C) Resulting array I-V (left) and P-V curves (right axis).

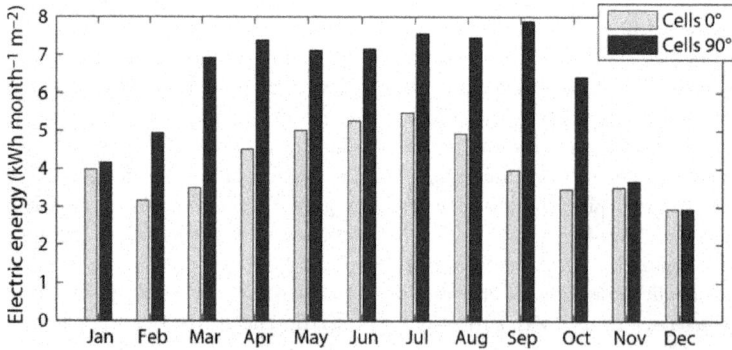

Figure 12. Monthly electricity generation of horizontal louver system for horizontal and vertical PV cell orientations with one integrated bypass diode per module.

the distance between PV modules reduces mutual shading and increases solar irradiance on the module. In addition, reducing the amount of shading leads to less mismatch between cells and improves the electrical conversion efficiency. On the other hand, increasing the distance between modules reduces the FCR (façade cover ratio). The FCR is a measure of the density of PV modules on the façade and defined as the PV module area relative to the total area of the façade. Based on the FCR, the performance of different system configurations can be compared at the same module density. Figure 13A shows the relation between number of modules, FCR and distance between

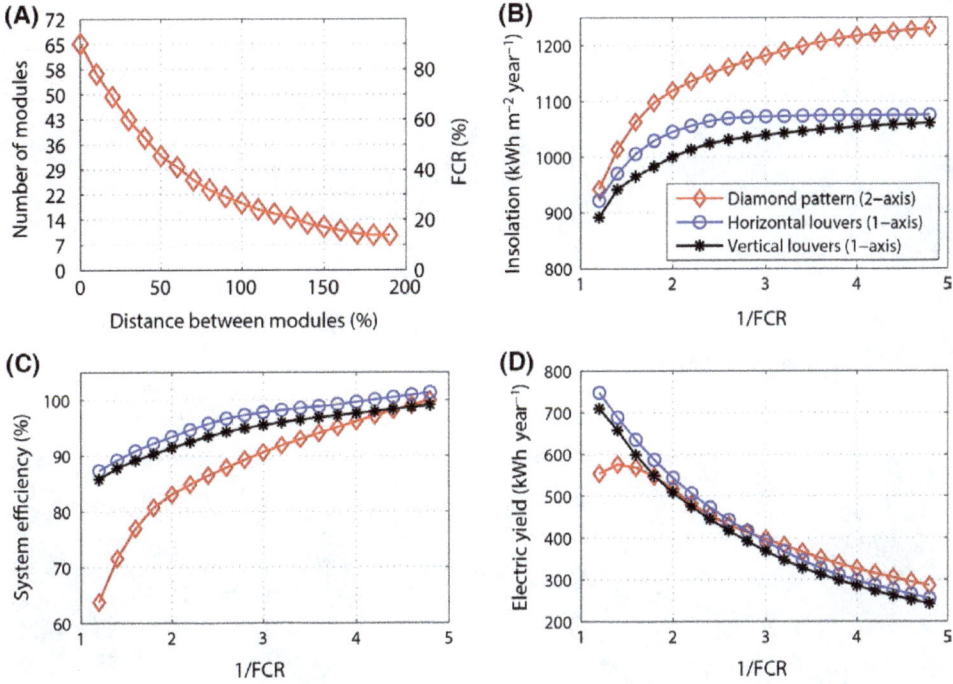

Figure 13. (A) Number of modules in the diamond pattern configuration and façade cover ratio as a function the distance between modules. (B) Annual solar insolation per module area as a function of the inverse façade cover ratio for three different system configurations. (C) Corresponding normalized electrical efficiency and (D) total electric yield of the façade shading systems.

modules for the diamond pattern configuration. In the analysis, the FCR is reduced from ca. 90% to 15% by varying the spacing between individual modules of the diamond pattern or rows and columns of the louver systems, but keeping the total façade area constant. Figure 13B–D illustrate corresponding performance characteristics for a south oriented façade as a function of the inverse FCR for the three different shading system configurations analyzed. Note that each module is fully laminated with thin-film CIGS cells (used in the most efficient orientation) and equipped with one external bypass diode.

Figure 13B shows solar insolation per module area as a function of 1/FCR. Due to mutual shading it decreases with increasing module density. Only at large module spacing, the full potential of the solar tracking system can be exploited, in particular for the two-axis system. Figure 13C shows the corresponding average efficiency for conversion of solar radiation to electric energy. It is normalized to the maximum efficiency of the diamond pattern system. With increased mutual shading at narrow module distance, electrical efficiency decreases. This is in particular the case for the diamond pattern system, due to inefficient alignment of the module shading pattern relative to cell orientation. At wide distance between

modules, the efficiency of all systems is similar. Total electric yield of the façade system is the product of specific solar insolation, PV module area, and electrical efficiency. It is depicted for constant façade area in Figure 13D. Electric yield continuously decreases with increasing module distance for the horizontal and vertical louver systems. In the diamond pattern configuration, it exhibits a maximum at ca. 70% FCR or 30% distance between modules. Up to this value, the increase in electrical efficiency and specific solar insolation offsets the reduction in active module area. Overall, the highest electric yield is achieved with the horizontal louver system up to a FCR of ca. 35% and with the diamond pattern system at lower FCR.

Note that the results shown in this section only apply to a south facing façade at the location of Zurich and modules covered fully by PV. Efficiency loss at narrow module distance can be reduced by PV module design with improved shading tolerance [41], the integration of multiple bypass diodes per module, or placement of PV cells in module areas with little shading. The efficiency degradation in the diamond configuration relative to the louver systems at narrow module distance is also less critical for east and west relative to south oriented facades.

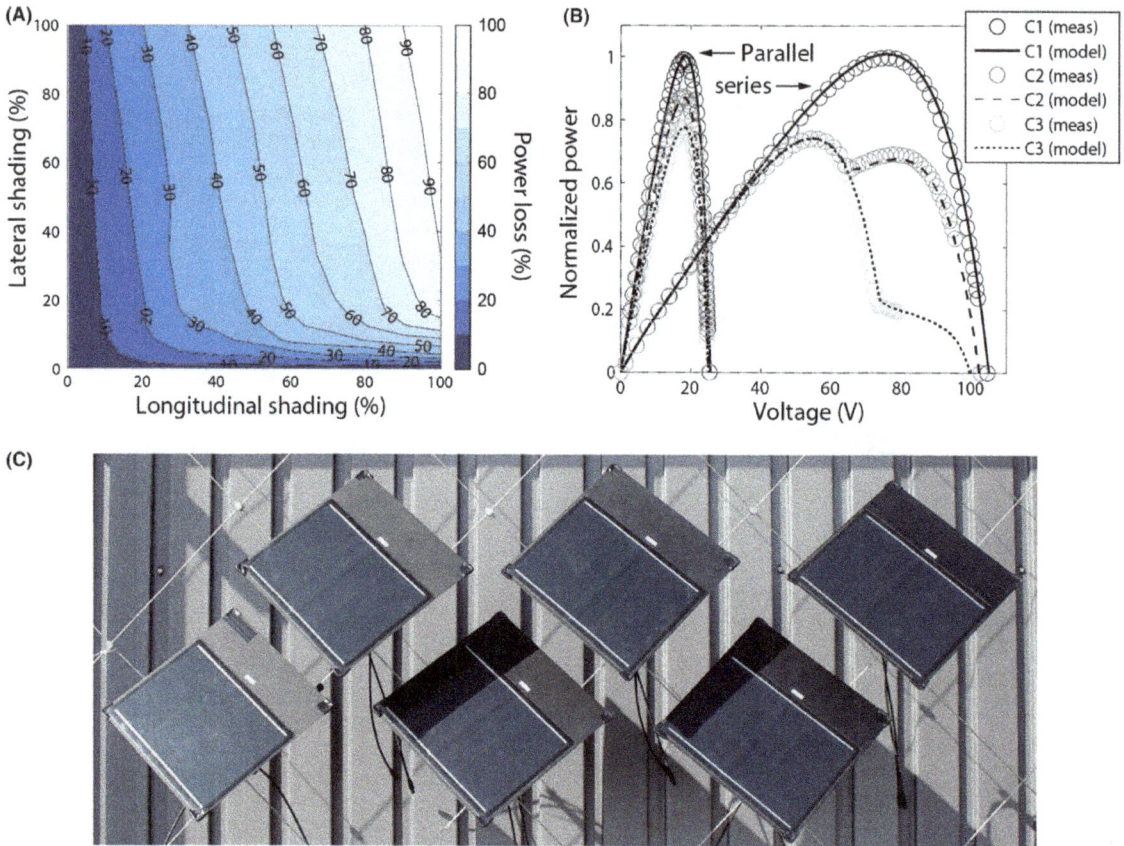

Figure 14. (A) Measured power loss of a single module as a function of longitudinal and lateral shading. (B) Normalized measured and modeled P-V curves of four modules in parallel and series connection for different shading conditions: C1 – no module shaded, C2 – one module shaded 100% in lateral and 50% in longitudinal direction, C3 – one module shaded 50% in lateral and 100% in longitudinal direction. (C) Thin-film copper indium gallium selenide module prototypes installed in diamond pattern on cable-net structure. The spacing between modules is increasing from left to right, leading to asymmetric mutual shading. Each module consists of three submodules connected with a bus bar in parallel.

Experimental Implementations

Measurement of I-V characteristics in partially shaded conditions

To validate the electrical model, outdoor measurements have been performed using six CIGS module prototypes produced by Flisom AG (www.flisom.ch). Each module has a size of 40 by 40 cm with one external bypass diode integrated in the junction box. Currently only approximately two-third of the module area is covered by PV cells and each module consists of three submodules connected in parallel (Fig. 14C). The next generation of modules is planned to be entirely laminated with PV cells that span the full width of the module.

I-V curves have been measured for different shading conditions with a PVPM2540C measurement device from PV Engineering. I-V curves were obtained in controlled

shading experiments for single modules at clear sky. In these experiments, a module was mounted horizontally on the ground and an appropriate shading object moved in approximately 30 cm distance from the module to incrementally adjust the lateral and longitudinal shading fraction with a step size of 5–10%, depending on the influence on output power. The irradiance in the shaded and unshaded region was measured using an external reference cell and a pyranometer. For each shading condition, the MPP was inferred from the P-V characteristic curve. The measured power loss relative to the unshaded condition is shown Figure 14A. In order to compare these measurements to simulation results, we applied nonlinear least squares fitting of the model to the measured, unshaded module I-V curve to extract parameters relevant for the employed equivalent circuit [43]. Based on this and the assumed cell reverse characteristic, the subcell I-V curve in the shaded and unshaded region is modeled and the module I-V curve reconstructed. Note that it is not possible to directly relate the measurement results of Figure 14A to the simulation results shown in Figure 4E, because the underlying module characteristics are different. For comparison of measurement and simulation, the specific design of the measured module prototype as a parallel connection of three submodules has been considered in the model. The deviation of measured and simulated MPP is below 5% for most shading conditions and largest for regions with strong variation in the MPP, that is, at low lateral and high longitudinal shading. The percentage root mean square deviation between measured and simulated values is 4.8% and the mean absolute deviation 4.6%. This deviation may be caused by the inaccuracy in positioning the shading object and potential unknown mismatch between submodules.

In addition to single module measurements, I-V curves of arrays under different partial shading conditions and module interconnections were obtained. To compare measurements and model, the array I-V curve was reconstructed based on a fit of each unshaded module I-V curve, the specific shading pattern, and the interconnection of modules. As an example, Figure 14B shows measured and simulated normalized P-V curves of four modules in series and parallel for three different shading conditions. As expected, power loss in shaded conditions is lower for parallel-connected than series connected modules, however, currents and resistive losses are higher. For both module interconnections, the power loss with one module shaded 50% in lateral and 100% in longitudinal direction is higher than in the opposite case of 100% lateral and 50% longitudinal shading. For series connected modules, this has no influence on the MPP though. Multiple extrema of the P-V curve are caused by bypass diodes, which limit the power loss of partially shaded strings.

Figure 14C shows CIGS module prototypes mounted in diamond pattern configuration on a cable-net structure. Each module is equipped with a novel soft robotic actuator, which allows the module to rotate along two axes based on pneumatic inflation [56]. This can be used for solar tracking and daylight control. Measurements of the PV modules in this configuration are currently performed.

Building integration

The framework presented in this study is currently being applied in a living lab experiment at the ETH House of Natural Resources (HoNR, www.honr.ethz.ch), in which an ASF (Adaptive Solar Façade) is designed, constructed, and operated [2, 4]. The design of the ASF is close to the diamond pattern façade discussed in this study. The HoNR has been inaugurated in June 2015 and the ASF installation is shown in Figure 15. The ASF consists of 50 modules on a modular frame placed in front of a south facing, two-person office, in order to evaluate its thermal and electric performance in an office environment. For reference, the adjacent office with a conventional shading system is monitored. Continuous monitoring will allow us to study the interconnection between power generation, solar intake, user preferences, and heating and ventilation systems energy demand.

Discussion and Conclusions

In this work we presented a parametric modeling framework of solar insolation and electric energy yield for building integrated dynamic PV systems. The methodology is applied to evaluate the performance of different façade-mounted PV shading system configurations with one and two-axis solar tracking. We furthermore study the influence of façade orientation and module distance.

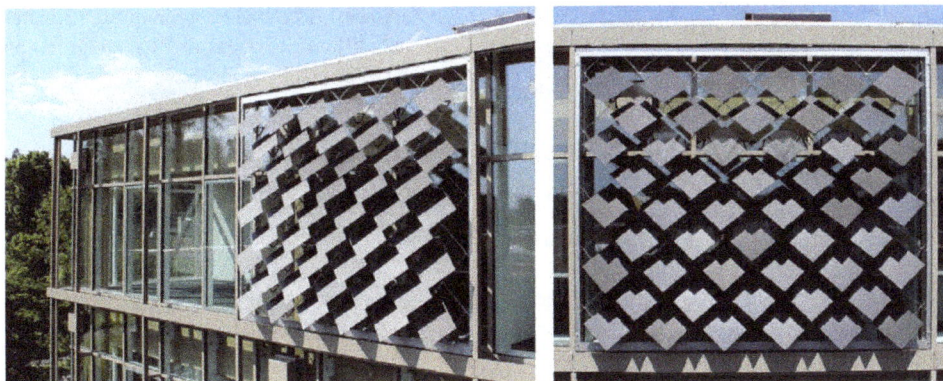

Figure 15. Adaptive Solar Façade installed at ETH House of Natural Resources. The right picture shows the shadow pattern on the modules for highest sun position, which can be compared to the calculated shadow pattern in Figure 8.

The analysis shows, that there is a trade-off between tracking performance, in terms of the amount of incident solar radiation perpendicular to module surface, and mutual shading between modules, in particular for closely positioned modules. The orientation of the tracking axis strongly influences timing and spatial distribution of shading patterns on the modules, which has immediate consequences on the optimal electrical configuration. The distance between modules is a critical parameter influencing the amount of mutual shading and hence limiting solar irradiation on shading modules using solar tracking. For a certain module spacing, highest solar insolation per module area is reached for two-axis tracking, followed by one-axis tracking, and finally fixed systems. The difference between the systems is negligible at very small distances, but increases for larger module distances. The relative performance between horizontal and vertical louver façade tracking systems depends on the season, façade orientation, and geographical location.

With regard to PV electricity generation, mutual shading between PV modules reduces the conversion efficiency from incident solar radiation to electrical energy. At narrow module spacing, the highest electric yield on a south facing façade is reached by horizontal louver systems. Only at wider module spacing with less shading, the full potential of the two-axis tracking system can be exploited. The methodology we have developed offers high spatial resolution, which helps to minimize electrical mismatch losses in case of strongly shaded conditions. We show that careful planning of module string configuration, PV cell orientation, and location of bypass diodes reduces electrical mismatch losses induced by partial shading and can result in more than 50% higher energy yield compared to uninformed design strategies. Knowledge of the precise shading pattern and electrical performance is furthermore useful to evaluate potential benefits of power electronic converters using advanced MPPT [30, 53–55] and for the development of thin-film PV cell geometries with improved shading tolerance [41].

The analysis presented in this study focuses on solar insolation and electric energy yield of different BIPV shading system configurations. However, these systems will be used as adaptive shading elements in exchange with users and the building climate and energy system. Analysis taking into account building lighting, heating, and cooling energy demand in addition to PV electricity production will be performed for a complete assessment of the different systems. In this regard, it is possible to couple dynamic shading system simulations in Rhinoceros/Grasshopper to building energy modeling in EnergyPlus [57]. Previous work of dynamic shading systems has shown that specific module control sequences minimize building energy demand for heating, cooling, and lighting [2].

While solar tracking achieves good glare protection and high insolation on PV modules, it does not take into account energy required for heating, cooling, and lighting. In future work, solutions balancing PV electricity production and building energy demand will be investigated.

Single and dual-axis tracking systems have the potential to significantly enhance power generation of PV modules [49–51]. While solar tracking is adequate to determine the potential of systems with large module distances, there may be other control strategies that are more suitable for panel configurations with smaller module distances and high mutual shading. The methodology presented in this study will help us to develop such advanced control methods minimizing power losses. One control strategy used for tracking PV plants is the so-called back-tracking method, for which shading is completely avoided and losses due to small angle of incidence are minimized. Solutions of back-tracking for one and two-axis tracking systems exist [10, 11], however, the parametric 3D modeling framework developed in this work may help to find solutions for complex system geometries. In addition to back-tracking, strategies with motion control of individual modules (or module clusters) and their impact on electricity generation can be evaluated. Further integration of 3D parametric, PV electric, and building system simulation will enhance integrated design and open new optimization possibilities.

The work presented in this study is currently applied in the context of the ASF project. The developed methodology is used to investigate the performance of different geometrical arrangements, advanced module control, and PV system electrical design.

Acknowledgments

This research has been financially supported by CTI within the SCCER FEEB&D (CTI.2014.0119) and by the Building Technologies Accelerator program of Climate-KIC. We gratefully acknowledge Flisom AG for provision of high-efficiency CIGS PV modules.

Conflict of Interest

None declared.

References

1. Perez, M. J., V. Fthenakis, H. C. Kim, and A. O. Pereira. 2012. Façade–integrated photovoltaics: a life cycle and performance assessment case study. Prog. Photovoltaics Res. Appl. 20:975–990.
2. Jayathissa, P., Z. Nagy, N. Offedu, and A. Schlueter. 2015. Numerical Simulation of Energy Performance, and

Construction of the Adaptive Solar Façade. *Proceedings of Advanced Building Skin Conference.*

3. Velasco, R., A. P. Brakke, and D. Chavarro. 2015. Computer-aided architectural design futures. The next city-new technologies and the future of the built environment. Springer, Berlin Heidelberg, 172–191.

4. Rossi, D., Z. Nagy, and A. Schlueter. 2012. Adaptive distributed robotics for environmental performance, occupant comfort and architectural expression. Int. J. Arch. Comp. 10:341–360.

5. Yoo, S. H., and E. T. Lee. 2002. Efficiency characteristic of building integrated photovoltaics as a shading device. Build. Environ. 37:615–623.

6. Yoo, S. H., and H. Manz. 2011. Available remodeling simulation for a BIPV as a shading device. Sol. Energy Mater. Sol. Cells 95:394–397.

7. Mandalaki, M., K. Zervas, T. Tsoutsos, and A. Vazakas. 2012. Assessment of fixed shading devices with integrated PV for efficient energy use. Sol. Energy 86:2561–2575.

8. Roche, D., H. Outhred, and R. J. Kaye. 1995. Analysis and control of mismatch power loss in photovoltaic arrays. Prog. Photovoltaics Res. Appl. 3:115–127.

9. D'Alessandro, V., A. Magnani, L. Codecasa, F. Di Napoli, P. Guerriero, and S. Daliento. 2015. Dynamic electrothermal simulation of photovoltaic plants. *International Conference on Clean Electrical Power (ICCEP),* 682–688.

10. Narvarte, L., and E. Lorenzo. 2008. Tracking and ground cover ratio. Prog. Photovoltaics Res. Appl. 16:703–714.

11. Lorenzo, E., L. Narvarte, and J. Munoz. 2011. Tracking and back-tracking. Prog. Photovoltaics Res. Appl. 19:747–753.

12. Castellano, N. N., J. A. G. Parra, J. Valls-Guirado, and F. Manzano-Agugliaro. 2015. Optimal displacement of photovoltaic array's rows using a novel shading model. Appl. Energy 144:1–9.

13. Diaz-Dorado, E., A. Suarez-Garcia, C. J. Carrillo, and J. Cidras. 2011. Optimal distribution for photovoltaic solar trackers to minimize power losses caused by shadows. Renewable Energy 36:1826–1835.

14. Mermoud, A. 2012. Optimization of Row-Arrangement in PV Systems, Shading Loss Evaluations According to Module Positioning and Connexions. *Proceedings of the 27th European Photovoltaic Solar Energy Conference.*

15. Tian, H., F. Mancilla-David, K. Ellis, E. Muljadi, and P. Jenkins. 2013. Determination of the optimal configuration for a photovoltaic array depending on the shading condition. Sol. Energy 95:1–12.

16. La Manna, D., V. Li Vigni, E. R. Sanseverino, V. Di Dio, and P. Romano. 2014. Reconfigurable electrical interconnection strategies for photovoltaic arrays: a review. Renew. Sustain. Energy Rev. 33:412–426.

17. Alonso-Garcia, M. C., J. M. Ruiz, and W. Herrmann. 2006. Computer simulation of shading effects in photovoltaic arrays. Renew. Energy 31:1986–1993.

18. Dongaonkar, S., C. Deline, and M. A. Alam. 2013. Performance and reliability implications of two-dimensional shading in monolithic thin-film photovoltaic modules. IEEE J. Photovol. 3:1367–1375.

19. Bishop, J. W. 1988. Computer simulation of the effects of electrical mismatches in photovoltaic cell interconnection circuits. Solar Cells 25:73–89.

20. Deline, C., A. Dobos, S. Janzou, J. Meydbray, and M. Donovan. 2013. A simplified model of uniform shading in large photovoltaic arrays. Sol. Energy 96:274–282.

21. Patel, H., and V. Agarwal. 2008. MATLAB-based modeling to study the effects of partial shading on PV array characteristics. IEEE Trans. Energy Convers. 23:302–310.

22. Karatepe, E., M. Boztepe, and M. Colak. 2007. Development of a suitable model for characterizing photovoltaic arrays with shaded solar cells. Sol. Energy 81:977–992.

23. Maeki, A., S. Valkealahti, and J. Leppaaho. 2012. Operation of series-connected silicon-based photovoltaic modules under partial shading conditions. Prog. Photovol. 20:298–309.

24. Celik, B., E. Karatepe, N. Gokmen, and S. Silvestre. 2013. A virtual reality study of surrounding obstacles on BIPV systems for estimation of long-term performance of partially shaded PV arrays. Renewable Energy 60:402–414.

25. Tsai, H.-L. 2010. Insolation-oriented model of photovoltaic module using matlab/simulink. Sol. Energy 84:1318–1326.

26. Ishaque, K., Z. Salam, and Syafaruddin. 2011. A comprehensive MATLAB simulink PV system simulator with partial shading capability based on two-diode model. Sol. Energy 85:2217–2227.

27. Quaschning, V., and R. Hanitsch. 1996. Numerical simulation of current-voltage characteristics of photovoltaic systems with shaded solar cells. Sol. Energy 56:513–520.

28. d'Alessandro, V., F., Di Napoli, P., Guerriero, and S. Daliento, et al. 2015. An automated high-granularity tool for a fast evaluation of the yield of PV plants accounting for shading effects. Renew. Energy 83:294–304.

29. Capdevila, H., A. Marola, and M. Herrerias. 2013. High resolution shading modeling and performance simulation of sun-tracking photovoltaic systems. *Proceedings of the 9th Int. Conf. on Concentrator Photovoltaic Systems.*

30. Poshtkouhi, S., V. Palaniappan, M. Fard, and O. Trescases. 2012. A General approach for quantifying the benefit of distributed power electronics for fine grained MPPT in photovoltaic applications using 3-D modeling. IEEE Trans. Power Electron. 27:4656–4666.

31. www.rhino3d.com (accessed 30 June 2015).

32. www.grasshopper3d.com (accessed 30 June 2015).

33. Häberlin, H. 2012. Photovoltaics system design and practice. John Wiley & Sons, Hoboken, New Jersey, USA.

34. Goswami, D. Y., F. Kreith, and J. F. Kreider. 2015. Principles of solar engineering, 3rd ed. Taylor and Francis, CRC Press, Boca Raton, Florida, USA.

35. Jakubiec, J. A., and C. F. Reinhart. 2011. DIVA 2.0: Integrating daylight and thermal simulations using Rhinoceros 3D, Daysim and EnergyPlus. *Proceedings of the 12th Conference of International Building Performance Simulation Association.*

36. Duffie, J. A., and W. A. Beckman. 2006. Solar engineering of thermal processes (Vol. 3). Wiley, New York, NY.

37. Salomon, D.. 2011. The computer graphics manual: oblique projections. Springer Science & Business Media, London, England.

38. www.luxrender.net (accessed 30 June 2015).

39. Pharr, M., and G. Humphreys. 2010. Physically based rendering: from theory to implementation. Morgan Kaufmann, Burlington, Massachusetts, USA.

40. www.meteonorm.com (accessed 30 June 2015).

41. Dongaonkar, S., and M. A. Alam. 2015. Geometrical design of thin film photovoltaic modules for improved shade tolerance and performance. Prog. Photovoltaics Res. Appl. 23:170–181. doi:10.1002/pip.2410.

42. Koishiyev, G. T., and J. R. Sites. 2009. Impact of sheet resistance on 2-D modeling of thin-film solar cells. Sol. Energy Mater. Sol. Cells 93:350–354.

43. Mermoud, A., and T. Lejeune. 2010. Performance assessment of a simulation model for PV modules of any available technology. *Proc. of the 25th European Photovoltaic Solar Energy Conference.*

44. Product datasheet Shell Solar ST40. Available at http://www.gehrlicher.com/fileadmin/content/pdfs/de/produktarchiv/Shell_ST40.pdf (accessed 30 June 2015).

45. www.pvsyst.com (accessed 30 June 2015).

46. Spirito, P., and V. Abergamo. 1982. Reverse bias power dissipation of shadowed or faulty cells in different array configurations, *Proc. of the 25th European Photovoltaic Solar Energy Conference*, 296–300.

47. Alonso-Garcia, M. C., and J. M. Ruiz. 2006. Analysis and modelling the reverse characteristic of photovoltaic cells. Sol. Energy Mater. Sol. Cells 90:1105–1120.

48. Mack, P., T. Walter, R. Kniese, D. Hariskos, and R. Schäffler. 2008. Reverse Bias and Reverse Currents in CIGS Thin Film Solar Cells and Modules, *Proc. of the 23rd European Photovoltaic Solar Energy Conference.*

49. Huld, T., M. Suri, and E. D. Dunlop. 2008. Comparison of potential solar electricity output from fixed-inclined and two-axis tracking photovoltaic modules in Europe. Prog. Photovoltaics Res. Appl. 16:47–59.

50. Koussa, M., A. Cheknane, S. Hadji, M. Haddadi, and S. Noureddine. 2011. Measured and modelled improvement in solar energy yield from flat plate photovoltaic systems utilizing different tracking systems and under a range of environmental conditions. Appl. Energy 88:1756–1771.

51. Drury, E., A. Lopez, P. Denholm, and R. Margolis. 2014. Relative performance of tracking versus fixed tilt photovoltaic systems in the USA. Prog. Photovoltaics Res. Appl. 22:1302–1315.

52. Perpinan, O. 2012. Cost of energy and mutual shadows in a two-axis tracking PV system. Renewable Energy 43:331–342.

53. Garcia, M., J. M. Maruri, L. Marroyo, E. Lorenzo, and M. Perez. 2008. Partial shadowing, MPPT performance and inverter configurations: observations at tracking PV plants. Prog. Photovoltaics Res. Appl. 16:529–536.

54. Sarvi, M., S. Ahmadi, and S. Abdi. 2015. A PSO-based maximum power point tracking for photovoltaic systems under environmental and partially shaded conditions. Prog. Photovoltaics Res. Appl. 23:201–214.

55. Ishaque, K., and Z. Salam. 2013. A review of maximum power point tracking techniques of PV system for uniform insolation and partial shading condition. Renew. Sustain. Energy Rev. 19:475–488.

56. Svetozarevic, B., Z. Nagy, D. Rossi, and A. Schlueter. 2014. Experimental Characterization of a 2-DOF Soft Robotic Platform for Architectural Applications. *Proceedings of the Robotics Science and Systems Conference*, UC Berkeley, California.

57. Roudsari, M., M. Pak, and A. Smith. 2014. Ladybug: A Parametric Environmental Plugin for Grasshopper to Help Designers Create an Environmentally-Conscious Design. *Proc. of Int. Conf. of IBPSA*, Chambery, France.

9

Irradiance adjustment system developed for various types of solar cells and illumination conditions

Yoshihiro Nishikawa[1], Kiyoshi Imai[1], Keiji Miyao[1], Satoshi Uchida[2], Daisuke Aoki[3], Hidenori Saito[3], Shinichi Magaino[3] & Katsuhiko Takagi[3]

[1]KONICA MINOLTA, INC., 3-91 Daisen Nishi-machi, Sakai-ku, Sakai City, Osaka 590-8551, Japan
[2]The University of Tokyo, 4-6-1 Komaba, Meguro-ku, Tokyo 153-8904, Japan
[3]Kanagawa Academy of Science and Technology, 3-2-1 Sakado, Takatsu-ku, Kawasaki-shi, Kanagawa 213-0012, Japan

Keywords

DSR method, I–V characterizations, organic solar cell, reference cell, spectral responsivity, spectroradiometer

Correspondence

Yoshihiro Nishikawa, KONICA MINOLTA, INC. Osaka 580-8551, Japan.
E-mail:
yoshihiro.nishikawa@konicaminolta.com
Shinichi Magaino, Kanagawa Academy of Science & Technology (KAST), Kanagawa 213-0012, Japan.
E-mail:
magaino@newkast.or.jp

Funding Information

The authors express their sincere appreciation to the Japan Science and Technology Agency (JST) Development of Advanced Measurement and Analysis Systems funding program for their financial support.

Abstract

A new irradiance adjustment system for the light source named "Programmable Reference cell system for Irradiance adjustment by Spectral Measurement (PRISM)" has been developed for photovoltaics evaluation. It can be applied for various kinds of solar cells and illumination conditions without the preparation of conventional reference cells. This system consists of a device for absolute spectral responsivity measurements, that is, the spectral responsivity unit, and a device for irradiance adjustment of the light source, that is, the spectral reference cell. First, the short circuit current of the target solar cell (I_{ref}) is calculated by convolution of the absolute spectral responsivity with the spectral irradiance under STC (AM1.5 G, 1 kW/m^2, 25°C) as defined in IEC 60904-3. The spectral irradiance of the light source is then adjusted for I–V measurements so that the short-circuit current of the spectral reference cell (I_{cal}), calculated by convolution of the absolute spectral responsivity with the spectral irradiance of the light source, agrees with I_{ref}. The difference between I_{ref} and the short circuit current, as determined at an internationally recognized test center under STC, was within ±2% for crystalline silicon solar cells and ±3% for organic solar cells. The difference between I_{cal} and the short-circuit current measured under a solar simulator or LED in which the irradiance was adjusted by this system was within ±4% when the irradiance was higher than 4 W/m^2.

Introduction

Crystalline silicon solar cells are used in the largest quantities for solar energy systems, however, their high manufacturing cost and heavy weight prevent the widespread use of solar energy. To address this problem, various types of solar cells, for example, compound semiconductors, organic thin films, dye-sensitized, and perovskite [1] solar cells are under investigation by various groups with some of them being already commercialized.

Further development of materials and cell structure as well as appropriate evaluation methods for cell functions will be required to gauge and improve the overall performance of solar cells. International standards for the evaluation of silicon solar cells have been established by the IEC as the 60904 Series, however, they define only test methods under standard test conditions (STC) although, when compared to crystalline silicon-based cells, higher performance can be expected for some types of solar cells under weak irradiation such as faint morning, evening, indoor or oblique light [2]. Moreover, new types of solar cells sometimes have characteristics different from that of silicon solar cells such as nonlinearity between the current and light intensity, a light soaking effect, and slow response time, which are not taken into consideration in the IEC 60904 standards [3, 4].

The reference cell method is widely used for irradiance adjustment of a solar simulator [5] and this method requires the preparation of a reference cell as defined in IEC 60904-2, 4 [6, 7]. However, preparation of a reference cell with a spectral response that matches the solar cell being tested is difficult and extremely time-consuming [8, 9], especially, for such unstable cells as dye-sensitized solar cells (DSCs). It is necessary to prepare a filtered reference cell with its relative spectral responsivity coincident with that of the organic PV to be tested, which should then be calibrated at a recognized institute. However, this may take several months to 1 year since the filtered reference cell must also be checked for long-term stability. In this work, we have developed a new irradiance adjustment system for the light source named, "Programmable Reference cell system for Irradiance adjustment by Spectral Measurement (PRISM)" which can be applied for any type of solar cell under various illumination conditions without the use of conventional reference cells [10–19]. In this work, the performance of this system has been investigated for various solar cells and illumination conditions.

Principle

Basic procedures

The PRISM system consists of a device for absolute spectral responsivity measurements (spectral responsivity unit: SK-1150) which meets IEC 60904-4, 8 standards [7, 8] and a device for light source irradiance adjustment (spec-

tral reference cell: SRC-1100), as shown in Figure 1. The basic procedures for measurements at STC (AM1.5G, 1 kW/m^2, 25°C) are as follows:

1 Spectral responsivity of the target solar cell ($S(\lambda)$) is measured with the SK-1150 unit.
2 The short-circuit current of the target solar cell (I_{ref}) is calculated by convolution of the absolute spectral responsivity $S(\lambda)$ with the spectral irradiance at STC (AM1.5G, 1 kW/m^2, 25°C) as defined in IEC 60904-3 ($E_{AM1.5G}(\lambda)$) [9], using the SRC-1100 reference cell device, as follows:

$$I_{ref} = \int S(\lambda) \cdot E_{AM1.5G}(\lambda)\, d\lambda \qquad (1)$$

3 The spectral irradiance of the solar simulator ($E_{ss}(\lambda)$) is adjusted until the short-circuit current of SRC-1100 (I_{cal}), calculated from equation (2), agrees with I_{ref}.

$$I_{cal} = \int S(\lambda) \cdot E_{SS}(\lambda)\, d\lambda \qquad (2)$$

4 Current-voltage characteristics (I–V curves) of the target cell are measured by a solar simulator, the irradiance of which is adjusted through the above procedures.

Measurement of nonlinear cells

The short-circuit current of some types of organic solar cells is not always proportional to the irradiance. In this case, equation (1) should be modified to

$$I_{ref} = \int S(\lambda, E_b) \cdot E_{AM1.5G}(\lambda)\, d\lambda, \qquad (3)$$

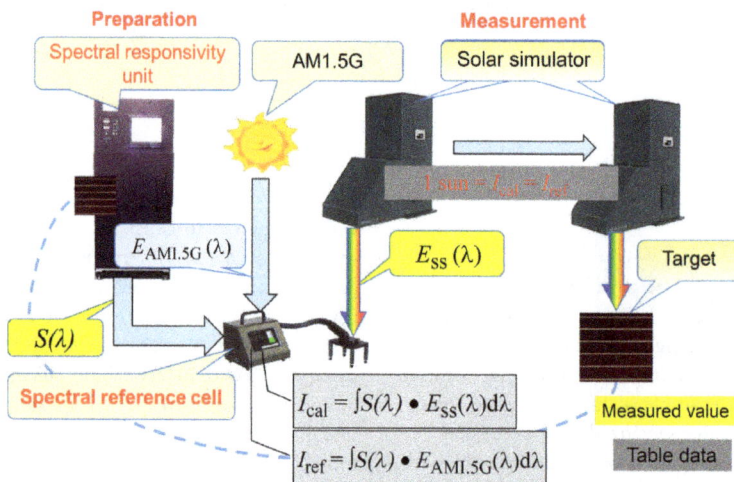

Figure 1. Schematic diagram of the PRISM system. PRISM, Programmable Reference cell system for Irradiance adjustment by Spectral Measurement.

where $S(\lambda, E_b)$ indicates the spectral responsivity which may vary with the irradiance ($E_b = \int E_{ss}(\lambda) \, d\lambda$). However, it is difficult to measure $S(\lambda, E_b)$ by commonly available devices since they are not able to regulate the irradiance level to desired values. To address this problem, we have developed the PRISM system which uses the differential spectral responsivity (DSR) method developed by Physikalisch-Technische Bundesanstalt (PTB) and defined in IEC 60904-4, 8 and ISO 15387 [7, 8, 20, 21].

Measurements under various illumination conditions

The PRISM system employing the DSR method can be applied for measurements under any illumination conditions. Thus, the short-circuit current (I_{sc}) for any solar cell under any illumination conditions can be calculated as follows:

$$I_{sc} = \int S(\lambda, E_b) \cdot E(\lambda) \, d\lambda, \qquad (4)$$

where $E(\lambda)$ is the spectral irradiation from any light source.

Key Devices for the PRISM System

Spectral responsivity unit (SK-1150)

The SK-1150 unit has been constructed to simultaneously illuminate the bias light irradiance (E_b) and monochromatic light irradiance onto a solar cell. Both lights are combined in the lens and irradiated on the solar cell. The irradiation unit consists of a halogen lamp, Xenon lamp, grating and chopping units, and light monitors to compensate for any instability of the lamps. The current measurement unit consists of a source meter and lock-in amplifier, both of which are controlled by a computer.

The absolute spectral responsivity of this device is measured using a standard detector calibrated at an internationally recognized test center. Differential spectral responsivity is measured at various irradiances of bias light and $S(\lambda, E_b)$ is then automatically calculated by a computer [7, 8, 20, 21].

When the current response to the applied light irradiation is slow, for example, in the case of dye-sensitized solar cells (DSCs), the chopping frequency of the monochromatic light should be low enough to obtain a steady-state current. Aoki et al. have reported that $S(\lambda, E_b)$ decreases as the chopping frequency is increased [3], indicating that it should be lower than 3.3 Hz to obtain accurate data under 1 sun bias light irradiation. However, spectral responsivity measurement using the lock-in amplifier is unsuitable for low-frequency measurements since long and repetitive

measurements are required to obtain accurate data, for example, several hours at 0.3 Hz.

To address this problem, this system has two methods to determine $S(\lambda, E_b)$, that is, the selective amplification of the current caused by monochromatic light irradiation using the lock-in amplifier and the calculation of the difference between currents measured when the shutter of the monochromatic light is open and closed using the DC current amplifier. The latter method is similar to the AC-a method reported by Guo et al. [22], and requires less measuring time because the lock-in amplifier method takes a chopping frequency of 10 cycles.

Spectral reference cell device (SRC-1100)

The SRC-1100 device consists of a spectroradiometer for spectral irradiance measurement of the light source, an optical receiver and computer. With this device, irradiance adjustment of the solar simulator can be performed according to the procedures described in the Basic procedures section.

The spectroradiometer equipped in this device has a high sensitivity image sensor of back-thinned CCD to measure very low irradiance, such as indoor light, as compared to AM 1.5 G (1 kW/m^2). Additionally, it has a unique optical system to reduce stray light which causes inaccurate readings. Figure 2 shows the transmittance level of the stray light measured in accordance with JIS Z8724 (Japan Industrial Standards, Methods of measurement Light-source colour) [23].

The irradiance measured by the spectroradiometer should have a linear relationship with the short-circuit current of the linear cell under a wide irradiation range, for example, from indoor light to 1 sun. The short-circuit current (I_{sc}) and linearity error (ε) were determined for the linear cell (Konica Minolta, AK-200) by the spectroradiometer and were plotted against the *Irradiance* ($\int E_{ss}(\lambda) d\lambda$), as shown in Figure 3 (Konica Minolta is hereafter referred to as KM). The ε was calculated as follows:

$$\varepsilon = (I_{sc}/I_{mes} - 1) \times 100, \qquad (5)$$

where

$$I_{sc} = \int S(\lambda) \cdot E_{ss}(\lambda) \, d\lambda.$$

The linearity error (ε) was within $\pm 0.2\%$ at an irradiation range of 10–1000 W/m^2, as shown in Figure 3.

The sensitivity of a spectroradiometer is known to change with time [24] by vibration and degradation of its optical components, etc., while instrument-to-instrument variations also exist [25, 26]. No change in the relative spectral responsivity of the standard lamp was measured, but about a 5% decrease in irradiance was measured by the spectroradiometer in 100 days.

The SRC-1100 is first calibrated at the manufacturing factory using a standard lamp traceable to that of a recognized national institute; however, as mentioned above, the measured value changes with time. To reduce these effects, SRC-1100 has a user calibration function to resolve changes before measurements. This apparatus has a stable reference cell (KM, AK-200) with absolute spectral responsivity ($S_{rc}(\lambda)$) traceable to the calibrated value of a national institute. First, the reference cell is placed under the light source and the short-circuit current (I_{mes_rc}) is measured. Next, the irradiance ($E_{ss}(\lambda)$) is measured and the absolute irradiance calculated by multiplying ($E_{ss}(\lambda)$) by a correcting factor ($K = I_{mes\_rc}/E_{ss}(\lambda) \cdot S_{rc}(\lambda) d \lambda$). If the sensitivity of SRC-1100 shows no change with time, K equals 1.

Intercomparison of Short Circuit Currents Among Calibration Centers

Intercomparisons of the short circuit currents were carried out at three internationally recognized calibration centers, that is, the National Institute of Advanced Industrial Science & Technology (AIST), National Renewable Energy Laboratory (NREL), and PTB [27] for multiple

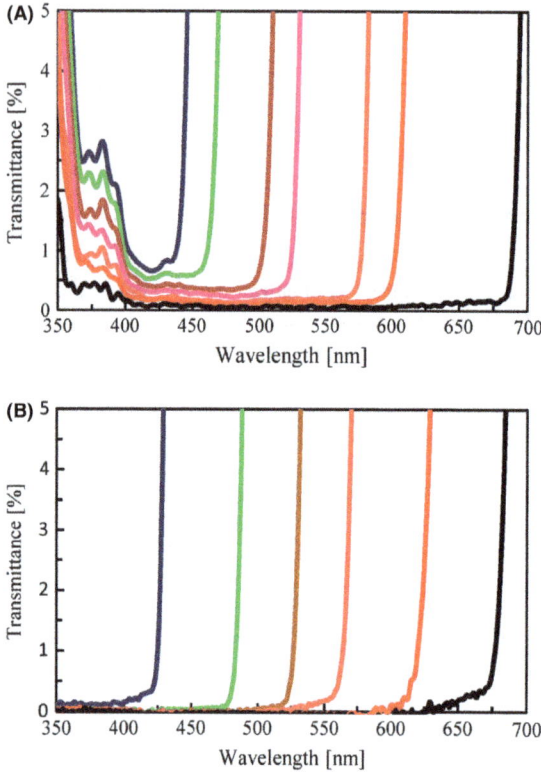

Figure 2. Transmittance level of stray light determined for: (A) a conventional spectroradiometer; and (B) the spectroradiometer equipped in this device.

Table 1. Types of reference cells.

Reference cells	Type	Centroid wavelength
AK-100	Amorphous Si-type	521 nm
AK-110	Microcrystalline Si-type	791 nm
AK-120	Top layer of triple junction	499 nm
AK-130	Middle layer of triple junction	664 nm
AK-140	Bottom layer of triple junction	831 nm
AK-200	Crystalline Si-type	772 nm
AK-300	DSC N719 type	553 nm

DSC, dye-sensitized solar cells.

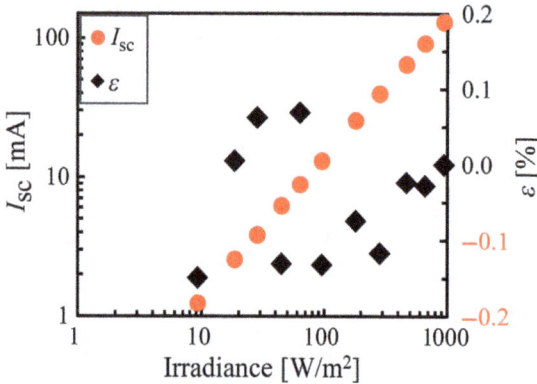

Figure 3. Short-circuit current (I_{sc}) and linearity error (ε) of SRC-1100 plotted against the irradiance. SRC, spectral reference cell

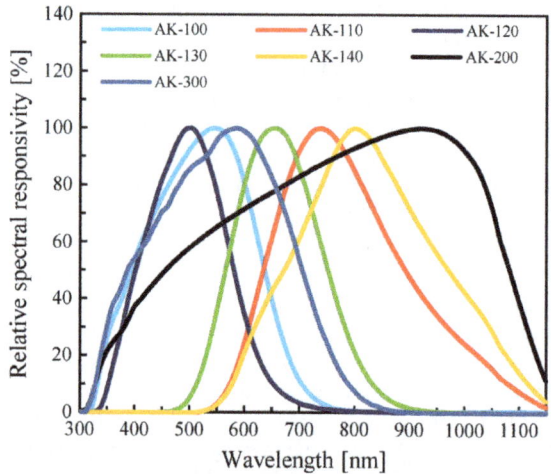

Figure 4. Relative spectral responsivities of the reference cells.

reference cells (KM, AK series) to investigate variations in their measured values. Table 1 and Figure 4 show the type of cells in the AK series and their relative spectral responsivities, respectively. Table 1 also summarizes the centroid wavelengths (W_c) of the reference cells for the AK series which were calculated by equation 6. Here, it should be noted that W_c is the representative wavelength of the spectral responsivity at STC.

$$W_c = \int S(\lambda) \cdot \lambda d\lambda / \int S(\lambda)\, d\lambda. \qquad (6)$$

The centroid wavelength can be employed as the index of each reference cell to indicate the representative spectral responsivity, thus, allowing comparisons of multiple reference cells. For example, the centroid wavelengths of AK-100 (Amorphous Si-type cell), AK-200 (Crystalline Si-type cell), and AK-110 (Microcrystalline Si-type cell) are 521 nm, 772 nm, and 791 nm, respectively. First, the short-circuit current of each cell was measured at KM, then the secondary reference solar cell was calibrated at AIST. The cells were then sent to PTB and NREL, and variations in the short circuit currents among the three recognized calibration centers were investigated. The difference in short-circuit current with the elapse of time was found to be within ±0.3% in a 2-year period from 2011 to 2013.

Figure 5 shows the differences in the short-circuit current among the test centers (ΔI_{sc_tc}) plotted against the centroid wavelength of the spectral responsivity (W_c) for each cell. The ΔI_{sc_tc} were calculated as follows:

$$\Delta I_{sc\_tc} = (I_{sc\_tc}/I_{sc\_km} - 1) \times 100, \qquad (7)$$

where I_{sc_tc} is the short-circuit current determined at PTB or NREL, and I_{sc_km} is that determined at KM with the secondary reference solar cell calibrated at AIST. Dirnberger et al. have reported intercomparison of the short-circuit current for crystal Si-type and amorphous Si-type modules performed by the Fraunhofer Institute for Solar

Energy Systems (ISE), European Solar Test Installation (ESTI of the European Commission, Joint Research Centre, Institute for Energy and Transport, Renewable Energy Unit), and the NREL and AIST [28]. They reported that deviations in the short-circuit current of the crystal Si-type module was 2% (−1.1 to 0.9%), but that of the amorphous Si-type module was −2.9% for the data measured at AIST and 1.6% for that measured at NREL. The value from NREL was found to be 4.5% higher than that of AIST. In this paper, deviation of the short-circuit current was 0.9% for crystalline Si-type cells and 2.8% (0–2.8%) for amorphous Si-type cells. Thus, the value from NREL was 2.8% higher than that of AIST. The tendency for greater deviation of the short-circuit current for amorphous Si-type cells than crystalline Si-type cells was also recognized in this paper. As shown in Figure 5, the ΔI_{sc_tc} values of cells having W_c in the long wavelength range were negative and those of cells having W_c in the short wavelength range were positive, although the reason for this tendency is not yet clear at present. Further investigation is required to analyze this tendency. On the other hand, intercomparison results of this paper were found to be reasonable since the total estimated uncertainty to determine the short-circuit current was 3.8%. The estimated uncertainty of a recognized calibration center was within 1% and within 1.8% at KM. Taking these results into consideration, the maximum deviation under STC for the PRISM system was set within 4%, equivalent to the error range allowance in the above results.

Experimental

Performance verification of SK-1150

The performance of the SK-1150 unit was verified by comparing the short circuit currents of the solar cells at STC with those determined at a recognized test center. That is, the difference in the short-circuit current (ΔI_{sc_stc}) between that determined by SK-1150 (I_{ref}) and

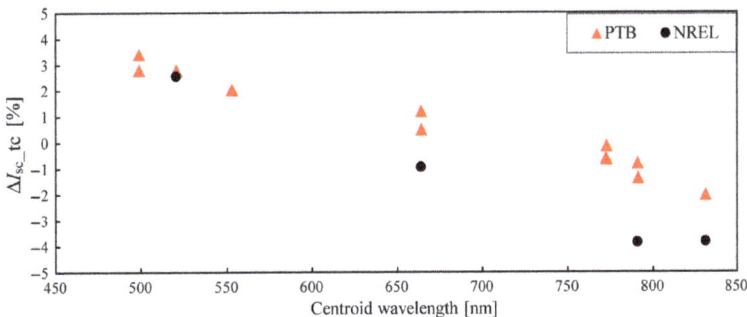

Figure 5. Difference in the short-circuit current among the calibration centers plotted against the centroid wavelength.

that determined at a test center (I_{ref_tc}) was calculated as follows:

$$\Delta I_{sc\_stc} = (I_{ref}/I_{ref\_tc} - 1) \times 100. \qquad (8)$$

Performance verification of SRC-1100

Performance of the SRC-1100 device was verified by comparing the short circuit currents calculated from equation (2) (I_{cal}) with those measured under a solar simulator (I_{mes}) or LED. The difference in the short-circuit current between I_{cal} and I_{mes} (ΔI_{sc}) was calculated as follows:

$$\Delta I_{sc} = (I_{cal}/I_{mes} - 1) \times 100. \qquad (9)$$

The three types of solar simulators and LED light source used in the evaluations are shown in Table 2. The solar simulators were all class AAA but with differing spectral irradiance distributions in order to examine the spectral irradiance dependence. The source meters were Agilent B2901A, ADCMT 6242, and Advantest R6246.

Spectral responsivity measurements of DSC

Two different types of dye-sensitized solar cells (DSCs) were tested: one cell containing 3-Methoxypropionitrile (MPN) as the solvent for the electrolyte (DSC-A) and the other containing an ionic liquid electrolyte (DSC-B). The viscosity of the ionic liquid electrolyte was about 30 times higher than that containing MPN at 20°C. A mesoscopic TiO_2 semiconductor electrode sensitized by cis-RuLL'-

(SCN)$_2$ (L = 2,2'-bipyridyl-4,4'-dicarboxylic acid, L' = 4, 4'-dinonyl-2,2'-bipyridyl) (Z907) dye was used for both cells. The active area for each cell was 0.196 cm^2 [3]. Additionally, DSCs supplied by the National Institute for Materials Science (DSC-NIMS-1 and DSC-NIMS-2) were also used. For these DSCs, a carbazole dye with hexyl-substituted oligothiophenes, MK-2 (Soken Chemical & Engineering Co., Ltd., Tokyo, Japan), was used as a sensitizer. An ionic liquid based on imidazolium iodide was used as the electrolyte. Transparent nanocrystalline TiO_2 films were prepared as follows: The films were formed by a screen printing method, that is, TiO_2 paste (Nanoxide-20N, Nikki Syokubai Kasei) was screen-printed onto a transparent conducting oxide (TCO) glass substrate (F-doped SnO_2, sheet resistance 10 ohm/cm^2, Nippon Sheet Glass), followed by sintering at 500°C for 60 min in air.

Before measurements, the cells were stabilized by light soaking at the same irradiance as that for the bias light. In the case of DSC-NIMS-1, the short-circuit current reached a steady-state value in about 96 h during light soaking at an irradiance of 1 sun. The cell was kept in the dark for about 100 h after 96 h light soaking, and then irradiated again under the same conditions. The short-circuit current reached a steady-state value immediately after irradiation.

The absolute spectral responsivity ($S(\lambda)$) was measured by SK-1150 at 25°C with bias light irradiance levels between 0 to 1.2 sun. Wavelength resolution was 10 nm and the chopping frequency of the monochro-

Table 2. Solar simulators and LED light source.

Type	Manufacturer	Irradiance level	Target cells
XES-40S	SAN-EI ELECTRIC	10–1000 W/m^2	Reference cells/Nonlinear cells/DSC-NIMS/OPV
94023A	NewPort	1000 W/m^2	Reference cells
YSS-T150A	YamashitaDenso	10–1000 W/m^2	DSC-A/DSC-B
LED	Imac	10–300 W/m^2	Reference cells/Nonlinear cells/DSC-NIMS/OPV

DSC-NIMS-1, dye-sensitized solar cells-National Institute for Materials Science; OPV, organic thin film solar cell.

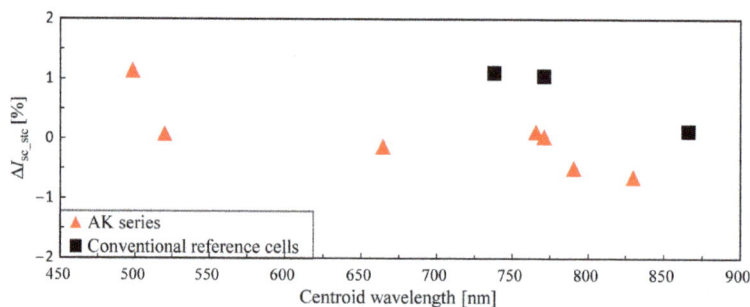

Figure 6. Difference in the short-circuit current (ΔI_{sc_stc}) between that determined by the PRISM SK-1150 unit and that determined at a calibration center for linear cells under STC plotted against the centroid wavelength of the spectral responsivity (W_c). PRISM, Programmable Reference cell system for Irradiance adjustment by Spectral Measurement; SK-1150, Spectral responsivity unit; STC, standard test conditions.

matic light was set low enough to obtain a steady-state current.

I–V measurements

Solar cell efficiencies are, at present, measured at STC, however, by adjusting the light irradiance using the PRISM method, I–V characteristics were measured under various irradiances with a solar simulator (San-ei Electric XES-40S) and LED light source (Imac). The voltage sweep rates were set so that an I–V curve determined by the forward scan (from short circuit to open circuit) agreed with

that by reverse scan. The solar cell efficiency (η) and fill factor (FF) were calculated as follows:

$$\eta = \frac{V_{oc} \times J_{sc} \times FF}{P_{rad}} \times 100, \tag{10}$$

$$FF = \frac{P_{max}}{V_{oc} \times J_{sc}}, \tag{11}$$

where V_{oc}, J_{sc}, P_{rad}, and P_{max} represent the open circuit voltage, short circuit current density, incident light intensity and maximum power, respectively.

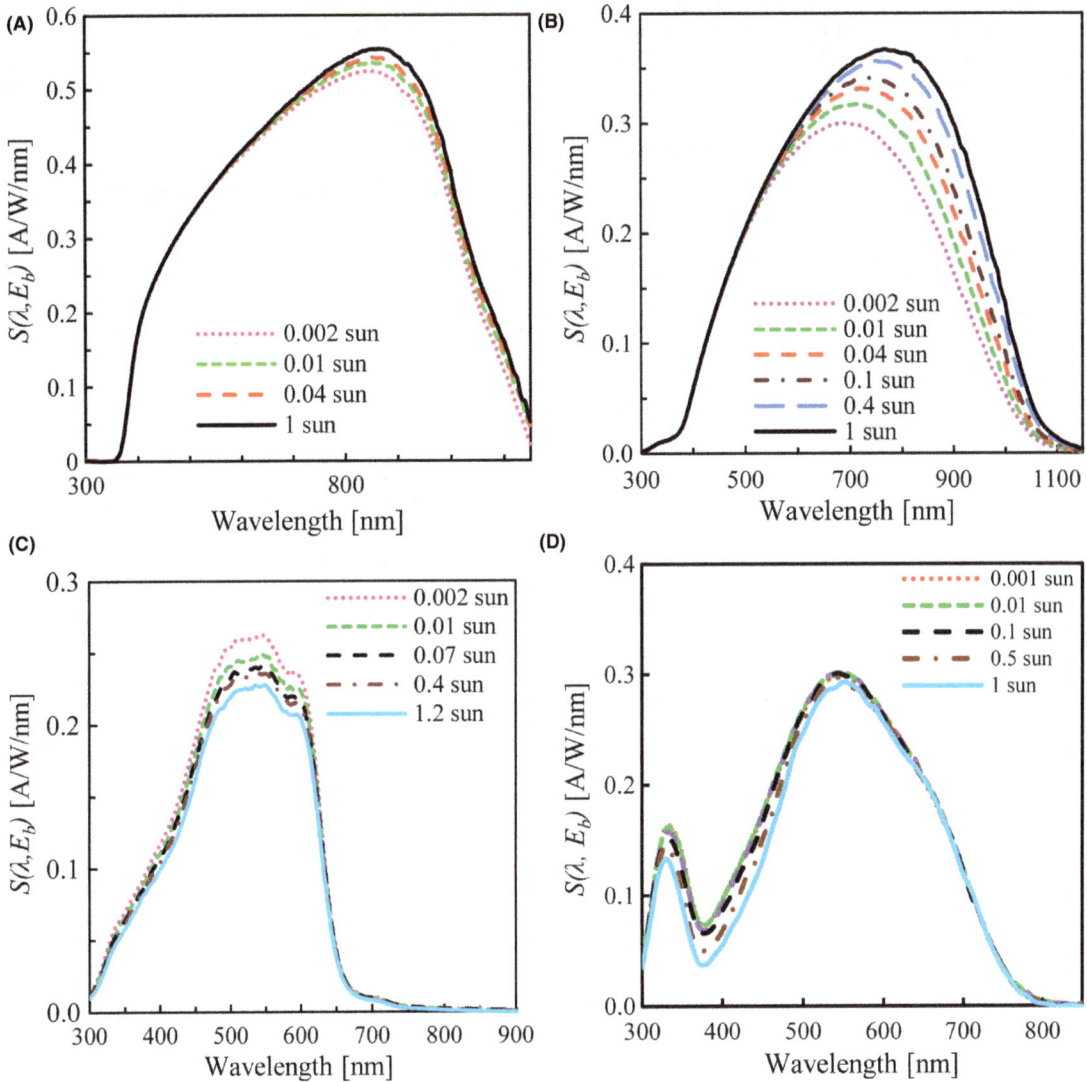

Figure 7. Spectral responsivity ($S(\lambda, E_b)$) determined for two nonlinear cells (cells A and B), OPV and DSC-B: (A) Cell A, (B) Cell B, (C) OPV, (D) DSC-B. DSC, dye-sensitized solar cells; OPV, organic thin film solar cell.

Results and Discussions

Performance of the SK-1150 unit

Figure 6 shows the differences in the short-circuit current (ΔI_{sc_stc}) between that determined by SK-1150 and that determined at a recognized calibration center for linear reference cells at STC plotted against the centroid wavelength (W_c). As shown in Figure 6, the ΔI_{sc_stc} was within $\pm 1.5\%$.

The spectral responsivity ($S(\lambda, E_b)$) determined for the two nonlinear cells (cells A and B), an organic thin film solar cell (OPV), and a dye-sensitized solar cell containing an ionic liquid electrolyte (DSC-B) are shown in Figure 7. The ΔI_{sc_stc} values determined for these cells are shown in Tables 3 and 4. Nonlinear cells are those in which the spectral responsivities change with the bias light intensity [8]. The $S(\lambda, E_b)$ of both cells increased with an increase in the bias light intensity in a wavelength range over

Table 3. ΔI_{sc_stc} determined for two nonlinear cells (Cell A and Cell B).

Cell	ΔI_{sc_stc} (%)
A	0.1
B	−0.3

Table 4. ΔI_{sc_stc} determined for OPV and DSC-B.

Cell	ΔI_{sc_stc} (%)
OPV	−1.1
DSC-B	−2.6

DSC, dye-sensitized solar cells; OPV, organic thin film solar cell.

500 nm, while the degree of increase was larger for cell B. The ΔI_{sc_stc} was 0.1% for cell A and −0.3% for cell B.

Unlike the nonlinear cells, the $S(\lambda, E_b)$ of the OPV and DSC-B decreased with an increase in the bias light intensity at wavelengths under 650 nm. Similar results have been reported for DSCs by Hara et al. [29], Guo et al. [22] and Aoki et al. [3]. Hara et al. have suggested that this decrease in the spectral responsivity may be related to light absorption in the wavelength range of 400–500 nm caused by I_3^- accumulation in the vicinity of the electrode due to oxidation of I^- and accelerated by bias light irradiation. Table 4 shows that ΔI_{sc_stc} was −1.1% for the OPV and −2.6% for DSC-B.

The maximum deviation under STC for the PRISM system was set within 4%, as mentioned before. The ΔI_{sc_stc} values determined for the linear and nonlinear reference cells, OPV, and DSC were found to be within this maximum deviation.

Performance of the SRC-1100 device

First, the differences in the short-circuit current (ΔI_{sc}) between I_{cal} and I_{mes} calculated from equation (9) for the linear reference cells under STC were determined, and the results are shown in Figure 8. Two sets of the SRC-1100 device with different solar simulators (San-ei Electric XES-40S and Newport Corp 94023A) were employed. As shown in Figure 8, ΔI_{sc} was within $\pm 3\%$, which is within the maximum deviation set for the PRISM system.

Next, the ΔI_{sc} of the nonlinear cells, OPV, and DSCs were determined by changing the irradiance. A San-ei Electric XES-40S solar simulator was used for the nonlinear cells and the OPV, and a Yamashita Denso YSS-T150A simulator was used for the DSCs. Figure 9 shows the ΔI_{sc} determined for these cells. The ΔI_{sc} of the nonlinear cells and the OPV was within $\pm 4\%$ and larger for cell B than cell A. The ΔI_{sc} of the DSCs was within $\pm 3\%$

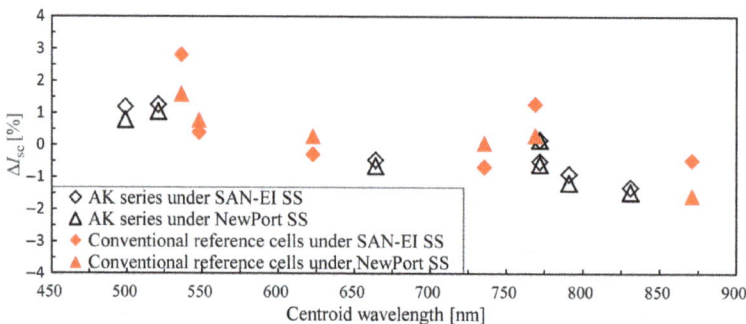

Figure 8. Difference in the short-circuit current (ΔI_{sc}) between I_{cal} and I_{mes} calculated from equation (9) for the reference cells (Konica Minolta AK series and conventional reference cells made by other companies) under the solar simulators.

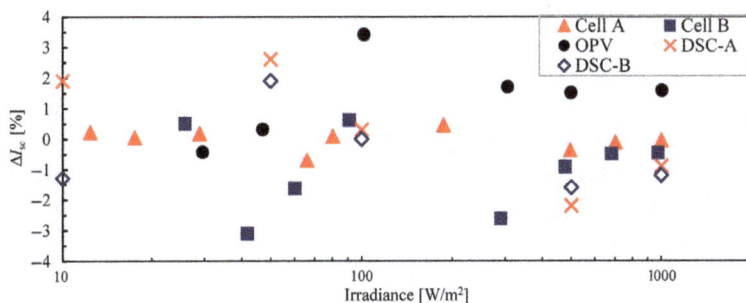

Figure 9. ΔI_{sc} determined for two nonlinear cells (cells A and B), OPV, DSC-A and DSC-B under the solar simulator. OPV, organic thin film solar cell; DSC, dye-sensitized solar cells.

and larger for DSC-A than DSC-B. The ΔI_{sc} determined for DSC-NIMS-1 and DSC-NIMS-2 were within ±4% under the solar simulator with an irradiance range of 10–1000 W/m². These results show that the ΔI_{sc} is within the maximum deviation set for the PRISM system even under such low irradiance as 10 W/m².

The ΔI_{sc} under LED was also investigated. The ΔI_{sc} for the reference cells under LED with an irradiance of 3.97 W/m² and that determined for nonlinear cells by changing the irradiance of LED are shown in Figure 10. The ΔI_{sc} of the reference cells was within ±4% which is within the maximum deviation set for the PRISM system. The ΔI_{sc} of the nonlinear cells was within ±3% when the irradiance was higher than 4 W/m². These results show that the PRISM system is applicable not only to sunlight but also LED light when the irradiance is higher than 4 W/m². Since an increase in ΔI_{sc} under low irradiance may result from the S/N ratio of the equipment, improvement in the equipment components will also need to be implemented for less uncertainty.

I–V characteristic measurements

Changes in solar cell efficiency and open circuit voltage upon irradiance with a light source

The PRISM system was applied to determine changes in the solar cell efficiency (η) and open circuit voltage (V_{oc}) upon irradiance with a solar simulator. The target cells were DSC-NIMS-2 and AK-100, and the solar simulator was a San-ei Electric XES-40S. Figure 11 shows the changes in η and V_{oc} upon irradiance with the solar simulator. The η of the DSC increased with an increase in irradiance at a range from 0.01 to about 0.2 sun, then gradually decreased, whereas the V_{oc} monotonously increased with an increase in irradiance. Contrarily, in the case of AK-100, both η and V_{oc} decreased with a monotonous decrease in irradiance. The rates of decrease in η

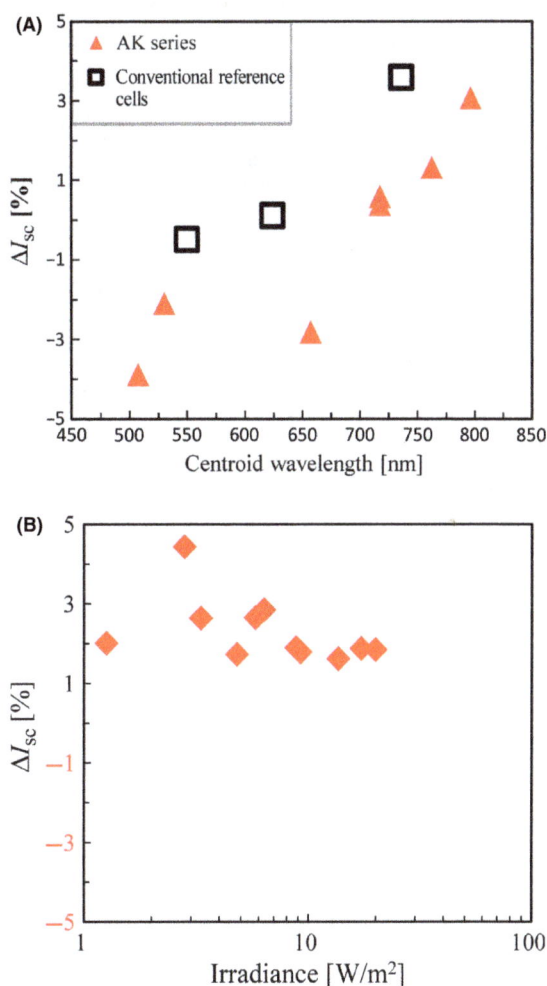

Figure 10. (A) ΔI_{sc} determined for the reference cells under LED with an irradiance of 3.97 W/m² and (B) ΔI_{sc} determined for nonlinear cells by changing the irradiance of LED.

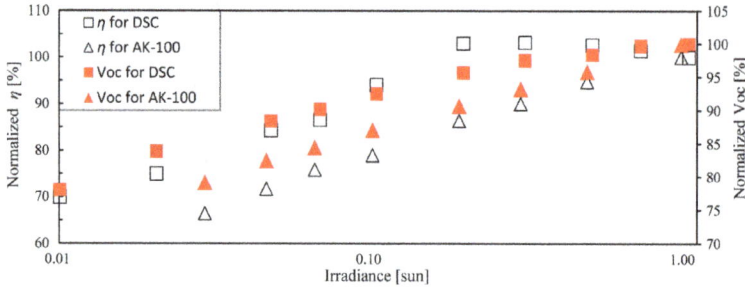

Figure 11. Changes in the solar cell efficiency (η) and open circuit voltage (V_{oc}) determined for DSC-NIMS-2 and AK-100 upon irradiance with a solar simulator. DSC-NIMS-2, dye-sensitized solar cells-National Institute for Materials Science.

Table 5. Performance of OPV under a solar simulator (SAN-EI ELEC-TRIC XES-40S) and LED (Imac).

Light source	J_{sc} mA/cm^2	V_{oc} V	P_{max} W/m^2	Irradiance W/m^2	η %
Solar simulator	7.05	0.46	18.0	1000 (1 sun)	1.8
LED	0.0815	0.34	0.176	4.57	3.9

OPV, organic thin film solar cell.

Table 6. Performance of DSC-NIMS-1 under a solar simulator (SAN-EI ELECTRIC XES-40S) and LED (Imac).

Light source	J_{sc} mA/cm^2	V_{oc} V	P_{max} W/m^2	Irradiance W/m^2	η %
Solar simulator	13.8	0.60	61.4	1000 (1 sun)	6.1
LED	0.120	0.47	0.345	4.10	8.4

DSC-NIMS-1, dye-sensitized solar cells-National Institute for Materials Science.

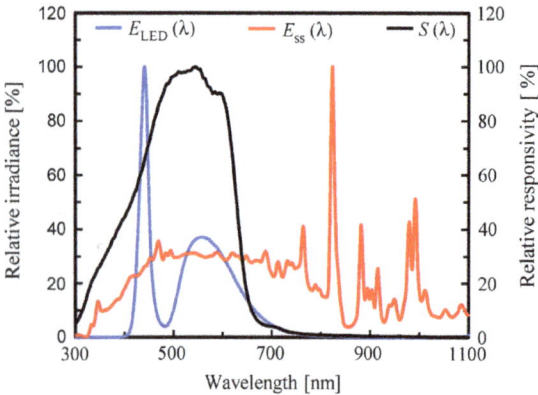

Figure 12. Relationship between the spectral responsivity of the OPV ($S(\lambda)$) and spectral irradiance of the light sources (($E\_LED(\lambda)$ for LED, $E\_ss(\lambda)$ for the San-ei Electric XES-40S (1 sun)). OPV, organic thin film solar cell.

and V_{oc} for the DSC with a decrease in irradiance were less than of those for AK-100.

Determination of solar cell efficiency under weak LED irradiation

The results shown in Figure 11 suggest that, under weak irradiation, higher performance can be expected for some types of solar cells when compared to crystalline solar cells. For example, organic solar cells can be specifically considered for indoor use due to their flexibility, lightness and high designability, however, a standard method to verify their performance has yet to be established.

The PRISM system was also applied to determine the photoelectric conversion efficiencies of OPV and DSC under LED which exhibits irradiance similar to indoor light. Table 5 shows the performance of the OPV under the solar simulator and LED. The solar cell efficiency (η) under LED was more than twice that obtained under 1 sun. Figure 12 shows the relationship between the spectral responsivity of the OPV and spectral irradiances of the light sources. As shown in Figure 12, the light emission wavelength region of the LED was approximately within the absorption wavelength region of the OPV, however, the wavelength region of natural sunlight is much wider than the absorption wavelength region of the OPV. This may explain why the η obtained under LED was greater than that obtained under 1 sun. The performance of DSC-NIMS-1 under the solar simulator and LED are shown in Table 6. As with the OPV, the η under LED was larger than that under 1 sun.

Conclusions

The newly developed PRISM system, consisting of a SK-1150 unit which measures the absolute spectral responsivity of a target cell and the SRC-1100 device which stores this data along with the reference spectral irradiance and that of the light source in order to adjust its irradiance, was found to be applicable for any type of solar cell under various illumination conditions. Significantly, the SRC-1100 acts as the reference cell, allowing comparison studies without the preparation of conventional reference cells as defined in IEC 60904-2, 4, thus, dramatically reducing the preparation time for I–V characterization measurements of a newly developed cell. The PRISM system makes it unnecessary to prepare a filtered reference cell and then confirm its stability over months. It is thus possible to avoid these tedious and lengthy procedures before measurement of organic solar cell performance. This is because the data of the absolute spectral responsivity itself serves the same function as a reference cell.

Regarding accuracy, differences in the short-circuit current between that determined by the SK-1150 and at a recognized test center under STC were within ±2% for the filtered reference cells and nonlinear cells, and within ±3% for organic solar cells. The typical calibration uncertainty at a secondary or working calibration institute is around 2%, meaning that SK-1150's accuracy is comparable to calibrations at these facilities. Upon irradiance adjustment with this PRISM system, the difference in the short-circuit current between that obtained by convolution of the absolute spectral responsivity with the spectral irradiance of a solar simulator or LED and that measured with the adjusted light source was within ±3% under STC and ±4% under various illumination conditions when the irradiance was higher than 4 W/m^2.

In the case of calibration of the light irradiance under STC conditions using a conventional method with a reference cell, the adjustment accuracy was also comparable the present PRISM method. However, in the case of the calibration of various kinds of light sources with even low irradiance, the PRISM system enables a traceable measurement method for irradiance adjustment which had so far been difficult due to the lack of a suitable reference cell with a certified value. This system will be particularly useful for the preparation and I–V measurement of nonlinear organic solar cells which have a variety of spectral responsivities.

Acknowledgments

The authors express their sincere appreciation to the Japan Science and Technology Agency (JST) Development of Advanced Measurement and Analysis Systems funding program for their financial support and encouragement. They thank Dr M. Yanagida of the National Institute for Materials Science for supplying the stable DSC cells used in this work. Sincere gratitude is also extended to the National Institute of Advanced Industrial Science & Technology (AIST, Japan), National Renewable Energy Laboratory (NREL, USA), and Physikalisch Technische Bundesanstalt (PTB, Germany) for their useful advice on the traceable calibration method.

Conflict of Interest

None declared.

References

1. Lee, M. M., J. Teuscher, T. Miyasaka, T. N. Murakami, and H. J. Snaith. 2012. Efficient hybrid solar cells based on meso-superstructured organometal halide perovskites. Science 338:643–647. doi: 10.1126/science.
2. Ito, S., H. Matsui, K. Okada, S. Kusano, T. Kitamura, Y. Wada, et al. 2004. Calibration of solar simulator for evaluation of dye-sensitized solar cells. Solar Energy Mater. Solar Cells 82:421–429. doi: 10.1016/j.solmat.
3. Aoki, D., T. Aoki, H. Saito, S. Magaino, and K. Takagi. 2012. Methods for spectral responsivity measurements of dye-sensitized solar cells. Electrochemistry 80:640–646. doi: 10.5796/electrochemistry.
4. Takagi, K., S. Magaino, H. Saito, T. Aoki, and D. Aoki. 2013. Measurements and evaluation of dye-sensitized solar cell performance. Photochem. Rev. 14:1–12. doi: 10.1016/j.photochemrev.
5. Emery, K. A., and C. R. Osterwald. 1989. Solar cell calibration methods. Solar Cells 27:445–453.
6. IEC 60904-2 2nd edition 2007–03: photovoltaic devices - part 2; Requirements for reference solar devices.
7. IEC 60904-4 1st edition 2009–06: photovoltaic devices - part 4; Reference solar devices - procedures for establishing calibration traceability.
8. IEC 60904-8 3rd edition 2014–05: photovoltaic devices - Part 8; Measurement of spectral responsivity of a photovoltaic (PV) device.
9. IEC 60904-3 2nd edition 2008–04: photovoltaic devices - Part 3; Measurement principles for terrestrial photovoltaic (PV) solar devices with reference spectral irradiance data.
10. Gevorgyan, S. A. 2011. An inter-laboratory stability study of roll to roll flexible polymer solar modules. Sol. Energy Mater. Sol. Cells 95:1398–1468.
11. Hinsch, A. 2013. Status of the dye solar cell technology (DSC) as a guideline for further research. Proceeding of the 28th European PV Solar Energy Conference and Exhibition 2013; 2239–2247.
12. Nishikawa, Y., K. Miyao, S. Tochimoto, K. Imai, and S. Winter. 2012a. The application of spectral reference cell to various photovoltaic cells. The Japan Society of Applied Physics Spring Meeting; March 2012: 16P-GP9-1.
13. Nishikawa, Y., K. Miyao, S. Tochimoto, K. Imai, and S. Winter. 2012b. The application of spectral reference cell to

various photovoltaic cells. The Japan Society of Applied Physics Spring Meeting; March 2012: 16P-GP9-2.

14. Nishikawa, Y., K. Imai, S. Tochimoto, K. Miyao, and S. Winter. 2013. The application of spectral reference cells to non-linear photovoltaic cell. Konica Minolta Technol. Rep. 10:81–87.

15. Uchida, S., and Y. Nishikawa. New evaluation method of photovoltaic by using spectral technology. International Conference on the Evaluation & Standardization of Organic Solar Cells 2013 (ICES2013); January 2013: IL3.

16. Nishikawa, Y. Organic photovoltaic evaluation method by utilizing spectral reference cell. Proceeding of the 93rd Annual Meeting of The Chemical Society of Japan; March 2013: 3H1-31.

17. Uchida, S., Y. Nishikawa, K. Miyao, K. Imai, S. Tochimoto, S. Magaino, et al. The new generation's PV evaluation method by utilizing spectral technology. Proceeding of the 28th European PV Solar Energy Conference and Exhibition 2014; 3552-3556.

18. Uchida, S., Y. Nishikawa, K. Miyao, K. Imai, S. Tochimoto, S. Magaino, et al. The differential spectral responsivity (DSR) technology for evaluation of new generation PV. Proceeding of the 23rd International Photovoltaic Science and Engineering Conference; October 2013: 5-O-5.

19. Nishikawa, Y. New traceable evaluation method for organic photovoltaics. International Conference on the Evaluation & Standardization of Organic Solar Cells 2014 (ICES2014); February 2014: IL5.

20. ISO 15387: 2005 space systems - single-junction solar cells - measurement and calibration procedures.

21. Metzdorf, J., W. Möller, T. Wittchen, and D. Hünerhoff. 1991. Principle and application of differential spectroradiometry. Metrologia 28:247–250.

22. Guo, X.-Z., Y.-H. Luo, Y.-D. Zhang, X.-C. Huang, D.-M. Li, and Q.-B. Meng. 2010. Study on the effect of measuring methods on incident photon-to-electron conversion efficiency of dye-sensitized solar cells by home-made setup. Rev. Sci. Instrum. 81:103106.

23. JIS Z 8724: 1997 methods of color measurement - Light-source colour.

24. Sperfeld, P., D. Dzafic, F. Plag, F. Haas, C. Kröber, H. Albert, et al. Usability of compact array spectroradiometers for the traceable classification of pulsed solar simulators. Proceeding of the 27th European Photovoltaic Solar Energy Conference and Exhibition 2012; 3060–3065.

25. Habte, A., A. Andreas, L. Ottoson, C. Gueymard, G. Fedor, S. Fowle, et al. Indoor and outdoor spectroradiometer intercomparison for spectral irradiance measurement. Technical Report NREL/TP-5D00-61476. 2014: i–21.

26. Galleano, R., W. Zaaiman, P. Morabito, A. Minuto, A. Spena, S. Bartocci, et al. 2012. Intercomparison of spectroradiometers for solar spectral irradiance measurements. Preliminary results. AIP Conf. Proc. 1477:139–142. doi: 10.1063/1. 4753853.

27. Green, M. A., K. Emery, Y. Hishikawa, and W. Warta. 2009. Solar cell efficiency tables (Version 33). Prog. Photovoltaics Res. Appl. 17:85–94.

28. Dirnberger, D., U. Kräling, H. Müllejans, E. Salis, K. Emery, Y. Hishikawa, et al. 2014. Progress in photovoltaic module calibration: results of a worldwide intercomparison between four reference laboratories. Meas. Sci. Technol. 25:105005.

29. Hara, K., S. Igari, S. Takano, and G. Fujihashi. 2005. Characterization of photovoltaic performance of dye-sensitized solar cells. Electrochemistry 73:887–896. doi: 10.5796/Electrochemistry.

10

Estimating the specific chemical exergy of municipal solid waste

Francis Chinweuba Eboh[1,2], Peter Ahlström[1] & Tobias Richards[1]

[1]Swedish Centre for Resource Recovery, University of Borås, 501 90 Borås, Sweden
[2]Department of Mechanical Engineering, Michael Okpara University of Agriculture, Umudike, Abia, Nigeria

Keywords
Higher heating value, municipal solid waste, specific chemical exergy, standard entropy, statistical model

Correspondence
Francis Chinweuba Eboh, Swedish Centre for Resource Recovery, University of Borås, 501 90 Borås, Sweden.
E-mail: eboh.francis_chinweuba@hb.se

Funding Information
TETfund Academic Staff Training and Development, Nigeria.

Abstract

A new model for predicting the specific chemical exergy of municipal solid waste (MSW) is presented; the model is based on the content of carbon, hydrogen, oxygen, nitrogen, sulfur, and chlorine on a dry ash-free basis (daf). The proposed model was obtained from estimations of the higher heating value (HHV) and standard entropy of MSW using statistical analysis. The ultimate analysis of 56 different parts of MSW was used for the derivation of the HHV expression. In addition, 30 extra parts were used for validation. One hundred and seventeen relevant organic substances that represented the main constituents in MSW were used for derivation of the standard entropy of solid waste. The substances were divided into different waste fractions, and the standard entropies of each waste fraction and for the complete mixture were calculated. The specific chemical exergy of inorganic matter in the waste was also investigated by considering the inorganic compounds in the ash. However, as a result of the extremely low value calculated, the exergy of inorganic matter was ignored. The results obtained from the HHV model show a good correlation with the measured values and are comparable with other recent and previous models. The correlation of the standard entropy of the complete waste mixture is less accurate than the correlations of each individual waste fraction. However, the correlations give similar results for the specific chemical exergy, indicating that HHV has a greater impact when estimating the specific exergy of solid waste than entropy.

Introduction

The quantities of solid wastes increase as the population continues to grow throughout the world, whereas the available space for disposal decreases. If this waste is not properly treated and handled, it will pollute the air, land, ground water, and soil as well as has a negative impact on the hygienic conditions of the people [1]. A waste-to-energy (WTE) plant is one of the most robust and effective posttreatment options to decrease the volume of produced waste, reduce greenhouse gas emissions, and utilize the energy content in nonrecyclable waste for the production of electricity and heat, thereby reducing dependence on fossil fuel.

For the efficient design of energy conversion processes, the chemical exergy and energy content of the fuel are basic properties to be considered to estimate the maximum available energy entering the system for performance analysis and optimization of the entire process. This can be done by detecting and evaluating quantitatively the thermodynamic imperfection (exergy loss) of the process under consideration, its main sources of loss and possible ways of improving such process can be indicated [2]. Estimating the chemical exergy of fuel is an important step when performing exergy analysis [3] in waste-to-energy plants. Exergy analysis is a method that uses the conservation of mass and energy together with the second law of thermodynamics. This analysis method is useful to achieve a more efficient energy-resource use because it enables the locations, types, and magnitudes of losses to be identified and to determine meaningful efficiencies [4].

However, because many solid fuels have unknown structures and chemical compositions, their exergy values cannot be calculated directly because of the lack of standard absolute entropy values [5]. Many models for the prediction and estimation of the chemical exergy of carbon-based fuels with complex bond interactions and unknown thermodynamics properties have been proposed based on the characteristics of the known homogeneous organic substances in the fuel. The first attempt was performed by Rant [6], involving the formulation of a semiempirical model to evaluate the availability (exergy) content of a structurally complicated material species. In that model, the chemical exergy of a fuel is evaluated from the computation of pure organic substances of known absolute entropies. Rant evaluated the ratios of the estimated chemical exergies and the higher heating values for seven gases and 12 liquid organic substances. Szargut and Styrylska [7] improved Rant's correlation by considering the chemical composition of the fuels. They obtained correlation formulas to express the dependence between the ratio of the standard chemical exergy to the lower heating value using mass ratios of hydrogen, oxygen, nitrogen and sulfur to carbon that describe the chemical composition of the fuel. Due to lack of thermodynamic data, sulfur was not considered in their model for solid fuels, and their correlations were theoretically limited to Szargut's reference environmental (R.E) model.

Using a model for estimating the thermodynamic properties of coal, char, tar, and ash, Eisermann et al. [8] approximated the standard entropy of coal by comparing the behavior of the standard entropies of a number of aliphatic and aromatic hydrocarbons as a function of several elemental ratios: $H/(C + N)$, $O/(C + N)$, $N/(C + N)$ and $S/(C + N)$. Shieh and Fan [9] estimated the specific chemical exergy of a structurally complicated material by adopting the concepts of the dead (or reference) state and the properties of the constituents in the material based on the first and second laws of thermodynamics. It was assumed that the entropy of a fuel is equal to the entropies of its constituent elements. This assumption is not accurate in many cases. Ikumi et al. [10] developed a method for estimating the entropies of coals based on the mole ratios of hydrogen, oxygen, nitrogen, and sulfur elements to the carbon element. Bilgen and Kaygusuz [11] used the entropy correlation proposed by Eisermann et al. [8] to improve the Shieh and Fan [9] model for the calculation of the chemical exergy of coals, and Stepanov [5] applied the entropy model developed by Ikumi et al. [10] to modify Shieh and Fan [9] to calculate the exergies of coke-oven gases of different metallurgical mills. These models are limited to coal fuels only because their constituent organic compounds have been derived from the standard entropies of the relevant organic substances of coals.

Song et al. [3] developed a model based on Shieh and Fan [9] to estimate the specific entropy of the organic

matter in biomass used for the exergy calculations. Although their model showed a high accuracy and was simpler than the Szagut and Styrylska's correlation, it has a limited application, as it is only applicable to biomass. Song et al. [12] also proposed a model for estimating the entropy of solid fuels and then extended the Shieh and Fan [9] model using the major organic constituents of solid fuel for the prediction of the specific chemical exergy of solid fuels. However, they combined the higher heating value derived on a dry basis (db) with values of the standard entropy obtained based on a dry ash-free basis (daf) for the estimation of the chemical exergy. Furthermore, their model cannot be used for estimating the chemical exergy of substances containing elements other than C, H, O, N, and S and for combustible materials, such as certain categories of leather, plastic, and rubber, that are part of municipal solid waste. To the author's knowledge, no model has been found in the literature that is derived for predicting the chemical exergy of MSW.

The objective of this work is to propose a model for calculating the specific chemical exergy of MSW containing the C, H, O, N, S, and Cl from its elemental compositions on a dry ash-free basis.

Derivation of the Estimated Model

Municipal Solid Waste consists of a complex, heterogeneous mixture of organic and inorganic substances. The organic elements in MSW are mainly C, H, O, N, S, and Cl, which can be obtained from ultimate analysis, whereas the inorganics are commonly Si, Ca, K, P, Al, Mg, Fe, S, Na, Zn, Cu, Mn, and Cr, from which their oxides can be obtained from ash analysis data. Previous reports have shown that the influence of inorganic matters on the exergy value can be neglected in solid fuel as a result of their relatively small value [3, 12].

The standard chemical exergy of a substance that is not present in the environment can be evaluated by considering a reaction of the substance with other substances for which the chemical exergies are known [13]. The exact calculation of the chemical exergy of a material with complicated structures is difficult [14]; as a result, the standard chemical exergy of the substance in the environment is not readily available.

Standard chemical exergy of a substance

The chemical exergy of a substance is equal to the maximum amount of work that can be obtained from the substance by taking it to chemical equilibrium with the reference environment [15]. The standard exergy of a substance can be evaluated by considering an idealized reaction of the substance with other substances (generally

Table 1. Standard chemical exergy and standard entropies of various compounds.

Substance	e^0 (kJ/mol)	s^0 (kJ/mol K)
CO_2	19.87	0.214
H_2O_l	0.95	0.070
O_2	3.97	0.205
N_2	0.72	0.192
SO_2	310.93	0.248
SiO_2	1.636	0.041
HCl	85.5	0.187
CaO	129.881	0.038
K_2O	412.544	0.102
P_2O_5	377.155	0.117
Al_2O_5	4.479	0.051
MgO	62.417	0.027
Fe_2O_3	17.656	0.087
SO_3	242.003	0.257
Na_2O	296.32	0.075
MnO	122.390	0.060
ZnO	37.080	0.042
Cr	538.610	0.024
Pb	226.940	0.065
As	477.040	0.035
Cd	290.920	0.052
Cl	163.940	0.166

l, liquid phase.

Source: [3, 17, 18].

reference substances) of known chemical exergies [16]. The known chemical exergies can be obtain from the table of standard chemical exergy based on Szargut's R.E model (Model II), as shown in Table 1. With consideration of the reversible reaction for chemical formation of a compound, Szargut et al. [2] expressed the standard chemical exergy of elements or compounds as

$$b^o_{ch} = \Delta G^o_f + \sum_i n_i b^o_{chi} \text{ (kJ/mol)} \tag{1}$$

where ΔG^o_f, n_i, and b^o_{chi} represent the standard Gibbs energy of formation, the mole fraction of component i in the mixture, and the standard chemical exergy of the constituent element i, respectively.

Calculation of the specific chemical exergy of municipal solid waste

For simplicity, suppose 1 kg of MSW (daf), expressed as $C_m H_n N_p O_q Cl_r S_t$, undergoes complete combustion at a standard state for the steady condition to produce carbon dioxide, water, nitrogen, hydrogen chloride, and sulfur dioxide as follows:

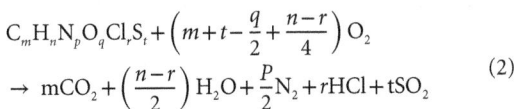

$$C_m H_n N_p O_q Cl_r S_t + \left(m+t-\frac{q}{2}+\frac{n-r}{4} \right) O_2$$
$$\rightarrow mCO_2 + \left(\frac{n-r}{2} \right) H_2O + \frac{p}{2}N_2 + rHCl + tSO_2 \tag{2}$$

All substances are assumed to enter and exit at the reference temperature, $T_0 = 298.15$ K, and reference pressure, $P_0 = 101.325$ kPa. The subscripts m, n, p, q, r, and t are the numbers of atoms of C, H, N, O, Cl, and S, respectively, in kmol/kg MSW or the molal compositions per kg of MSW expressed as:

$$m = \frac{0.01C}{12.011} \text{ kmol/kg} \quad \text{or} \quad \frac{10C}{12.011} \text{ mol/kg} \tag{3}$$

$$n = \frac{0.01H}{1.008} \text{ kmol/kg} \quad \text{or} \quad \frac{10H}{1.008} \text{ mol/kg} \tag{4}$$

$$p = \frac{0.01N}{14.007} \text{ kmol/kg} \quad \text{or} \quad \frac{10N}{14.007} \text{ mol/kg} \tag{5}$$

$$q = \frac{0.01O}{15.999} \text{ kmol/kg} \quad \text{or} \quad \frac{10O}{15.999} \text{ mol/kg} \tag{6}$$

$$r = \frac{0.01Cl}{35.45} \text{ kmol/kg} \quad \text{or} \quad \frac{10Cl}{35.45} \text{ mol/kg} \tag{7}$$

$$t = \frac{0.01S}{32.066} \text{ kmol/kg} \quad \text{or} \quad \frac{10S}{32.066} \text{ mol/kg} \tag{8}$$

where the elements in Equations (3–8) are expressed in wt% (daf). For the steady state, under the standard condition, the energy balance of the reaction in Equation (2) is given by

$$W = Q + h^0_{MSW} + \left(m+t-\frac{q}{2}+\frac{n-r}{4} \right) h^0_{O_2} - mh^0_{CO_2}$$
$$- \left(\frac{n-r}{2} \right) h^0_{H_2O} - \frac{p}{2}h^0_{N_2} - rh^0_{HCl} - th^0_{SO_2} \tag{9}$$

The entropy balance is expressed as

$$0 = \frac{Q}{T_o} + s^0_{MSW} + \left(m+t-\frac{q}{2}+\frac{n-r}{4} \right) s^0_{O_2} - ms^0_{CO_2}$$
$$- \left(\frac{n-r}{2} \right) s^0_{H_2O} - \frac{p}{2}s^0_{N_2} - rs^0_{HCl} - ts^0_{SO_2} + S_{gen} \tag{10}$$

where W and Q are the work and heat transfer, respectively. S_{gen} is the entropy generated by the irreversibility in the reaction, and s^0 and h^0 represent the standard entropy and enthalpy, respectively.

Eliminating the heat transfer Q between Equation (9) and (10) gives the following:

$$W = \left[h^0_{MSW} + \left(m + t - \frac{q}{2} + \frac{n-r}{4} \right) h^0_{O_2} - h s^0_{CO_2} \right.$$
$$- \left(\frac{n-r}{2} \right) h^0_{H_2O} - \frac{p}{2} h^0_{N_2} - r h^0_{HCl} - t h^0_{SO_2} \right]$$
$$- T_o \left[s^0_{MSW} + \left(m + t - \frac{q}{2} + \frac{n-r}{4} \right) s^0_{O_2} \right.$$
$$\left. - h s^0_{CO_2} - \left(\frac{n-r}{2} \right) s^0_{H_2O} - \frac{p}{2} s^0_{N_2} - r s^0_{HCl} - t s^0_{SO_2} \right] - T_o S_{gen} \tag{11}$$

The maximum work, W_{max}, will occur when there is no irreversibility in the system. Hence, Equation (11) can be expressed as

$$W_{max} = \left[h^0_{MSW} + \left(m + t - \frac{q}{2} + \frac{n-r}{4} \right) h^0_{O_2} - m h^0_{CO_2} \right.$$
$$- \left(\frac{n-r}{2} \right) h^0_{H_2O} - \frac{p}{2} h^0_{N_2} - r h^0_{HCl} - t h^0_{SO_2} \right]$$
$$- T_o \left[s^0_{MSW} + \left(m + t - \frac{q}{2} + \frac{n-r}{4} \right) s^0_{O_2} - m s^0_{CO_2} \right.$$
$$\left. - \left(\frac{n-r}{2} \right) s^0_{H_2O} - \frac{p}{2} s^0_{N_2} - r s^0_{HCl} - t s^0_{SO_2} \right] \tag{12}$$

or

$$W_{max} = -\Delta H^o_r - T_o \left[s^0_{MSW} + \left(m + t - \frac{q}{2} + \frac{n-r}{4} \right) s^0_{O_2} \right.$$
$$\left. - m s^0_{CO_2} - \left(\frac{n-r}{2} \right) s^0_{H_2O} - \frac{p}{2} s^0_{N_2} - r s^0_{HCl} - t s^0_{SO_2} \right] \tag{13}$$

where ΔH^o_r represents the heat of reaction of the combustion process, which is equal to the negative higher heating value [9], that is

$$\Delta H^o_r = -\text{HHV} \tag{14}$$

then

$$W_{max} = \text{HHV} - T_o \left[s_{msw} + \left(m + t - \frac{q}{2} + \frac{n-r}{4} \right) s^0_{O_2} - m s^0_{CO_2} \right.$$
$$\left. - \left(\frac{n-r}{2} \right) s^0_{H_2O} - \frac{p}{2} s^0_{N_2} - r s^0_{HCl} - t s^0_{SO_2} \right] \tag{15}$$

Assume that the reaction in Equation (2) at 298.15 K and 101.325 kPa is an adiabatic process with no irreversibility. The exergy balance equation, in absence of changes in the kinetic and potential energy for reacting systems, is given by

$$0 = -W_{max} + e_{MSW} + \left(m + t - \frac{q}{2} + \frac{n-r}{4} \right) e^0_{O_2} - m e^0_{CO_2}$$
$$- \left(\frac{n-r}{2} \right) e^0_{H_2O} - \frac{p}{2} e^0_{N_2} - r e^0_{HCl} - t e^0_{SO_2} \tag{16}$$

where e represents the specific chemical exergy. Substituting Equation (15) into Equation (16), the specific chemical exergy of MSW (daf), e_{MSW}, is presented as

$$e_{MSW} = \text{HHV} - T_o \left[s^0_{MSW} + \left(m + t - \frac{q}{2} + \frac{n-r}{4} \right) s^0_{O_2} - m s^0_{CO_2} \right.$$
$$\left. - \left(\frac{n-r}{2} \right) s^0_{H_2O} - \frac{p}{2} s^0_{N_2} - r s^0_{HCl} - t s^0_{SO_2} \right] + m e^0_{CO_2}$$
$$+ \left(\frac{n-r}{2} \right) e^0_{H_2O} + \frac{p}{2} e^0_{N_2} + r e^0_{HCl}$$
$$+ e^0_{SO_2} - \left(m + t - \frac{q}{2} + \frac{n-r}{4} \right) e^0_{O_2} \tag{17}$$

or

$$e_{MSW} = m \left[\left(e^0_{CO_2} + T_o s^0_{CO_2} \right) - \left(e^0_{O_2} + T_o s^0_{O_2} \right) \right]$$
$$+ \frac{n}{2} \left[\left(e^0_{H_2O} + T_o s^0_{H_2O} \right) - \frac{1}{2} \left(e^0_{O_2} + T_o s^0_{O_2} \right) \right]$$
$$+ \frac{q}{2} \left(e^0_{O_2} + T_o s^0_{O_2} \right) + \frac{p}{2} \left(e^0_{N_2} + T_o s^0_{N_2} \right)$$
$$+ t \left[\left(e^0_{SO_2} + T_o s^0_{SO_2} \right) - \left(e^0_{O_2} + T_o s^0_{O_2} \right) \right]$$
$$- \frac{r}{2} \left[\left(e^0_{H_2O} + T_o s^0_{H_2O} \right) - \frac{1}{2} \left(e^0_{O_2} + T_o s^0_{O_2} \right) \right]$$
$$- 2 \left(e^0_{HCl} + T_o s^0_{HCl} \right) + \text{HHV}_{MSW} - T_o s^0_{MSW} \ (\text{kJ/kg}) \tag{18}$$

where e^0 is the standard exergy in kJ/mol and s^0 is the standard entropy in kJ/mol K, as tabulated in Table 1. s^0_{MSW} is the standard entropy of municipal solid waste, in kJ/K kg (daf), and HHV_{MSW} is the higher heating value (HHV) of MSW, in kJ/kg (daf). The specific chemical exergy of MSW can be calculated once the standard chemical exergies of CO_2, $H_2O(l)$, N_2, O_2, SO_2, and HCl; the higher heating value; and the absolute entropies are known.

Estimating the higher heating value of municipal solid waste

In the absence of a measured value, the HHV of fuel can be estimated from their elemental composition [19, 20]. In this study, the HHV estimate was performed by considering 56 data points and 30 data points of MSW samples for the derivation and validation of the

Table 2. Assumed correlations used for the selection of the proposed model for HHV (daf).

S. No.	Assumed expression	Criteria for selection	Reference
1.	$HHV = aC + bH + cO + dN + eS + fCl$	Assuming fuel HHV to be a linear function of it constituents.	Current model
2.	$HHV = aC + bH + cO + dN + eS$	Based on Gumz's criteria	[19]
3.	$HHV = a^0 + bH + cO + dN + eS + fCl$	Based on Chang's criteria	[23]
4.	$HHV = aC + b(H - O/8) + eS$	Based on Dulong's criteria	[19]
5.	$HHV = aC + bH + cO + eS$	Based on modified version of Dulong's criteria	[19]
6.	$HHV = a^0 + aC + bH + cO^2$	Based on Seyler's criteria	[19]
7.	$HHV = a(C - (3/8)O) + b(3/8)O) + c(H - (1/6)O) + eS$	Based on Steuer's criteria	[19]
8.	$HHV = a(C - 0.75(O/2)) + b(H - 0.125(O/2)) + eS$	Based Sumegi's criteria	[19]
9.	$HHV = aC + bH + c((N + O - 1)/8) + eS$	Dulong-Berthelot's criteria	[19]

where C, H, O, N, S, and Cl represents carbon, hydrogen, oxygen, nitrogen, sulfur, and chlorine, respectively, in % by mass on a dry ash free basis. a^0, a, b, c, d, e, and f are constants of correlation.

correlation, respectively; in addition, the chemical composition and HHV of each sample was collected from the published literature and presented in Tables A1 and A2 (Appendix 1). These data cover six categories of combustible MSW fractions, namely, food, wood, paper, textiles, plastics, and rubber waste [21, 22]. For the selection of a suitable model, 9 assumed algebraic expressions from previous work based on the correlation of the HHV and ultimate analysis of solid fuel (daf) were used, as shown in Table 2. Using regression analysis based on the generalized method of least squares [19] on the 56 data points, the constant terms of these algebraic expressions were evaluated. The correlation that has the least error and highest coefficient of determination, as described in Selection of the best correlation, was selected. The newly estimated correlation was compared with the experimental values of HHV and the results of previous models collected from the open literature, for further validation.

Estimating the standard entropy of municipal solid waste

Municipal solid waste (MSW) contains mainly organic polymers in plastics, wood, paper, textile, rubber, and food waste. The entropies of these polymers in the organic waste are estimated or evaluated by the entropies of their organic monomers structures as there is no significant difference between the entropies of the solid organic monomers and their polymers [24]. The difference ranges from 0.1 to 12.5% (Table 3).

The standard entropy of MSW was derived from organic substances with known standard entropies. In this work, 117 samples of organic compounds relevant to MSW were collected from the published literature [3, 8, 10, 12, 17, 24] and tabulated in Table A3 (Appendix 1). The data points where selected based on the molecular structures of the organic substances that are associated or

Table 3. Standard entropies of some solid organic polymers and monomers at 298.15 K.

S. No.	Substance	S^0 (J/mol K)
1.	$C_6H_{11}NO$	173.21
	$(C_6H_{11}NO)_n$	173.0
2	$C_4H_4O_4$	157.2
	$(C_4H_4O_4)_n$	151.4
3	$C_{15}H_{10}N_2O_2$	332.5
	$(C_{15}H_{10}N_2O_2)_n$	294
4	$C_{13}H_{24}O_2$	401.9
	$(C_{13}H_{24}O_2)_n$	351.6

Source: [24].

linked with the formation of larger molecular structure network of municipal solid waste. The organic compounds were grouped into the six categories of waste fractions, as previously used for the higher heating value, namely: food, plastic, textile, rubber, wood, and paper. This was accomplished by considering the molecular structures of the organic substances that can be found in each of the molecular structures of the waste fractions. For wood, it contains three major chemical components: cellulose, hemicelluloses, and lignin [25]. Each of the chemical structure of the wood constituents [26, 27] was studied and organic compounds (monomers) that can be made or found from these structures are selected. In the food, the main structural elements identified are proteins, carbohydrate and lipids [28]. The molecular structures of these food components [29, 30] were also investigated and organic monomers that are linked with the structure are selected. The same method was carried on chemical structures of plastic [31], textile [32, 33], and rubber [34] materials with identifications of biologically important molecules which form the building structure of their polymers. Based on the absolute entropies and elemental compositions of the selected organic substances, a first-order polynomial correlation was derived statistically for the standard entropy of the waste fractions and the mixture.

Selection of the best correlation

Three statistical parameters were used as evaluating parameters for both HHV and the standard entropy of MSW, which are computed as follows:

$$\text{Average absolute error (AAE)} = \frac{1}{n}\sum_{i=1}^{n}\left|\frac{Z_{est}-Z_{exp}}{Z_{exp}}\right|\times 100\% \quad (19)$$

$$\text{Average bias error (ABE)} = \frac{1}{n}\sum_{i=1}^{n}\frac{Z_{est}-Z_{exp}}{Z_{exp}}\times 100\% \quad (20)$$

$$\text{Coefficient of determination } (R^2) = 1-\sum_{i=1}^{n}\frac{\left(Z_{est}-Z_{exp}\right)^2}{\left(Z_{exp}-\overline{Z}_{exp}\right)^2} \quad (21)$$

where Z_{est} and Z_{exp} denote the estimated and experimental values, respectively. $\overline{Z}_{exp}$ is the experimental average value. AAE is the average error of a correlation. A smaller error of correlation will occur when AAE is low, which indicates higher accuracy. ABE denotes the average bias error of correlation. A positive value of ABE indicates an overall overestimation, whereas a negative value implies an overall underestimation. The smaller the absolute value of ABE, the smaller the bias of correlation. R^2 is used as a comprehensive parameter to measure the accuracy of the model. A higher R^2 value means a better estimation and fitting [20]. These three parameters are the important statistical criteria and are primarily employed to assess correlations [3, 12, 19].

Specific chemical exergy of inorganic matter in municipal solid waste

Inorganic substances of waste materials are contained in the ash and obtained from complete combustion of solid fuel; ash is mainly contained in various metallic oxides and has a high thermal stability [35, 36]. The specific chemical exergy of inorganic matter in kJ/kg MSW was calculated from the major ash compositions data in Table 4 from a stoker-type incinerator [37] as follows [3, 12, 36]:

$$e_{ioc} = 0.01A\left(\sum n_i x_i e_{ioc}^0 + RT_o\sum n_i x_i \ln x_i\right) \text{ (kJ/kg)} \quad (22)$$

where n_i represents number of moles of the component in inorganic matter, in mol/kg. e_{ioc}^0 and x_i are the standard chemical exergy and mole fraction of components i in inorganic matter, respectively. R is the universal gas constant, 0.0083145 kJ/mol K, and A is the ash content of MSW in wt%.

Table 4. Chemical composition of MSW ashes.

Component	Bottom ash, BA (wt%)	Fly ash, FA (wt%)
SiO_2	37.8	2.47
CaO	20.79	44.5
K_2O	0.85	3.01
P_2O_5	3.63	0.26
Al_2O_3	13.4	0.55
MgO	2.91	0.57
Fe_2O_3	7.46	0.32
SO_3	1.01	1.61
Na_2O	5.38	4.39
ZnO	0.52	2.25
CuO	0.51	0.096
MnO	0.17	0.04
Cr	0.63	0.008
Pb	0.22	0.51
As	0.021	0.062
Cd	0.0003	0.003
Cl	3.51	35.15

Source: [37].

Results and Discussion

Correlation based on the higher heating value

For the higher heating value of MSW, the correlation derived that showed the minimum error with a higher accuracy among the nine assumed correlations used in Table 2 was expressed as

$$\text{HHV} = 0.364C + 0.863H - 0.075O$$
$$+ 0.028N - 1.633S + 0.062Cl \text{ (MJ/kg)} \quad (23)$$

$$35.8\% \leq C \leq 86.1\%, 4.1\% \leq H \leq 13.9\%, 0.0\%$$
$$\leq O \leq 54.9\%, 0.0\% \leq N \leq 20.3\%, 0.0\%$$
$$\leq S \leq 2.7\%, 0.0\% \leq Cl \leq 56.4\%,$$
$$13.0 \text{ MJ/kg} \leq \text{HHV} \leq 43.2 \text{ MJ/kg}.$$

The results of the validation of the derived model and the comparison with published correlations using the experimental values of 30 samples of MSW in the different categories of food, wood, plastic, textile, rubber, and paper waste are shown in Table 5 and represented in Figures 1–5. Figures 1, 2, and 4 show the best correlation with experimental data (highest coefficient of determination), representing the model developed in this work, model by Channiwala and Parikh [19] and Dulong's correlation. However, the proposed model shows significantly better estimations when considering the errors (AAE and ABE) compared to the other models. This is not surprising, as these models have been derived from mixed solid fuel and coal. Figure 3 shows a correlation proposed by Sheng and Azevedo [20]. Although the correlation has a good coefficient of determination ($R^2 = 0.92$), it has a higher error and

Table 5. Derived correlation compared with previous models.

S No.	Name	Correlation (MJ/kg)	Application	AAE (%)	ABE (%)	R^2	Reference
1.	Proposed Model	HHV = 0.364C + 0.863H − 0.075O + 0.028N − 1.633S + 0.062Cl	MSW	5.738	0.032	0.95	Current model
2.	Channiwala and Parikh	HHV* = 0.3491C* + 1.1783H* + 0.1005S* − 0.1034O* − 0.0151N* − 0.0211A*	Mixed waste	6.456	2.254	0.95	[19]
3.	Sheng and Azevedo	HHV* = −1.3675 + 0.3137C* + 0.7009H* + 0.0318 (100 − C* − H* − A*)	Biomass	9.657	−3.650	0.92	[20]
4.	Dulong	HHV = 0.3383C + 1.443 (H − (O/8)) + 0.0942S	Coal	11.822	−4.832	0.95	[19]
5.	Chang	HHV = 35.8368 + 0.7523H − 0.2674S − 0.4654O − 0.3814Cl − 0.2802N	MSW	7.234	3.067	0.93	[19]

(*) shows the correlations obtained in % by mass on a dry basis, whereas the others are on dry ash-free basis.

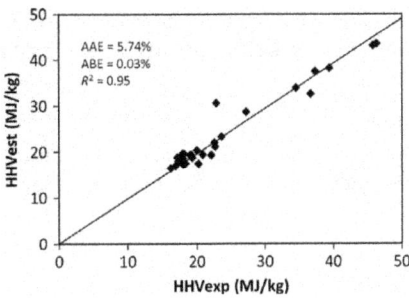

Figure 1. Comparison between the experimental and the estimated HHV by the developed correlation.

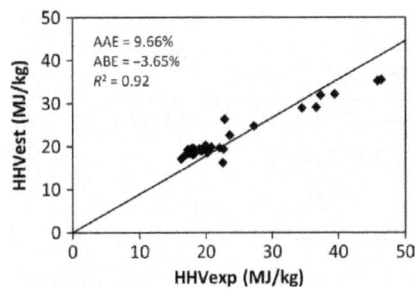

Figure 3. Comparison between the experimental and the estimated HHV by the Sheng and Azevedo [20] correlation.

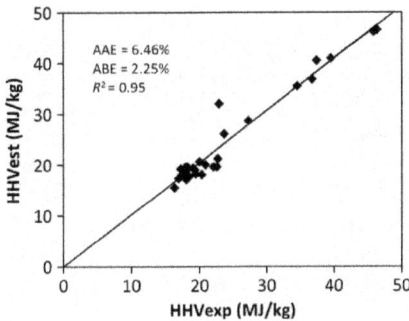

Figure 2. Comparison between the experimental and the estimated HHV by the Channiwala and Parikh [19] correlation.

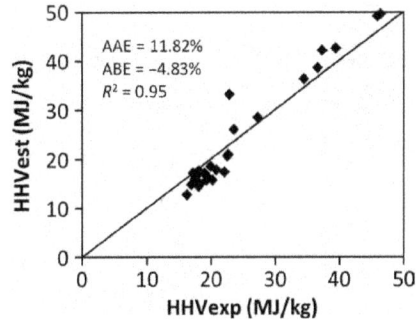

Figure 4. Comparison between the experimental and the estimated HHV by Dulong's correlation.

underestimated the HHV. In addition, the correlation is limited to biomass. The correlation proposed by Chang (Table 5 and Fig. 5) has a considerable accuracy, with a coefficient of determination of $R^2 = 0.93$. Although this correlation was derived from MSW, it overestimated the correlation and has a higher error value when compared with the present model.

Standard entropy of municipal solid fuel

For the prediction of the standard entropy of the organic substance in MSW, a correlation in the form of the

first-order polynomial was used. The five correlations derived for estimating the standard entropy of waste fractions and the mixture of waste were expressed as follows:

For Plastic waste

$$s^0_{pl} = 0.0087C + 0.0753H + 0.0134O$$
$$+ 0.0077N + 0.0084Cl \ (kJ/K \ kg) \tag{24}$$

$$10.3\% \leq C \leq 94.7\%, 0.0\% \leq H \leq 14.3\%, 0.0\% \leq O \leq 54.2\%,$$
$$0.0\% \leq N \leq 66.7\%.$$

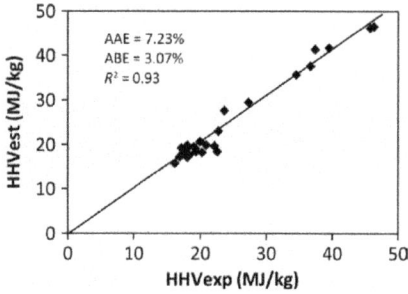

Figure 5. Comparison between the experimental and the estimated HHV by Chang's correlation.

With ABE, AAE, and R^2 of 0.722, 7.314, and 0.7674, respectively.

For Textile/Rubber waste

$$s_{tr}^0 = 0.0097C + 0.0635H + 0.0128O + 0.0136N \\ + 0.0165S \, (kJ/K\,kg) \tag{25}$$

$$15.8\% \le C \le 95.1\%, 3.0\% \le H \le 9.7\%, 0.0\% \le O \le 55.2\%, \\ 0.0\% \le N \le 66.7\%, 0.0\% \le S \le 42.1\%$$

With ABE, AAE, and R^2 of 0.714, 6.476, and 0.5457, respectively.

For Wood/Paper waste

$$s_{wp}^0 = 0.0162C + 0.0116H + 0.0081O + 0.00691Cl \, (kJ/K\,kg)$$

$$26.7\% \le C \le 77.8\%, 0.4\% \le H \le 7.7\%, 5.1\% \le O \le 71.1\%, \\ 0.0\% \le Cl \le 66.3\% \tag{26}$$

With ABE, AAE, and R^2 of 0.329, 5.215, and 0.728, respectively.

For Food waste

$$s_{fo}^0 = 0.0065C + 0.0808H + 0.0127O \\ + 0.0101N + 0.0100S \, (kJ/K\,kg) \tag{27}$$

$$19.2\% \le C \le 92.3\%, 1.4\% \le H \le 14.1\%, 0.0\% \le O \\ \le 59.7\%, 0.0\% \le N \le 51.9\%, 0.0\% \le S \le 34.0\%$$

With ABE, AAE, and R^2 of 0.414, 5.886, and 0.6922, respectively.

For Mixed waste

$$s_{msw}^0 = 0.0101C + 0.0630H + 0.0106O + 0.0108N \\ + 0.0155S + 0.0084Cl \, (kJ/K\,kg) \tag{28}$$

$$10.3\% \le C \le 95.1\%, 0.00\% \le H \le 14.3\%, 0.0\% \le O \le 71.1\%, 0.0\% \\ \le N \le 66.7\%, 0.0\% \le S \le 42.1\%, 0.0\% \le Cl \le 89.7\%,$$

with ABE, AAE, and R^2 of 1.118, 8.293, and 0.5414, respectively.

Comparing the five equations obtained, the results show that the standard entropy correlations for each waste fraction in MSW are more accurate than the standard entropy correlation for the waste mixture. This is as a result of complicated mixture, heterogeneous molecule structure and variation in municipal solid waste chemical compositions and properties. Nevertheless, because the standard entropy of plastic, textile/Rubber, wood/paper, food, and waste mixture gave similar average values for the specific exergy of the MSW estimation of 24,359, 24,364, 24,426, 24,393, and 24,387 (kJ/kg), respectively, the correlation of the standard entropy of the waste mixture can be used for the derivation of exergy.

Specific chemical exergy of municipal solid fuel (daf) and specific chemical exergy of ash

By substituting Equation (3)–(8), (23), and (24)–(28) into Equation (18), along with the standard chemical exergy data from Table 1, the specific chemical exergy of solid waste on a dry ash-free basis can be expressed as follows:

For Plastic waste:

$$e_p = 376.879C + 787.351H - 58.654O + 46.398N \\ - 1533.261S + 100.981Cl \, (kJ/kg) \tag{29}$$

For Textile/Rubber waste:

$$e_{TR} = 376.580C + 790.869H - 58.475O + 44.639N \\ - 1538.180S + 98.566Cl \, (kJ/kg) \tag{30}$$

For Wood/Paper waste

$$e_{WP} = 374.642C + 806.343H - 57.074O + 48.693N \\ - 1533.261S + 101.425Cl \, (kJ/kg) \tag{31}$$

For Food waste

$$\text{Food } e_F = 377.535C + 785.711H - 58.446O + 45.682N \\ - 1536.242S + 103.486Cl \, (kJ/kg) \tag{32}$$

For mixed waste

$$e_{msw} = 376.461C + 791.018H - 57.819O + 45.473N \\ - 1536.242S + 100.981Cl \, (kJ/kg) \tag{33}$$

The minimum, maximum, and average specific exergy values of municipal solid waste calculated were 17,602, 43,396, and 24,387 in (kJ/kg), respectively. Although Equation (33) slightly underestimated the specific chemical exergy calculated by Equations (29) and (30), that is, an ABE of −0.139 and −0.113, respectively, and slightly overestimated the exergy estimated by Equations (31) and (32), that is, an ABE of 0.179 and 0.009, respectively, when compared, the coefficient of determination shows that Equations (29–32) are similar to Equation (33) (i.e., a value of 1 was achieved in all cases). This result indicates that Equation (33) can be used to estimate the specific exergy of municipal solid waste, that is, the HHV has more impact on the exergy.

The overall average ratio of the specific exergy of MSW developed with the higher heating value was obtained as 1.036, showing that the value of exergy is slightly higher than the HHV. The ratio of exergies to heating values obtained in this work is similarly when compared with Szargut and Styrylska [7] model with ratio of 1.047. As their methods were commonly used for evaluating the chemical exergy of solid fuels. This result indicates that the present model is reliable and accurate. However, the slight variation in the ratios is due to different types of fuel used.

The specific exergies of inorganic matter in MSW calculated from Equation (22) using the chemical ash composition data in Table 4 are 0.86 and 1.79 kJ/kg for bottom ash and fly ash, respectively. These values are very small when compared with the average specific chemical exergy values, 24,387 kJ/kg of MSW (daf) estimated, demonstrating that the specific chemical exergy of inorganic matter can be neglected.

Conclusions

Following the evaluations of the previous equations for estimating the specific chemical exergy of solid fuels, the present proposed models in this study were found to be more accurate when using municipal solid waste as a fuel. All other methods have either ignored the inclusion of chlorine from the elemental compositions of waste or have used other solid fuels with a limited amount of MSW. In this work, a simple method for estimating the specific exergy of municipal solid waste on (daf) from their ultimate analysis based on HHV, standard entropy, and exergy equation of reaction was proposed.

The higher heating values of the estimated MSW showed a good correlation and a higher accuracy compared with previous models. It is calculated as

$$HHV = 0.364C + 0.863H - 0.075O + 0.028N$$
$$- 1.633S + 0.062Cl\,(MJ/kg)$$

The standard entropy of the estimated waste mixture has a rather low accuracy when compared with the waste fractions. However, the standard entropy can be used for the estimation of the specific chemical exergy of a solid, as it showed a similar result with the standard entropy of waste fractions; the standard entropy is expressed as

$$s^0_{msw} = 0.0101C + 0.0630H + 0.0106O + 0.0108N$$
$$+ 0.0155S + 0.0084Cl\,(kJ/K\,kg).$$

This result indicates that a higher heating value has more impact on the derivation of the specific chemical exergy of solid waste than entropy. In other words, the specific exergy of MSW mainly depends on the values of HHV.

Due to very low calculated values of specific chemical exergy of inorganic matter in MSW, the specific chemical exergy developed in this work is equal to the specific chemical exergy of the organic matter in MSW and is presented as

$$E_{msw} = 376.461C + 791.018H - 57.819O$$
$$+ 45.473N - 1536.242S + 100.981Cl\,(kJ/kg).$$

The results obtained demonstrate that the specific chemical exergy is always slightly higher than the highest heating value, indicating the validity and accuracy of the model.

The present correlation can be accepted for estimating the specific chemical exergy of MSW using the elemental compositions of the fuel within the range specified based on a dry ash-free basis. The model is applicable for the efficient modeling of a combustion system in a waste-to-energy plant.

Acknowledgments

The authors acknowledge the Nigerian Government and Michael Okpara University of Agriculture Umudike, Abia State, Nigeria, for supporting this work through TETfund Academic Staff Training and Development.

Conflict of Interest

None declared.

Nomenclature

AAE   average absolute error
A ash content in the waste (%)
ABE   average bias error
E chemical exergy (kJ)
e specific chemical exergy (kJ/kg) or (kJ/mol)
FC   fixed carbon (%)
G Gibbs energy (kJ/kg) or (kJ/mol)
H enthalpy (kJ/kg)
HHV higher heating value (kJ/kg)
MSW municipal solid waste
P pressure (kpa)
R^2 coefficient of determination

S	entropy (kJ/K)
S_{gen}	entropy generated
s	specific entropy (kJ/kg K) or (kJ/mol K)
T	Temperature (K)
V	volatile matter (%)

Subscripts

ba	bottom ash
daf	dry ash-free basis
est	estimate
exp	experiment
fa	fly ash
f	formation
fo	food
ioc	inorganic compound
max	maximum
msw	municipal solid waste or mixed solid waste
o	standard state
pl	plastic
R	reaction
tr	textile/rubber
wp	wood/paper

Superscripts

0	reference state

Greek Symbols

Δ	change
Σ	summation

References

1. Georgieva, K., and K. Varma. 1999. Municipal solid waste incineration. *World Bank Technical Guidance Report.* Washington D.C.
2. Szargut, J., D. R. Morris, and F. R. Steward. 1988. Exergy analysis of thermal, chemical, and metallurgical processes. Hemisphere, New York, NY.
3. Song, G., L. Shen, and J. Xiao. 2011. Estimating specific chemical exergy of biomass from basic analysis data. Ind. Eng. Chem. Res. 50:9758–9766.
4. Dincer, I., and M. A. Rosen. 2007. Exergy, energy, environment and sustainable development. Wiley, New York, NY.
5. Stepanov, V. S. 1995. Chemical energies and exergies of fuels. Energy 20:235–242.
6. Rant, Z. 1961. Towards the estimation of specific exergy of fuels (in German). Allg. Wärmetech 10:172–176.
7. Szargut, J., and T. Styrylska. 1964. Approximate evaluation of the exergy of fuels (in German). Brennst. Wärme Kraft 16:589–596.
8. Eisermann, W., P. Johnson, and W. L. Conger. 1980. Estimating thermodynamic properties of coal, char, tar and ash. Fuel Process. Technol. 3:39–53.
9. Shieh, J. H., and L. T. Fan. 1982. Estimation of energy (enthalpy) and exergy (availability) contents in structurally complicated materials. Energy Sources 6:1–46.
10. Ikumi, S., C. D. Luo, and C. Y. Wen. 1982. Method of estimating entropies of coals and coal liquids. Can. J. Chem. Eng. 60:551–555.
11. Bilgen, S., and K. Kaygusuz. 2008. The calculation of the chemical exergies of coal-based fuels by using the higher heating values. Appl. Energy 85:776–785.
12. Song, G., J. Xiao, H. Zhao, and L. Shen. 2012. A unified correlation for estimating specific chemical exergy of solid and liquid fuels. Energy 40:164–173.
13. Bejan, A., G. Tsatsaronis, and M. Moran. 1996. Thermal design and optimization. John Wiley, New York, NY.
14. Bilgen, S. 2014. The estimation of chemical availability (Exergy) values for various types of coals in geographical regions of Turkey. Energy Sources 36:830–842.
15. Rivero, R., and M. Garfias. 2006. Standard chemical exergy of elements updated. Energy 31:3310–3326.
16. Kaygusuz, K. 2009. Chemical exergies of some coals in Turkey. Energy Sources 31:299–307.
17. Standard thermodynamic properties of chemical substances. 2000. CRC Press. Available at http://www.update.uu.se/~jolkkonen/pdf/CRC_TD.pdf (accessed 3 July 2015).
18. Kotas, T. J. 1995. The exergy method of thermal plant analysis. Krieger Publishing Company, Malabar, FL.
19. Channiwala, S. A., and P. P. Parikh. 2002. A unified correlation for estimating HHV of solid, liquid and gaseous fuels. Fuel 81:1051–1063.
20. Sheng, C., and J. L. T. Azevedo. 2005. Estimating the higher heating value of biomass fuels from basic analysis data. Biomass Bioenergy 28:499–507.
21. Zhou, H., A. Meng, Y. Long, Q. Li, and Y. Zhang. 2014. Classification and comparison of municipal solid waste based on thermochemical characteristics. J. Air Waste Manag. Assoc. 64:597–616.
22. Zhou, H., Y. Long, A. Meng, Q. Li, and Y. Zhang. 2015. Classification of municipal solid waste components for thermal conversion in waste-to-energy research. Fuel 145:151–157.
23. Kathiravale, S., M. N. M. Yunus, K. Sopian, A. H. Samsuddin, and R. A. Rahman. 2003. Modeling the heating value of municipal solid waste. Fuel 82:1119–1125.
24. Domalski, E. S., and E. D. Hearing. 1996. Heat capacities and entropies of organic compounds in the condensed phase. Volume III. J. Phys. Chem. Ref. Data 25:1–525.

25. Pandey, K. K. 1999. A study of chemical structure of soft and hardwood and wood polymers by FTIR spectroscopy. J. Appl. Polym. Sci. 17:1969–1975.

26. Dietrich, F., and W. Gerd. 1983. Wood: chemistry, ultrastructure, reactions. De Gruyter, Berlin, pp. 66–181.

27. Bledzki, A. K., V. E. Sperber, and O. Faruk. 2002. Natural wood and fiber reinforcement in polymers. Smithers Rapra, Shropshire, United Kingdom. pp. 4–18.

28. Fundo, J. F., M. A. C. Quintas, and C. L. M. Silva. 2015. Molecular dynamics and structure in physical properties and stability of food systems. Food Eng. Rev. 7:384–392.

29. Wrolstad, R. E. 2011. Food carbohydrate chemistry. Wiley-Blackwell, West Sussex, United Kingdom.

30. Coultate, T. P. 2002. Food: the chemistry of its components. Royal Society of Chemistry, Cambridge, United Kingdom, pp. 7–408.

31. Brydson, J. A. 1999. Plastics materials, 7th ed. Butterworth-Heinemann, Oxford, United Kingdom. pp. 205–813.

32. Fan, Q. 2005. Chemical testing of textile, 1st ed. Woodhead Publishing, Cambridge, England.

33. Gordon, S., and Y. L. Hsieh. 2007. Cotton: science and technology. Woodhead Publishing, Cambridge, England.

34. Simpson, R. B. 2002. Rubber basics. Smithers Rapra, United Kingdom.

35. Wang, S. 2008. Application of solid ash based catalysts in heterogeneous catalysis. Environ. Sci. Technol. 42:7055–7063.

36. Song, G., L. Shen, J. Xiao, and L. Chen. 2013. Estimation of specific enthalpy and exergy of biomass and coal ash. Energy Sources 35:809–816.

37. Park, K., J. Hyun, S. Maken, S. Jang, and J. W. Park. 2005. Vitrification of municipal solid waste incinerator fly ash using Brown's gas. Energy Fuels 19:258–262.

38. Hla, S. S., and D. Roberts. 2015. Characterisation of chemical composition and energy content of green waste and municipal solid waste from Greater Brisbane, Australia. Waste Manag. 41:12–19.

39. Xiao, G., M. Ni, Y. Chi, B. Jin, R. Xiao, Z. Zhong, et al. 2009. Gasification characteristics of MSW and an ANN prediction model. Waste Manage. (Oxford) 29:240–244.

40. Xiao, G., M. J. Ni, H. Huang, Y. Chi, R. Xiao, Z. P. Zhong, et al. 2007. Fluidized-bed pyrolysis of waste bamboo. J. Zhejiang Univ. Sci. A 8:1495–1499.

41. Meraz, L., A. Domínguez, I. Kornhauser, and F. Rojas. 2003. A thermochemical concept-based equation to estimate waste combustion enthalpy from elemental composition. Fuel 82:1499–1507.

42. Zheng, J., Y. Jin, Y. Chi, J. Wen, X. Jiang, and M. Ni 2009. Pyrolysis characteristics of organic components of municipal solid waste at high heating rates. Waste Manage. (Oxford) 29:1089–1094.

Appendix

Table A1. Chemical characteristics of MSW (daf) used for derivation.

MSW groups, Subgroup and variety	Proximate analysis (wt %)			Ultimate analysis (wt %)						HHV (MJ/kg)	Reference
	A	V	FC	C	H	O	N	S	Cl		
Food waste											
1. Flour	–	–	–	42.78	6.19	48.39	2.48	0.15	–	18.157	[21]
2. Rice	0.40	84.42	15.18	45.97	6.35	45.74	1.67	0.25	–	18.213	[22]
3. Peanut shell	–	–	–	53.6	6.70	38.3	1.20	0.20	–	20.598	[21]
4. Pea	–	–	–	42.13	5.88	48.62	3.14	0.22	–	16.533	[21]
5. Scallion	–	–	–	48.12	6.27	41.74	3.09	0.78	–	18.042	[21]
6. Potato	3.15	79.52	17.33	44.41	5.33	47.82	1.81	0.64	–	17.656	[22]
7. Spinach	15.97	65.26	18.77	47.58	6.48	43.93	1.57	0.43	–	20.326	[22]
8. Celery	14.58	65.36	20.06	38.46	6.16	54.52	0.21	0.65	–	15.886	[22]
9. Pakchoi	18.44	63.97	17.59	43.37	5.93	48.64	1.25	0.81	–	23.173	[22]
10. Tangerine peel	2.91	76.49	20.6	48.74	5.92	43.83	1.43	0.08	–	19.024	[22]
11. Banana peel	10.85	64.38	24.77	35.8	4.79	54.93	4.37	0.10	–	18.385	[22]
12. Orange peel	2.44	76.27	21.29	43.93	5.64	48.93	1.34	0.07	0.08	18.550	[21]
13. Rib	–	–	–	52.92	8.83	25.63	2.29	0.32	–	17.277	[21]
14. Fish bone	39.82	56.25	3.93	63.87	8.01	19.08	8.39	0.64	–	26.245	[21]
15. Food waste	6.1	82.11	18.00	51.54	7.14	37.06	3.13	0.21	0.92	21.619	[38]
Wood waste											
16. Poplar wood	7.54	73.85	18.61	51.36	5.86	41.00	1.52	0.22	–	20.009	[22]
17. Poplar leaf	15.69	68.74	15.57	49.54	5.24	43.30	1.32	0.59	–	19.986	[22]
18. Chinar leaf	9.23	69.74	21.03	52.95	4.88	40.51	1.01	0.65	–	21.064	[22]
19. Gingko leaf	11.62	73.19	15.19	41.35	5.54	50.88	1.36	0.87	–	17.289	[22]
20. Pine wood	0.95	83.5	15.54	50.51	5.95	43.39	0.11	0.03	–	19.834	[21]
21. Sawdust	0.42	81	18.58	49.42	7.26	42.92	0.39	0.01	–	21.267	[21]
22. Wood	1.00	81.62	18.38	50.10	6.16	43.47	0.17	0.02	0.07	19.697	[38]
23. Wood chips	1.95	82.66	15.4	49.54	6.21	44.06	0.12	0.04	0.03	19.544	[21]
24. Bamboo	0.69	81.03	18.27	50.46	6.32	42.73	0.22	0.1	0.16	19.716	[21]
25. Leaves	8.92	73.7	17.38	47.25	5.57	46.26	0.19	0.73	–	18.882	[21]
26. Pine needles	–	–	–	52.57	6.3	40.44	0.54	0.16	–	20.843	[21]
27. King grass	7.44	74.12	18.43	46.91	5.89	46.3	0.7	0.21	–	19.428	[21]
Paper waste											
28. Blank printing paper	10.69	79.33	9.98	45.12	5.31	48.91	0.38	0.28	–	15.127	[22]
29. Tissue paper	0.52	90.47	9.01	45.18	6.13	48.32	0.25	0.11	–	17.340	[22]
30. Newspaper	8.07	79.54	12.39	48.01	5.71	45.86	0.33	0.09	–	18.666	[22]
31. Magazine	29.49	62.44	8.07	41.04	8.99	49.15	0.42	0.4	–	16.771	[21]
32. Writing paper	–	–	–	43.66	5.84	50.16	0.16	0.18	–	13.69	[21]
33. Cardboard	5.27	81.75	12.97	46.09	5.36	48.02	0.32	0.21	–	18.239	[21]
34. Carton	7.22	83.95	8.82	48.97	6.14	44.52	0.21	0.16	–	18.430	[21]
35. Printing paper	9.70	82.83	17.17	47.51	5.98	46.25	0.14	0.03	0.09	18.051	[38]
36. Packaging paper	12.2	85.88	14.12	46.92	5.92	46.74	0.22	0.09	0.10	17.654	[38]
Textile											
37. Absorbent cotton gauze	0.14	94.85	5.01	46.74	5.69	47.23	0.27	0.08	–	14.664	[22]
38. Cotton cloth	3.09	78.71	18.21	56.49	5.87	33.3	3.52	0.18	0.65	14.664	[21]
39. Wool	1.24	84.76	14.00	60.07	4.24	31.48	2.65	1.55	–	21.183	[21]
40. Acrylic fiber	0.14	75.25	24.61	66.78	5.2	7.31	20.26	0.45	–	29.812	[21]
41. Chemical fiber	–	–	–	48.09	7.16	34.06	9.43	1.26	–	21.959	[21]
42. Polyester taffeta	0.44	90.63	8.93	60.1	4.5	35.11	0.28	0.01	–	22.178	[21]
43. Terylene	0.49	88.6	10.91	62.16	4.14	33.12	0.29	0.28	–	20.963	[22]
44. Textiles	1.40	82.86	17.14	52.54	6.19	39.26	1.76	0.20	1.42	21.197	[38]
Plastics waste											
45. PS	0.04	99.57	0.39	86.06	6.27	1.93	5.73	–	–	38.946	[22]
46. LDPE	–	99.98	0.02	85.98	11.20	2.61	0.21	–	–	46.480	[22]
47. HDPE	0.18	99.57	0.25	85.35	12.70	1.90	0.05	0.14	–	46.444	[22]

Table A1. Continued.

MSW groups, Subgroup and variety	Proximate analysis (wt %)			Ultimate analysis (wt %)						HHV (MJ/kg)	Reference
	A	V	FC	C	H	O	N	S	Cl		
48. PVC	–	94.93	5.07	38.34	4.47	–	0.23	0.61	56.35	20.830	[22]
49. PET	0.09	90.44	9.47	63.01	4.27	32.69	0.04	–	–	23.111	[22]
50. PE	0.15	99.85	–	85.45	14.32	–	0.16	0.07	–	46.388	[21]
51. PP	0.02	99.97	0.01	85.41	12.51	1.85	0.23	–	–	46.248	[21]
52. Packaging plastic	3.90	95.21	4.79	75.75	9.78	12.00	0.35	0.03	2.08	26.951	[38]
53. Other plastic	1.30	99.09	0.91	84.90	9.63	0.97	3.35	0.03	1.11	41.135	[38]
Rubber waste											
54. Rubber	8.36	84.77	6.86	77.72	10.12	7.42	0	2.66	2.08	25.474	[21]
55. Tire	25.70	68.05	6.25	79.19	8.45	11.38	0.69	0.28	–	35.654	[21]
Other combustibles											
56. Other combustibles	20.40	90.83	9.17	70.48	8.79	17.53	1.63	0.83	0.74	32.161	[38]

All the proximate, ultimate analysis data and HHV on dry basis are converted to dry ash-free basis. Also all HHV are converted to MJ/kg.

Table A2. Chemical characteristics of MSW (daf) used for validation.

MSW groups, Subgroup and variety	Proximate analysis (wt %)			Ultimate analysis (wt %)						HHV (MJ/kg)	Reference
	A	V	FC	C	H	O	N	S	Cl		
Food waste											
1. Rice	0.42	87.74	11.84	44.2	5.73	48.75	1.20	0.1	0.02	18.048	[21]
2. Potato	–	–	–	42.09	6.5	49.06	2.12	0.23	–	16.912	[21]
3. Orange peel	2.91	76.49	20.6	48.74	5.92	43.72	1.43	0.19	–	19.024	[21]
4. Rib	38.22	61.56	0.23	51.61	6.38	31.91	9.48	0.69	–	22.716	[21]
Wood waste											
5. Wood	0.82	81.64	17.54	48.35	6.62	44.7	0.04	0.29	–	20.868	[21]
6. Wood chips	3.45	81.5	15.05	49.03	5.69	44.98	0.22	0.07	–	19.255	[21]
7. Wooden chopsticks	2.18	83.45	14.37	48.79	5.16	45.7	0.3	0.04	–	19.355	[39]
8. Bamboo	1.79	81.36	16.84	51.42	6.01	41.92	0.36	0.29	–	19.974	[40]
9. Leaves	9.43	74.32	16.25	47.18	5.61	46.35	0.18	0.68	–	20.278	[21]
10. King grass	–	–	–	48.37	6.30	44.58	0.49	0.25	–	22.127	[21]
Paper waste											
11. Newspaper	5.43	85.04	9.53	45.24	7.17	47.1	0.25	0.23	–	17.204	[21]
12. Printing paper	12.3	87.65	0.04	44.93	4.55	50.43	0.09	–	–	16.233	[21]
13. Cardboard	–	–	–	46.71	5.31	47.35	0.32	0.32	–	18.367	[21]
14. Toilet paper	0.52	90.47	9.01	45.18	6.13	48.32	0.25	0.11	–	17.337	[21]
15. Paper food cartons	6.93	–	–	48.07	6.55	45.04	0.16	0.17	–	18.137	[41]
16. Magazine stock	29.26	–	–	46.55	6.56	46.44	0.16	0.30	–	17.967	[41]
17. Plastic-coated paper	2.77	–	–	44.53	6.35	46.80	0.19	0.08	–	17.556	[41]
Textile											
18. Cotton cloth	1.52	84.53	13.95	46.51	5.8	46.98	0.43	0.28	–	17.699	[22]
19. Cotton	1.45	86.7	11.85	46.19	6.12	47.07	0.54	0.08	–	17.500	[21]
20. Wool	–	–	–	58.53	6.48	18.23	15.12	1.65	–	23.632	[22]
21. Shoe heel and sole	30.09	–	–	76.13	10.14	11.10	0.72	1.92	–	36.676	[41]
22. leather	10.1	–	–	66.74	8.90	12.79	11.12	0.44	–	22.892	[41]
23. Upholstery	2.80	–	–	48.46	6.28	44.86	0.31	0.10	–	17.891	[41]
Plastics waste											
24. PE	0.15	99.85	–	85.45	14.32	–	0.16	0.07	–	46.388	[42]
25. PP	0.16	99.84	–	84.3	14.44	1.05	0.18	0.03	–	45.842	[21]
26. PVC	0.04	95.16	4.8	38.75	5.21	–	0.22	–	55.82	22.575	[21]
27. Polyurethane	4.38	87.29	8.32	66.17	6.55	18.46	6.26	0.02	2.53	27.300	[41]
28. Plastic film	6.72	–	–	72.05	10.42	16.96	0.49	0.08	–	34.519	[41]
Rubber waste											
29. Rubber	15.38	65.26	19.36	89.18	8.54	–	1.23	1.05	–	39.473	[21]
30. Tire	19.27	63.11	17.61	88.56	8.52	0.88	0.75	1.29	–	37.364	[21]

All the proximate, ultimate analysis data and HHV on dry basis are converted to dry ash-free basis. Also all HHV are converted to MJ/kg.

Table A3. Standard entropies at 298.15K of organic compounds relevant to MSW.

Name	Formula	S^0 (kJ/kg K)
Food		
1. Allantoin	$C_4H_6N_4O_3$	1.233
2. Alloxan	$C_4H_2N_2O_4$	1.314
3. Arginine	$C_6H_{14}N_4O_2$	1.439
4. Asparagine	$C_4H_8N_2O_3$	1.322
5. Aspartic acid	$C_4H_7NO_4$	1.279
6. Citric acid	$C_6H_8O_7$	1.312
7. Creatine	$C_4H_9N_3O_2$	1.445
8. Cystine	$C_6H_{12}N_2O_4S_2$	1.347
9. D-Glutamic acid	$C_5H_9NO_4$	1.230
10. L-Lactic acid	$C_3H_6O_3$	1.579
11. L-Phenylalanine	$C_9H_{11}NO_2$	1.293
12. L-Proline	$C_5H_9NO_2$	1.425
13. Maleic acid	$C_4H_4O_4$	1.373
14. Malic acid	$C_4H_6O_5$	1.199
15. Methionine	$C_5H_{11}NO_2S$	1.552
16. Phenanthrene	$C_{14}H_{14}$	1.207
17. Trytophan	$C_{11}H_{12}N_2O_2$	1.229
18. Tyrosine	$C_9H_{11}NO_3$	1.181
19. Uric acid	$C_5H_4N_4O_3$	1.030
20. Valine	$C_5H_{11}NO_2$	1.527
21. Xanthine	$C_5H_4N_4O_2$	1.059
22. Stearic acid	$C_{18}H_{36}O_2$	1.531
23. Taurine	$C_2H_7NO_3S$	1.231
24. Urea	CH_4N_2O	1.742
25. Hexadecanoic acid	$C_{16}H_{32}O_2$	1.764
26. Adenine	$C_5H_5N_5$	1.118
27. Creatinine	$C_4H_7ON_3$	1.483
28. L-Serine	$C_3H_7O_3N$	1.419
29. L-Glutamine	$C_5H_{10}O_3N_2$	1.335
30. DL-Alanyl glycine	$C_5H_{10}O_3N_2$	1.460
31. Glycylglycine	$C_4H_8N_2O_3$	1.438
32. Alanine	$C_3H_7NO_2$	1.450
33. Cysteine	$C_3H_7NO_2S$	1.402
34. Dimethyl sulfone	$C_2H_6O_2S$	1.509
35. D-Lactic acid	$C_3H_6O_3$	1.593
36. Fumaric acid	$C_4H_4O_4$	1.447
37. Guanine	$C_5H_5N_5O$	1.061
38. Gycine	$C_2H_5NO_2$	1.379
39. Isoleucine	$C_6H_{13}NO_2$	1.586
40. Leucine	$C_6H_{13}NO_2$	1.586
41. L-Glutamic acid	$C_5H_9NO_4$	1.279
42. 1-Hexadecanol	$C_{16}H_{34}O$	1.864
43. Hypoxanthine	$C_5H_4ON_4$	1.070
44. Glycolide	$C_4H_4O_4$	1.354
Plastic		
45. 1,3,5-Trioxane	$C_3H_6O_3$	1.476
46. Benzophenone	$C_{13}H_{10}O$	1.346
47. Biphenyl	$C_{12}H_{10}$	1.358
48. Hexachloroethane	C_2Cl_6	1.002
49. Diphenyl carbonate	$C_{13}H_{10}O_3$	1.300
50. Diphenyl ether	$C_{12}H_{10}O$	1.372
51. Diphenylcarbinol	$C_{13}H_{12}O$	1.330
52. Polypropylene, isotatic	$(C_3H_6)_n$	1.662
53. Polypropylene, syndiotic	$(C_3H_6)_n$	1.798
54. Pyromellitic dianhydride	$C_{10}H_2O_6$	1.087

Table A3. Continued,

Name	Formula	S^0 (kJ/kg K)
55. Naphthalene	$C_{10}H_8$	1.306
56. Succinic acidic	$C_4H_6O_4$	1.417
57. Cyanuric acid	$C_3H_3N_3O_3$	1.947
58. Acetamide	C_2H_5NO	1.840
59. Durene	$C_{10}H_{14}$	1.166
60. Hexamethlenetetramine	$C_6H_{12}N_4$	1.116
61. Triphenylene	$C_{18}H_{12}$	1.273
62. Hydroquinone	$C_6H_6O_2$	1.182
63. Melamine	$C_3H_6N_6$	1.251
64. Phthalic acid	$C_8H_6O_4$	1.215
65. Phathalic anhydide	$C_8H_4O_3$	1.405
66. Triethylenediamine	$C_6H_{12}N_2$	1.947
67. 4,4'-diphenylmethane diisocyanate	$C_{15}H_{10}N_2O_2$	1.329
68. Polyisocyanurate	$(C_{15}H_{10}N_2O_2)_n$	1.175
69. Tridecanolactone	$C_{13}H_{24}O_2$	1.893
70. Polytridecanolactone	$(C_{13}H_{24}O_2)_n$	1.656
71. Polyvinylidene chloride	$(C_2H_2Cl_2)_n$	0.894
72. Polyvinyl chloride	$(C_2H_3Cl)_n$	1.042
73. poly(1-butene), isotactic	$(C_4H_8)_n$	1.836
74. Polystyrene	$(C_8H_8)_n$	1.294
Textile		
75. 3-Nitrobenzoic acid	$C_7H_5NO_4$	1.227
76. 1,2-Diphenylethene	$C_{14}H_{12}$	1.39
77. Adipic acid	$C_6H_{10}O_4$	1.504
78. 2-Methlnaphthalene	$C_{11}H_{10}$	1.547
79. Acenaphthene	$C_{12}H_{10}$	1.225
80. Anthracene	$C_{14}H_{10}$	1.164
81. 1,4-Benzoquinone	$C_6H_4O_2$	1.506
82. Diphenylamine	$C_{12}H_{11}N$	1.666
83. Pyrene	$C_{16}H_{10}$	1.112
84. Thiourea	CH_4N_2S	1.523
85. Ammonium thiocyanate	CH_4N_2S	1.842
86. 3-Nitroaniline	$C_6H_6N_2O_2$	1.276
87. Resorcinol	$C_6H_6O_2$	1.341
88. Triphenylmethane	$C_{19}H_{16}$	1.277
89. Triphenylmethanol	$C_{19}H_{16}O$	1.265
90. Isoquinoline	C_9H_7N	1.324
91. Acridine	$C_{13}H_9N$	1.161
92. 2-Nitrobenzoic acid	$C_7H_5O_4N$	1.247
93. 1,3-Phenylenediamine	$C_6H_8N_2$	1.429
94. Dicyanodiamide	$C_2H_4N_4$	1.538
95. ε-Caprolactam	$C_6H_{11}NO$	1.531
96. Poly-ε-Caprolactam	$(C_6H_{11}NO)_n$	1.529
97. Polyglycolide	$(C_4H_4O_4)_n$	1.304
Wood		
98. L-Sorbose	$C_6H_{12}O_6$	1.226
99. o-Cresol	C_7H_8O	1.530
100. Oxalic acid	$C_2H_2O_4$	1.220
101. p-Cresol	C_7H_8O	1.547
102. Sucrose	$C_{12}H_{22}O_{11}$	1.052
103. D-Mannitol	$C_6H_{14}O_6$	1.309
104. Pentachlorophenol	C_6HCl_5O	0.946
105. Galactose	$C_6H_{12}O_6$	1.14
106. Phenol	C_6H_6O	1.53
107. 2-Hydroxybenzoic acid	$C_7H_6O_3$	1.29
108. Glucose	$C_6H_{12}O_6$	1.161

Table A3. Continued,

Name	Formula	S^0 (kJ/kg K)
109. Xylose	$C_5H_{10}O_5$	0.956
110. 3-Hydroxybenzoic acid	$C_7H_6O_3$	1.281
111. 4-Hydroxybenzoic acid	$C_7H_6O_3$	1.272
112. Benzoic acid	$C_7H_6O_2$	1.372
113. Catechol	$C_6H_6O_2$	1.364
114. Lactose	$C_{12}H_{22}O_{11}$	1.128
115. o-Hydroxybenzoic acid	$C_7H_6O_3$	1.29
116. o-Hydroxybenzoic acid	$C_7H_6O_3$	1.281
117. o-Hydroxybenzoic acid	$C_7H_6O_3$	1.272

Source: [3, 8, 10, 12, 17, 24].

11

Hydrotreating reactions of tall oils over commercial NiMo catalyst

Jinto M. Anthonykutty[1], Juha Linnekoski[1], Ali Harlin[1] & Juha Lehtonen[2]

[1]VTT Technical Research Centre of Finland, Biologinkuja 7, Espoo, FI-02044 VTT, Finland
[2]Department of Biotechnology and Chemical Technology, School of Chemical Technology, Aalto University, PO Box 16100, Aalto FI-00076, Finland

Keywords

Commercial NiMo catalyst, decarboxylation, hydrogenation, hydrotreating, tall oil

Correspondence

Jinto M. Anthonykutty, VTT Technical Research Centre of Finland, Biologinkuja 7, Espoo, FI-02044 VTT, Finland.
E-mail: jinto.manjalyanthonykutty@vtt.fi

Funding Information

Jinto Manjaly Anthonykutty acknowledges the financial support from VTT Graduate School. The authors also acknowledge Stora Enso for supporting this research and their vision on wood-based olefins.

Abstract

Catalytic hydrotreating is an attractive method for upgrading bio-derived oils into renewable feedstocks with less oxygen content, suitable for producing valuable hydrocarbons through various petro-refinery processes. This study evaluates the catalytic activity of a commercial alumina (Al_2O_3) supported NiMo catalyst for hydrotreating tall oil feeds such as crude tall oil (CTO), distilled tall oil (DTO), and tall oil fatty acid (TOFA). Catalytic experiments carried out in a bench-scale fixed bed reactor set-up at different process conditions [space velocity (1–3 h^{-1}), temperature (325–450°C), and H_2 pressure (5 MPa)] produced a wide-range of products from tall oil feeds. Hydrotreating of TOFA produced highest yield of n-alkanes (>80 wt%) compared to DTO and CTO hydrotreating. A high conversion of fatty acids and resin acids was obtained in DTO hydrotreating. In CTO hydrotreating, a drop in conversion of fatty acids and resin acids was observed especially at the lowest temperature tested (325°C). The study revealed that there are various deoxygenation pathways preferential at different hydrotreating temperatures. As an example for TOFA, the decarboxylation route is dominant over the hydrodeoxygenation route at high temperatures (>400°C).

Introduction

Catalytic hydrotreating is an essential petro-refinery method to reduce the content of heteroatoms (S, N, and O) from raw materials, which has been widely employed to upgrade refinery feeds prior to such processes as catalytic reforming, catalytic cracking, and steam cracking [1]. In petro-refineries that use sulfur-rich fossil-fuel-based feeds, the catalytic hydrotreating technology is fully developed specifically for the hydrodesulfurization (HDS) of petroleum products and naphtha streams [2–4]. However, the recent research developments prompted by the increasing demand of bio-based fuels and chemicals reinvent the importance of developing optimal hydrotreating (hydrodeoxygenation, HDO) methods for reducing the amount of oxygenates from complex bio-derived feeds before the thermochemical conversion process into fuels or chemicals.

The amount of oxygenates in bio-derived oils vary depending on the origin of raw materials. Bio-oils produced by pyrolysis of lignocellulosic biomass are rich in oxygenates (35–40%). Therefore, deep hydrotreatment conditions are required to achieve complete deoxygenation from such bio-oils, inducing high processing cost [5]. Plant-based oils or vegetable oils are another class of bio-derived feedstock. Vegetable oils have been extensively studied as a feedstock for the production of bio-chemicals and biofuels through a catalytic HDO route [6–9]. The patented bio-synfining process produces naphtha range hydrocarbons for conventional steam cracking process from vegetable oils and fats [10, 11]. However, most of the widely studied vegetable oils are edible and remain as less favorable feedstocks as their selection to a refinery may create several environmental, economic, and societal issues. Therefore, a proper selection of a sustainable feedstock is vital. Tall oil, a by-product of the Kraft-pulping process, is nonfood chain affecting, economically feasible and low oxygen content feedstock [12]. Tall oil upgrading has already been received substantial attention as a sustainable bio-refinery method for producing biofuels and refinery feeds [12–14]. Tall oil comprises fatty acids, resin acids, and sterols. Different tall oil feeds such as crude tall oil (CTO), distilled tall oil (DTO), and tall oil fatty acid (TOFA) are commercially available, and can be used as valuable raw materials for chemical production. CTO is produced by means of the acidulation of tall oil soap skimmings from the black liquor which is obtained from pulping. DTO and TOFA are obtained from CTO distillation along with other fractions. Among the aforementioned tall oil feeds, CTO is regarded as the most cost-competitive raw material [15].

Sulfided molybdenum catalysts on alumina supports with nickel or cobalt as promoter metals are the most widely employed industrial catalysts for hydrotreating [16]. Supported CoMo and NiMo catalysts are more active in sulfided forms. There have been many studies reported on the origin of catalytic synergy between two main group elements in hydrotreating catalysts [17]. Different theories on the origin of catalytic activity and the nature of active sites can be found in literature [17, 18]. Kubicka et al. [19] reports that during hydroprocessing of vegetable oils over a sulfided NiMo catalyst, nickel sulfide phases are more active for decarboxylation than molybdenum sulfide phases. Therefore, the extent of HDO (hydrogenation/dehydration) versus decarboxylation reactions is highly dependent on factors such as promotor metal (Ni) concentrations and the dispersion of sulfidic phases. It is already learnt from previous studies that in comparison with CoMo catalysts, NiMo catalysts are more active for the HDO of aliphatic oxygenates [20]. Moreover, NiMo catalysts are found to be effective for hydrodeoxygenating cyclic oxygenates to cycloalkanes and also for causing ring opening reactions from cyclics at high temperatures, which produces valuable hydrocarbons for further refinery processes [12, 21]. Based on these observations, it can be suggested that NiMo catalyst is more applicable than a CoMo catalyst for hydrodeoxygenating bio-derived oils such as tall oil which contains aliphatic oxygenates as well as cyclic oxygenates. During catalytic hydrotreating of tall oil, oxygenates are removed by various deoxygenation routes which produces a wide range of hydrocarbons mainly paraffin range hydrocarbons [22, 23].

Among tall oil fractions, catalytic deoxygenation of TOFA has been studied with considerable interest for producing a hydrocarbon fraction suitable as a diesel fuel [24]. Egeberg et al. [2] reports that hydroconversion of TOFA can be slightly different than that of triglycerides (vegetable oils) as the former produces more methane by means of methanation reaction, whereas the latter produce more propane by the scission of glycerol backbone. Catalytic deoxygenation of TOFA investigated by other researchers reveal that a significant yield of n-heptadecane (C_{17}) with high selectivity can be obtained from TOFA using a palladium mesoporous carbon catalyst [24]. Low temperatures are found to be more favorable for the formation of n-octadecane from fatty acids through a HDO route. Hydrotreating of CTO has been discussed in the literature through several patents which mainly focuses the application of hydrotreated products as a diesel fuel [25–27]. Previously, we successfully reported on the use of hydrotreated DTO and CTO as a steam cracker feed [22, 23]. A high degree of deoxygenation was obtained at low temperatures (325–400°C) and space velocity of weight hourly space velocity (WHSV) = 1 h^{-1}, yielding a maximum in paraffins. Our group has also studied the hydrotreating chemistry of sterols in CTO and proposed a reaction scheme for the HDO of sterols at low temperatures in line with studies on the HDO of phenolic compounds and saturated alicyclic alcohols over sulfided catalysts [23]. Importantly, in our study, a high conversion of sterols was obtained at the tested conditions irrespective of space time and temperature.

In this study, a comparative study is presented using different tall oil feedstocks (TOFA, DTO, and CTO) The effect of space time and temperature on the hydrotreating of DTO and CTO is already been reported [22, 23], therefore, they are not discussed further in this study. More specifically, this study presents the results of TOFA hydrotreating under the applied conditions, which are then compared with the results of DTO and CTO hydrotreating. This comparison has carried out on the basis of achieved product distribution, composition of organic phase samples and the conversion of acid fractions at the most favorable space time (WHSV = 1 h^{-1}). The influence of feedstock type and reaction severity (temperature) during hydrotreating of tall oils (6 h of time on stream) over a sulfided NiMo catalyst is thus studied in this article.

Materials and Methods

Commercially available TOFA (SYLFAT® 2) and DTO (SYLVATAL® 25/30 S); and CTO obtained from Stora Enso pulping facilities in Finland were used as feeds for hydrotreating experiments. Detailed chemical and elemental composition and total acid number of the employed feeds are represented in Tables 1 and 2. A commercial alumina (Al_2O_3) supported NiMo catalyst in sulfided form was employed for hydrotreating studies. The pre-sulfidation was carried out by using a H_2S/H_2 mixture for 5 h at 400°C (H_2S/H_2 = 5 vol. %). The test runs (6 h) with each tall oil feed were performed in a continuous down flow fixed bed reactor (stainless steel tube, length: 450 mm long, internal diameter: 15 mm) at different process conditions; temperature = 325–450°C, H_2 partial pressure = 5 MPa, weight hourly space velocity (whsv) = 1–3 h^{-1} and H_2/feedstock molar ratio = 17.4. Duplicate experiments were carried out for certain test runs so as to check the reproducibility of yield. In a typical experiment, a known amount of the catalyst (6, 3, and 2 g corresponds to WHSV 1 h^{-1}, WHSV 2 h^{-1}, and WHSV 3 h^{-1}) was loaded in the reactor which was then placed in an oven (Oy Meyer Vastus, Monninkylä, Finland). The catalyst bed temperature was monitored by a temperature controller (TTM-339 series). A set amount of feedstock was fed to the reactor along with hydrogen (62 mol/kg of CTO). TOFA and DTO were used as received for hydrotreating experiments, whereas CTO was preheated in order to reduce the viscosity, and thus to enhance the feeding of the material to the reactor. Gas chromatographic tools such as such as GC-MS (HP-5 MS column), GC×GC-MS (ZB-5 HT inferno and ZB-35 HT inferno columns)and GC×GC-FID/(TOF-MS) (Rtx-1 PONA and BPX-50 columns) were used for quantitatively (using external calibration method) and qualitatively analyzing the organic phase samples obtained from hydrotreating of tall oils [22, 23]. GC-FID (DB-23 and Zebron ZB-1 columns) analyses were carried out to calculate the residual amounts of fatty acids, resin acids, and sterols from organic phase samples. The amount (wt%) of water in aqueous phase samples was determined by Karl Fischer (KF) titration (method: ASTM E 203). Analysis of gas phase components was carried out by FT-IR and micro-GC instruments. The residual oxygen composition of organic phase samples from hydrotreated tall oils was determined by elemental analysis using CHN equipment based on ASTM D 5291 method. Sulfur content was determined by means of ASTM D 4239 method.

Terminology and calculations

For the sake of clarity, the term paraffins in this study correspond to n-alkanes and i-alkanes. Nonaromatics denote cycloalkanes or naphthenes (-mono, -di, or -tri) and other cyclic oxygenates as well as fatty alcohols and unsaturated hydrocarbons. The reactions which mention in this study under hydrotreating conditions are hydrogenation of

Table 1. Feedstock composition.

Composition (wt.%)	TOFA	DTO	CTO	
Free and bonded Fatty Acids	**92.6**	**71.3**	**48.1**	
			Free	Bonded
(16:0) Palmitic acid	0.40	0.20	2.1	0.04
(17:0) Margaric acid	0.80	0.30	0.4	0.01
(18:0) Stearic acid	0.90	0.70	0.8	0.01
(18:1) Oleic acid	27.4	15.3	8.6	0.16
(18:1) 11-octadecenoic acid	0.70	0.50	0.3	0.01
(18:2) 5.9-octadecadienoic acid	0.50	0.30	0.2	0.02
(18:2) conj. octadecadienoic acid	6.00	8.30	5.5	0.36
(18:2) Linoleic acid	23.1	24.3	19.1	2.4
(18:3) Pinolenic acid	8.00	4.4	1.6	0.5
(18:3) Linolenic acid	4.30	0.60	0.6	0.02
(18:3) conj. octadecatrienoic acid	0.20	1.80	0.4	0
(20:0) Arachidic acid	0.80	0.40	0.7	0.01
(20:3) 5.11.14-eicosatrienoic acid	7.90	7.60	1.4	0.08
(20:3) 7.11.14-eicosatrienoic acid	1.30	0.60	0.2	0.2
Other fatty acids	10.3	6.00	1.8	0.28
Free and bonded Resin Acids	**1.3**	**23**	**28.5**	
			Free	Bonded
8,15-isopimaradien-18-oic acid	0.1	0.5		
Pimaric acid	0.7	4.8	2.6	0.03
Sandaracopimaric acid	–	0.3	0.8	0.03
Diabietic acid	–	0.5	–	–
Palustric acid	–	2.2	3.9	0.01
Isopimaric acid	–	1.1	1.3	0.01
7. 9 (11) -abietic acid	–	0.4	0.1	0
13-B-7.9(11)-abietic acid	–	0.3	0.2	0.02
Abietic acid	0.1	7.7	9.1	0.1
Dehydroabietic acid	–	3.6	2.4	0.04
Neoabietic acid	–	0.4	3.3	0.02
Other resin acids	0.4	1.3	2.8	0.7
Neutrals		**23**		
Campesterol	–	–	0.02	–
Campestanol	–	–	0.17	–
β-Sitosterol	–	–	2.5	–
β-Sitostanol	–	–	0.5	–
Lupeol	–	–	0.5	–
Cycloartenol	–	–	0.2	–
24-methylenecycloartenol	–	–	2.5	–
Prenol-7	–	–	0.07	–
α-Sitosterol	–	–	1.1	–
Methyl betulinate	–	–	0.02	–
Betulin	–	–	0.3	–
Betulinic acid	–	–	0.5	–
Prenol-8	–	–	0.9	–
Other	–	–	12.6	–

Table 2. Elemental composition total acid number (TAN).

Composition	TOFA	DTO	CTO
Elemental composition			
C (wt%)	76.8	77.4	78.6
H (wt%)	11.6	11.1	11.2
O (wt%)	11.6	11.4	9.9
N (ppm)	12	30	100
S (ppm)	53	161	1800
P (ppm)	–	–	36
Alkali metals and alkaline earth metals (ppm)	–	–	56
Total acid number (TAN) mg/kg	326	300	129.5

double bonds, decarboxylation, decarbonylation, HDO, isomerization, hydrocracking of alkane and cyclic structures, and thermal cracking reactions. HDO is represented as a reaction resulted from complete hydrogenation with saturation of double bonds as the first step. Selective deoxygenation denotes the catalytic deoxygenation reactions at low temperatures (Mainly HDO, removal of oxygen as water). Nonselective deoxygenation represents the deoxygenation reactions which occur at high temperatures (thermal or catalytic cracking, removal of oxygen as CO_2).

The mass based yield (%) of various product streams in this study is calculated as a relative yield based on the amount (g/h) of feedstock entering to the reactor. Conversion of fatty acids and resin acids was calculated by assuming that residual acid fractions are mostly present in the organic phase. The equation used for calculating the conversion is

$$\text{Conversion (\%)} = \frac{n_{A}, \text{Feed} - n_{A}, \text{O.P.}}{n_{A}, \text{Feed}} \times 100,$$

n_{A}, Feed, is the total mol of acids (fatty acid/resin acid) present in the feed, n_{A}, O.P., is the total mol of acids (fatty acid/resin acid) present in the organic phase.

Product yield of paraffins (n-alkanes + i-alkanes) was calculated by assuming that paraffins are solely formed from fatty acid fraction in tall oils. Product yield was calculated by using the following equation.

$$\text{Product yield (\%)} = \left[\frac{\text{Total mol of paraffins in O.P}}{\text{Total mol of fatty acids in Feed}} \right] \times 100$$

The reaction steps discussed in this study in association with the hydrogenation and dexoygenation of fatty acids (linoleic acid and oleic acid) in tall oils over a sulfided NiMo catalyst are shown below by Equations (1), (2), (3), (4), (5), and (6)

Hydrogenation:

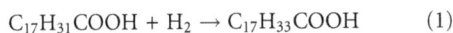

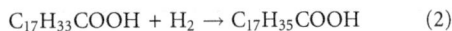

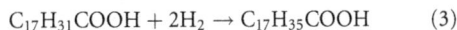

$$C_{17}H_{31}COOH + H_2 \rightarrow C_{17}H_{33}COOH \quad (1)$$

$$C_{17}H_{33}COOH + H_2 \rightarrow C_{17}H_{35}COOH \quad (2)$$

$$C_{17}H_{31}COOH + 2H_2 \rightarrow C_{17}H_{35}COOH \quad (3)$$

Deoxygenation:

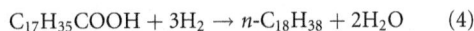

$$C_{17}H_{35}COOH + 3H_2 \rightarrow n\text{-}C_{18}H_{38} + 2H_2O \quad (4)$$

$$C_{17}H_{35}COOH + H_2 \rightarrow n\text{-}C_{17}H_{36} + CO + H_2O \quad (5)$$

$$C_{17}H_{35}COOH \rightarrow n\text{-}C_{17}H_{36} + CO_2 \quad (6)$$

Experimental Results

Hydrotreating of TOFA

Hydrotreating experiments with TOFA on a NiMo catalyst resulted in liquid product streams which differ in color and yield depending on the conditions in the reactor. Importantly, the samples from TOFA hydrotreating at low temperatures (<350°C) appeared as oily samples with the tendency of solidification at room temperature. It was also observed that the temperature, 300°C, tested with TOFA resulted the excessive deposition of wax; thus, reactor plugging. In line with earlier studies [22, 23], mass balance error (±5%) was higher in experiments at elevated temperatures (>400°C) in comparison with low-temperature experiments. Liquid products obtained from TOFA hydrotreating were separated into organic and aqueous phases. Table 3 shows the detailed product distribution in organic phases obtained from TOFA hydrotreating at various process conditions. As hydrotreating of TOFA produces more paraffins than other tall oil feeds, the product distribution of major paraffinic hydrocarbons such as n-octadecane and n-heptadecane obtained in TOFA hydrotreating at different process conditions is presented in this study as a separate section.

Product distribution of C_{17} and C_{18} hydrocarbons in TOFA hydrotreating

The amount of n-octadecane and n-heptadecane obtained in TOFA hydrotreating on a sulfided NiMo catalyst is shown in Table 3. It is clear from Table 3 that a significant drop in the production n-octadecane occurred with TOFA from 325 to 450°C especially at the longest space time tested (WHSV = 1 h^{-1}), whereas n-heptadecane production steadily increased until 400°C and then sharply decreased at 450°C. As already noted from earlier studies [22, 23], low temperatures (<400°C) are more favorable for the formation of n-octadecane through a HDO route. As the temperature increases, routes consuming less hydrogen such as decarboxylation and decarbonylation are

Table 3. Product distribution of the organic phase obtained from the hydrotreating of TOFA at different process conditions.

| Temp (°C) | 325 | 350 | 400 | 450 | 325 | 350 | 400 | 450 | 325 | 350 | 400 | 450 |
WHSV (h^{-1})	1	1	1	1	2	2	2	2	3	3	3	3
Composition (wt%)												
Paraffins												
nC_7-C_9	0.4	1.2	6.8	10.1	0.9	1.8	4.4	8.3	0.2	0.8	3.1	9.2
nC_{10}-C_{16}	1.4	2.2	4.3	8.6	4.2	2.9	3	14.1	2.8	1.9	4.4	8.3
nC_{17}	27.7	36	36	27	24	31	31	20	21.5	29	30	23
nC_{18}	53	44	32	22	49.8	49	35	29	43	41	33	33
nC_{19}	1.2	1	1.3	2.1	3.8	2.4	0.9	1.1	0.8	0.7	0.6	0.4
nC_{20}	2	0.8	1.4	1.5	2.4	2.1	0.2	0.9	2.9	1.5	1.4	0.2
*i*alkanes	2.1	2.5	4.1	6	1.4	1.7	3.4	3	1.4	2.1	2.7	2.9
Total	87.6	87.7	86	78.1	86.5	90.9	77.9	76.4	72.6	77	75.2	77
Nonaromatics												
Cycloalkanes	0.08	0.2	1.4	3	0.9	0.4	0.4	2.5	0	0.03	0.9	0.3
Olefins	0.07	0.4	0	0	0.08	0.03	0.01	0.6	0.1	0.2	0.5	0.6
Alcohols	2.3	3.5	2.7	0.3	0.4	0.4	1.8	1	0.3	1	1	0.9
Other	9.5	7.9	4.9	15.8	5.1	7	9	9.1	13.9	5.9	6.8	6
Total	11.9	12	13	19.1	6.4	7.83	11.2	13.2	14.3	7.13	9.2	7.8
Aromatics												
Monoaromatics	0	0	1.7	4.1	0.07	0.04	0.9	1.4	0	0.04	0.8	1.2
Overall wt%	99.5	99.7	96.7	91.3	93	98.7	90	90	86.9	84	85.2	86
Residual oxygen	0.8	0.8	0.46	0.4	1.3	0.9	0.7	0.9	1.6	1.9	1.1	1

prominent, producing a significant amount of *n*-heptadecane at these temperatures. Higher concentration of *n*-heptadecane over *n*-octadecane was observed beyond 350°C at the longest space time tested (WHSV = 1 h^{-1}). This observation is in agreement with the earlier report on the promotion of decarboxylation route over HDO as a function of increasing temperature on a sulfided NiMo catalyst [28]. It may also imply that decarboxylation + decarbonylation reactions predominate over HDO reaction solely at longer space times, that is, WHSV = 1 h^{-1} with TOFA. Importantly, in comparison with hydroprocessing of vegetable oils such as rapeseed oil and sunflower oil over a NiMo catalyst [29], a similar yield of *n*-alkanes from TOFA hydrotreating at lower temperature (350°C) was achieved, especially for the *n*-C_{17} + *n*-C_{18} fraction. It is noteworthy that the amount of oleic acid ($C_{18:1}$) and linoleic acid ($C_{18:2}$) is considerably lower in TOFA compared to the aforementioned vegetable oils. Interestingly, in TOFA hydrotreating, irrespective of space time a similar product distribution of *n*-heptadecane and *n*-octadecane was observed at 400°C.

Comparison of TOFA, DTO, and CTO hydrotreating

Overall product distribution

Figure 1 shows the yield of various product streams obtained as a function of temperature at WHSV = 1 h^{-1} in hydrotreating of tall oils. As noted from Figure 1,

among the three tall oil feeds tested, TOFA was found to give the highest yield of liquid products and organic products at different temperatures. It was also observed with the employed tall oil feedstocks that the lowest temperature tested (T = 325°C) resulted to a maximum yield of liquid products. As expected, maximum gas product yield was obtained at 450°C with each tall oil feeds. At this temperature, the highest yield of gas products was obtained from CTO in comparison with DTO and TOFA. The FT-IR detected gas phase components were CO_2, CO, methane, ethane, ethylene, propane, propylene, and butane. Importantly, lighter alkanes and alkenes were obtained solely at high temperatures especially from CTO and DTO. As can be seen from Figure 1, among three tall oil feedstocks the highest yield of water was obtained from TOFA. The trend with the production of water from tall oils followed the decreasing order as TOFA–DTO–CTO. Water is obtained as a by-product of HDO (hydrogenation/dehydration) reaction during hydrotreating.

Composition of organic phase

Figure 2 shows a comparison of the product composition of organic phase samples as a function of temperature obtained in TOFA, DTO, and CTO hydrotreating at WHSV 1 h^{-1}. As can be seen from Figure 2, maximum amount (wt%) of paraffins was obtained from TOFA among three tall oil feeds. This observation appear as an

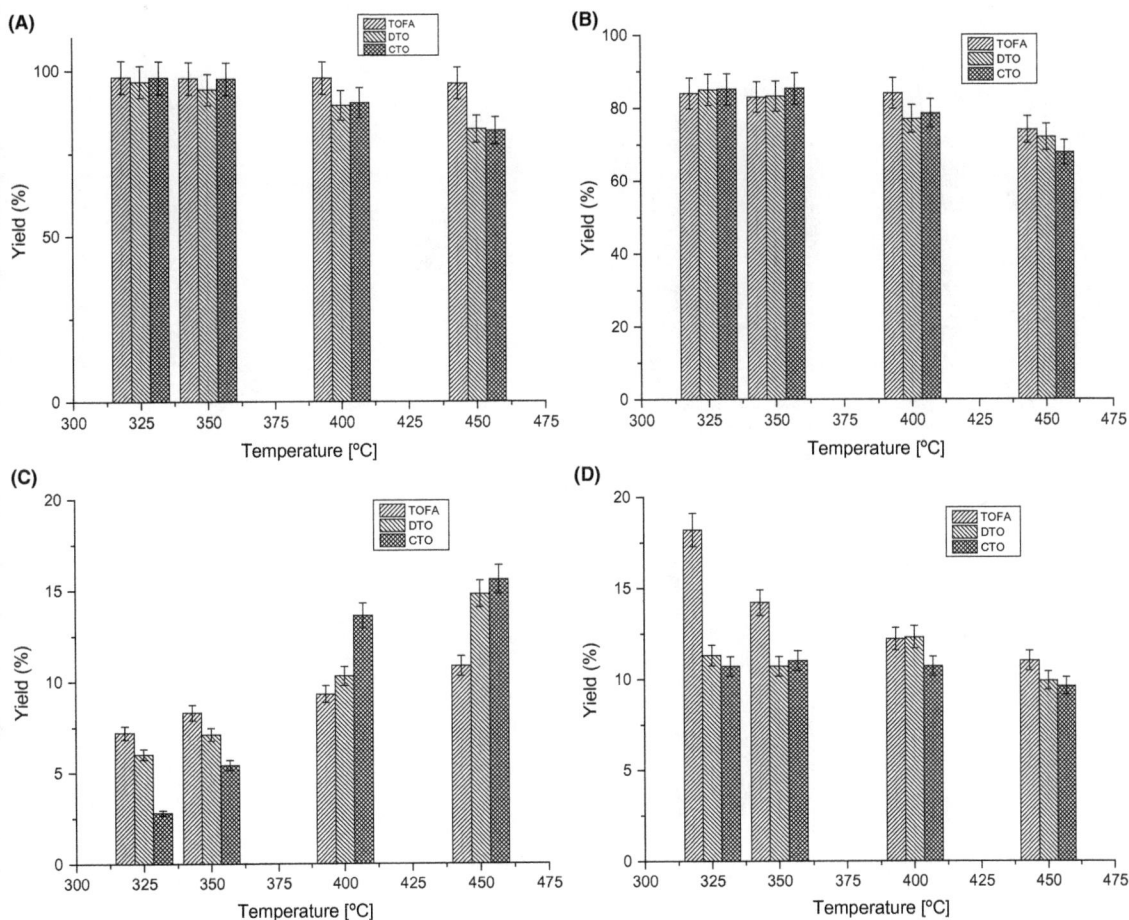

Figure 1. Yield (%) of various product streams obtained in hydrotreating of tall oils over a NiMo catalyst, WHSV = 1 h^{-1}, T = 325–450°C (A) liquid product yield, (B) organic phase product yield, (C) gas product yield, and (D) water yield.

obvious result in relation to the higher amount (wt%) of fatty acids in TOFA.

Product distribution of aromatic and nonaromatic hydrocarbons

As can be seen from Figure 2, nonaromatics such as cyclic structures (monocyclic and polycyclic hydrocarbons), alcohols, unsaturated hydrocarbons, and sterol derivatives were obtained in higher amount with CTO in comparison with other tall oil feeds. Among the three tall oil feeds, maximum production of aromatics was obtained from DTO. It has been reported previously that the aromatics from DTO hydrotreating comprised primarily of monoaromatics such as substituted benzenes in addition to polyaromatics [22]. Organic phases obtained from three feedstocks especially from

CTO also comprised some sulfur (100–500 ppm) and nitrogen (10–20 ppm) [23].

Fatty acid and resin acid conversion

Fatty and resin acid conversions were evaluated based on GC-FID analysis. Figure 3 shows the fatty and resin acid conversions obtained with different tall oil feedstocks at the most favorable space time (WHSV = 1 h^{-1}) as a function of temperatures. Figure 3A shows that fatty acid conversion achieved in a high extent with TOFA and DTO. A similar conversion level achieved with fatty acids in TOFA and DTO in low-temperature hydrotreating experiments (325–400°C). In comparison with TOFA and DTO, a drop in conversion of fatty acids was obtained in CTO hydrotreating under the tested conditions. Figure 3B shows that elevated conversion level was reached for resin

(A)

(B)

(C)

(D)

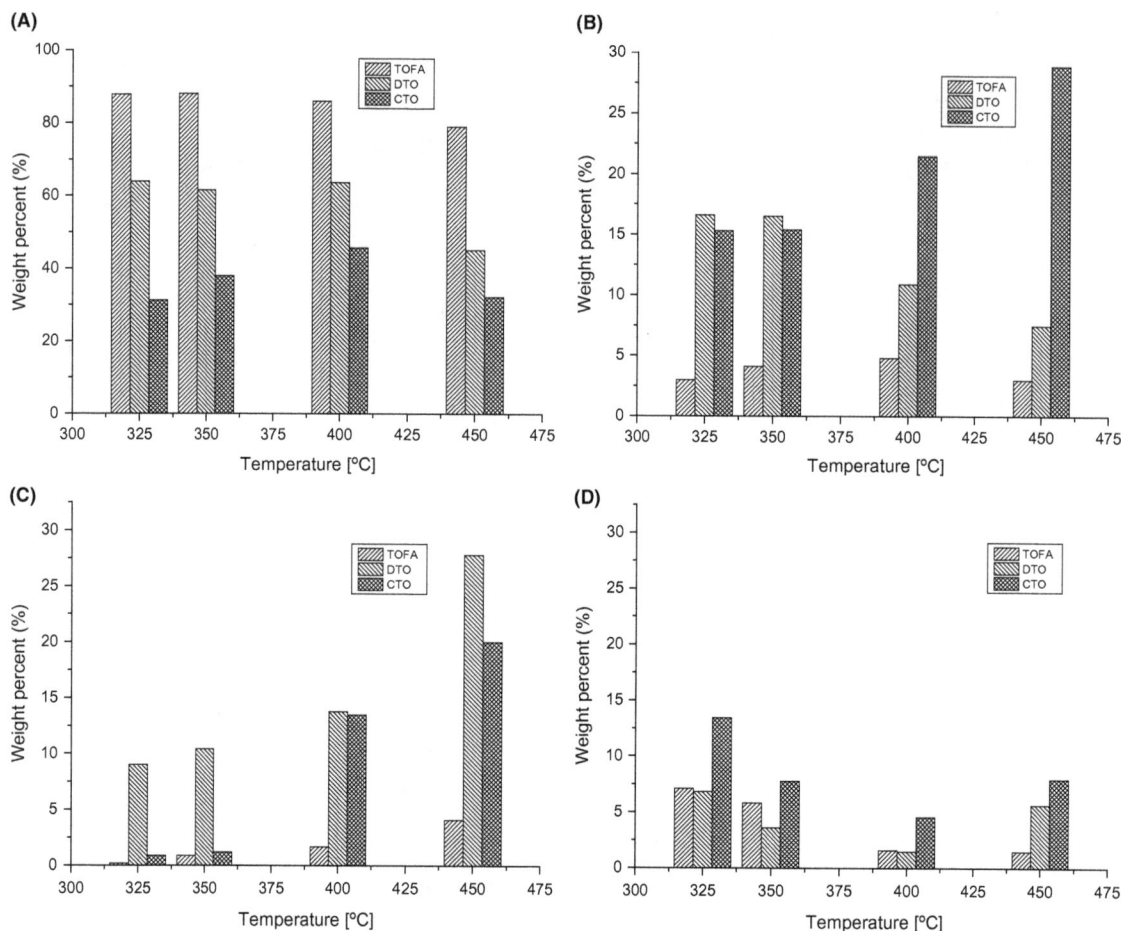

Figure 2. Product composition (wt%) of organic phase samples in tall oil hydrotreating (A) paraffins,(B) nonaromatics, (C) aromatics, and (D) residual fractions; WHSV =1 h^{-1}, T = 325–450°C.

acids as well in DTO hydrotreating. In CTO hydrotreating, a low conversion level of resin acids was obtained in line with the conversion of fatty acids from the same feedstock.

Furthermore, in order to get more insight into the behavior of fatty acids and resin acids during hydrotreating of tall oils, the conversion of individual fatty acids and resin acids in each tall oil feedstock was assessed. Figure 4A–I shows the conversion of linoleic acid, oleic acid, and dehydroabietic acid obtained during the hydrotreating of TOFA, DTO, and CTO at different space times (WHSV = 1–3 h^{-1}) as a function of temperature. It should be stated that all fatty acids and resin acids other than the aforementioned ones have achieved more or less full conversion levels during the hydrotreating of tall oils under the tested conditions; therefore, they are not included in the results.

Discussion

Based on distribution of gaseous products obtained from tall oil hydrotreating experiments, it can be assumed that in TOFA and DTO hydrotreating at high temperatures (>400°C), mainly nonselective deoxygenation occurs with a sulfided NiMo catalyst by means of cracking (thermal or catalytic) [12, 28] which requires no hydrogen and produces short-chain oxygenates and CO_2 as well as hydrocarbons (cycloalkanes and aromatics) [30]. With CTO, it is suggested that nonselective deoxygenation can be lower at high temperatures (>400°C) from sterols [23] although sterol hydrotreating chemistry at high temperature has not been elucidated yet. The gas product composition with high concentration (~60% yield of total gaseous products) of lighter alkanes and alkenes from CTO hydrotreating may imply that sterols in CTO are mainly converted by thermal

(A)

(B)

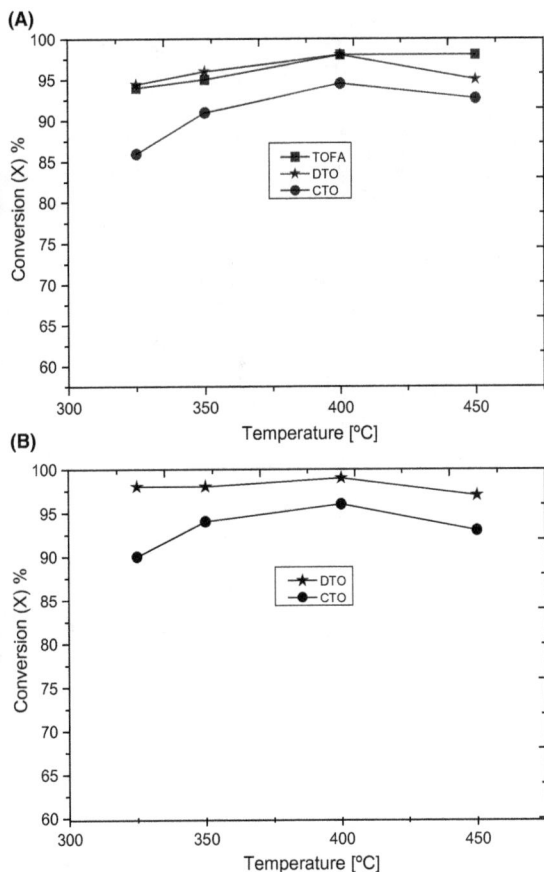

Figure 3. Conversion of fatty acids (A) and resin acids (B) obtained in hydrotreating of TOFA, DTO, and CTO at WHSV = 1 h^{-1}, T = 325–350°C.

decomposition (cracking) of side chains as well as the ring opening at high temperatures (>400°C). The thermal decomposition and ring opening reactions from sterols produce gas phase light hydrocarbons along with oxygenates and unsaturated cyclic structures [31]. A ring opening and thermal decomposition route is also expected from resin acids in CTO and DTO at high temperatures, producing lighter alkanes and alkenes [32] On the other hand, the increased yield of gaseous products in low temperature (<400°C) TOFA and DTO hydrotreating in comparison with the CTO hydrotreating at similar conditions is attributed to the increased formation of CO_2 (~40%), CO (~20%), and methane (~20%) through selective deoxygenation routes. Interestingly, in contrast with methane, propane was obtained only in minor yields in low temperature (<400°C) tall oil hydrotreating.

n-Octadecane and n-heptadecane were obtained as the major paraffinic hydrocarbons from tall oil hydrotreating

through HDO, decarboxylation and decarbonylation reactions. In order to further evaluate the extent of decarboxylation + decarbonylation reactions versus HDO reaction in tall oil hydrotreating, the C_{17}/C_{18} ratio of the liquid products was calculated. Figure 5 shows the comparison of the C_{17}/C_{18} ratio obtained from the hydrotreating of TOFA, DTO, and CTO at different process conditions (space time and temperature). It is interesting to note from Figure 5 that in agreement with TOFA hydrotreating, DTO hydrotreating at different WHSV's also results an increase in C_{17}/C_{18} ratio as a function of increasing temperature until 400°C. Surprisingly with CTO, a decrease in C_{17}/C_{18} ratio was observed from 325 to 400°C in all tested cases. The higher C_{17}/C_{18} ratio obtained in low temperature (325°C and 350°C) experiments irrespective of space time is attributed to the decrease in formation of n-octadecane. This result suggests that the complex nature of CTO feedstock plays a significant role during its hydrotreating over a NiMo catalyst. The selective deoxygenation reactions may be reduced with CTO at lower temperatures (<400°C) due to a number of reasons, which is related to the reactivity of fatty acids in CTO and discussed in a separate section in this study. At 450°C, it appears that a comparison of C_{17}/C_{18} ratio as a function of WHSV is challenging for the tested tall oil feeds. Presumably, the formation n-octadecane and n-heptadecane can be significant at shorter space times (WHSV = 2 h^{-1} and WHSV = 3 h^{-1}), whereas cracking reactions which produce shorter chain alkanes from C_{17} and C_{18} hydrocarbons can also be significant as a function of increasing space time.

Regarding the other fractions in organic phase, monoaromatics are mainly formed from fatty acid fractions through intermediate n-alkanes by means of cyclization and aromatization reaction routes [30]. In addition to this, a direct route for the formation of cycloalkanes and aromatics prior to deoxygenation has also been proposed in the literature [33]. Monoaromatics can also be resulted from resin acids and sterols in tall oil by selective ring opening and dehydrogenation reactions which are valid on the employed catalyst at high temperatures. As fatty acids act as the major precursor for monoaromatics formation at high temperatures, it is obvious that lower amount of fatty acids in CTO in comparison with DTO causes a low yield of monoaromatics [30]. Interestingly, the ratio of polyaromatics to monoaromatics was higher with CTO compared to DTO, and it can be credited to the supplement formation of polyaromatics from the sterol fraction in CTO, either prior to or after deoxygenation.

Based on the obtained composition of reaction products in organic phase samples, a reactivity scale is proposed for the formation of major products from each tall oil feeds as a function of temperature (325–450–°C), and represented as Figure 6. It should be noted that the

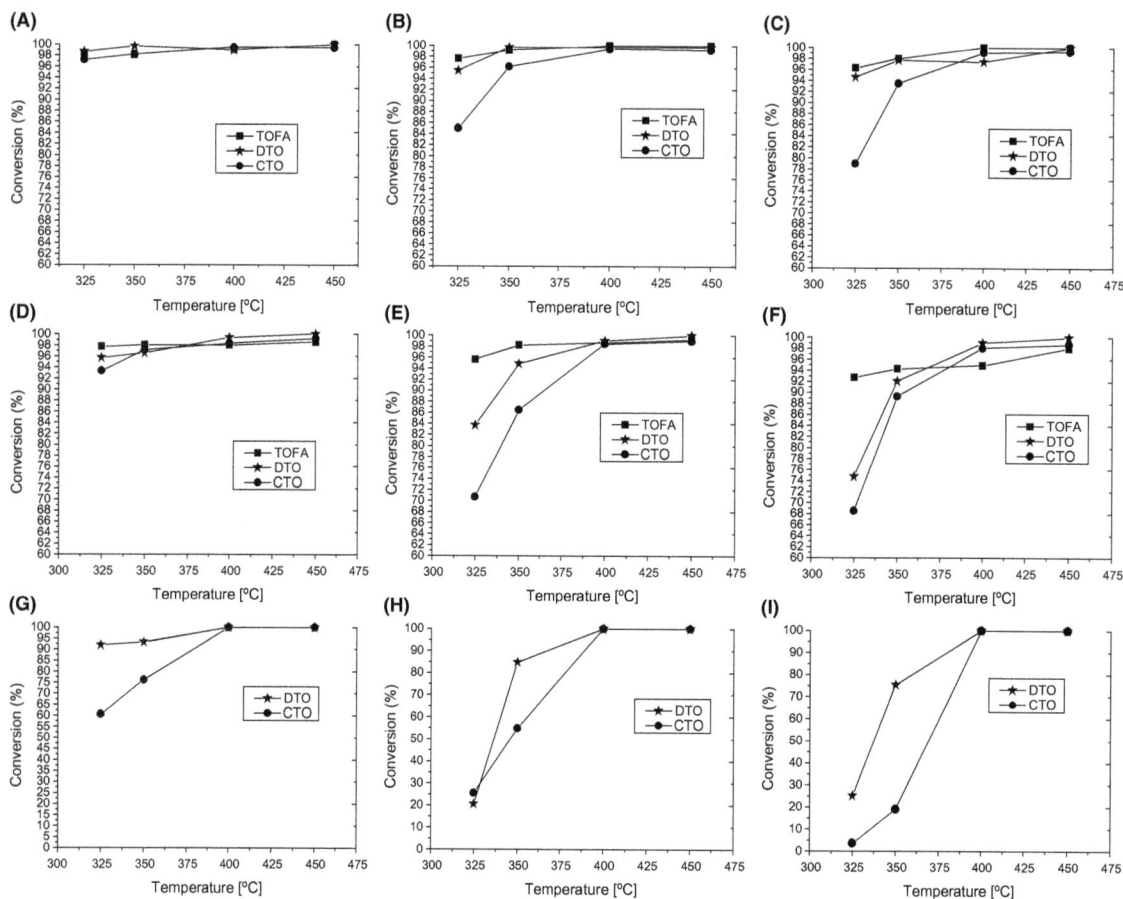

Figure 4. Conversion of individual fatty acids and resin acids; linoleic acid: (A) WHSV =1 h^{-1}, (B) WHSV =2 h^{-1}, (C) WHSV = 3 h^{-1}; oleic acid: (D) WHSV = 1 h^{-1}, (E) WHSV = 2 h^{-1}, (F) WHSV = 3 h^{-1}; dehydroabietic acid: (G) WHSV = 1 h^{-1}, (H) WHSV = 2 h^{-1}, and (I) WHSV = 3 h^{-1}.

proposed reactivity scale excludes the possible thermal or catalytic cracking reactions which occur directly from tall oil fractions in high-temperature experimental runs.

Conversion and reactivity of acid fractions in tall oil

Fatty acids

As noted from Figure 3A, a decrease in conversion of fatty acids was obtained in CTO hydrotreating. Therefore, it is important to assess the effect of other fractions on the conversion of fatty acids in CTO. It is already learnt from DTO hydrotreating that resin acids are not inducing any significant effect on fatty acids in terms of inhibiting its reactivity. In addition to resin acids CTO also contains a considerable amount of neutrals (sterols). In view of the earlier study on the HDO of phenolic compounds and alicyclic alcohols [20], it can be proposed that the sulfur

anion vacancy (CUS) site which is responsible for the HDO of acid fractions are also responsible for the HDO of sterols (direct C–O hydrogenolysis). Interestingly, in our previous study on CTO hydrotreating [23], a high conversion range of sterols occurred at the tested conditions irrespective of space time and temperature. Therefore, based on these observations we may only conclude that a high conversion of sterols is achievable in spite of the competitive adsorption on CUS between acid fractions and sterols in CTO.

The effect of sulfur on fatty acid conversion

It is widely accepted that metal impurities present in the feed influences the hydroprocessing reactions on a sulfided NiMo catalyst [3]. However, the effects of metal impurities on the conversion of fatty acids are ruled out in this research study as the test runs were carried out

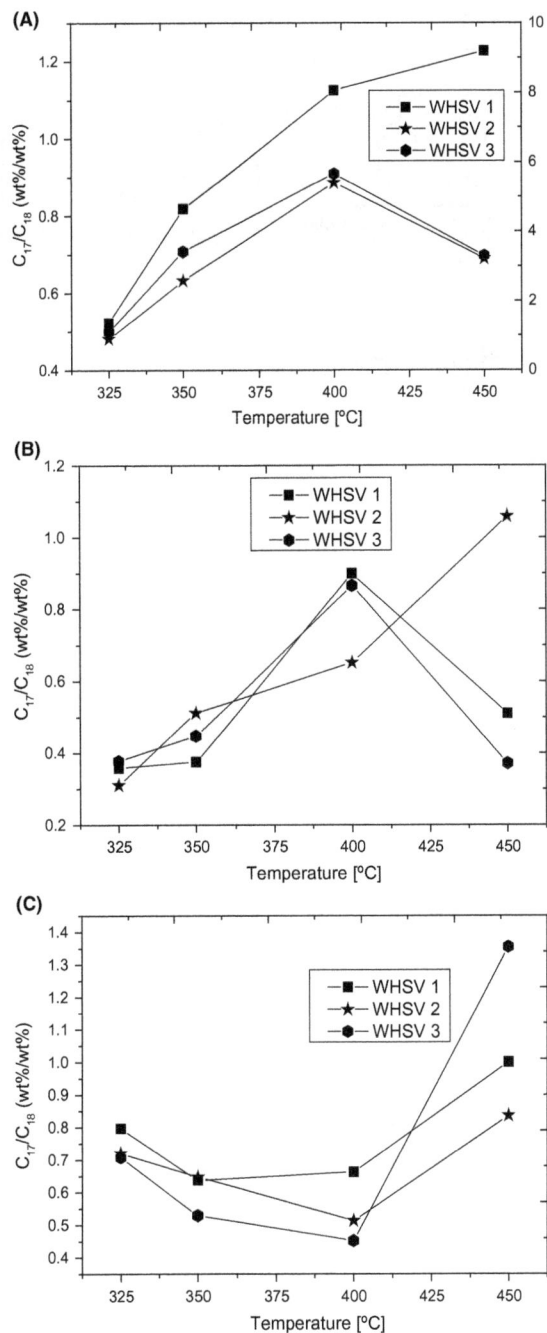

Figure 5. C_{17}/C_{18} (wt%/wt%) ratio obtained from tall oil hydrotreating (A) TOFA (B) DTO (C) CTO; WHSV = 1–3 h^{-1}, T = 325–450°C.

only for 6 h of time-on-stream. The drop in conversion of fatty acids in CTO hydrotreating is considered in this study mainly based on the sulfur content in the feedstock.

It has been reported that the sulfur content in the feed, for instance, elemental sulfur, preferably in a range 2000–5000 w ppm is able to enhance the extent of decarboxylation reactions over a sulfided NiMo catalyst [26]. This increase in the magnitude of decarboxylation reactions is attributed to the increased catalyst acidity by means of adsorption of sulfur species onto the catalyst surface. Senol et al. [34] and Ryymin et al. [35] studied the effects of sulfur additives (H_2S and dimethyl disulfide (DMDS)) in the HDO of aliphatic oxygenates and published that in the presence of sulfur additives, reductive reactions such as hydrogenation (saturation of double bonds), and HDO (addition of hydrogen adjacent to carbonyl carbon, which results the step-wise formation of alkanes and water through intermediate alcohol) are suppressed on a sulfided NiMo catalyst. On the basis of this aspect, we can conclude that the unexpected higher C_{17}/C_{18} ratio obtained in CTO hydrotreating at low temperature (<350°C) (Fig. 5) is presumably due to the higher concentration (1800 ppm) of sulfur in CTO compared to DTO and TOFA, and their influence on selective deoxygenation routes from fatty acids. However, a significant increase in the decarboxylation route (C_{17} formation) in comparison with HDO route (C_{18} formation) is not observed in CTO hydrotreating by virtue of the sulfur content in the feed, but rather it seemed like both HDO and decarboxylation are occurred in a similar extent in CTO hydrotreating. It may presume that in comparison with TOFA and DTO hydrotreating, decarboxylation route was more prominent in CTO hydrotreating in low-temperature experiments apparently due to the effects of sulfur content. The drop in HDO of fatty acids in CTO at low temperatures compared to acids in other feedstocks can be attributed to the decreased reductive reactions due to the effect of sulfur. However, no conclusive comments are drawn here associating the structure–activity relationship of the employed catalyst during hydrotreating of tall oils.

Reaction mechanism assessment of fatty acids

The reaction mechanism of fatty acids is discussed here based on the conversion of individual fatty acids presented in Figure 4. Figure 4A–C shows that linoleic acid, one of the main fatty acids present in tall oil (present in similar concentrations in TOFA, DTO, and CTO), was converted in similar extent with all feedstocks at longer space time (WHSV = 1 h^{-1}). However, at shorter space times (WHSV = 2 and 3 h^{-1}) a prominent drop in reactivity of linoleic acid especially at low temperatures (325–350°C) is observed with CTO compared to other feedstocks. It is also observed from Figure 4D–F that other major fatty acid, oleic acid, converted in a slightly lesser extent in CTO hydrotreating even at WHSV 1 h^{-1} compared to the hy-

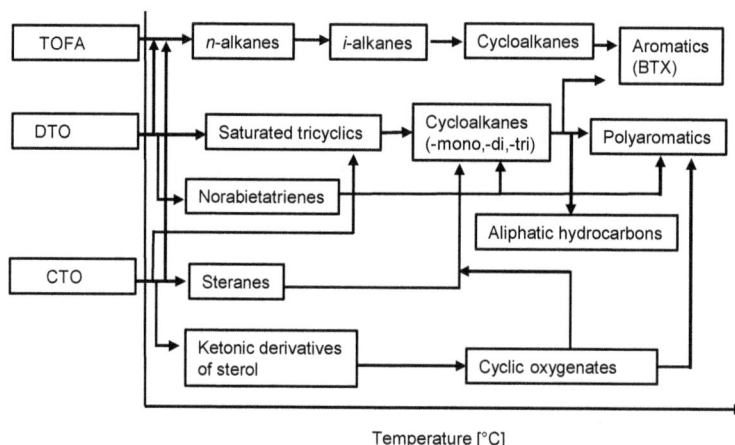

Figure 6. Proposed reactivity scale for the formation of major products from tall oil feeds in hydrotreating (*T* = 325–450°C).

drotreating of TOFA and DTO. At shorter space times (WHSV = 2 and 3 h^{-1}), a significant drop in the conversion of oleic acid was also observed in DTO hydrotreating. Oleic acid can be formed from linoleic acid by partial hydrogenation [36], therefore, we perceive the possibility of oleic acid formation from linoleic acid at least favorable space times such as WHSV = 2 and 3 h^{-1}. The formation of oleic acid at these conditions indirectly results a drop in the conversion of oleic acid. Furthermore, Figure 7 is plotted to show the molar amount (moles) of stearic acid (a saturated fatty acid (C$_{18}$) which is reported to form as an intermediate acid during the hydrotreating of C$_{18}$ unsaturated fatty acids) in feedstock and hydrotreated products, respectively, at WHSV=1 h^{-1}. It is evident from Figure 7 that high concentrations of stearic acid is obtained in low temperature (<400°C) hydrotreating experiments. This observation confirms that fatty acids (linoleic acid and oleic acid) undergo deoxygenation through an initial fatty acid chain double-bond hydrogenation as proposed by other researchers [8, 36, 37]. The saturated fatty acid formed; stearic acid in this case, can then undergo deoxygenation via different routes.

It can be suggested that hydrogenation is prominent at elevated temperatures as well, and produce saturated fatty acids. However, as shown by reactions (4), (5), and (6), *n*-alkanes are formed (up to an optimum temperature level) at the expense of intermediate fatty acids, which in turn results a drop in the concentration of saturated fatty acids (stearic acid) in the product stream from high temperatures (>400°C); as evident from Figure 7. Interestingly for all feeds short-chain fatty acids as well as partially hydrogenated fatty acids were detected in product streams, where for CTO the concentration of partially hydrogenated fatty acids especially at low temperatures

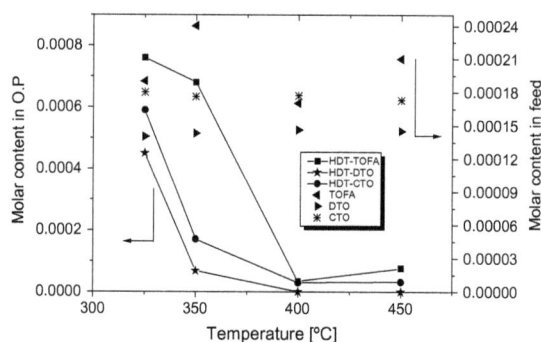

Figure 7. Concentration (molar content) of stearic acid in feeds and hydrotreated products at WHSV = 1 h^{-1}.

was markedly higher than for DTO and TOFA. With TOFA and DTO, a high yield (wt%) of short-chain fatty acids at high temperatures are resulted by means of cracking reactions.

Fatty acid conversion versus paraffin yield

The product yield (%) of paraffins was calculated in this study in order to correlate fatty acid conversion and paraffin formation. It is important to observe that product yield of paraffins corresponds to the amount of fatty acids in feedstocks especially in DTO and CTO hydrotreating experiments. Palanisamy et al. [38] proposes a route for the formation of *n*-alkanes and *i*-alkanes from resin acids (abietic acid) through a ring opening mechanism over a sulfided NiMo catalyst. This ring opening mechanism through C–C bond cleavage may be valid in our experimental conditions producing acyclic paraffinic hydrocar-

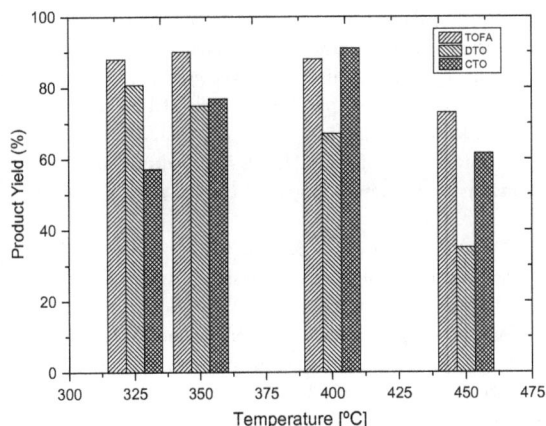

Figure 8. Product yield of paraffins (%) from TOFA, DTO, and CTO as a function of hydrotreating temperature (WHSV = 1 h^{-1}).

bons (*n*-alkanes and *i*-alkanes). Figure 8 shows the yield of paraffins as a function of temperature, obtained from TOFA, DTO, and CTO hydrotreating at WHSV = 1 h^{-1}. It should be taken into consideration that product yield of paraffins can be lower at a high temperature (>400°C) due to the successive formation of cycloalkanes and aromatics from paraffins. Considering this, it can be seen from Figure 8 that a high yield (%) of paraffins was obtained from TOFA hydrotreating, which is no surprise as fatty acids were converted in high extent in that case. A high yield of paraffins was obtained from low-temperature experimental runs in DTO hydrotreating. Remarkably, a high yield of paraffins was obtained from CTO hydrotreating at temperatures 350 and 400°C. As already discussed, fatty acids in CTO are converted in a lesser extent than fatty acids in TOFA and DTO. Furthermore, fatty alcohols and fatty acid esters were detected in significant concentrations in samples from low temperature (325–400°C) hydrotreating of CTO. These fatty alcohols and fatty acids are resulted from the incomplete HDO of fatty acids. This observation also signifies that even at low temperatures all converted fatty acids in CTO were not turned into paraffins. Therefore, a possibility of paraffin formation from fractions other than fatty acids is proposed in CTO hydrotreating. It is known that CTO feed contains a considerable fraction of fatty alcohols (6–8 wt %) along with other fractions; consequently, one route can be suggested as paraffin formation from fatty alcohols via hydrogenation. Ring opening activity of sulfided NiMo catalyst with resin acids as proposed by Palanisamy et al. [38] can also be extend to sterols in CTO. Nonetheless, it is suggested that detailed studies are needed in order to further evaluate the possible routes for the formation of paraffins from cyclic structures in tall oils.

Resin acids and their reaction mechanism assessment

From DTO hydrotreating, it is known that fatty acids did not impose any inhibiting effects on resin acid conversions over a sulfided NiMo catalyst, and vice versa [22]. Figure 3 shows that conversion of resin acids in DTO and CTO follows the same trend as of fatty acids. Therefore, it is proposed that resin acid conversion is also altered by the same effects which are related to fatty acid conversion. Figure 4G–I shows that the conversion of resin acid, dehydroabietic acid, was lower in the case of DTO and CTO hydrotreating at low temperatures (<400°C), which implies that a temperature above 350°C may be needed for achieving a complete conversion of dehydroabietic acid. Earlier, it has been reported by our group that 18-norabietane ($C_{19}H_{34}$) is formed from resin acids [e.g., abietic acid ($C_{19}H_{29}COOH$)] via complete hydrogenation and deoxygenation [22]. Coll et al. propose that deoxygenation from resin acids can occur by means of complete hydrogenation which produces intermediate aldehydes and alcohols and finally a hydrocarbon structure with same number of carbon atoms as the parent resin acids [12]. However, the formation of 18-norabietane as a major product from resin acids at low-temperature hydrotreating [22] implies that deoxygenation was mainly occurred through a hydrogenation and decarboxylation/decarbonylation step and not through an intermediate aldehyde–alcohol route as proposed by Coll et al. [12] The formation of a small fraction of abietane from resin acids also implies that hydrogenation/dehydration mechanistic route is still valid from resin acids. Furthermore, it was reported [22] that norabietatrienes and monoaromatic tricyclic structures are formed even at 350°C from abietic-type resin acids presumably through a parallel dehydrogenation and decarboxylation reaction route as proposed by Dutta et al. [21] The decarboxylation step involved in this reaction route can be catalytic (in the presence of a HDO catalyst under H_2 pressure) or noncatalytic (thermal reaction in the absence of a HDO catalyst). Interestingly, as evidence to noncatalytic route, it has been reported that norabietatrienes are formed during the thermal treatment of abietic acid at 350°C [22, 39]. Therefore, different reaction routes from resin acids are proposed for low-temperature hydrotreating (<400°C) as in the case of fatty acids. Moreover, it may infer that during the hydrotreating of DTO at low temperatures (<400°C), most of the resin acids are consumed through a decarboxylation or decarbonylation as well as through the noncatalytic step (direct decarboxylation + dehydrogenation) without inducing much interaction with the active site responsible for HDO (hydrogenation/dehydration) reaction.

Conclusions

The activity of a commercial alumina supported NiMo catalyst was evaluated for hydrotreating of different tall oil feedstocks. In this study, sulfided NiMo catalyst was found to be active for producing wide-range of products from tall oil feedstocks through various reaction routes existing at different temperatures. At lower temperatures (<400°C), selective deoxygenation routes such as, HDO, decarboxylation, and decarbonylation routes were prominent. At higher temperatures reactions such as cracking (thermal or catalytic) and dehydrogenation were prominent, which produced cycloalkanes and aromatics as main products. Importantly, the trend obtained for the formation of n-octadecane and n-heptadecane in this study especially from TOFA is found to be in well agreement with the behavior of sulfided NiMo catalyst reported in the literature for hydrotreating of vegetable oils. Moreover, fatty acids were found to convert in high extent in TOFA and DTO hydrotreating. In CTO hydrotreating, a decrease in conversion of acid fractions was observed at low temperatures, which is attributed to the complex nature of CTO with significant amount of sulfur compounds in it. Sulfur compounds are proposed to alter the deoxygenation reactions particularly decarboxylation reactions from acid fractions at low temperatures in CTO hydrotreating. Furthermore, it is proposed based on the results in this study that deoxygenation reaction from resin acids at low temperatures mainly occurs through a hydrogenation and decarboxylation/decarbonylation step.

Acknowledgments

Jinto Manjaly Anthonykutty acknowledges the financial support from VTT graduate school. The authors also acknowledge Stora Enso for supporting this research and their vision on wood-based olefins.

Conflict of Interest

None declared.

References

1. Furimsky, E. 2000. Catalytic hydrodeoxygenation. Appl. Catal. A: Gen. 199:147–190.
2. Egeberg, R., N. Michaelsen, L. Skyum, and P. Zeuthen. 2010. Hydrotreating in the production of green diesel. PTQ Q2 Available at www.digitalrefining.com/article/1000156 (accessed 31 May 2014).
3. Furimsky, E., and F. E. Massoth. 1999. Deactivation of hydroprocessing catalysts. Catal. Today 52:381–495.
4. Furimsky, E., and F. E. Massoth. 1993. Regeneration of hydroprocessing catalysts. Catal. Today 17:537–659.
5. Elliott, D. C. 2007. Historical developments in hydroprocessing bio-oils. Energy Fuels 21:1792–1815.
6. Kubička, D., P. Šimáček, and N. Žilková. 2009. Transformation of vegetable oils into hydrocarbons over mesoporous-alumina-supported CoMo catalysts. Top. Catal. 52:161–168.
7. Kikhtyanin, O. V., A. E. Rubanov, A. B. Ayupov, and G. V. Echevsky. 2010. Hydroconversion of sunflower oil on Pd/SAPO-31 catalyst. Fuel 89:3085–3092.
8. Guzman, A., J. E. Torres, L. P. Prada, and M. L. Nunez. 2010. Hydroprocessing of crude palm oil at pilot plant scale. Catal. Today 156:38–43.
9. Kovács, S., T. Kaszaa, A. Thernesz, I. W. Horváthb, and J. Hancsóka. 2011. Fuel production by hydrotreating of triglycerides on NiMo/Al2O3/F catalyst. Chem. Eng. J. 176–177:237–243.
10. Abhari, R., and Havlik P. Z. 2009. Hydrodeoxygenation process. US patent publication 2009/0163744.
11. Abhari, R., H. L. Tomlinson, and E. G. Roth. 2009. Biorenewable naphtha. US patent publication 2009/0300971) A1.
12. Coll, R., S. Udas, and W. A. Jacoby. 2001. Conversion of the rosin acid fraction of crude tall oil into fuels and chemicals. Energy Fuels 15:1166–1172.
13. Sharma, R. K., and N. N. Bakhshi. 1991. Catalytic conversion of crude tall oil to fuels and chemicals over HZSM-5: effect of co-feeding steam. Fuel Process. Technol. 27:113–130.
14. Furrer, R. M., and N. N. Bakhshi. 1988. Catalytic conversion of tall oil to chemicals and gasoline range hydrocarbons. Res. Thermochem. Biomass Convers 956.
15. Kirshner, M. 2005. Chemical profile: tall oil. Chemical Market Reporter: 34.
16. Furimsky, E. 1998. Selection of catalysts and reactors for hydroprocessing. Appl. Catal. 17:177–206.
17. Furimsky, E. 1983. Chemistry of catalytic hydrodeoxygenation. Catal. Rev-Sci. Eng. 25:421–458.
18. Topsøe, H., R. G. Egeberg, and K. G. Knudsen. 2004. Future challenges of hydrotreating catalyst technology. Prepr. Pap-Am. Chem. Soc. Div. Fuel Chem. 49: 568–569.
19. Kubička, D., and L. Kaluza. 2010. Deoxygenation of vegetable oils over sulfided Ni, Mo, and NiMo catalysts. Appl. Catal. A: Gen. 372:199–208.
20. Şenol, O. I., E. M. Ryymin, T. R. Viljava, and A. O. I. Krause. 2007. Effect of hydrogen sulphide on the hydrodeoxygenation of aromatic and aliphatic oxygenates on sulphided catalysts. J. Mol. Catal. A: Chem. 277:107–112.
21. Dutta, R. P., and H. H. Schobert. 1993. Hydrogenation/dehydrogenation reactions of rosin. Fundam. Studies Coal Liquefaction 38:1140–1146.

22. Anthonykutty, J. M., K. M. Van Geem, R. D. Bruycker, J. Linnekoski, A. Laitinen, J. Räsänen, et al. 2013. Value added hydrocarbons from distilled tall oil via hydrotreating over a commercial NiMo catalyst. Ind. Eng. Chem. Res. 52:10114–10125.

23. Anthonykutty, J. M., J. Linnekoski, A. Harlin, A. Laitinen, and J. Lehtonen. Catalytic upgrading of crude tall oil into a paraffin-rich liquid. Biomass Conv. Bioref. doi: 10.1007/s13399-014-0132-8

24. Rozmysłowicz, B., P. Mäki-Arvela, S. Lestari, O. A. Simakova, K. Eränen, I. L. Simakova, et al. 2010. Catalytic deoxygenation of tall oil fatty acids over a palladium-mesoporous carbon catalyst: a new source of biofuels. Top. Catal. 53:1274–1277.

25. Knuuttila, P., P. Kukkonen, and Ulf Hotanen. 2010. Method and apparatus for preparing fuel components from crude tall oil. US patent publication WO 2010/097519 A2.

26. Stigsson, L., and V Naydenov. 2009. Conversion of crude tall oil to renewable feedstock for diesel range fuel compositions. European patent publication WO 2009/131510.

27. Diaz, M. A. F., R. A. Markovits, and S. A. Markovits. 2005. Process for refining a raw material comprising black liquor soap, crude tall oil or tall oil pitch. EP1568 760 A1.

28. Kubičková, I., and D. Kubička. 2010. Utilization of triglycerides and related feedstocks for production of clean hydrocarbon fuels and petrochemicals: a review. Waste Biomass Valorization 1:293–308.

29. Mikulec, J., J. Cvengros, L. Jorı́kova, M. Banic, and A. Kleinova. 2010. Second generation diesel fuel from renewable sources. J. Clean. Prod. 18:917–926.

30. da Rocha Filho, G. N., D. Brodzki, and G. Djéga-Mariadassou. 1993. Formation of alkanes, alkylcycloalkanes and alkylbenzenes during the catalytic hydrocracking of vegetable oils. Fuel 72:543–549.

31. Wagner, J. L., V. P. Ting, and C. J. Chuck. 2014. Catalytic cracking of sterol-rich yeast lipid. Fuel 130:315–323.

32. Severson, R. F., and W. H. Schuller. 1972. The thermal behavior of some resin acids at 400–500°C. Can. J. Chem. 50:2224.

33. Scharmann, H., W. R. Eckert, and A. Zeman. 1969. Elucidation of structure of the methyl esters of cyclic fatty acids V: mass-spectrometry of isomeric methyl esters of phenyl undecanoic acid. Fette Seifen Anstrichmiftel 71:118.

34. Senol, O. I., T.-R. Viljava, and A. O. I. Krause. 2007. Effect of sulfiding agents on the hydrodeoxygenation of aliphatic esters on sulphided catalysts. Appl. Catal. A: Gen. 326:236–244.

35. Ryymin, E.-M., M. L. Honkela, T.-R. Viljava, and A. O. I. Krause. 2010. Competitive reactions and mechanisms in the simultaneous HDO of phenol and methyl heptanoate over sulphided NiMo/γ-Al2O3. Appl. Catal. A: Gen. 389:114–121.

36. Monnier, J., H. Sulimma, A. Dalai, and G. Caravaggio. 2010. Hydrodeoxygenation of oleic acid and canola oil over alumina-supported metal nitrides. Appl. Catal. A 382:176–180.

37. Snåre, M., I. Kubic̆kova, P. Ma1ki-Arvela, K. Era1nen, and D. Y. Murzin. 2006. Heterogeneous catalytic deoxygenation of stearic acid for production of biodiesel. Ind. Eng. Chem. Res. 45:5708–5715.

38. Palanisamy, S. 2013. Co-processing fat-rich material into diesel fuel. Doctoral thesis, Chalmers University

39. Bernas, A., T. Salmi, Y. D. Murzin, J. Mikkola, and M. Rintola. 2012. Catalytic transformation of abietic acid to hydrocarbons. Top. Catal. 55:673–679.

12

An old solution to a new problem? Hydrogen generation by the reaction of ferrosilicon with aqueous sodium hydroxide solutions

Paul Brack[1], Sandie E. Dann[1], K. G. Upul Wijayantha[1], Paul Adcock[2] & Simon Foster[2]

[1]Energy Research Laboratory, Department of Chemistry, Loughborough University, Loughborough, Leicestershire LE11 3TU, United Kingdom
[2]Intelligent Energy Ltd, Charnwood Building, Holywell Park, Ashby Road, Loughborough, Leicestershire LE11 3GB, United Kingdom

Keywords
Chemical hydrogen storage, ferrosilicon, hydrogen generation, hydrolysis

Correspondence
Paul Brack, Department of Chemistry, Loughborough University, Loughborough, Leicestershire LE11 3TU, United Kingdom.
E-mail: p.brack@lboro.ac.uk

Funding Information
The authors would like to thank the Engineering and Physical Sciences Research Council (EPSRC) and Intelligent Energy Ltd for funding this project.

Abstract

The chemical hydrogen storage properties of ferrosilicon were investigated. A hydrogen yield of ~4.75 wt.% (with respect to the mass of ferrosilicon) was estimated by the reaction of varying quantities of ferrosilicon with 5 mL of 40 wt.% sodium hydroxide solution. The reaction of ferrosilicon with aqueous sodium hydroxide solution to form hydrogen was found to have an activation energy of 90.5 kJ mol^{-1} by means of an Arrhenius plot. It was observed that the induction period of the hydrogen generation reaction varies exponentially with temperature. Although this combination of high activation energy and a lengthy induction period at low temperatures reduces the attractiveness of ferrosilicon for portable hydrogen storage applications unless methods can be developed to accelerate the onset and rate of hydrogen generation, its low cost and widespread availability make it attractive for further studies focused on higher temperature stationary applications.

Introduction

In recent years, there has been a great surge of research activity in the area of energy. This is primarily driven by two conflicting factors; the need to satisfy an ever increasing demand for energy while simultaneously reducing greenhouse gas emissions to sustainable levels. Broadly speaking, there are two ways that these goals can be achieved. The first, typically termed carbon capture and storage (CCS), is to capture the emissions from the fuels which provide our energy, and store them in such a way that the levels released to the atmosphere are substantially reduced [1]. This type of technology is well suited to large industrial plants or power stations, but its deployment in vehicular or portable applications would be problematic. The second solution is to switch to using an alternative fuel to hydrocarbons. Hydrogen is arguably the most promising option, given its large amount of chemical energy per unit mass (142 MJ kg^{-1}) and the benign nature of its combustion product, water [2]. However, due to its low density (0.0824 g L^{-1} at 298 K [3]), the storage and transportation of hydrogen gas is problematic as it must be heavily compressed. Compressed gases must be stored in heavy steel containers, partially nullifying the benefit in terms of high gravimetric energy density of using hydrogen in the first place. Alternatively, hydrogen can be stored in cryogenic containers, but these are costly and accumulate considerable "boil off" losses [4]. Both of these approaches are also hazardous and present considerable risks to the user and the public at large.

Hence for the deployment of hydrogen in vehicular and portable applications to be viable, alternative storage methods must be developed [5]. A particularly attractive approach is that of chemical hydrogen storage materials, which release their hydrogen when required by means of a chemical reaction [6, 7]. While at first consideration, this may seem a relatively new approach, it has in fact been heavily utilized before. In the first three decades of the 20th century, Zeppelins, huge airships filled with hydrogen, flew regularly across the Atlantic Ocean. They were also deployed in World War I. The filling of the airships required a supply of a substantial volume of hydrogen gas in a form which could be safely stored for months at a time and transported across large distances of uneven, battle torn terrain. To obtain this, the US military made use of a material known as ferrosilicon [8, 9]. Ferrosilicon is an alloy formed of various phases of iron and silicon. It has been used extensively in the steel industry for ~125 years to remove oxygen from melted steel during the casting process, thus improving its physical properties [10]. Ferrosilicon is produced by the reduction of quartz with coke in the presence of iron or iron ore [10]. Ferrosilicon up to 15% silicon content is produced in a blast furnace, while higher silicon content can be reached by the use of a submerged electric arc furnace [10–13]. Each batch has a fixed amount of quartz and a variable amount of coal and iron oxide to attain the desired iron to silicon ratio [14]. The most widely available composition is ferrosilicon 75, which consists of ~75 wt.% silicon, the remaining ~25 wt.% consisting principally of iron (in the form of iron disilicide, $FeSi_2$) with a small amount of aluminium (1–2 wt.%) [14]. To obtain hydrogen for their Zeppelins, the US military reacted ferrosilicon with sodium hydroxide solutions in large batch reactors [8, 9, 15].

It is perhaps surprising that despite the present volume of literature on chemical hydrogen storage materials, there have been no published investigation of the hydrogen generating properties of ferrosilicon since the reports by the US military some three quarters of a century ago. In this article we examine the chemical hydrogen storage properties of ferrosilicon for portable applications.

Experimental

Ferrosilicon and sodium hydroxide were purchased from Castree Kilns (St. Clears, Carmarthenshire, Wales) and Sigma Aldrich (Gillingham, Dorset, UK), respectively, and used as received. The evolved hydrogen was collected and quantified using the water displacement method [16, 17].

For the evaluation of the volume of hydrogen generated per unit mass of ferrosilicon, 5 mL of 40 wt.% sodium hydroxide solution was added to a 50 mL round bottomed flask in a water bath at 60°C and left for 10 mins to equilibrate. To this was then added the desired mass of ferrosilicon,

and the volume of hydrogen evolved recorded over a 60-min period. Each reaction was performed in triplicate.

To determine the activation energy of the reaction, 5 mL of 40 wt.% sodium hydroxide solution was heated in a water bath in a 50 mL round bottomed flask and left to equilibrate for 10 mins. To this was then added 1.00 g of ferrosilicon, and the volume of hydrogen evolved recorded over a 10 min period. Each reaction was performed in triplicate.

Results and Discussion

Hydrogen yield

The generation of hydrogen from silicon in alkali solutions is well known and proceeds by the following reaction:[18, 19]

$$2NaOH + Si + H_2O \rightarrow Na_2SiO_3 + 2H_2 \quad (1)$$

In this way, silicon can theoretically produce 14% of its own weight in hydrogen. The literature on ferrosilicon hydrolysis suggests that its hydrogen generation reactions proceed in a similar way [8, 9, 15]. To estimate experimentally the hydrogen yield with respect to the mass of ferrosilicon, 0.75, 1.00, 1.25, and 1.50 g of ferrosilicon were added to a 40 wt.% solution of sodium hydroxide at 60°C, and the hydrogen generation recorded over a period of an hour as shown in Figure 1. After an induction period, hydrogen generation increased rapidly at first before gradually slowing.

As shown in Figure 2, the total volume of hydrogen evolved can be plotted against the mass of ferrosilicon added to the reaction mixture to obtain a linear trend. The total volume of hydrogen evolved can be converted to a mass, and from this the hydrogen yield with respect to the mass of ferrosilicon can be estimated as a

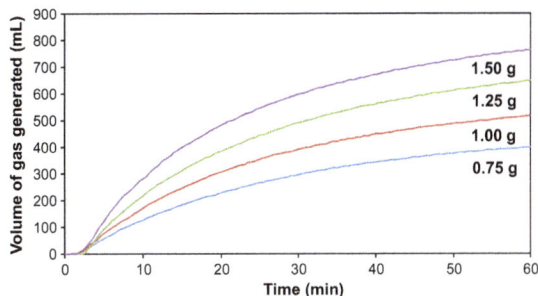

Figure 1. A plot of volume of hydrogen generated over time during the reaction of 0.75 g (blue), 1.00 g (red), 1.25 g (green), and 1.50 g (purple) of ferrosilicon with 5 mL of 40 wt.% sodium hydroxide solution in a water bath at 60°C.

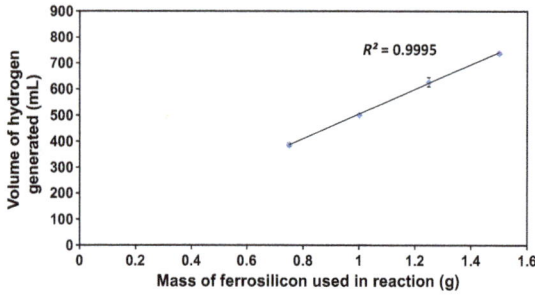

Figure 2. A plot of total volume of hydrogen evolved against mass of ferrosilicon used in the reaction.

reasonably consistent ~4.75 wt.% across these experiments (Fig. 3). This is considerably lower than the theoretical maximum for silicon.

The order of reaction with respect to ferrosilicon can be obtained from the gradient of a plot of the natural log of the maximum or initial hydrogen generation rate (HGR) against the natural log of the mass of ferrosilicon used in the reaction (Fig. 4), and in this case is estimated to be ~0.7.

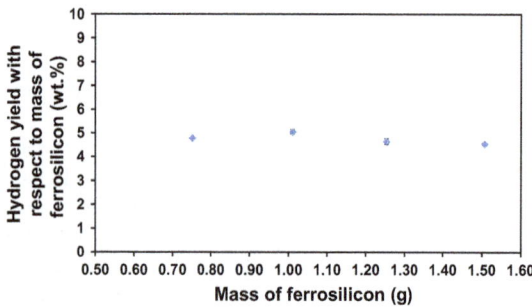

Figure 3. A plot of the mass of ferrosilicon used in the reaction against the hydrogen yield with respect to the mass of ferrosilicon.

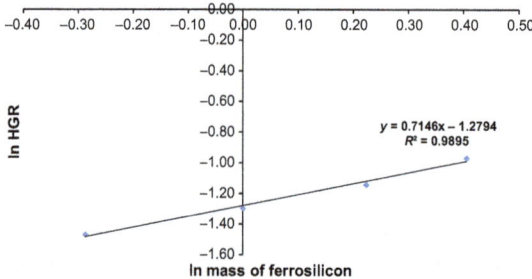

Figure 4. A plot of ln HGR against ln mass of ferrosilicon used in each reaction to obtain the order of reaction with respect to ferrosilicon.

Arrhenius and Eyring plot

To obtain an estimate of the activation energy for the reaction of ferrosilicon with sodium hydroxide solution to form hydrogen, an Arrhenius plot can be constructed [20, 21]. About 1.00 g of ferrosilicon was added to 5 mL of 40 wt.% sodium hydroxide solution in a water bath at 331, 335, 339, 344, and 348 K, and the volume of gas evolved recorded for 10 mins (Fig. 5). This is sufficient to calculate the maximum or initial rate of reaction commonly referred to as the hydrogen generation rate. The Arrhenius equation can be written in the following form:

$$\ln k = \ln A - (E_a/RT) \tag{2}$$

where k is the hydrogen generation rate (mL s^{-1}), A is the pre-exponential factor, E_a is the activation energy (kJ mol^{-1}), R is the gas constant (8.314 J mol^{-1} K^{-1}) and T is the hydrolysis temperature (K). Hence, the gradient of a plot of ln HGR versus $1/T$ will give a gradient equal to E_a/R, and from this the activation energy can be determined (Fig. 6). In this case, the activation energy is estimated to

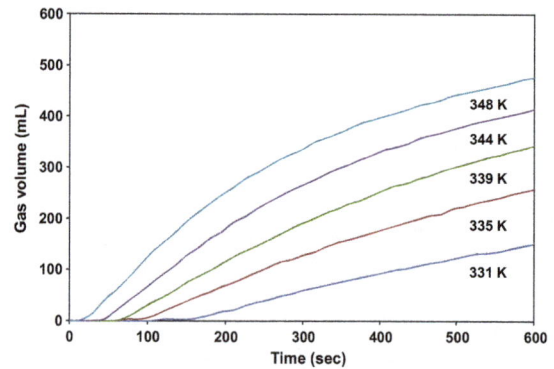

Figure 5. Hydrogen generation curves at 331 K (blue), 335 K (red), 339 K (green), 344 K (purple), and 348 K (turquoise) from the reaction of 1.00 g of ferrosilicon with 5 mL of a 40 wt.% sodium hydroxide solution.

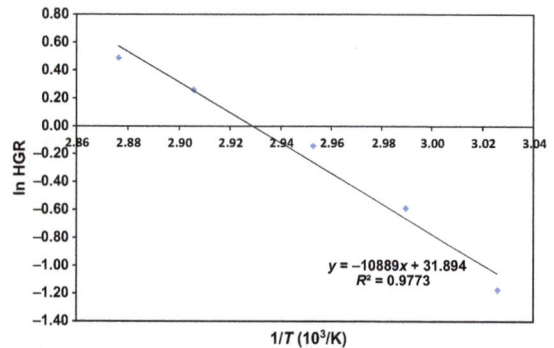

Figure 6. Arrhenius plot for the reaction of ferrosilicon with sodium hydroxide solution.

Table 1. Activation energy of various hydrogen generation reactions calculated using the Arrhenius equation.

Catalyst	E_a (kJ/mol)	References
NaBH$_4$		
Co–P	23.9	22
Co films	27.6	23
Ru nanoclusters	28.51	24
Co–Ni–P–B	28.7	25
Co–W–B	29	26
Co–Cu–B	29.9	27
Co–Fe–B	31	27
Pd–Ni–B	31.1	28
Co–B	33	29
Co–Ni–B	34	27
Co–Cr–B	37	27
Co–Ni–P	38	30
Co–PVP	38.7	31

Hydrogen generating mixture	E_a (kJ/mol)	References
Al based systems		
Al–Bi–Li–NaCl	22.57	32
Al–Li–In–Zn	22.91	33
Al–Li$_3$AlH$_6$	33	34
Al–Ga–In–Sn	43.8	35
Al powder	64.2	35
Activated aluminium	69	36
Al/NaAlO$_2$	71	37

Figure 7. Eyring plot for the reaction of ferrosilicon with sodium hydroxide solution.

be 90.5 kJ mol^{-1}. As can be seen from Table 1, this is substantially higher than the activation energy determined for the hydrolysis reactions of other chemical hydrogen storage materials, such as sodium borohydride and aluminium. We believe that this is due to the presence of iron disilicide in the ferrosilicon matrix, which renders the silicon less available for, and thus slows the rate of, reaction. This sluggish rate of reaction of ferrosilicon is a major disadvantage in the context of portable applications.

Further information about the reaction kinetics can be obtained from an Eyring plot [20, 21]. The Eyring equation can be expressed as follows:

$$\ln k/T = -\Delta H^{\neq}/R.1/T + \ln(k_b/h) + \Delta S^{\neq}/R \qquad (3)$$

where k is the hydrogen generation rate (mL s^{-1}), R is the gas constant (8.314 J mol^{-1} K^{-1}), T is the hydrolysis temperature (K), k_b is the Boltzmann constant, h is the Planck constant, and $\Delta H^{\neq}$ and $\Delta S^{\neq}$ are the activation enthalpy and entropy of the hydrolysis reaction, respectively. Thus, a plot of ln k/T against $1/T$ yields a straight line graph with a gradient of $\Delta H^{\neq}/R$ and a y axis intercept of ln $(k_b/h) + \Delta S^{\neq}/R$. This is presented in Figure 7, and gives a value for the activation enthalpy of 87.7 kJ mol^{-1} and a value for the activation entropy of 10.9 J K^{-1} mol^{-1}. The sign of the activation entropy gives an indication of

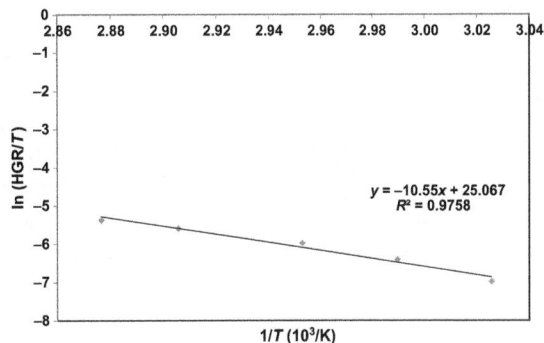

whether the rate determining step is an associative (negative value) or dissociative (positive value) step. Previous studies on the etching of bulk silicon in hydroxide solutions show that that reaction proceeds by the following steps:[18, 38]

(1) Hydroxide attack of bulk silicon

$$Si + 2OH^- + 4h^+ \rightarrow Si(OH)_2^{2+}$$

(2) Reduction of water

$$4H_2O \rightarrow 4OH^- + 2H_2 + 4h^+$$

(3) Formation of a water-soluble complex

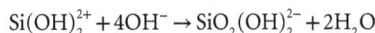

$$Si(OH)_2^{2+} + 4OH^- \rightarrow SiO_2(OH)_2^{2-} + 2H_2O$$

Step (1) was found to be the rate determining step, and, as it is an associative step, a negative activation entropy is expected for this reaction. As we obtained a positive activation entropy, our results suggest that the etching of silicon in ferrosilicon powders follows a slightly different mechanism to the etching of bulk silicon, wherein a dissociative step, possibly the separation of silicon from the ferrosilicon matrix, is rate determining. However, given the closeness to zero of the activation entropy and the uncertainty in the trend line, it is difficult to be certain.

The onset of hydrogen evolution is not instant. The time required for the onset of hydrogen evolution is found to vary exponentially with temperature (Fig. 8). In the reaction of silicon with sodium hydroxide solutions, a similar induction period is observed, which is typically ascribed to the etching of the surface silicon dioxide layer which must be accomplished before hydrogen generation can commence [38]. This has been found to be an Arrhenius process, and it seems likely that the induction period for ferrosilicon is caused by this. Extrapolation of the trend line from Figure 8 towards room temperature indicates that ferrosilicon is unsuitable as a hydrogen storage material for low temperature portable applications as, notwithstanding the

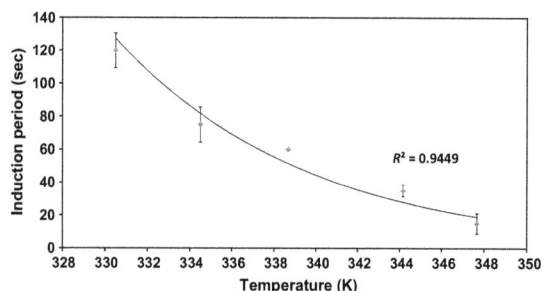

Figure 8. The variation in induction period with temperature for the reaction of 1.00 g of ferrosilicon with 5 mL of 40 wt.% sodium hydroxide solution in a water bath.

slow kinetics of the reaction once it had commenced, the delay period before the onset of hydrogen production would be much too long for practical application.

Conclusions

Ferrosilicon is cheap, commercially available and safe to transport and handle. However, reactions of ferrosilicon with solutions of sodium hydroxide are dogged by slow kinetics (E_a = 90.5 kJ mol^{-1}) and long induction periods at low temperatures. This suggests that although ferrosilicon is potentially useful for higher temperature, for example, stationary, applications, its utility as a chemical hydrogen storage material for portable applications is limited unless methods can be found to accelerate the onset and rate of hydrogen generation.

Acknowledgements

The authors thank the EPSRC and Intelligent Energy Ltd for funding this project. PB also thanks the SCI for the award of a Messel Scholarship.

References

1. Hammond, G. P., and J. Spargo. 2014. The prospects for coal-fired power plants with carbon capture and storage: a UK perspective. Energ. Convers. Manage. 86:476–489.

2. Schlapbach, L., and A. Züttel. 2001. Hydrogen-storage materials for mobile applications. Nature 414:353–358.

3. Haynes, W. M. 2014. CRC Handbook of Chemistry and Physics, 95th ed. CRC Press, UK.

4. Sarkar, A., and R. Banerjee. 2005. Net energy analysis of hydrogen storage options. Int. J. Hydrogen Energy 30:867–877.

5. Dalebrook, A. F., W. Gan, M. Grasemann, S. Moret, and G. Laurenczy. 2013. Hydrogen storage: beyond conventional methods. Chem. Commun. (Camb.) 49:8735–8751.

6. Marrero-Alfonso, E. Y., A. M. Beaird, T. A. Davis, and M. A. Matthews. 2009. Hydrogen generation from chemical hydrides. Ind. Eng. Chem. Res. 48:3703–3712.

7. Weidenthaler, C., and M. Felderhoff. 2011. Solid-state hydrogen storage for mobile applications: Quo Vadis? Energy Environ. Sci. 4:2495–2502.

8. Weaver, E. R. 1920. The generation of hydrogen by the reaction between ferrosilicon and a solution of sodium hydroxide. J. Ind. Eng. Chem. 12:232–240.

9. Teed, P. L.. 1919. The chemistry and manufacture of hydrogen. Edward Arnold, London, UK.

10. Farzana, R., R. Rajarao, and V. Sahajwalla. 2014. Synthesis of ferrosilicon alloy using waste glass and plastic. Mater. Lett. 116:101–103.

11. Hauksdottir, A. S., A. Gestsson, and A. Vesteinsson. 2002. Current control of a three-phase submerged arc ferrosilicon furnace. Control Eng. Pract. 10:457–463.

12. Buø, T. V., R. J. Gray, and R. M. Patalsky. 2000. Reactivity and petrography of cokes for ferrosilicon and silicon production. Int. J. Coal Geol. 43:243–256.

13. Hauksdottir, A. S., A. Gestsson, and A. Vesteinsson. 1998. Submerged-arc ferrosilicon furnace simulator: validation for different furnaces and operating ranges. Control Eng. Pract. 6:1035–1042.

14. Ingason, H. T., and G. R. Jonsson. 1998. Control of the silicon ratio in ferrosilicon production. Control Eng. Pract. 6:1015–1020.

15. Taylor, H. S. 1921. Industrial hydrogen. The Chemical Catalog Company, Inc., New York, USA.

16. Macanás, J., L. Soler, A. M. Candela, M. Muñoz, and J. Casado. 2011. Hydrogen generation by aluminum corrosion in aqueous alkaline solutions of inorganic promoters: the AlHidrox process. Energy 36:2493–2501.

17. Fan, M.-Q., S. Liu, W.-Q. Sun, Y. Fei, H. Pan, C.-J. Lv, et al. 2011. Hydrogen generation from Al/NaBH4 hydrolysis promoted by Li-NiCl2 additives. Int. J. Hydrogen Energy 36:15673–15680.

18. Goller, B., D. Kovalev, and O. Sreseli. 2011. Nanosilicon in water as a source of hydrogen: size and pH matter. Nanotechnology 22:305402.

19. Erogbogbo, F., T. Lin, P. M. Tucciarone, K. M. Lajoie, L. Lai, G. D. Patki, et al. 2013. On-demand hydrogen generation using nanosilicon: splitting water without light, heat, or electricity. Nano Lett. 13:451–456.

20. Akdim, O., R. Chamoun, U. B. Demirci, Y. Zaatar, A. Khoury, and P. Miele. 2011. Anchored cobalt film as stable supported catalyst for hydrolysis of sodium borohydride for chemical hydrogen storage. Int. J. Hydrogen Energy 36:14527–14533.

21. Zhu, J., R. Li, W. Niu, Y. Wu, and X. Gou. 2012. Facile hydrogen generation using colloidal carbon supported cobalt to catalyze hydrolysis of sodium borohydride. J. Power Sources 211:33–39.

22. Guo, Y., Q. Feng, Z. Dong, and J. Ma. 2013. Electrodeposited amorphous Co–P catalyst for hydrogen generation from hydrolysis of alkaline sodium borohydride solution. J. Mol. Catal. A Chem. 378:273–278.

23. Li, H., J. Liao, X. Zhang, W. Liao, L. Wen, J. Yang, et al. 2013. Controlled synthesis of nanostructured Co film catalysts with high performance for hydrogen generation from sodium borohydride solution. J. Power Sources 239:277–283.

24. Özkar, S., and M. Zahmakıran. 2005. Hydrogen generation from hydrolysis of sodium borohydride using Ru(0) nanoclusters as catalyst. J. Alloys Compd. 404–406:728–731.

25. Fernandes, R., N. Patel, and A. Miotello. 2009. Efficient catalytic properties of Co–Ni–P–B catalyst powders for hydrogen generation by hydrolysis of alkaline solution of NaBH4. Int. J. Hydrogen Energy 34:2893–2900.

26. Dai, H. B., Y. Liang, P. Wang, X. D. Yao, T. Rufford, M. Lu, et al. 2008. High-performance cobalt–tungsten–boron catalyst supported on Ni foam for hydrogen generation from alkaline sodium borohydride solution. Int. J. Hydrogen Energy 33:4405–4412.

27. Patel, N., R. Fernandes, and A. Miotello. 2010. Promoting effect of transition metal-doped Co–B alloy catalysts for hydrogen production by hydrolysis of alkaline NaBH4 solution. J. Catal. 271:315–324.

28. Liu, W., H. Cai, P. Lu, Q. Xu, Y. Zhongfu, and J. Dong. 2013. Polymer hydrogel supported Pd–Ni–B nanoclusters as robust catalysts for hydrogen production from hydrolysis of sodium borohydride. Int. J. Hydrogen Energy 38:9206–9216.

29. Dai, H.-B., Y. Liang, P. Wang, and H.-M. Cheng. 2008. Amorphous cobalt–boron/nickel foam as an effective catalyst for hydrogen generation from alkaline sodium borohydride solution. J. Power Sources 177:17–23.

30. Guo, Y., Q. Feng, and J. Ma. 2013. The hydrogen generation from alkaline NaBH4 solution by using electroplated amorphous Co–Ni–P film catalysts. Appl. Surf. Sci. 273:253–256.

31. Metin, O., and S. Özkar. 2009. Hydrogen generation from the hydrolysis of ammonia-borane and sodium borohydride using water-soluble polymer-stabilized cobalt(0) nanoclusters catalyst. Energy Fuels 23:3517–3526.

32. Liu, S., M.-Q. Fan, C. Wang, Y.-X. Huang, D. Chen, L.-Q. Bai, et al. 2012. Hydrogen generation by hydrolysis of Al–Li–Bi–NaCl mixture with pure water. Int. J. Hydrogen Energy 37:1014–1020.

33. Fan, M. Q., S. Liu, C. Wang, D. Chen, and K. Y. Shu. 2012. Hydrolytic hydrogen generation using milled aluminum in water activated by Li, In, and Zn additives. Fuel Cells 12:642–648.

34. Wu, T., F. Xu, L.-X. Sun, Z. Cao, H.-L. Chu, Y.-J. Sun, et al. 2014. Al-Li3AlH 6: a novel composite with high activity for hydrogen generation. Int. J. Hydrogen Energy 39:10392–10398.

35. Gai, W.-Z., W.-H. Liu, Z.-Y. Deng, and J.-G. Zhou. 2012. Reaction of Al powder with water for hydrogen generation under ambient condition. Int. J. Hydrogen Energy 37:13132–13140.

36. Rosenband, V., and A. Gany. 2010. Application of activated aluminum powder for generation of hydrogen from water. Int. J. Hydrogen Energy 35:10898–10904.

37. Soler, L., A. M. Candela, J. Macanás, M. Muñoz, and J. Casado. 2009. In situ generation of hydrogen from water by aluminum corrosion in solutions of sodium aluminate. J. Power Sources 192:21–26.

38. Seidel, H., L. Csepregi, A. Heuberger, and H. Baumgartel. 1990. Anisotropic etching of crystalline silicon in alkaline solutions. J. Electrochem. Soc. 137:3612–3626.

13

Liquinert quartz crucible for the growth of multicrystalline Si ingots

Kozo Fujiwara[1], Yukichi Horioka[2] & Shiro Sakuragi[3]

[1]Institute for Materials Research (IMR), Tohoku University, Katahira 2-1-1, Aoba-ku, Sendai 980-8577, Japan
[2]Frontier Technology Business Research Institute Co. Ltd., Yamazaki Umenodai 11-9, Noda 278-0024, Japan
[3]Union Materials Inc., Oshido 1640, Tone-machi, Kitasoma-gun, Ibaraki 300-1602, Japan

Keywords
Casting, liquinert, multicrystalline silicon, quartz crucible

Correspondence
Kozo Fujiwara, Institute for Materials Research (IMR), Tohoku University, Katahira 2-1-1, Aoba-ku, Sendai 980-8577, Japan.

E-mail: kozo@imr.tohoku.ac.jp

Funding Information
This work was supported by a Kakenhi Grant-in-Aid (No. 26246016) from the Ministry of Education, Culture, Sports, Science and Technology (MEXT) of Japan.

Abstract

The growth of a multicrystalline silicon (mc-Si) ingot for solar cell applications was attempted using a Liquinert quartz crucible. A mc-Si ingot was also grown in a quartz crucible coated with Si_3N_4 powder for comparison with that from the Liquinert quartz crucible. The mc-Si ingot grown in the Liquinert quartz crucible had a shinier surface which has few impurity particles and higher minority carrier lifetime than the mc-Si ingot grown in a quartz crucible coated with Si_3N_4 powder. These results indicate that contamination with impurities was reduced with the Liquinert quartz crucible; therefore, this crucible has the potential to be a powerful tool for the production of high-quality mc-Si ingots for solar cell applications.

Introduction

Multicrystalline Si (mc-Si) ingots for solar cell applications are produced by casting based on unidirectional solidification. It has been demanded to reduce the defects, such as impurities, dislocations, sub-grain boundaries, and grain boundaries, to improve the efficiency of mc-Si solar cells. Quartz crucibles are typically used in the casting process and are coated with silicon nitride (Si_3N_4) powder, which serves as a mold release agent to inhibit adherence of the mc-Si ingot to the crucible. However, it is difficult to avoid contamination of the mc-Si ingot with impurities during the melting/solidification process, where impurities included in the Si_3N_4 powder are dissolved into the Si melt or where Si_3N_4 powder is detached

from the crucible and dissolved into the Si melt. Such impurities act as lifetime killer and can be the origin of dislocations, which degrades the quality of mc-Si ingots. Therefore, other materials have been considered as substitutes for the Si_3N_4 powder coating layer or for the crucible itself [1–8].

In this study, the use of a Liquinert quartz crucible was attempted for the growth of a mc-Si ingot. The concept of Liquinert, the name of which comes from "liquid is in an inert state," was proposed by Sakuragi [9, 10], whereby a liquid is in a state that is non-wetting and non-reactive with a crucible at high temperature. There are three requirements so as to realize the liquinert state; (1) high-purity raw material, (2) high-purity atmosphere, and (3) less-reactive crucible. When all of them are satisfied, molten raw material becomes round shape without wetting with crucible wall. In this situation, we can take out a crystal from a crucible without sticking after crystal growth. For example, in the crystal growth of NaI(Tl) scintillation material which has strong hygroscopic nature, a tiny amount of residual water in raw materials or in atmosphere becomes the cause of wetting and sticking even after dehydration procedure by vacuum pumping at high temperature. Therefore, we had to remove residual water perfectly from raw materials/atmosphere to realize a liquinert state for producing a high quality crystal [10]. Union Materials Inc. (UM) have developed this technology and have produced a variety of shaped crystals, such as needle-shaped BiSbTe single crystals, sheet-shaped mc-Si, and window-shaped CaF_2 single crystals [9]. In a growth of a mc-Si ingot, the required essentials of above (1) and (2) are generally satisfied because we can use high-purity Si raw materials (the purity is 11N) and a high-purity Argon gas (G1 grade, the purity is 99.9999%). Therefore, if we could use a suitable crucible and/or coating layer materials, it would be possible to realize the liquinert state even for silicon. If a mc-Si ingot could be grown in the liquinert condition, the problem of impurity contamination into the melt/crystal during the growth processes would be solved. Recently, according to this concept, FTB Research Institute, Co., Ltd., and Sakuragi have developed a Liquinert quartz crucible for the growth of single crystal Si by the Czochralski method [11]. If this crucible is available for the growth of a mc-Si ingot by casting, then the quality of a mc-Si ingot could be significantly improved, which would lead to an improvement in the energy conversion efficiency of solar cells.

In this study, we report on the growth of a mc-Si ingot by casting with a Liquinert quartz crucible to determine whether this crucible has the potentiality for the production of high-quality mc-Si ingot or not.

Si Crystal Grown Under Nonwetting and Nonreacting Conditions

Before the development of the Liquinert quartz crucible for the growth of a mc-Si ingot, preliminary experiments were conducted using a small amount of Si raw materials. Si melt typically reacts with SiO_2, so that Si crystals adhere to a silica container. Sakuragi et al. showed that the chemical reaction between a Si melt and a crucible at a high temperature was triggered by a minute amount of residual water in the atmosphere [10]; therefore, the liquinert condition, which is in a nonwetting and nonreacting condition, would be obtained if a water-free atmosphere could be achieved.

Figure 1 shows tiny Si crystals grown in a small SiC crucible coated with Si_3N_4 powder under flowing high-

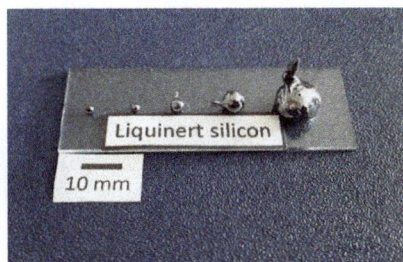

Figure 1. Si crystals grown under the liquinert condition.

purity Ar gas containing $SiCl_4$ as a liquinert condition. Spherical Si crystals were obtained without adherence of the Si to the above crucible. The surfaces of the crystals grown with the liquinert condition were shiny, as shown in Figure 1. To create a liquinert condition in a casting furnace, a quartz crucible must be coated with a material that is less reactive with the Si melt, because it is not practical to enclose $SiCl_4$ gas in a casting furnace, and a SiC crucible for much larger amount of Si raw materials is also not practical. Therefore, FTB Research Institute, Co., Ltd., and UM have been developing the Liquinert quartz crucible for larger size of Si ingots.

Growth of mc-Si Ingot in a Liquinert Quartz Crucible

The Liquinert quartz crucible used in this study was produced by FTB Research Institute, Co., Ltd., and consists of a quartz crucible with the coating layer. We expected to form a barium oxide (BaO) as a coating layer because it is stable at high temperature (the melting temperature of BaO is 2013°C [12]). Furthermore, a quartz crucible coated by BaO is often used to reduce the pollution from the crucible wall for the growth of single crystal Si by the Czochralski method [13]. Therefore, in this study, barium hydroxide was spread all over the inner surface of a

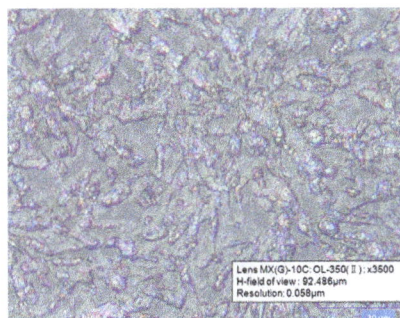

Figure 2. Optical image of inner surface of a liquinert quartz crucible.

quartz crucible (GE214), and then it was heated to form a coating layer, as shown in Figure 2. The size of the crucible used in this study was $186 \times 186 \times 250$ mm^3. For comparison, a similar quartz crucible (GE214) coated with Si$_3$N$_4$ powder (purity >98%) was also used. Many mc-Si ingots were grown in both crucibles under the same growth conditions to evaluate the effectiveness of the Liquinert quartz crucible for improvement in the quality of mc-Si.

For the growth experiments, 3.8 kg of high-purity raw Si materials (11N) and 5.25 g of B-doped Si wafer (0.016 Ω cm) were mixed in a crucible, and the crucible was set in a furnace. The temperature of the furnace was elevated to 1430°C to melt the materials in the crucible. When the materials were completely melted, the unidirectional growth process was started to grow a mc-Si ingot. The mc-Si ingot was grown by the dendritic casting method, in which dendrite growth is promoted along the bottom wall of the crucible during the earlier stage of growth [14, 15]. In this method, the bottom of the crucible was quickly cooled at the earlier stage of growth to promote dendrite growth. Then, the mc-Si ingot was grown at a constant rate of 0.4 mm/sec until the melt was completely solidified. All mc-Si ingots were grown under similar conditions with the same procedure. During the heating and the crystal growth processes, a high-purity Ar gas (6N) was flowing inside of the furnace.

Figure 3 shows the top surfaces of as-grown mc-Si ingots grown in Liquinert and Si$_3$N$_4$-coated quartz crucibles. The difference in the shininess of both ingot surfaces was evident. The surface of the ingot grown in the Liquinert quartz crucible was much shinier owing to the disuse of Si$_3$N$_4$ powder for the coating. According to this result, it is expected that contamination with impurities is restrained with the Liquinert quartz crucible. Figure 4 shows the side and bottom surfaces of both as-grown mc-Si ingots. The surfaces of the ingot grown in the Liquinert quartz crucible were very smooth and shiny, so that the grain structures were not visible (Fig. 4A), while the

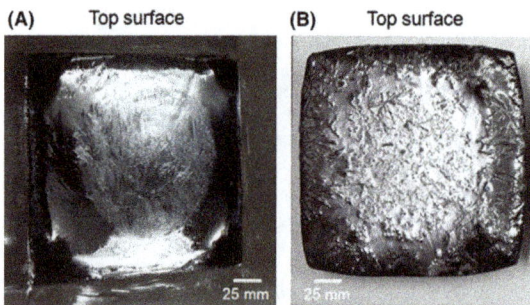

Figure 3. Top surfaces of as-grown mc-Si ingots grown in (A) a liquinert quartz crucible and in (B) a Si$_3$N$_4$-coated quartz crucible.

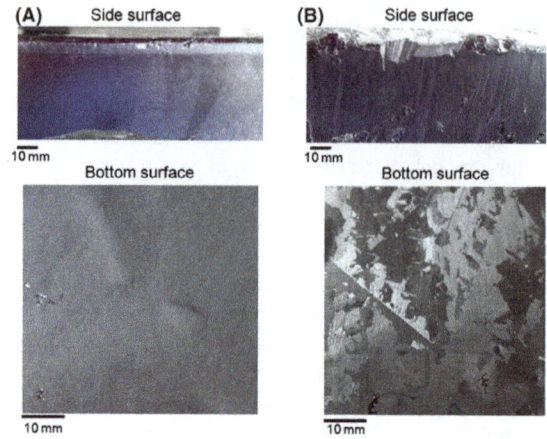

Figure 4. Side and bottom surfaces of as-grown mc-Si ingots grown in (A) a liquinert quartz crucible and in (B) a Si$_3$N$_4$-coated quartz crucible.

Table 1. Comparison of the resistivity and minority carrier lifetime of wafers cut from mc-Si ingots grown in liquinert and Si$_3$N$_4$-coated quartz crucibles.

Crucible	Liquinert quartz crucible	Si$_3$N$_4$ coated crucible
Wafer size (mm^3)	$125 \times 125 \times 0.2$	$125 \times 125 \times 0.2$
Average resistivity ($\Omega \cdot$cm)	2.42	1.21
Average lifetime (μsec)	67.7	20.8

surfaces of the ingot grown in the Si$_3$N$_4$-coated quartz crucible were rough and the grain structures were visible (Fig. 4B). From those results, we can conclude that wetting and reaction between the Si melt and crucible were restrained in the Liquinert quartz crucible.

Wafers with a size of $125 \times 125 \times 0.2$ mm^3 were cut from the middle part of both ingots, and the resistivity and minority carrier lifetime were measured. The average values of resistivity and minority carrier lifetime for each wafer are presented in Table 1. The resistivity of the wafer taken from the ingot grown in the Si$_3$N$_4$-coated quartz crucible was lower than that of the wafer taken from the ingot grown in the Liquinert quartz crucible, although the growth conditions for both ingots and the portions of the ingot from which the wafers were cut were similar. This suggests that impurities were dissolved into the Si melt/ingot from Si$_3$N$_4$ powder. The higher resistivity and minority carrier lifetime shown for the wafer taken from the ingot grown in the Liquinert quartz crucible confirm that the impurity contamination was reduced by using the Liquinert quartz crucible.

This was the first step to develop the Liquinert quartz crucible for the growth of mc-Si ingots, and we obtained positive results. Therefore, we will continue to develop this technology for the improvement of the quality of mc-Si ingots.

Summary

An attempt was made to use a Liquinert quartz crucible for the growth of mc-Si ingot for solar cell applications. It was shown that contamination with impurities from the crucible/coating layer was reduced with the Liquinert quartz crucible and a shiny ingot was successfully grown. The minority carrier lifetime was improved compared to that of a mc-Si ingot grown in a Si_3N_4-coated quartz crucible. These results indicate that the Liquinert quartz crucible has the potential to be a powerful tool in the near future for the production of high quality mc-Si ingots for solar cell applications.

Acknowledgments

The minority carrier lifetime measurements were performed at Kyocera Co. Ltd. This work was supported by a Kakenhi Grant-in-Aid (No. 26246016) from the Ministry of Education, Culture, Sports, Science and Technology (MEXT) of Japan.

Conflict of Interest

None declared.

References

1. Saito, T., A. Shimura, and S. Ichikawa. 1983. A reusable mold in directional solidification for silicon solar-cells. Sol. Energy Mater. 9:337–345.
2. Minster, O., J. Granier, C. Potard, and N. Eustathopoulos. 1987. Molding and directional solidification of solar-grade silicon using an insulating molten-salt. J. Cryst. Growth 82:155–161.
3. Prakash, P., P. K. Singh, S. N. Singh, R. Kishore, and B. K. Das. 1994. Use of C silicon oxynitride as a graphite mold releasing coating for the growth of shaped multicrystalline silicon-crystals. J. Cryst. Growth 144:41–47.
4. Khattak, C. P., and F. Schmid. Reusable crucible for silicon ingot growth. US Patent US 2004/0211496 A1.
5. Julsrud, S., and T. L. Naas. Method and crucible for direct solidification of semiconductor grade multi-crystalline silicon ingot. International Patent WO 2007/148987 A1.
6. Roligheten, R., G. Rian, and S. Julsrud. Reusable crucibles and method of manufacturing them. International Patent WO 2007/148986 A1.
7. Huguet, C., V. Brizé, S. Bailly, H. Lignier, E. Flahaut, B. Drevet, et al. Releasing coatings for PV-Si processing by liquid routes: comparison between the conventional and a high-purity coating. 26th European Photovoltaic Solar Energy Conference and Exhibition, Hamburg, Germany, 2011.
8. Hsieh, C. C., A. Lan, C. Hsu, and C. W. Lan. 2014. Improvement of multi-crystalline silicon ingot growth by using diffusion barriers. J. Cryst. Growth 401:727–731.
9. Sakuragi, S. "LIQUINERT": a new concept for shaped crystal growth. 19th European Photovoltaic Solar Energy Conference, Paris, France, 2004.
10. Sakuragi, S., T. Shimasaki, G. Sakuragi, and H. Nanba. Poly-silicon sheet for solar cell prepared by die-casting. 19th European Photovoltaic Solar Energy Conference, Paris, France, 2004.
11. Horioka, Y., and S. Sakuragi. Method of coating quartz crucible for growing silicon crystal, and quartz crucible for growing silicon crystal. JP4854814/KR101325628.
12. Okamoto, H. 2000. Phase diagrams for binary alloys, Desk handbook, vol. 1. ASM International, Materials Park, Ohio.
13. Hansen, R. L., L. E. Drafall, R. M. McCutchan, J. D. Holder, L. A. Allen, and R. D. Shelley. Surface-treated crucibles for improved zero dislocation performance. US005976247A.
14. Fujiwara, K., W. Pan, N. Usami, K. Sawada, M. Tokairin, Y. Nose, et al. 2006. Growth of structure-controlled polycrystalline silicon ingot for solar cells by casting. Acta Mater. 54:3191–3197.
15. Nakajima, K., K. Kutsukake, K. Fujiwara, K. Morishita, and S. Ono. 2011. Arrangement of dendrite crystals grown along the bottom of Si ingots using the dendritic casting method by controlling thermal conductivity under crucibles. J. Cryst. Growth 319:13–18.

14

Response surface methodology for optimization of bio-lubricant basestock synthesis from high free fatty acids castor oil

Venu Babu Borugadda & Vaibhav V. Goud

Department of Chemical Engineering, Indian Institute of Technology Guwahati, Guwahati, Assam 781039, India

Keywords

Biolubricant, epoxidation, free fatty acid, optimization, oxirane oxygen content

Correspondence

Vaibhav V. Goud, Department of Chemical Engineering, Indian Institute of Technology Guwahati, Guwahati, Assam 781039, India.
E-mail: vvgoud@iitg.ernet.in

Funding Information

No funding information provided.

Abstract

In this paper, an eco-friendly single-step process for the synthesis of biolubricant basestock from high free fatty acid (FFA) castor oil (CO) via epoxidation reaction was investigated. Influence of various process parameters on the structural modification of CO and their interaction with the maximum oxirane oxygen content (OOC) was optimized. Central composite design (CCD) as one of the tools in response surface methodology (RSM) was used to evaluate the effects of process variables on maximum OOC. Iodine value (IV) and OOC was used to monitor the progress of epoxidation. From the RSM study, the optimal condition inferred was H_2O_2, 1.65 mol; catalyst loading, 15.14 wt%; temperature, 52.81°C; and reaction time, 2.81 h. At this optimum condition, OOC was found to be 3.85 mass%. Further, the epoxide product was confirmed by ^1H, ^{13}C NMR spectral technique and OOC was determined by the standard HBr method. Finally, the significant physico-chemical properties for the prepared epoxide were determined and compared with the castor oil.

Introduction

Currently, plant seed oil/vegetable oil-derived renewable products have been gaining much importance to replace the conventional source of energy due to their depletion at a faster rate [1, 2]. In addition, the use of fossils has stimulated the search for eco-friendly alternatives to conventional resources. Nevertheless, inexhaustible renewable resources can supply a raw material basis for day-to-day life products, and this can avoid contribution to green house effects due to CO_2 discharge minimization [3]. In addition to that, proper utilization of renewable raw materials such as plant seed oils can bridge the gap between the fossil reserves' demand and consumption in the future. Above all, renewable materials could meet the principles of green chemistry in terms of easy degradation and lower toxicity [4]. Moreover, the plant oils

offer a wide number of advantages discussed by various technocrats [1, 2, 4]. In this regard, lubricants are one of the areas, which demand an alternative to the conventional lubricant basestocks due to diverse environmental issues reported by many researchers [5, 6]. Therefore, an effective utilization of the bio-based feedstocks for the lubricant basestock synthesis could bring down the dependency on imported petroleum as well as promote the sustainable agricultural initiative [6].

In general, liquid lubricants are the most common form of lubricants; their composition consists of 70–99% basestock and 30–1% of additives to improve the performance properties. However, the ultimate performance of lubricant depends on the basestock which can be synthesized by using plant seed oils [7]. Replacement of lubricant basestocks with the plant-based resources offers a wide range of advantages [5–8]. Even though the plant seed oils offer many advantages, they restrict their direct use due to the inadequate thermal and oxidative stability, hydrolytic stability, and poor cold flow properties [9]. These negative impacts are due to the presence of bis-allylic protons in plant seed oils structure which are highly susceptible to free radical attack, and thereby it undergoes oxidative degradation to form polar oxy compounds [10–13]. Plant seed oil oxidizes similarly to the hydrocarbon mineral oil by following the same free radical oxidation mechanism, but the oxidation rate of plant seed oils is faster than the hydrocarbon mineral oils [14, 15]. This fast rate of oxidation can be attributed to the presence of unsaturated fatty acids in its composition [16]. On the other hand, thermo-oxidative stability, hydrolytic stability, and low temperature performance of nonedible plant oils are much low and poor [17]. However, chemical modification of triglycerides can eliminate the poly unsaturation (for better thermo-oxidative stability), and an optimal extent of chemical change can ameliorate low temperature behavior [18].

Till date, many studies are available in the literature on structural modification of different plant seed oils and their methyl esters to prepare lubricant basestock by various methods, such as epoxidation (chemical or structural modification) [19], genetic modification [20], blending with additives [21], and hydrogenation of double bonds [22]. Among these methods, one of the most significant methods is hydrogenation of unsaturated double bonds in vegetable oils and its methyl esters [23, 24]. Hydrogenation yields total saturation of double bonds, resulting in miserable cold flow properties [23]. Genetic modification approach is tried either to reduce the saturated fatty acid content of the plant seed oils or to reduce the polyunsaturated fatty acid content, as these constituents have a negative impact on the thermo-oxidative stability [23, 25]. Genetically modified oils consistently exhibited improved oleic acid content ranging from 84 to 88% during multiple environmental conditions [26, 27]. However, among all these methods, structural alteration of unsaturated bonds via epoxidation has gained much attention due to high reactivity of three-member oxirane rings. Chemical modification of plant oils at the double bond sites' results in improved thermo-oxidative stability of the modified product [28].

The present study is focused on epoxidation of castor oil (nonedible oil) which is abundantly available all over India. Castor plant has the botanical name of *Ricinus Communis* of the family *Eurphorbiacae* [29]. Castor plant is primitively a tree or shrub that can grow in most tropical and subtropical countries above 10 m high, reaching an age up to 4 years [30]. Especially, castor plant needed a temperature between 15 and 38°C with lower humidity throughout the growing season

in order to receive maximum oil yields [30, 31]. Castor oil is a colorless to pale yellow liquid with mild/no odor or no taste, on an average CO seed contains about 46–55% oil by weight [29]. Oil fraction of castor seeds contains a higher amount of the ricinoleic acid as a hydroxylated fatty acid, and this unique structure has given a unique identity to this biological source for industrial synthesis of the variety of compounds [32]. Around the globe, India is the world's largest exporter of castor oil, and the other major producers are China and Brazil [30]. The total world production of castor seeds is estimated around one million tons, and the oil extracted is about 500,000 tons with productivity of 470 kg of oil per hectare [30, 33]. Irrespective of castor seed origin, season in which it has grown, its fatty acid composition remains unique [34].

Goud et al. [35] reported the epoxidation of castor oil with acetic acid and formic acid, using Amberlite IR-120 as heterogeneous acid catalysts [36, 37]. Further, Salimon et al. [38] also discussed the synthesis of ricinoleic acid epoxide and its characterization. Likewise, Salih et al. [14] described the synthesis of biolubricant basestocks from chemically modified ricinoleic acid-based tetra-esters via epoxidation, ring opening, and esterification reactions [39]. Recently, Hajar et al. [32] reported the production of biolubricant from castor oil substrate using novozyme 435. But very scanty information is available in the literature on the optimization of high FFA castor oil epoxidation process. Therefore, in the current study, an attempt has been made to examine the behavior of the epoxidation reaction and the physicochemical properties of epoxide prepared from high FFA raw material. However, very scanty information is available in the literature on optimization of high FFA castor oil epoxidation to understand the effect of reaction variables on oxirane oxygen content. Therefore, the present communication is aimed to bridge this gap by conducting the study on the modification of high FFA castor oil structure via epoxidation. The reaction was aimed at higher OOC and the main focus was to understand the effect of process parameters and interaction among them. RSM was adopted to optimize the epoxidation parameters and CCD was applied to understand the effect of process variables, using analysis of variance (ANOVA).

Materials and Methods

Materials and analytical techniques

Castor seeds were collected from Cherukupalli (Andhra Pradesh, India). Hydrogen peroxide (purity = 50% v/v) was purchased from Rankem, ion-exchange resin

(Amberlite IR-120, strong acid), glacial acetic acid (purity = 99–100%) was obtained from Merck India Ltd. All other reagents used for analysis were of analytical grade and used as received.

Extraction of CO was carried out according to the standard AOAC method. Similarly, physico-chemical characterization of all the samples was carried out as per the standard method reported earlier in some of our studies [16, 40].

Thin layer chromatography

The silica coated aluminum thin layer chromatography (TLC) plates were used for the analysis. The solvent system used was a mixture of hexane and dichloromethane (CH_2Cl_2) at 2:7 ratio to which few drops of glacial acetic acid were added, and the sample coated plates were visualized in an iodine chamber.

Epoxidation experimental design

Interaction between the process variables on maximum OOC was judged by RSM. A full factorial CCD technique was used for the optimization of the CO epoxidation process variables. Four reaction variables such as substrate molar ratio, that is C=C bonds to H_2O_2 mole ratio, catalyst loading (wt%), temperature (°C), and reaction time (h) were chosen to understand the effect of these variables on CO epoxidation reaction, as these variables are highly responsible for maximum OOC. The range of epoxidaton reaction variables involved in this study are shown in Table 1 along with lower (−1), medium (0), and higher (+1) levels of the variables. Value of α (alpha) is fixed at level 2 ($\alpha = 2^{4/4}$). A 2^4 full factorial CCD for four independent variables was used by giving a total number of 30 ($=2^n + 2n + 6$) experiments, where 'n' is the number of independent variables. During the optimization study, eight axial experimentations and 16 factorial runs were carried out with six extra replications at the center of design to estimate the pure error.

Table 1. Epoxidation process variables and their levels for response surface design of the castor oil epoxidation.

| Independent variables | Symbol | Unit | Variable levels | | |
			−1	0	+1
Time	A	h	2	3	4
Temperature	B	°C	50	60	70
Catalyst Loading	C	wt%	10	15	20
Oil: Hydrogen peroxide (H_2O_2)	D	mol	1	1.5	2

Epoxidation reaction procedure

Epoxidation of CO was carried out in 250 mL three necked glass reactor equipped with five blade glass stirrer and condenser; the entire setup was immersed in a heating oil bath. During the experiment CO, hydrogen peroxide and acetic acid were measured in a molar ratio; ion-exchange resin (Amberlite, IR-120) was added in weight% based on the organic phase. Initially, 20 g of CO (0.07 mol) was transferred into the reactor and heated to the desired reaction temperature (60°C). Then glacial acetic acid 2.1 g (0.035 mol) and other reactants, that is, hydrogen peroxide and catalysts were added to the reaction mixture. Processing time and temperature conditions used are mentioned in an experimental design matrix (Table S1). Addition of hydrogen peroxide was carried out drop wise for the first-half hours, when the temperature was 5°C below the reaction temperature to avoid the explosion. During epoxidation 14 g stirring speed was maintained to ensure consistent mixing. After complete addition of hydrogen peroxide, the reaction was continued for the desired time duration as mentioned in Table S1. Upon completion of the reaction and prior to the analysis the samples were washed repeatedly with warm Millipore water (40°C) to make it neutral. The sample was concentrated by rotary evaporator. The OOC was determined for the final epoxide product after each run. All the measurements were carried out in duplicate, and the average values are reported.

Preliminary study of process variables

Single parameter optimization process was followed to decide the optimum range of variables. Initially, the reaction time was optimized by continuing the reaction up to 9 h and samples were drawn at regular 1 h intervals to estimate the OOC. Linear increases in the epoxide content were observed up to the reaction time of 3 h, beyond which a gradual decrease was noticed in OOC (Fig. 1). A similar trend was noticed for all other reaction variables. Hence, depending on the preliminary studies, ranges of all the process variables were chosen and coded using the reported expression [41].

Statistical analysis

Based on experimental data shown in Table S1, the regression coefficient was determined by design expert software 8.0.7.1 trial version to predict the process response as a function of independent variables and their interactions were used to understand the system behavior. The mathematical relationship between the process variables

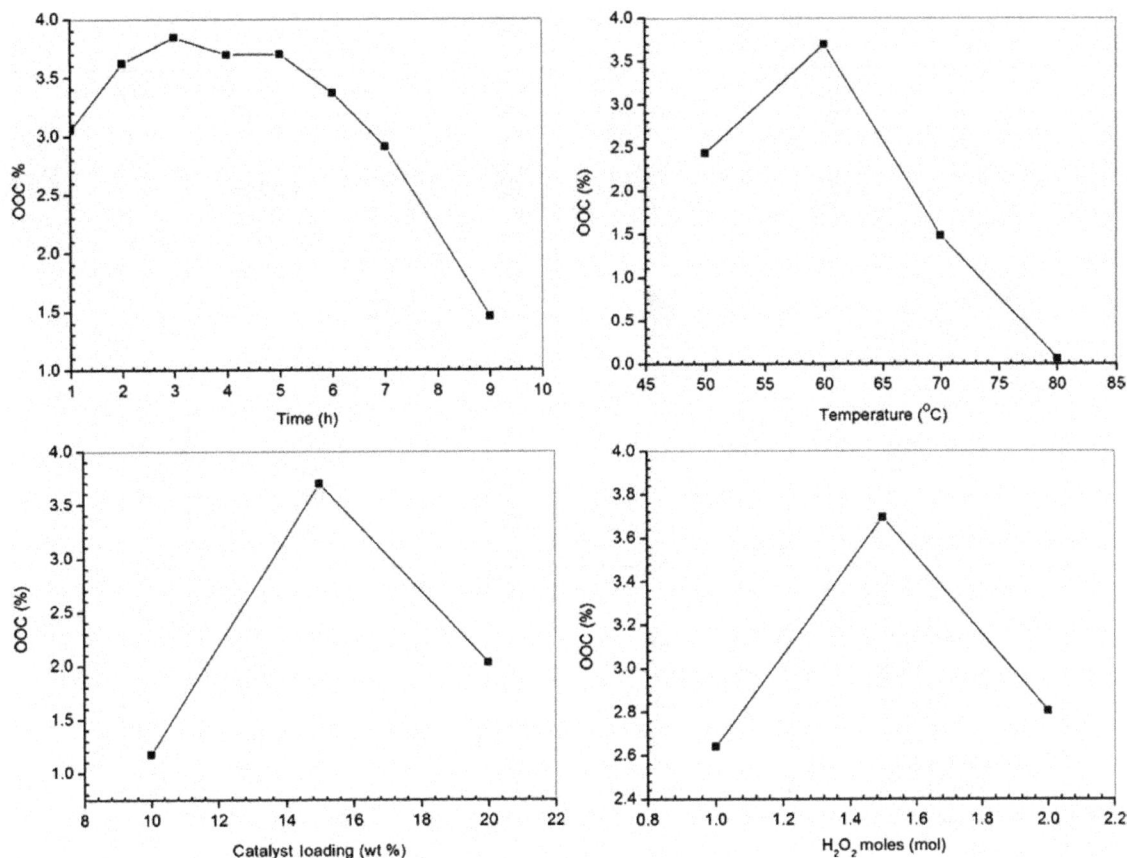

Figure 1. Preliminary study data for castor oil epoxidation to find the effects of individual reaction variables (time, catalyst loading, substrate ratio and temperature) on response.

and response was calculated by the following quadratic polynomial expression:

$$Y = \beta_0 + \sum_{i=1}^{n} \beta_i x_i + \sum_{i=1}^{n} \beta_{ii} x_i^2 + \sum_{i=1}^{n} \sum_{j<1}^{n} \beta_{ij} x_i x_j \quad (1)$$

where Y is the response, that is, the epoxide content, X_i and X_j represent the independent variables, β_0 is constant, β_i is linear term coefficient, β_{ii} is the quadratic term coefficient, β_{ij} is cross-term coefficient and 'n' is the number of process variables studied and optimized during the study. ANOVA was carried out to estimate the effects of process variables and their possible interaction effects on the maximum OOC in the response surface regression procedure. The goodness and best fit of the model was evaluated by a regression coefficient R^2. The response surface and counter plots are obtained using the fitted quadratic polynomial equation generated from regression analysis by keeping two of the independent variables at central value (0) and varying the other two.

Results and Discussion

Model fitting and ANOVA analysis

In order to optimize the epoxidation process variables for maximum epoxide content, a three level, four-factorial CCD was favored. ANOVA as a multivariate technique was studied to determine optimum reaction conditions. All the 30 designed experimental runs (Table S1) were performed and the results were analyzed by multiple regression analysis (Table 2). A quadratic polynomial equation was obtained from the experimental data to predict the epoxide content as shown below in terms of coded variables.

Response (higher OOC) = 3.82 + 0.038A − 0.82B − 0.19C

+ 0.068D − 0.37AB − 0.16AC

+ 0.022AD − 0.15BC − 0.33BD

+ 0.16CD − 0.21A^2

− 0.56B^2 − 0.56C^2 − 0.58D^2 2

Table 2. Regression coefficients of the predicted, quadratic polynomial model for response variable (maximum epoxide content), and analysis of a surface quadratic model (ANOVA) for epoxidized castor oil.

Source	Sum of squares	Degrees of freedom	Mean square	F value	P-value (prob > F)
Model	42.74	14	3.05	2261.27	<0.0001
A-time	0.035	1	0.035	25.56	0.0001
B-temp	15.99	1	15.99	11843.23	<0.0001
C-cat Load	0.83	1	0.83	611.11	<0.0001
D-H_2O_2	0.11	1	0.11	81.99	<0.0001
AB	2.24	1	2.24	1660.91	<0.0001
AC	0.41	1	0.41	305.74	<0.0001
AD	7.656E-003	1	7.656E-003	5.67	0.0309
BC	0.35	1	0.35	255.64	<0.0001
BD	1.71	1	1.71	1266.18	<0.0001
CD	0.39	1	0.39	291.64	<0.0001
A^2	1.16	1	1.16	858.43	<0.0001
B^2	8.70	1	8.70	6440.64	<0.0001
C^2	8.54	1	8.54	6326.76	<0.0001
D^2	9.32	1	9.32	6906.31	<0.0001
Lack-of-fit	0.018	10	1.812E-003	4.26	0.0616
Pure error	2.128E-003	5	4.255E-04		

In order to ensure a thorough model fit, to measure the analysis of variance on individual model coefficients test for lack-of-fit need to be estimated. The lack-of-fit is an assessment of failure of a model to represent the data that cannot be reported by random error [42]. Generally, the significant process variables are decided based on the F-value or P-value (also known as a probability of error value or "prob > F" value) [42]. Greater the magnitude of F-value and correspondingly smaller the "prob > F" value, more important is the corresponding coefficient [42]. The results of second-order response surface model in the form of ANOVA for the maximum epoxide content are summarized in Table 2. From the table, it can be seen that F-value of the model is 2261.27, and the corresponding P-value (prob > F) is very small, that is <0.0001 implying that the model is highly significant. The P-values are adopted as a tool to ensure the importance of each co-efficient. In this study, the main linear effects of time (A), temperature (B), catalyst loading (C), and hydrogen peroxide molar ratio (D), cross and quadratic effects of all four process variables (i.e., AB, AC, BC, BD, A^2, B^2, C^2, and D^2) are highly significant, as the P-value is very less (<0.0001). The other model term (cross variable) time and hydrogen peroxide molar ratio (AD) are also significant variables, since the P-value is 0.0309 (<0.05). In the present model, the absence of insignificant parameters intending that all the liner, cross, and quadratic terms are highly important for maximum epoxide content. The insignificant lack-of-fit F-value of 4.26 indicates that lack-of-fit is considerably significant relative to the pure error, which signifies that the model is extremely accurate without any noise, and the results

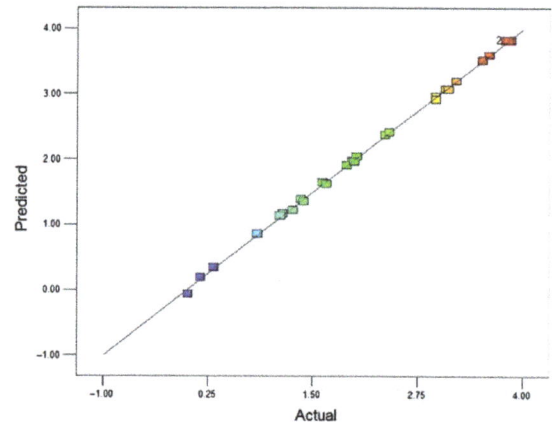

Figure 2. Predicted versus actual plot of response (maximum OOC) on castor oil epoxidation.

are reproducible. Furthermore, the actual values are very much close to the predicted as shown in Figure 2.

The precision of a model is judged by the regression coefficient (R^2). The R^2 value is always in between 0 and 1, and its order of magnitude suggests the aptness of the model [8]. For a good statistical model, the R^2 value should be close to one and the regression value for higher epoxide content is presented as 0.9996, which is close to 1, and it signifies that the 99.96% model behavior can be interpreted for higher epoxide content while only about 0.04% of the full variance cannot be explained by the model. Regression coefficient R^2 represents that the accuracy and general ability of the polynomial model is good. The predicted R^2 value (0.9982) is in reasonable agreement with the adjusted R^2 (0.9993) which recommends prominent corelational statistics between the remarked values and the predicted data. Thus, the regression model provides an excellent explanation of the relationship between the independent process variables and the response variable [42].

Influence of various process variables on maximum OOC

In order to estimate the best reaction condition for maximum epoxide content, the effects of linear, cross and quadratic reaction variable on the epoxidation was studied. The three-dimensional response surface plots and the two-dimensional counter plots, which is the graphical histrionics of the regression equation, are obtained by employing design expert software [26, 43]. These graphical representations of the plots are shown by varying two process variables at a time while keeping the other two variables at a central level (0). During each experiment, samples were withdrawn at regular intervals, washed, neutralized, and analyzed for OOC and α-Glycol content (i.e., oxirane cleavage).

Effect of time and temperature on OOC

Figure S1a and b describes the two-dimensional counter plot and corresponding three-dimensional response surface plot expressing the cross effect of time (A) and temperature (B) on the maximum OOC with respect to other process variables. Figure S1a and b shows that OOC increased with an increase in reaction time from 2 to 4 h, the increasing trend was noticed up to a certain time period of 3 h, beyond which gradual depletion in OOC was noticed (Table S1). Similarly, temperature was varied from 40 to 80°C in order to study the effect of temperature on OOC with respect to time. Results in Figure S1a and b show that OOC content increased with an increase in temperature up to 60°C, further increase in temperature, say 70°C OOC decreases (Table S1). The behavior of 2D and 3D figures revealed that longer reaction time and elevated temperature (beyond 60°C) lowers the OOC which might be due to oxirane cleavage (Fig. S2) [44]. The maximum OOC of 3.85 mass% was attained at the medium temperatures of 60°C and 3 h reaction time. Similar behavior was noticed by Salimon et al. [38] during their study on epoxidation of ricinoleic acid. In the current work, less OOC was obtained, which may be due to higher acid value of castor oil. Okieimen et al. [45] observed similar results during their study on epoxidation of high acid value rubber seed oil. Hence, 60°C temperature was considered as suitable for epoxidation of CO, when

molar ratio of unsaturation: acetic acid: hydrogen peroxide was 1:0.5:1.5 and catalysts loading 15 wt%.

Effect of time and catalyst loading on OOC

Keeping time (A) and catalyst loading (C) at central levels, that is, 3 h and 15 wt%, respectively, combined the effect of temperature (B) and hydrogen peroxide (D) was studied aiming at the higher epoxide content. From Figure S3a and b, it can be seen that OOC content increased almost linearly with a certain time period along with catalyst loading, thereafter OOC decreases (Table S1). The obtained results attributed to the hypothesis that availability of enough active surface area of catalyst for longer reaction time during the reaction results in cleavage of oxirane oxygen [44]. Hwang and Erhan [46] reported that presence of higher FFA content led to hydrolysis reaction in acid media thereby decreases the epoxide content due to oxirane cleavage. Hence, within the experimental conditions used in this study, the most favorable catalyst loading and reaction duration appeared to be 15 wt% and 3 h.

Effect of time and substrate molar ratio on OOC

Figure S4a and b shows the effect of varying substrate molar ratio and reaction time on the maximum epoxide content. Increase in the substrate molar ratio from 0.5 to 1.5 mol (Table S1) leads to linear increased in OOC content due to the formation of more peracetic acid and thereby OOC. The maximum OOC was attained with the mole ratio, 1.5 mol and the stability of oxirane observed with this ratio was similar to that observed with 1 molar ratio. However, an increase in the final oxirane value was relatively less when the ratio was further increased to 2.5 mol [47]. The use of a higher substrate molar ratio raised additional problem of agitation and decreases the mass transfer rate thereby decreases the OOC. The maximum substrate ratio and higher reaction time provided an opportunity to react oxirane rings with excess hydrogen peroxide, acetic acid, and by-product water [48]. Therefore, for subsequent experimentation, optimum reaction time, H_2O_2 moles considered are 3 h and 1.5 mol.

Effect of temperature and catalyst loading on OOC

Effect of reaction temperature (B) and catalyst loading (C) on the maximum OOC was investigated by varying catalyst loading from 5 to 25 wt% (Table S1) and temperature from 40 to 80°C with an interval of 10°C. Increase in the temperature (Fig. S5a and b) showed a favorable effect on epoxidation reaction and maximum OOC was obtained

(3.85 mass%) at 60°C and catalyst loading 15 wt%. From Figure S5a and b, it can also be seen that beyond 60°C decrease in OOC was observed. Epoxidation at higher temperatures (>60°C) acts as a medium to oxirane cleavage thereby decreasing the OOC value [49]. Likewise, altering the catalyst loading from 5 to 25% roaring in the epoxide content was observed up to 15 wt%. Beyond which inadequate OOC was noticed due to excess catalyst loading, which leads to oxirane cleavage [39]. However, under a given reaction condition and 15 wt% catalyst loading, it is assumed that reaction is free from mass transfer resistance and the maximum OOC can be obtained at a moderate reaction temperature of 60°C [39]. Hence, 15 wt% catalyst loading and 60°C temperature was considered as the optimum parameters for subsequent experimentation.

Effect of temperature and substrate molar ratio on OOC

The effect of the substrate molar ratio (D) on reaction temperature (B) and their combined interaction during the epoxidation process at constant catalyst loading of 15 wt% and 3 h of reaction time is represented by 3D response surface plot and corresponding 2D contour plot in Figure S6a and b. From the figures, it can be depicted with an increase in both temperature and substrate molar ratio, there was a progressive increase in the OOC which indicates that both the variables have a significant interaction between each other. However, the decrease in OOC was noticed beyond 60°C [39] and substrate molar ratio 1.5. Derawi et al. [50] reported that the rate of epoxidation increased as the concentration of hydrogen peroxide increased in the system, but the stability of oxirane rings was poor at higher substrate molar ratios [51]. On the other hand, Dinda et al. [39] have shown that oxirane ring was quite stable at lower hydrogen peroxide concentrations. Finally from the interaction of these variables, it can be seen that at a medium reaction temperature, the OOC increases with an increase in substrate molar ratio. However, higher reaction temperature and molar ratio showed a negative impact on OOC. Hence, a temperature 60°C and 1.5 substrate molar ratio was considered as optimum condition for further experimentations.

Effect of catalyst loading and substrate molar ratio on OOC

The influence of catalyst loading on varying hydrogen peroxide molar ratio was investigated at various catalyst loadings and various H_2O_2 molar ratios. Enhanced rate of per acid formation was anticipated with an increase catalyst loading and hydrogen peroxide concentration followed by an increase in the OOC [49, 52]. Figure S7a

and b describes the effect of catalyst loading and hydrogen peroxide on the course of epoxide formation. Higher selectivity was achieved at 1.5 moles of hydrogen peroxide per mol of ethylenic unsaturation and catalyst loading 15 wt%. Concentrations of hydrogen peroxide higher than 1.5 mol leads to a higher rate of epoxy ring decomposition as resolved by most of the researchers [53] and also discussed in the previous section. This observation was in agreement with the results shown in Figure S7a and b. Above all, lower epoxide content for high FFA CO was due to the hydrolysis reaction in the aqueous phase and its mechanism is well explained by Blee et al. [54].

Process optimization for epoxidation of high FFA CO

In the current study, this model is aspired to find the epoxidation process for the best fit of variables that gives maximum OOC. The predicted values of epoxidation reaction obtained from the model equation for a maximum amount of epoxide content are substrate ratio 1.65 mol, catalyst loading 15.14 wt%, and 2.81 h of reaction time at 52.8°C reaction temperature (Table S2). These values were chosen from the optimum solutions proposed by RSM optimization tool. The model expects the maximum OOC that can be obtained at these optimum conditions is 4.09 mass%. To confirm the model prediction, the best response variables are tested at the theoretical condition by conducting the laboratory experiment. At the optimized process condition epoxide content was found to be 3.85 mass% which agrees well with the model predicted value suggests that the formulated model was believed to be accurate and reliable (Table S2). In the present study, all the 30 experiments (Table S1) including confirmatory experiments (Table S2) were performed in duplicate at laboratory conditions and the average values are reported.

Physico-chemical characterization of prepared high FFA CO epoxide product

NMR spectroscopy

Epoxidation of high FFA CO was confirmed by ^{1}H NMR spectroscopy; it is one of the significant techniques to confirm the product formation and monitor the progress of reaction. ^{1}H NMR spectrum of the starting material (CO) and final product (CO epoxide) are shown in Figure S8a and b. Significant signals in the ^{1}H NMR spectra of CO shows the presence of unsaturation bands (-CH=CH-) from 5.25 to 5.58 ppm range (Fig. S8a). However, the partial disappearance of unsaturation bands have been noticed (Fig. S8b) after completion of the reaction, which signifies the presence of unsaturation in the end product at optimum condition. Similarly, the peaks (-CH-O-CH-) at 2.8 to 3.1 ppm range indicate the formation of the epoxide product, which are absent in the oil spectra. Furthermore, characterization of the epoxide product by ^{1}H NMR revealed the presence of the ring open product, which was confirmed by the additional peak at 3.41 ppm. Although ^{1}H NMR confirmed the complete conversion of unsaturation, but minimal oxirane cleavage was noticed due to the high FFA content of oil. Padmasiri et al. [47]; Derawi and Salimon [50]; Blee et al. [54] also noticed similar results during their study on mee and rubber oil.

Similarly, Figure 3A and B show ^{13}C NMR spectra of CO and its epoxide. Appearance of peaks at 125–133 ppm represents the olefinic carbons in CO (Fig. 3A). Comparison of Figure 3A and B, revealed the formation of epoxy protons at 53–59 ppm (Fig. 3B) and disappearance of olefinic carbons in the epoxide product at 125–133 ppm (Fig. 3A). The appearance of additional peaks at 53 and 59 ppm in the CO confirms the epoxidation of high FFA CO. Similar results are noticed by Madankar et al. [55] during their study on canola oil.

TLC analysis

Thin layer chromatography analysis of CO and its epoxide was carried out in order to confirm the product and mixture of compounds present in the final product (Fig. 4).

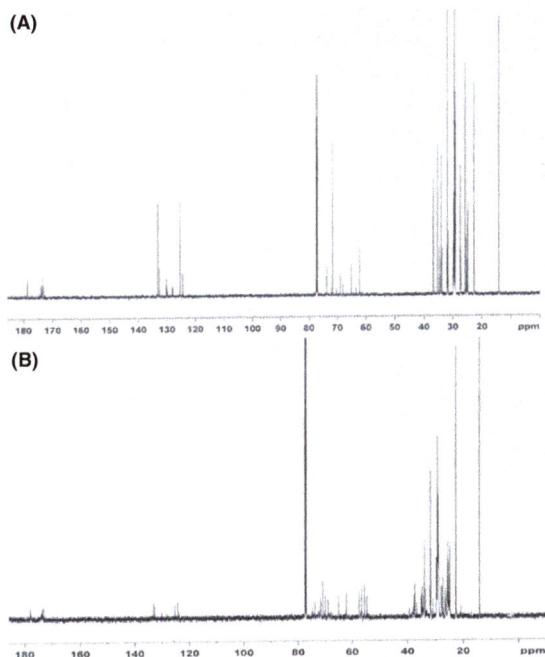

CO spot ←   → CO epoxide spot

Figure 4. Thin layer chromatography of castor oil and its epoxide.

The TLC spectra of the epoxide product indicate the presence of polar compounds, which was found to be absent in the CO confirms the formation of epoxide. From the CO epoxide TLC spectra, it was clear that some starting material is still present in the end product, which indicates an incomplete reaction. The same can be noticed from IV of the epoxide product (Table 3) and ^{1}H NMR spectrum (Fig. S8a and b) [47].

Pour point determination by DSC

Estimation of pour point is important in order to determine the low-temperature flow behavior of the modified (i.e. epoxide) and unmodified CO. It is a rough indication of the lowest temperature at which the epoxide is promptly pumpable [14]. In the present communication, cold flow properties of CO and its epoxide are determined using the method followed by Borugadda et al. [16]. DSC thermograms of CO and its epoxide are shown in Figure 5A and B. From the thermograms pour point of CO and CO epoxide was found to be −20 and −15 °C, respectively, which revealed that after epoxidation (i.e., the structural modification of CO), the pour point reduced significantly (5°C). This behavior is attributed to the fact that conversion of unsaturation content into epoxide

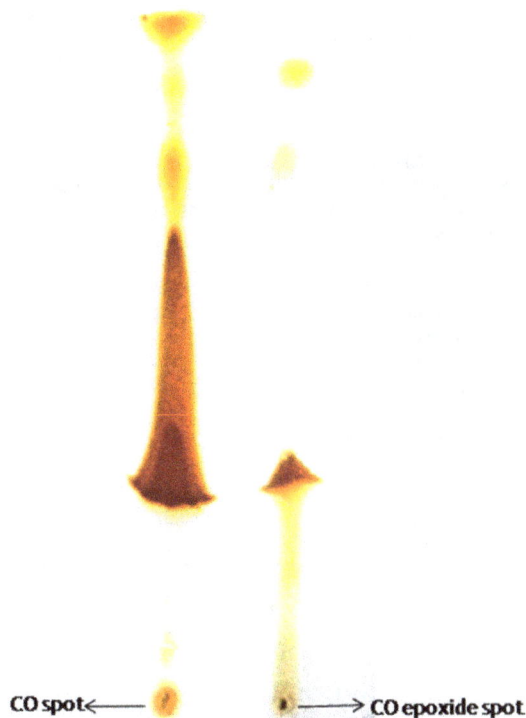

Figure 3. ^{13}C NMR spectrum of castor oil (A) and its epoxide (B).

Table 3. Physicochemical properties of castor oil (CO) and its epoxide.

Physicochemical properties	CO epoxide	CO
Acid value (mg KOH/g)	1.46	45.6
Density (kg/m³)	837.24	790.74
Free fatty acid (mg KOH/g)	0.73	22.8
Iodine value (gI$_2$/100 g of oil)	51.86	89.69
Kinematic viscosity (cSt) at 40°C	249.84	193.13
Moisture content (wt %)	0.21	0.9
Pour point (°C)	−15	−20
Refractive index (at 27.6°C)	1.479	1.477
Viscosity index (VI)	188.92	99.52
Oxirane content (Experimental, mass %)	3.85	–
Oxirane content (Theoretical, mass %)	5.35	–
Relative percentage conversion of oxirane (%)	71.96	–
Glycol content (Theoretical, mol/100 g)	0.31	–
Glycol content (Experimental, mol/100 g)	0.18	–
Relative percentage conversion to α-Glycol (mol/100 g)	40.67	–

altered the cold flow properties (PP). However, Soriano et al. [56] also observed similar kind of results during PP analysis of oils and esters. Elaborate literature survey on vegetable oils cold flow properties revealed that the cold flow property of plant oils is extremely inadequate, and this restricts their use at lower operating temperatures [10]. Plant oils have a tendency to form macro-crystalline structures at low temperatures through uniform stacking of the triglyceride backbone [10]. Formation of the macro-crystals restricts a free flow of the fluid due to loss of kinetic energy of individual molecules during the self-stacking [10]. In general, the pour point should be low enough to ensure that the epoxide is pumpable at lower temperatures in an application point of view [8]. Finally from this study, it can be seen that the modification in the structure of CO can alter the PP which was regarded as one of the major concerns for the plant oils to be used as an alternative to fossil resources. However, further improvement in the low-temperature properties can be done by additivation.

Thermo-oxidative stability

Thermo-oxidative stability of CO and its epoxide was estimated by onset (under nitrogen atmosphere) and oxidative onset (under oxygen atmosphere) temperatures. TGA thermo grams of (Fig. S9a and b) CO and its epoxide furnish the detailed information related to its degradation behavior in the presence of the inert gas (N$_2$) and reactive gas (O$_2$). Onset temperature is reported as the first perceptible temperature at which the degradation of CO and its epoxides starts. From Figure S9a, it can be seen that CO and its epoxide are stable up to 310 and 308°C (onset temperatures), respectively, in an inert atmosphere. Similarly, the maximum decomposition temperature represents the temperature at which utmost weight loss of CO and epoxide samples occurred, that is, 371.5 and 350.5°C, respectively. Lower thermal stability of epoxide signifies the degradation of epoxide at lower temperature compared to the unmodified CO. This may be due to the presence of high acid media in the mixture which is responsible for the lower thermal stability of CO epoxide [54].

Park et al. [57] reported that oil with saturated fatty acids, and mono unsaturation content has a positive influence and thermally more stable than polyunsaturation. Since the double bonds are converted to epoxide, hence higher thermal stability of epoxide was anticipated compared to unmodified CO, but due to the aforementioned reason, lower thermal stability was noticed. Study on thermal stability of epoxide is of great importance for various applications, and it depends on the chemical composition and structure of the epoxide.

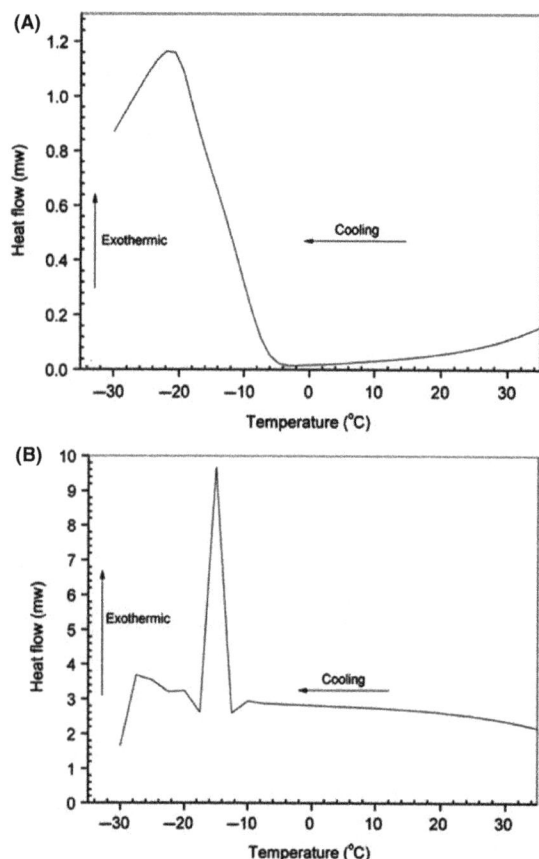

Figure 5. DSC thermogram of castor oil (A) and its epoxide (B) for pour point determination.

Similarly, oxidative stability was found as a quality indicative parameter under oxygen atmosphere. In the present communication, oxidative stability was specified as the resistance of the epoxide against oxidation in the air atmosphere. Figure S9b describes the TGA plots of CO and its epoxide, from the oxidative onset temperature of the thermogram, it was found to be 320°C for both modified and unmodified CO (Fig. S9b). Similarly, the maximum decomposition temperature was found to be 382 and 354°C, respectively, for CO and its epoxide. In addition to the aforementioned elucidation, this attitude can be justified by the presence of unsaturation in the epoxide sample. Finally, it can be concluded that the CO showed more thermo-oxidative stability than the structurally modified CO epoxide.

Viscosity and viscosity index

In the current study, kinematic viscosities of CO and its epoxide were found to be 193.13 and 249.84 cSt, respectively at 40°C (Table 3). Increase in the epoxide molecular weight can be attributed to the addition of an oxygen molecule at the unsaturation sites thereby forming the oxirane ring [45]. The results of the study depict that CO epoxide viscosity (249.84 cSt) has been improved over the unmodified CO, and this signifies that the synthesized CO epoxide can serve as an acceptable lubricant base stock with enhanced viscosity to reduce friction. The ability of substance to resist free flow is one of the highly significant attributes for many heavy duty and industrial materials such as fuels, lubricants, and surfactants [58]. Therefore, it is desirable that the viscosity must be high enough all the time to keep good oil film between the moving parts to reduce friction [58]. Otherwise, due to loss in the lubricant, there is a tendency to increase the friction thereby resulting in power loss and rapid wear on the machine parts [58]. Viscosity is one of the crucial parameters while selecting a lubricant for a specific application, and failure to use the right lubricant with the required viscosity results in extreme temperature that may result in poor lubrication, equipment failure, and damage [59]. As discussed above, viscosity of acceptable range cannot be covered by conventional vegetable oils, hence in this study, structural modification was attempted to enhance the viscosity and molecular weight. However, viscosity can also be improved by addition of a viscosity enhancer [60].

Similarly, it is also desirable to check the viscosity in terms of the viscosity index (VI) which indicates the lower sensitivity of viscosity variations at higher temperatures. During this study, CO and its epoxide VI was found to be 99.52 and 188.92, respectively. Higher VI of CO epoxide signifies that the addition of an oxygen molecule at the unsaturation sites leads to an overall increase in the molecular weight of the final epoxide product and the

viscosity index [14]. The higher VI of epoxide (188.92) almost doubles compared with the viscosity index of castor oil (99.52) indicating that the prepared epoxide can act as a high-temperature lubricant basestock.

Physico-chemical characterization

Significant physico-chemical properties of CO and its epoxide are estimated and given in Table 3. Acid value (AV) of CO and its epoxide was found to be 45.6 and 1.46 mg KOH/g, respectively. Lower acid value of CO epoxide signifies the smooth operation and functioning of the equipment during its usage. FFA is always considered as half of the AV, which signifies the formation of soap when mixed with water. The obtained FFA value was very less (0.73 mg KOH/g) and signifies trouble-free performance of the epoxide. The density of CO was found to be 790.74 kg/m^3 and corresponding epoxide density was 837.24 kg/m^3. Improved density of CO epoxide was assigned to increase molecular weight of epoxide by addition of an oxygen molecule in the midst of unsaturation sites. Determination of the IV (i.e., quantity of the double bonds) after epoxidation reaction is one of the best ways to confirm and support the completion of the epoxidation reaction. Initial IV of CO was found to be 89.69 (g I$_2$/100 g) whereas after epoxidation the value was 51.86 (g I$_2$/100 g). Higher IV of epoxide signifies incomplete epoxidation reaction and the same has been observed from ^{1}H NMR and TLC spectral analysis (Table 3). The presence of hydroxyl groups was confirmed by α-Glycol content analysis. The theoretical and experimental values of α-glycol content along with the relative percentage conversion of α-glycol is calculated and reported in Table 3.

Another significant property of epoxide is moisture content indicating the presence of water in the epoxide. Presence of moisture in the epoxide supports the bacterial growth which leads to an undesired performance upon usage thereby increasing AV, viscosity, and formation of free radical compounds via oxidation [21]. In this study, the CO epoxide moisture content was found to be 0.21 wt% (Table 3), which indicates safe performance of epoxide during usage. Refractive index for both the samples was found to be almost similar 1.477 and 1.479 (Table 3), which conveys that a very small quantity of heat energy can pass through the CO and epoxide samples, which helps to avoid the thermal degradation of end products during its usage and storage.

Conclusion

This is the first comprehensive report on the synthesis of castor oil epoxide from high FFA castor oil. The optimum condition for CO epoxidation occurred at temperature, 52.81°C; hydrogen peroxide to ethylenic un-saturation molar ratio, 1.65; catalyst loading, 15.14 wt %; and reaction time, 2.81 h. At this optimal condition, maximum epoxide content and relative conversion to oxirane was found to be 3.85mass% and 0.71%. Further, the product was confirmed by ^{1}H-NMR, ^{13}C-NMR, IV, and oxirane analysis. Quadratic polynomial model and ANOVA has well explained the interaction between the process variables. Further the model was examined and validated for the best fit. Comparative evaluation of physico-chemical properties of castor oil and its epoxide revealed surprising results. Among all the evaluated properties, only a few properties showed significant improvement, whereas thermo-oxidative stability and cold flow property results were unsatisfactory which may be due to high FFA content in CO or higher oxirane cleavage. However, the obtained viscosity index of epoxide indicates that the product can be used as high-temperature lubricant basestocks. However, from the present study, it can be concluded that low FFA CO could act as a suitable feedstock for lubricant synthesis with improved physico-chemical properties and anticipated to give improved thermo-oxidative stability and cold flow properties.

Conflict of Interest

None declared.

References

1. Kotwal, M., A. Kumar, and S. Darbha. 2013. Three-dimensional, mesoporous titanosilicates as catalysts for producing biodiesel and biolubricants. J. Mol. Catal. A. Chem. 377:65–73.
2. Borugadda, V. B., and V. V. Goud. 2014a. Synthesis of waste cooking oil epoxide as a bio-lubricant base stock: characterization and optimization study. J. Bioproc. Eng. Bioref. 3:57–72.
3. Oyedepo, S. O. 2012. Energy and sustainable development in Nigeria: the way forward. Energy Sustain. Soc. 2:15.
4. Eissen, M., J. O. Metzger, E. Schmidt, and U. Schneidewind. 2002. 10 years after Rio - concepts on the contribution of chemistry to a sustainable development. Angew. Chem. Int. Ed. Engl. 41:414–436.
5. Salimon, J., N. Salih, and E. Yousif. 2011b. Chemically modified biolubricant basestocks from epoxidized oleic acid: Improved low temperature properties and oxidative stability. J. Saudi Chem. Soc. 15:195–201.
6. Pawan, D. M., G. P. Ravindra, and V. P. Harshal. 2011. Epoxidation of wild safflower (Carthamus oxyacantha) oil with peroxy in of strongly cation exchange resin IR-122 as catalyst. Int. J. Chem. Tech. Res. 3:1152–1163.
7. Dietrich, H. 2002. Recent trends in environmentally friendly lubricants. J. Synthetic Lubrication 4:327–347.
8. Manivannan, P., and M. Rajasimman. 2011. Optimization of process parameters for the osmotic dehydration of beetroot in sugar solution. J. Food Process Eng 34:804–825.
9. Aguieiras, E. C. G., C. O. Veloso, J. V. Bevilaqua, D. O. Rosas, M. A. P. Silva, and M. A. P. Langone. 2011. Estolides synthesis catalyzed by immobilized lipases. Enzyme Res. 432746:1–7.
10. Adhvaryu, A., Z. Liu, and S. Z. Erhan. 2005. Synthesis of novel alkoxylated triacylglycerols and their lubricant base oil properties. Ind. Crops Prod. 21:113–119.
11. Sharma, B. K., J. M. Perez, and S. Z. Erhan. 2007. Soybean oil-based lubricants: a search for synergistic antioxidants. Energy Fuels 21:2408–2414.
12. Singh, C. P., and V. K. Chhibber. 2013. Chemical modification in karanja oil for biolubricant industrial applications. J. Drug Deliv. Therapeutics 3:117–122.
13. Erhan, S. Z., B. K. Sharma, and J. M. Perez. 2006. Oxidation and low temperature stability of vegetable oil-based lubricants. Ind. Crops Prod. 24:292–299.
14. Salih, N., J. Salimon, E. Yousif, and B. M. Abdullah. 2013. Biolubricant basestocks from chemically modified plant oils: ricinoleic acid based tetra-esters. Chem. Central J. 7:128.
15. Hamblin, P. 1999. Oxidative stabilisation of synthetic fluids and vegetable oils. J. Synthetic Lubrication 2:157–181.
16. Borugadda, V. B., and V. V. Goud. 2013a. Comparative studies of thermal, oxidative and low temperature properties of waste cooking oil and castor oil. J. Renew. Sustain. Energy 5:063104.
17. Fox, N. J., B. Tyrer, and G. W. Stachowiak. 2004. Boundary lubrication performance of free fatty acids in sunflower oil. Tribol. Lett. 4:275–281.
18. Bart, J. C. J., E. Gucciardi, and S. Cavallaro. 2012. Biolubricants: science and technology. Woodhead publishing series in energy No.46, 249. Woodhead Publishing Limited, UK.
19. Borugadda, V. B., and V. V. Goud. 2014b. Epoxidation of castor oil fatty acid methyl esters (COFAME) as a lubricant base stock using heterogeneous ion-exchange resin (IR-120) as a catalyst. Energy Procedia 54:75–84.
20. Adhvaryu, A., and S. Z. Erhan. 2002. Epoxidised soybean oil as a potential source of high temperature lubricants. Ind. Crops Prod. 15:247–254.
21. Doll, K. M., B. K. Sharma, and S. Z. Erhan. 2007. Synthesis of branched methyl hydroxyl stearates including an ester from bio-based levulinic acid. Ind. Eng. Chem. Res. 46:3513–3519.
22. Cermak, S. C., K. B. Brandon, and T. A. Isbell. 2006. Synthesis and physical properties of estolides from lesquerella and castor fatty acid esters. Ind. Crops Prod. 23:54–64.

23. Wadumesthrige, K., S. O. Salley, and K. Y. S. Ng. 2009. Effects of partial hydrogenation, epoxidation, and hydroxylation on the fuel properties of fatty acid methyl esters. Fuel Process. Technol. 90:1292–1299.

24. Cadenas, F. A. P., F. Kapteijn, M. P. Martijn, A. Jacob, and Z. Moulijn. 2007. Selective hydrogenation of fatty acid methyl esters over palladium on carbon-based monoliths: structural control of activity and selectivity. Catal. Today 128:13–17.

25. Suliman, E. T., Z. Meng, J. W. Li, J. Jiang, and Y. Liu. 2013. Optimisation of sunflower oil deodorising: balance between oil stability and other quality attributes. Int. J. Food Sci. Technol. 48:1822–1827.

26. Kinney, A. J., and T. E. Clemente. 2005. Modifying soybean oil for enhanced performance in biodiesel blends. Fuel Process. Technol. 86:1137–1147.

27. Kinney, A. J. 1996. Development of genetically engineered soybean oils for food application. J. Food Lipids 3:273–292.

28. Holser, R. A. 2008. Transesterification of epoxidized soybean oil to prepare epoxy methyl esters. Ind. Crops Prod. 1:130–132.

29. Castor oil plant. Available at http://en.wikipedia.org/wiki/Castor_oil_plant (accessed 14 November 2014).

30. Borugadda, V. B., and V. V. Goud. 2012. Biodiesel production from renewable feedstocks: status and opportunities. Renew. Sustain. Ene. Rev. 16:4763–4784.

31. Ogunniyi, D. S. 2006. Castor oil: a vital industrial raw material. Bioresour. Technol. 97:1086–1091.

32. Hajar, M., and F. Vahabzadeh. 2014. Artificial neural network modelling of biolubricant production using Novozym 435 and castor oil substrate. Ind. Crops Prod. 52:430–438.

33. Santana, G. C. S., P. F. Martins, D. L. N. Da Silva, C. B. Batistella, M. R. Filho, and W. M. R. Maciel. 2010. Simulation and cost estimate for biodiesel production using castor oil. Chem. Eng. Res. Design 88:626–632.

34. Salimon, J., N. Salih, and E. Yousif. 2010. Biolubricants: raw materials, chemical modifications and environmental benefits. Eur. J. Lipid Sci. Technol. 5:519–530.

35. Goud, V. V., A. V. Patwardhan, and N. C. Pradhan. 2006. Strongly acidic cation exchange resin of sulphonated polystyrene type used as catalyst for epoxidation of castor oil with peracetic acid and performic acid. Solid State Sci. Technol. 14:62–68.

36. Jankovic, M., O. Borota, and S. S. Fiser. 2012. Epoxidation of castor oil with peracetic acid formed in situ in the presence of an ion exchange resin. Chem. Eng. Process: Process Intensif. 62:106–113.

37. Leveneur, S., D. Y. Murzin, and T. Salmi. 2009. Application of linear free-energy relationships to perhydrolysis of different carboxylic acids over homogeneous and heterogeneous catalysts. J. Mol. Catal. A: Chem. 303:148–155.

38. Salimon, J., N. Salih, and E. Yousif. 2011c. Synthetic biolubricant basestocks from epoxidized ricinoleic acid: improved low temperature properties. Chem. Industry 60:127–134.

39. Dinda, S., A. V. Patwardhan, V. V. Goud, and N. C. Pradhan. 2008. Epoxidation of cottonseed oil by aqueous hydrogen peroxide catalysed by liquid inorganic acids. Bioresour. Technol. 99:3737–3744.

40. Borugadda, V. B., and V. V. Goud. 2014c. Thermal, oxidative and low temperature properties of methyl esters prepared from oils of different fatty acids composition: a comparative study. Thermochim. Acta 577:33–40.

41. Sagiroglu, A., S. S. Isbilir, H. M. Ozcan, H. Paluzar, and N. M. Toprakkiran. 2011. Comparison of biodiesel productivities of different vegetable oils by acidic catalysis. Ind. Eng. Chem. Res. 17:53–58.

42. Tabrizi, S. A. H., and E. T. Nassaj. 2011. Modeling and optimization of densification of nanocrystalline Al2O3 powder prepared by a sol–gel method using response surface methodology. J. Sol-Gel. Sci. Technol. 57:212–220.

43. Majumdar, S., and V. K. Lukka. 2011. Pectin based multiparticulate formulation of ketoprofen for bacterial enzyme dependent drug release in colon. Trends Biomater. Artif. Organs 25:133–143.

44. Chavan, V. P., A. V. Patwardhan, and P. R. Gogate. 2012. Intensification of epoxidation of soybean oil using sonochemical reactors. Chem. Eng. Process: Process. Intensif. 54:22–28.

45. Okieimen, F. E., C. Pavithran, and I. O. Bakare. 2005. Epoxidation and hydroxlation of rubber seed oil: one-pot multi-step reactions. Eur. J. Lipid Sci. Technol. 107:330–336.

46. Hwang, H. S., and S. Z. Erhan. 2011. Modification of epoxidized soybean oil for lubricant formulations with improved oxidative stability and low pour point. J. Am. Oil Chem. Soc. 78:1179–1184.

47. Padmasiri, K., B. M. O. Gamage, and L. Karunanayake. 2009. Epoxidation of some vegetable oils and their hydrolysed products with peroxyformic acid - optimised to industrial scale. J. Natl Sci. Found Sri Lanka 37:229–240.

48. Suarez, P. A. Z., M. S. C. Perreira, K. M. Doll, B. K. Sharma, and S. Z. Erhan. 2009. Epoxidation of methyl oleate using heterogeneous catalyst. Ind. Eng. Chem. Res. 48:3268–3270.

49. Goud, V. V., A. V. Patwardhan, S. Dinda, and N. C. Pradhan. 2007. Kinetics of epoxidation of jatropha oil with peroxyacetic and peroxyformic acid catalysed by acidic ion exchange resin. Chem. Eng. Sci. 62:4065–4076.

50. Derawi, D., and J. Salimon. 2010. Optimization of epoxidation of palm olein by using performic acid. J. Chem. 7:1440–1448.

51. Dinda, S., V. V. Goud, A. V. Patwardhan, and N. C. Pradhan. 2011. Selective epoxidation of natural

triglycerides using acidic ion exchange resin as catalyst. APJ Chem. Eng. 6:870–878.

52. Mungroo, R., N. C. Pradhan, V. V. Goud, and A. K. Dalai. 2008. Epoxidation of canola oil with hydrogen peroxide catalyzed by acidic ion exchange resin. J. Am. Oil Chem. Soc. 85:887–896.

53. Petrovic, Z., A. Zlatanic, C. C. Lava, and S. S. F. Fiser. 2002. Epoxidation of soybean oil in toluene with peroxoacetic and peroxoformic acids-kinetics and side reactions. Eur. J. Lipid Sci. Technol. 104:293–299.

54. Blee, E., S. Summerer, M. Flenet, H. Rogniaux, A. V. Dorsselaer, and F. Schuber. 2005. Soybean epoxide hydrolase. Identification of the catalytic residues and probing of the reaction mechanism with Secondary kinetic isotope effects. J. Biol. Chem. 280:6479–6487.

55. Madankar, C. S., A. K. Dalai, and S. N. Naik. 2013. Green synthesis of biolubricant base stock from canola oil. Ind. Crops Prod. 44:139–144.

56. Soriano, N. U., V. P. Migo, and M. Matsumura. 2006. Ozonized vegetable oil as pour point depressant for neat biodiesel. Fuel 85:25–31.

57. Park, J. Y., D. K. Kim, J. P. Lee, S. C. Park, Y. J. Kim, and J. S. Lee. 2008. Blending effects of biodiesels on oxidation stability and low temperature flow properties. Bioresour. Technol. 99:1196–1203.

58. Salimon, J., N. Salih, and E. Yousif. 2011a. Characterization and physicochemical properties of oleic acid ether derivatives as biolubricant basestocks. J. Oleo Sci. 60:613–618.

59. Lubricant Failure = Bearing Failure. Available at http://www.machinerylubrication.com/Read/1863/lubricant-failure (Accessed 14 November 2014).

60. Quinchia, L. A., M. A. Delgado, C. Valencia, J. M. Franco, and C. Gallegos. 2009. Viscosity modification of high oleic sunflower oil with polymeric additive for the design of new biolubricant formulations. Environ. Sci. Technol. 43:2060–2065.

15

University of Texas study underestimates national methane emissions at natural gas production sites due to instrument sensor failure

Touché Howard

Indaco Air Quality Services, Inc., Durham, North Carolina

Keywords
Greenhouse gases, methane, natural gas

Correspondence
Touché Howard, Indaco Air Quality Services, Inc., Durham, NC.
E-mail: touche.howard@indacoaqs.com

Funding Information
This paper used data that are publicly available and did not rely on external funding.

Abstract

The University of Texas reported on a campaign to measure methane (CH_4) emissions from United States natural gas (NG) production sites as part of an improved national inventory. Unfortunately, their study appears to have systematically underestimated emissions. They used the Bacharach Hi-Flow® Sampler (BHFS) which in previous studies has been shown to exhibit sensor failures leading to underreporting of NG emissions. The data reported by the University of Texas study suggest their measurements exhibit this sensor failure, as shown by the paucity of high-emitting observations when the wellhead gas composition was less than 91% CH_4, where sensor failures are most likely; during follow-up testing, the BHFS used in that study indeed exhibited sensor failure consistent with under-reporting of these high emitters. Tracer ratio measurements made by the University of Texas at a subset of sites with low CH_4 content further indicate that the BHFS measurements at these sites were too low by factors of three to five. Over 98% of the CH_4 inventory calculated from their own data and 41% of their compiled national inventory may be affected by this measurement failure. Their data also indicate that this sensor failure could occur at NG compositions as high as 97% CH_4, possibly affecting other BHFS measurement programs throughout the entire NG supply chain, including at transmission sites where the BHFS is used to report greenhouse gas emissions to the United States Environmental Protection Agency Greenhouse Gas Reporting Program (USEPA GHGRP, U.S. 40 CFR Part 98, Subpart W). The presence of such an obvious problem in this high profile, landmark study highlights the need for increased quality assurance in all greenhouse gas measurement programs.

Introduction

The climatic benefits of switching from coal to natural gas (NG) depend on the magnitude of fugitive emissions of methane (CH_4) from NG production, processing, transmission, and distribution [12, 13, 27]. This is of particular concern as the United States increasingly exploits NG from shale formations: a sudden increase in CH_4 emissions due to increased NG production could trigger climate "tipping points" due to the high short-term global warming potential of CH_4 (86× carbon dioxide on a 20-year time scale) [19]. The United States Environmental

Protection Agency (USEPA) estimates CH_4 emissions from the NG supply chain by scaling up individual ground-level measurements, mostly collected by reporting from industry [26]. However, some recent studies have questioned whether these "bottom-up" inventories are too low, since airborne measurements indicate that CH_4 emissions from NG production regions are higher than the inventories indicate [5, 14, 17, 20, 21].

In order to help determine the climate consequences of expanded NG production and use, and to address the apparent discrepancy in top-down and bottom-up measurements, the University of Texas (UT) at Austin and the

Environmental Defense Fund launched a large campaign to measure CH_4 emissions at NG production sites in the United States [1]. This study used both existing EPA GHG inventory data and new measurements to compile a new national inventory of CH_4 emissions from production sites. Forty-one percent of this new inventory was based on measurements made by [1], which included measurements of emissions from well completion flowbacks as well as measurements of emissions from chemical injection pumps, pneumatic devices, equipment leaks, and tanks at 150 NG production sites around the United States already in routine operation (measurements from tanks were not used for inventory purposes). However, the measurements of emissions at well production sites already in operation (which comprised 98% of the new inventory developed by [1]) were made using the Bacharach Hi-Flow Sampler (BHFS; Bacharach, Inc., New Kensington, PA) and recent work has shown that the BHFS can underreport individual emissions measurements by two orders of magnitude [10]. This anomaly occurs due to sensor transition failure that can prevent the sampler from properly measuring NG emission rates greater than ~0.4 standard cubic feet per minute (scfm; 1 scfm = 1.70 m^3 h^{-1} or 19.2 g min^{-1} for pure CH_4 at 60°F [15.6°C] and 1 atm; these are the standard temperature and pressure used by the U.S. NG industry). Although this failure is not well understood, it does not seem to occur when measuring pure CH_4 streams, but has been observed in four different samplers when measuring NG streams with CH_4 contents ranging from 66% to 95%. The sampler's firmware version and elapsed time since last calibration may also influence the occurrence of this problem [10, 18].

This paper presents an analysis of the UT [1] emissions measurements that were made with the BHFS, and shows that high emitters (>0.4 scfm [0.7 m^3 h^{-1}]) were reported very rarely at sites with a low CH_4 content in the wellhead gas (<91%), consistent with sensor transition failure. It also details testing of the exact BHFS instrument used in that study and shows the occurrence of this sensor failure at an NG production site with a wellhead composition of 91% CH_4 (the highest CH_4 concentration site available during testing). Finally, the downwind tracer ratio measurements made by [1] at a subset of their test sites are reexamined and indicate that the BHFS measurements made at sites with low wellhead CH_4 concentrations were too low by factors of three to five.

Evidence of BHFS Sensor Transition Failure in the UT Dataset

The Allen et al. [1] UT dataset is unique due to the large number of BHFS measurements made across a wide geographic range, the variety of emissions sources

(equipment leaks, pneumatic devices, chemical injection pumps, and tanks) and the wide range of NG compositions (67.4–98.4% CH_4) that were sampled. As such, the UT study provides an important opportunity to evaluate the occurrence of sensor transition failure in the BHFS as well as the impact of this issue on emission rates and emissions factors based on measurements in other segments of the NG supply chain.

The BHFS uses a high flow rate of air and a loose enclosure to completely capture the NG-emitting from a source, with the emission rate calculated from the total flow rate of air and the resulting sample NG concentration, after the background NG concentration is subtracted. The sampler uses a catalytic oxidation sensor to measure sample concentrations from 0% to 5% NG in air, but must transition to a thermal conductivity sensor in order to accurately measure sample concentrations higher than 5%. It is the failure of the sampler to transition to the higher range that has been previously observed by Howard et al. [10] and which can prevent the sampler from correctly measuring emission rates larger than 0.3–0.5 scfm (0.5–0.9 m^3 h^{-1}) (corresponding to sampler flow rates of 6–10 scfm [10–17 m^3 h^{-1}]). Figure 1 summarizes data

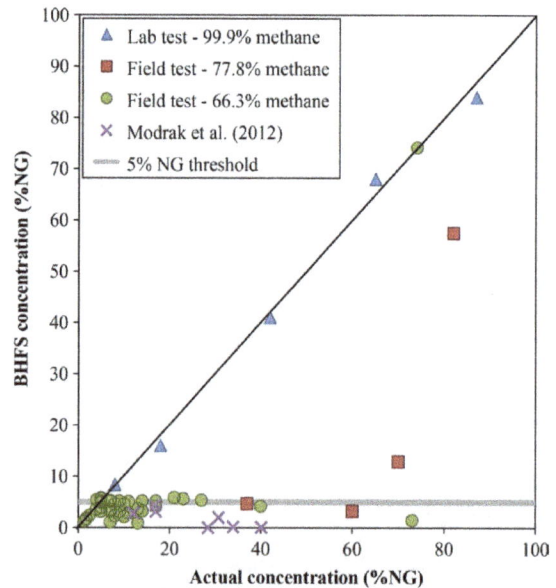

Figure 1. Occurrence of sensor transition failure in BHFS instruments with natural gas of varying CH_4 content from field and lab testing and from emission measurement studies (data from [10, 18]). NG concentrations in the BHFS sampling system measured by the BHFS internal sensor are compared to independent measurements of the sample NG concentrations. The 5% NG sample concentration threshold is the approximate concentration above which sensors should transition from catalytic oxidation to thermal conductivity. BHFS, Bacharach Hi-Flow Sampler; NG, natural gas.

showing the occurrence of sensor transition failure in several BHFS instruments during both field and laboratory testing as well as an example of the failure that occurred during an emission measurement study [10, 18].

Figure 2 presents the BHFS emission measurements from [1] as a function of percent CH_4 in wellhead gas at each site. Figure 2 also shows a line corresponding to emission rates of 0.3–0.5 scfm (0.5–0.9 m^3 h^{-1}), which represents the range of emission rates that would require transition from the catalytic oxidation sensor to the thermal conductivity sensor at sample flows ranging from 6 to 10 scfm (10–17 m^3 h^{-1}).

As seen in Figure 2, there are very few measurements in the thermal conductivity sensor range (above ~0.4 scfm [0.7 m^3 h^{-1}]) at sites where the wellhead gas composition of CH_4 is less than 91%, and this is true across all source categories. Raw data for sample flow and concentration from the BHFS were not provided in [1] supplemental information, so for this analysis, an average BHFS sample flow rate of 8 scfm (14 m^3 h^{-1}) has been assumed, which is the lower of the two sampling flows specified by the Bacharach operating manual [4]. At this sample flow rate, an emission source of 0.4 scfm (0.7 m^3 h^{-1}) corresponds with a sample concentration of 5% NG in air, above

which point the sampler would need to transition to the thermal conductivity sensor to allow for accurate measurements. For sites with CH_4 concentrations less than 91%, only four out of 259 measurements (1.5%) exceeded 0.4 scfm (0.7 m^3 h^{-1}), while for sites with CH_4 concentrations greater than 91%, 68 out of 510 measurements (13.3%) exceeded 0.4 scfm (0.7 m^3 h^{-1}). Consequently, there were almost nine times fewer measurements in the thermal conductivity range at sites with wellhead gas compositions of <91% CH_4 (Fig. 2). If the sample flow rate were 6 scfm (10 m^3 h^{-1}) (due to a flow restriction or reduced battery power), the threshold for transition to the thermal conductivity range would be 0.3 scfm (0.5 m^3 h^{-1}); this would still mean that there were almost seven times fewer measurements in the thermal conductivity range at sites with wellhead gas compositions of <91% CH_4 than at sites with >91% CH_4. Although it is well known that a small percentage of NG emission sources account for most of the total emissions from any given population [9, 15, 25], it is unlikely that almost all the significant emitters at NG production sites would occur only at sites with well head gas compositions >91% CH_4. It is also unlikely that the emission rates of all of the source categories surveyed by [1], which had diverse emission mechanisms such as equipment leaks, pneumatic controllers, chemical injection pumps, and tanks, would all have a ceiling of ~0.4 scfm (0.7 m^3 h^{-1}) at sites with lower wellhead gas CH_4 concentrations. Consequently, the low occurrence of high emitters at sites with lower wellhead gas CH_4 concentrations in [1] indicates that sensor transition failure occurred at sites with CH_4 content <91% and is consistent with the BHFS sensor failure found by Howard et al. [10].

Alternative Theories for the Emission Rate Pattern

Other possible causes of the emission rate pattern in the UT BHFS measurements were considered, including: regional operating differences at production sites; lighter gas densities resulting in higher emission rates; and improved detection of emissions by auditory, visual, and olfactory (AVO, e.g., [24]) methods at sites with heavier hydrocarbon concentrations.

Regional operating differences

Allen et al. [1] point out that air pollution regulations in Colorado which required installation of low bleed pneumatic devices in ozone nonattainment areas after 2009 might have led to lower emission rates in the Rocky Mountain region, which also had the lowest average concentration of CH_4 in the wellhead gas. However, if the

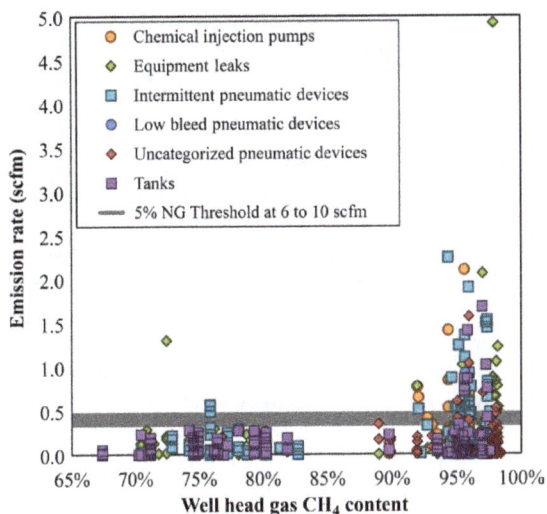

Figure 2. Emission rates of various sources measured by BHFS at NG production sites versus CH_4 concentration of the wellhead gas (data from [1]). The solid line indicates the maximum emission rate that could be measured by the catalytic oxidation sensor only (i.e., in the case of sensor transition failure). For sites with a NG composition greater than 91% CH_4, 13.3% of the measurements are in the TCD sensor range, assuming a sampler flow rate of 8 cubic feet per minute. For sites with less than 91% CH_4, only 1.5% of the measurements are in the TCD range. BHFS, Bacharach Hi-Flow Sampler; NG, natural gas; TCD, thermal conductivity detector.

Rocky Mountain region is removed from the analysis, the occurrence of emitters >0.4 scfm (0.7 m³ h⁻¹) at sites with wellhead gas <91% CH_4 was still only four out of 129 measurements (3.1%), while for sites with CH_4 concentrations greater than 91%, there remain 68 out of 510 measurements (13.3%) that exceeded 0.4 scfm (0.7 m³ h⁻¹) (there were no Rocky Mountain sites with CH_4 >91%). Consequently, even if the Rocky Mountain region is removed from consideration, the occurrence of emitters >0.4 scfm (0.7 m³ h⁻¹) was almost four times less at sites with less than 91% CH_4 than at sites with greater than 91% CH_4, so air quality regulations in Colorado do not appear to be the cause of the emission rate trend shown in Figure 2.

Beyond air pollution regulations, other unknown regional operating practices unrelated to CH_4 concentration might coincidentally cause the apparent relationship of site CH_4 concentrations with the occurrence of high emitters. However, as shown in Figure 3, the increase in leaks >0.4 scfm (0.7 m³ h⁻¹) directly correlates with the increase in the average regional CH_4 concentration. Because there are four regions and two variables (site CH_4 concentration and the percent of leaks >0.4 scfm [0.7 m³ h⁻¹]), the likelihood that regional operating characteristics would coincidentally cause the increase in occurrence of leaks >0.4 scfm (0.7 m³ h⁻¹) to mirror the increasing regional site CH_4 concentration is only one in 24 (four factorial), or ~4%.

Other known operating characteristics of the regions, such as average site pressure and average site age, are not related to the occurrence of equipment leaks >0.4 scfm (0.7 m³ h⁻¹): average site pressures show no correlation, and average site age is negatively correlated with the occurrence of equipment leaks >0.4 scfm (0.7 m³ h⁻¹).

Another argument against regional differences comes from the air quality study conducted by the City of Fort Worth ([6]; or the Ft. Worth study). Ft. Worth is part of the Mid-Continent region defined by [1], where the occurrence of equipment leaks only (as opposed to all BHFS measurement categories) >0.4 scfm (0.7 m³ h⁻¹) observed by [1] was 2.0% of the total equipment leaks in that region. However, equipment leaks >0.4 scfm (0.7 m³ h⁻¹) were 9.9% of the equipment leaks measured in the Ft. Worth study. This was determined using the Ft. Worth study categories of valves and connectors; their remaining category of "other", which included pneumatic control devices, had an even higher occurrence of sources >0.4 scfm (0.7 m³ h⁻¹) of 27.0%. Previous work [10] has shown that although sensor transition failure likely occurred in the Ft. Worth study, these incidents were limited compared to those in [1]. Consequently, the much lower occurrence of leaks >0.4 scfm (0.7 m³ h⁻¹) in the Mid-Continent region in [1] compared to the Ft. Worth study indicates that sensor transition failure was responsible for the low occurrence of emitters <0.4 scfm (0.7 m³ h⁻¹) as opposed to regional differences.

Gas density

Wellhead gas with a lower CH_4 and a greater heavier hydrocarbon content will be denser than gas with higher CH_4 content. Since gas flow through an opening is inversely related to the square root of the gas density, streams with lower CH_4 content would have a lower flow rate if all other conditions were the same. However, this would cause at most a 20% decrease for the lowest CH_4/highest heavier hydrocarbon streams compared to the highest CH_4/lowest heavier hydrocarbon streams observed in the UT study. This would also result in a gradual increase in emissions as CH_4 content increased, as opposed to the dramatic increase in emissions observed over a very narrow range of CH_4 concentrations (Fig. 2).

AVO detection

AVO methods might improve for gas streams with a greater proportion of heavier hydrocarbons, since those streams would have greater odor and might leave more visible residue near a leak. However, Figure 4 presents the occurrence of emitters >0.4 scfm (0.7 m³ h⁻¹) as a function of site CH_4 concentrations in the Appalachia

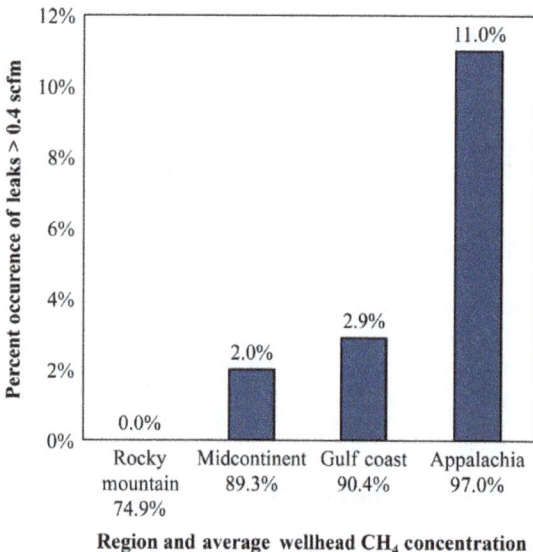

Figure 3. Occurrence of equipment leaks >0.4 scfm in each region of the [1] equipment leak data set. The odds of the occurrence of leaks >0.4 scfm being positively correlated with site CH_4 concentration are one in 24, which makes it unlikely this trend is due to regional operating effects.

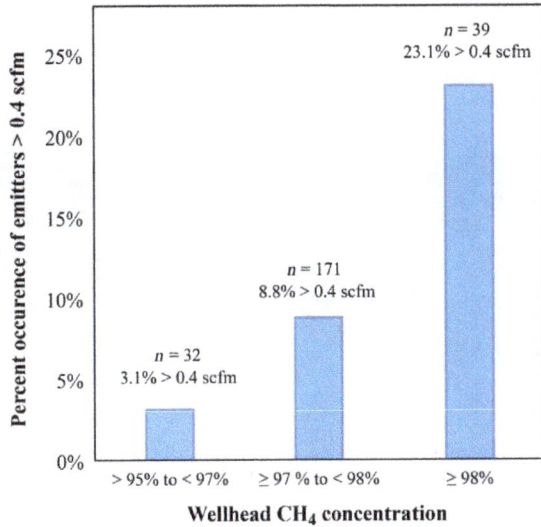

Figure 4. Occurrence of emitters >0.4 scfm as a function of site wellhead gas composition in [1] for the Appalachia region. An emission rate of greater than 0.4 scfm would require the transition from catalytic oxidation sensor to the thermal conductivity sensor for an average sample flow rate of 8 scfm. The dramatic increase in emitters >0.4 scfm over a narrow concentration range argues against the possibility that auditory, visual, and olfactory leak detection is the cause of the emission rate pattern seen in the [1] data set.

region alone. This region had the highest average CH_4 concentration in wellhead gas of any of the regions sampled in [1]. As seen in Figure 4, even over a very narrow range of site CH_4 concentrations (from 95% to >98% CH_4), there is a dramatic increase in emitters >0.4 scfm (0.7 m^3 h^{-1}) with increasing CH_4 concentration. It is unlikely that AVO methods would become so much more efficient over such a narrow range of high CH_4 concentrations where the gas streams are likely odorless and would leave little residue. This dramatic increase in high emitters at sites with high CH_4 concentrations within the Appalachia region alone also argues against the previously discussed regional operating differences hypothesis in general, since this trend is within a single region. Additionally, although the Rocky Mountain region surveyed by UT [1] had the lowest average site CH_4 concentration (74.9%) and heaviest hydrocarbon content, it actually had the highest number of equipment leaks (of any size) per well of all the regions, and there were 25% more leaks per well in that region than in the Appalachia region, which had the highest average site CH_4 concentration (97.0%) and therefore the lowest heavier hydrocarbon content. If AVO methods were more effective due to the presence of heavier hydrocarbons, it seems unlikely the region with the heaviest hydrocarbon concentrations would have the highest rate of overall leak occurrences.

Field Testing of the UT BHFS

Because the trend in the [1] data was consistent with sensor transition failure in the BHFS and no other explanation seemed plausible, I partnered with UT to test the sampler used by [1]. During that field program, the UT sampler had a version of firmware earlier than version 3.03, and older firmware versions have been shown to exhibit sensor transition failure [10]. However, the possible effect of the sampler's firmware version on the sensor failure was not known before this testing of the UT sampler, and at the time of my testing its firmware had been upgraded to a custom version (3.04).

As previously explained, the BHFS uses a catalytic oxidation sensor to measure sample stream concentrations from 0% to ~5% NG, and a thermal conductivity sensor for concentrations from ~5% to 100% NG. The catalytic oxidation sensor is typically calibrated with 2.5% CH_4 in air and the thermal conductivity sensor is calibrated with 100% CH_4 [4]. The manufacturer recommends sensor calibration every 30 days, a process which adjusts the response of the instrument. The calibration may also be checked ("bump-tested") periodically by the user, which does not adjust the instrument response. It is important to note that the description of the BHFS sensor operation in the supplemental information of [1] is incorrect, as they state that:

[A] portion of the sample is drawn from the manifold and directed to a combustibles sensor that measures the sample's methane concentration in the range of 0.05–100% gas by volume. The combustibles sensor consists of a catalytic oxidizer, designed to convert all sampled hydrocarbons to CO_2 and water. A thermal conductivity sensor is then used to determine CO_2 concentration.

However, the BHFS manual [4] clearly states that the catalytic oxidation sensor is used to measure concentrations from 0% to 5% CH_4 and the thermal conductivity sensor from 5% to 100% CH_4. This is a critical distinction because understanding that the BHFS uses a different sensor for each range and that it must transition from the catalytic oxidation sensor to the thermal conductivity sensor in order to conduct accurate measurements is critical to understanding the problem of sensor transition failure.

I initially conducted field testing of the UT sampler in conjunction with the UT team at a NG production site with a wellhead gas CH_4 concentration of 90.8%. NG composition analysis (via gas chromatograph-flame ionization detector) of wellhead gas at this site was conducted by the host company just prior to the sampler testing. The tests were conducted by metering known flow rates of NG into the BHFS inlets through a rotameter (King Instrument Company, Garden Grove, CA; 0–10 scfm air scale). The sample concentration indicated by the internal BHFS sensor was recorded and compared to an external gas concentration monitor used to measure the actual NG concentration at the sampler exhaust (Bascom-Turner Gas Sentry CGA 201, Norwood, MA). The Gas Sentry unit was calibrated with 2.5% and 100% CH_4 prior to the testing; exhaust concentrations measured using this unit agreed with concentrations calculated using the sampler flow rate and amount of NG metered into the inlet to within an average of ±6%.

This field testing was conducted in March of 2014 and is described by [10]; the UT sampler is identified therein as BHFS No. 3. At the time of this testing, the UT BHFS had firmware Version 3.04 (September 2013); this sampler had been calibrated 2 weeks prior to the field test and had been used for emission measurements at production sites since that time. The response of the sensors was checked ("bump-tested") by the UT field team but not calibrated prior to the start of testing. This was apparently consistent with the UT field program methodology: the sampler had been used for measurements with only sensor bump tests, but without the actual calibration unless the sensors failed the bump tests (as was acceptable according to the manufacturer's guidelines) during their ongoing field measurement program and was provided to me for these measurements "ready for testing".

Although the UT sampler's internal sensors initially measured the sample concentration correctly, after ~20 min of testing the sampler's sensors failed to transition from the catalytic oxidation scale (<5% NG) to the thermal conductivity scale (>5% NG), resulting in sample concentration measurements that were 11–57 times lower than the actual sample concentration (Fig. 5). Because sample concentration is directly used to calculate emission measurements made by the sampler, this would result in emission measurements that are too low. After this sensor transition failure occurred, the UT BHFS was calibrated (not simply "bump-tested") and thereafter did not exhibit any further sensor transition failures even during a second day of testing at sites with wellhead CH_4 concentrations as low as 77%. Two other BHFS that were not part of the UT program were also tested using the same procedure; these instruments had the most updated firmware commercially available (Version 3.03) and were put through an actual calibration sequence by the instrument distributor's representative prior to any testing. Neither of these instruments exhibited sensor transition failure at any of the sites. These results combined with the sensor transition failure previously observed in instruments with earlier versions of firmware suggest that the combination of updated firmware and frequent actual calibrations might reduce sensor failure, although this has not been proved conclusively [10, 11].

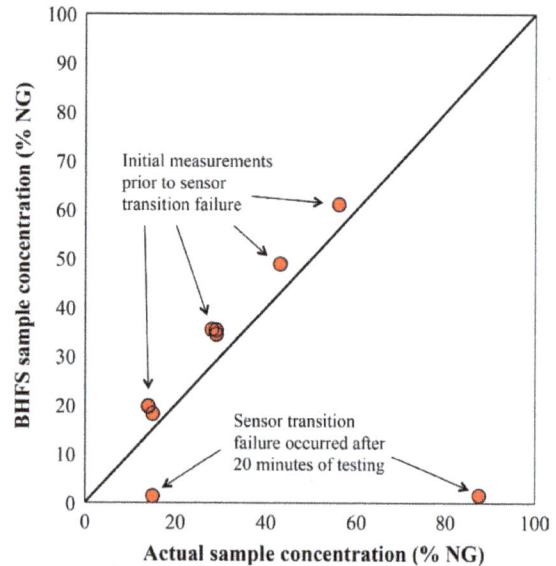

Figure 5. Performance of the BHFS used during the [1] study with NG composed of 90.8% CH_4; instrument firmware had been upgraded to version 3.04 after that study but before this testing; calibration was 2 weeks old. Sensor transition failure set in after ~20 min of testing; this failure was eliminated once the BHFS was put through a calibration sequence (as opposed to just a response test). BHFS, Bacharach Hi-Flow Sampler; NG, natural gas.

The UT recently published a follow-up study of pneumatic device emissions [2]. As part of this work, Allen et al. [2] conducted laboratory testing of the UT BHFS by making controlled releases of both 100% CH_4 and a test gas of 70.5% CH_4 mixed with heavier hydrocarbons into the UT BHFS and did not report any sensor transition failures during these tests, but during this laboratory testing the sampler (with the updated firmware version 3.04) was calibrated (not 'bump-tested") immediately prior to any testing. Consequently, the absence of sensor failure during their laboratory testing is consistent with the results observed during the March 2014 field tests, where calibrating the instrument eliminated the sensor failure.

Allen et al. [3] have suggested that the protocol during their field campaign was to check the calibration of the UT BHFS anytime it was turned on and that not following this protocol led to the sensor transition failure observed during this testing. However, in this instance, the sensor failure occurred both prior to and after the instrument was restarted. Additionally, the UT team observing the testing process did not suggest a calibration check when the instrument was turned back on for further testing. It was only after the sensor failure was observed that they checked and calibrated the instrument, so it does not appear that their protocol was to check the instrument calibration anytime it was turned on.

In summary, because the firmware for the UT sampler was updated prior to this testing (and therefore not the same as the version used during the UT field campaign [1]), and updated firmware may be a factor in reducing sensor failure, it is not expected that these test results are representative of how frequently sensor transition failure might have occurred during the UT study [1]. However, these results do clearly demonstrate that sensor transition failure could occur while using the UT BHFS.

Comparison With Other Pneumatic Device Studies

Two other recent studies have measured emission rates from pneumatic devices by installing meters into the supply gas lines of the devices, as opposed to measuring emissions using the BHFS as was done by Allen et al. [1]. Prasino [22] used the meter installation technique to study emissions from pneumatic controllers in British Columbia, and the UT follow-up study [2] installed meters to measure emission rates from pneumatics in the four regions surveyed in the previous UT study [1].

Unfortunately, it is not possible to compare the pneumatic device emission factors from [1] to those from either the Prasino study, or from [2], because even though [1] sought to randomly sample pneumatic devices, the result was clearly an emitter data set (measurements focused on pneumatic devices that were emitting), while the Prasino data set was made with a random selection of devices and [2] made comprehensive measurements of all devices that could be measured safely at each site. This difference can be demonstrated by comparing the percentage of emitting intermittent pneumatic devices occurring in [1] to that in [2]. In [1], 95.3% (123 out of 129 intermittent devices) were greater than zero, with the smallest nonzero emitter equal to 0.12 scfh (0.0034 m^3 h^{-1}). In [2], only 57.5% (184 out of 320 intermittent devices) were greater than zero. This percentage of nonzero measurements drops further if the lowest nonzero emitter (0.12 scfh; 0.0034 m^3 h^{-1}) observed by [1] is used as a threshold, in which case only 21.3% (68 out of 320) would be considered emitters. Since this threshold of 0.12 scfh (0.0034 m^3 h^{-1}) is 25 times lower than the typical minimum range of the Fox FT2A meters by [2], the reported emitters below this threshold are most likely instrument noise caused by the meter's thermal elements inducing convection currents [7].

Consequently, although the intent of [1] was to survey randomly selected devices, their approach actually resulted in a data set comprised almost exclusively of emitting devices; this possibility is acknowledged by [2]. Therefore, average emissions and emission factors for pneumatic devices calculated from [1] cannot be compared to those calculated from data collected by random or comprehensive sampling, such as presented in [22] or [2], because the emitter data set removes almost all the zero emitters and would result in much higher average emissions.

However, both [1] and [2] provide the CH_4 composition of the wellhead gas at the sites surveyed. This allows a comparison of emission rate patterns as a function of CH_4 concentration between devices measured by the BHFS [1] and by installed meters [2]. If the scarcity of high emitters measured by BHFS at sites with lower CH_4 concentrations in the initial UT study [1] was not an artifact caused by sensor transition failure, then the same concentration pattern should be present whether measured by the BHFS or by installed meters.

For this analysis, I removed the Rocky Mountain region to eliminate any bias from current or impending regulations that might have affected emission rates. Additionally, I focused on emissions from intermittent pneumatics because that provides the most complete data set from the two studies. Finally, as noted previously, the pneumatic device measurements from [1] apparently focused on emitting devices, whereas the devices surveyed in [2] were sampled as comprehensively as possible so the occurrences of high emitters in each study cannot be directly compared. Consequently, it is the ratio of the occurrences of high emitters at low CH_4 sites compared to high CH_4 sites within each study that must be compared.

As seen in Table 1, when measured by [1] via BHFS, the occurrence of emitters >0.4 scfm (0.7 m^3 h^{-1}) (on a percentage basis) at sites with wellhead gas compositions <91% CH_4 is almost a factor of five less than at sites with CH_4 >91%, consistent with BHFS sensor failure. Conversely, when measured via installed meters [2], the occurrence of emitters >0.4 scfm (0.7 m^3 h^{-1}) at sites with wellhead gas compositions <91% CH_4 is almost a factor of three higher than at sites with >91% CH_4, indicating a complete reversal in this trend. This stark difference between BHFS measurements and installed meter measurements corroborates that the scarcity of high emitters at sites with lower wellhead gas CH_4 content present in [1] was an artifact due to sensor failure in the BHFS.

Focused Analysis of the UT Study Equipment Leaks

In order to better understand the threshold of wellhead gas CH_4 concentrations at which sensor transition failure might occur, I conducted further analysis focused only on the equipment leak measurements in [1]. Equipment leaks were targeted because they are expected to be short term, steady state measurements, whereas emissions

Table 1. Occurrence of intermittent pneumatic device high emitters as a function of wellhead gas composition, measured by Bacharach Hi-Flow Sampler (BHFS) and installed meters (Rocky Mountain region excluded).

	No. of devices measured	No. of devices with emissions >0.4 scfm	% of devices with emissions >0.4 scfm
Allen et al. [1] (Measured by BHFS sampler)			
Wellhead gas composition >91% CH_4	85	28	32.9
Wellhead gas composition <91% CH_4	44	3	6.8
Ratio of frequency of high emitters at sites with wellhead gas compositions <91% CH_4 to sites with wellhead gas compositions >91% CH_4			0.21
Allen et al. [2] (Measured by installed meters)			
Wellhead gas composition >91% CH_4	106	3	2.8
Wellhead gas composition <91% CH_4	97	8	8.2
Ratio of frequency of high emitters at sites with wellhead gas compositions <91% CH_4 to sites with wellhead gas compositions >91% CH_4			2.9

reported from pneumatic devices and chemical injection pumps are likely to be an average of several measurements, and emissions from tanks may have an NG composition different from the reported wellhead composition.

Figure 6 presents the occurrence of equipment leaks in [1] that are >0.4 scfm (0.7 m^3 h^{-1}) as a function of site CH_4 concentrations. At sites with gas compositions of >97% CH_4, 11.7% of the leaks were >0.4 scfm (0.7 m^3 h^{-1}). At sites with wellhead compositions between 90% and 97% CH_4, only 2.7% of the leaks were >0.4 scfm

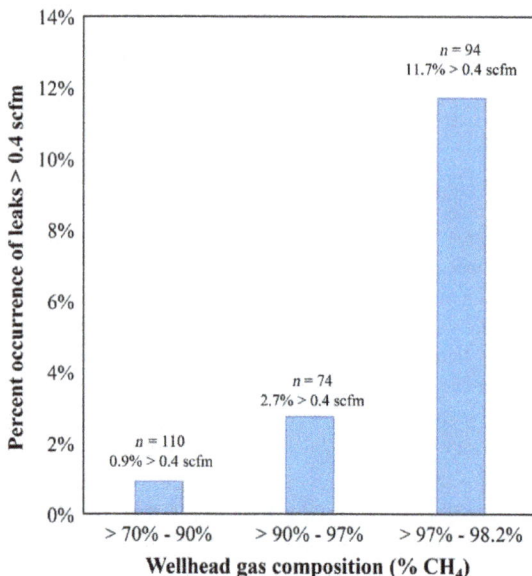

Figure 6. Occurrence of equipment leaks >0.4 scfm as a function of site well head gas CH_4 content in the [1] study. Leaks >0.4 scfm would require the transition from catalytic oxidation sensor to the thermal conductivity sensor for an average sample flow rate of 8 scfm. The large increase in the occurrence of leaks >0.4 scfm at sites with CH_4 content >97% indicates sensor transition failure below that threshold.

(0.7 m^3 h^{-1}), and this occurrence dropped to less than 1% at sites with wellhead gas compositions of <90% CH_4, indicating that the sampler's ability to measure leaks >0.4 scfm (0.7 m^3 h^{-1}) declined dramatically with decreasing concentrations of CH_4 in the wellhead gas (Fig. 6). This analysis indicates the BHFS may underreport emitters >0.4 scfm (0.7 m^3 h^{-1}) even when making measurements of NG streams with CH_4 content up to 97%, and provides a valuable refinement of the possible CH_4 concentration threshold where sensor failure may occur, since the highest CH_4 wellhead content available for direct field testing of the BHFS was only 91.8%.

Comparison of the UT Study Downwind Tracer Ratio Measurements to On-Site Measurements

Allen et al. [1] also made emission measurements using a downwind tracer ratio method at 19 sites for comparison to their on-site measurements. Their emissions from on-site measurements were calculated by using direct measurements of equipment leaks and pneumatic devices that were made by the UT team combined with estimates of emissions from any sources at the well pad that were not measured. These unmeasured sources included all tanks and compressors (compressors were a small source in comparison to all other sources) as well as any pneumatics that was not directly measured during the site survey. For CH_4 emissions from tanks and compressors, the authors used "standard emissions estimation methods" [1]. For pneumatic devices that were not surveyed, they applied their own emission factors based on the measurements of pneumatic devices collected during the UT study.

The tracer ratio measurements were made by releasing a tracer gas at a known rate to simulate the emissions from the site being measured. Simultaneous downwind measurements were then made of the concentrations of both the tracer gas and CH_4, and then the emission rate of CH_4 was calculated after correcting for background CH_4 and tracer concentrations. The tracer ratio method allows for the calculation of CH_4 emissions from the entire production site by accounting for the dilution of CH_4 as it is transported into the atmosphere from the source to the receptor.

In summarizing their tracer ratio measurements, [1] state: "For the production sites, emissions estimated based on the downwind measurements were also comparable to total on-site measurements; however, because the total on-site emissions were determined by using a combination of measurements and estimation methods, it is difficult to use downwind measurements to confirm the direct source measurements." However, upon further examination, I found that the downwind tracer measurements do in fact indicate the occurrence of sensor transition failure in their BHFS measurements.

Table 2 summarizes the characteristics of the sites surveyed by [1] using both the BHFS and the tracer ratio method. As described above, the on-site total is a combination of the measurements made by BHFS and estimates for any sources not actually measured by the UT team. I calculated the ratio of actual BHFS measurements to the total reported on-site emissions (estimated and measured) using the supplemental information provided by [1]. Actual measured emissions ranged from 1% to 79% of the total reported on-site emissions and the on-site total emissions range from 13% to 3500% of the downwind tracer ratio measurements (Table 2).

Table 3 compares the tracer ratio measurements to the on-site emissions, categorized by CH_4 content in the wellhead gas and by the fraction of actual BHFS measurements that comprise the on-site emissions. As shown in Table 3, when comparing all sites without separating them into these categories, the total of the tracer ratio measurements does agree closely to the on-site emissions, as [1] concluded. However, four of the sites had wellhead gas compositions of $\geq 97\%$ CH_4, at which the BHFS would be expected to make accurate measurements. The remaining 15 sites had wellhead gas compositions of $<82\%$ CH_4, at which sensor transition failure might occur and the BHFS would underreport emissions measurements.

Once the sites are categorized by these wellhead gas compositions, a deficit between the on-site emissions and the tracer ratio measurements appears in sites with lower CH_4 concentrations, and this deficit becomes more

Table 2. Sites surveyed by Allen et al. [1] using both Bacharach Hi-Flow Sampler (BHFS) and downwind tracer methods.

Tracer site name[1]	BHFS site name[1]	Wellhead gas CH_4 concentration (%)	On-site total[2] (BHFS measurements and estimates) (scfm CH_4)	BHFS measurements/on-site total[3]	Leaks measured by BHFS/on-site total[3]	Tracer ratio emission rate (scfm CH_4)	On-site total/ tracer ratio emission rate
MC-1	MC-1	70.9	1.89	0.12	0.12	2.32	0.815
MC-2	MC-14	78.1	0.99	0.34	0.01	2.00	0.495
MC-3	MC-20	77.2	1.63	0.45	0.18	2.95	0.552
MC-4	MC-5	74.2	2.31	0.19	0.14	3.36	0.687
MC-5	MC-16	79.3	1.85	0.56	0.18	4.16	0.445
RM-1	RM-7	81.9	0.22	0.11	0.09	0.584	0.368
RM-2	RM-8	74.5	4.43	0.02	0.02	1.70	2.60
RM-3	RM-1	76.4	0.13	0.67	0.69	0.442	0.303
RM-4	RM-3	74.9	0.11	0.21	0.00	0.839	0.137
RM-5	RM-2	74.5	0.09	0.35	0.33	0.240	0.392
RM-6	RM-5	74.5	0.74	0.41	0.42	0.421	1.75
RM-7	RM-14	74.5	0.27	0.26	0.26	0.368	0.736
RM-8	RM-19	76.2	0.29	0.82	0.79	1.08	0.266
RM-9	RM-12	74.5	0.38	0.05	0.05	0.864	0.436
RM-10	RM-4	76.2	2.86	0.01	0.00	0.080	35.7
AP-2	AP-23	97.6	1.28	0.68	0.35	0.270	4.74
AP-3	AP-43	97.0	4.75	0.62	0.59	4.12	1.15
AP-4	AP-37	97.0	1.36	0.44	0.42	0.709	1.92
AP-5	AP-18	97.0	0.39	0.74	0.69	0.288	1.37

[1]MC, Midcontinent; RM, Rocky Mountain; AP, Appalachia. Different site numbers were used to identify the same sites in the [1] supplemental information depending on whether BHFS or tracer ratio measurements were under discussion.
[2]On-site totals were calculated by [1] by combining measurements made by the BHFS with estimates of any sources not measured; these estimates were made using mathematical models for tanks as well as emission factors for compressors and any pneumatic controllers not directly measured.
[3]Calculated by this author from [1] supplemental information.

pronounced as the amount of the on-site emissions actually measured by the BHFS becomes a larger fraction of the total on-site emissions (measured and estimated). As seen in Table 3, for the high CH_4 sites where the sampler should function properly, the on-site measurements and estimates exceed the tracer measurements, but approach a ratio of one (complete agreement) as the amount of actual measurements increases. For the two sites with wellhead gas compositions ≥97% where the measured equipment leaks (which should produce steady emissions as compared to pneumatic devices which might be intermittent) averaged 64% of the total on-site measurements and estimates, the on-site total still exceeds the tracer measurements but are within 17% (Table 3). However, for the sites with wellhead gas CH_4 concentrations <82%, there is a clear trend of increasing deficit of the on-site emissions compared to the tracer ratio measurements as the actual BHFS measurements become a larger part of the on-site total. For instance, for the nine sites with at least 20% of on-site emissions from BHFS measurements (for an average of 45% of the total on-site emissions measured by the BHFS), the on-site emissions are only 49% of the tracer measurements (Table 3). For the two sites that had greater than 67% of on-site emissions data actually measured by the BHFS (for an average of 75% of on-site emissions data measured by the BHFS), the on-site emissions are only 28% of the tracer measurements (Table 3).

Comparing the on-site data to the downwind tracer measurements provides two valuable insights. First, there were six sites in the Rocky Mountain region for which at least 20% of the on-site emissions were measured by the BHFS (for an average of 45% actual BHFS measurements) (Table 2). For these six sites, the on-site emissions average 48% of the tracer data. For the two sites in this

region with at least 67% of on-site emissions from actual BHFS measurements (and with BHFS measurements averaging 75% of the total on-site data), the on-site emissions were only 28% of the tracer measurements (Table 2). This provides clear evidence that the sampler actually did fail in the Rocky Mountain region, as opposed to any possible regional differences (discussed previously) that might have created an emission pattern of no high emitters at sites with lower CH_4 concentrations in the wellhead gas.

Additionally, the tracer measurements provide a method to estimate the magnitude of errors introduced in the data collected by [1] due to BHFS sensor transition failure. For all of the sites with wellhead gas compositions ≥97% CH_4 (where the sampler should operate correctly), the emission rates determined by on-site measurements exceeded those determined by the downwind tracer ratio measurements. Assuming that the tracer method accurately measured the total emissions from the sites surveyed (e.g., [8, 15, 16]), I concluded that the methods used in [1] overestimated the on-site sources that were not directly measured. Therefore, I calculated the error in BHFS measurements at sites with low CH_4 wellhead gas composition by assuming the tracer ratio measurements are correct. I have also assumed for this analysis that the estimates of any on-site sources made by [1] are also correct, even though the tracer data indicate they may be too high, because this is conservative in the sense that correcting for this overestimate would increase the BHFS error calculated below. Given these assumptions, subtracting the on-site estimated emissions from the tracer ratio emissions gives the expected measurement total that should have been reported from the BHFS measurements. Comparing this expected measurement total to the actual

Table 3. Comparison of on-site measurements to tracer ratio measurements made by Allen et al. [1] categorized by wellhead gas CH_4 concentration.

Site category (number of sites in parentheses)	Average percentage of on-site emissions reported by BHFS	Total on-site emissions (reported by BHFS and estimated) (scfm CH_4)	Total emissions measured by tracer (scfm CH_4)	Ratio of on-site emissions to emissions measured by tracer
All sites (19)	37	26.0	26.8	0.97
Sites where BHFS measurements are expected to be accurate (wellhead gas composition ≥97% CH_4)				
All sites (4)	62	7.78	5.39	1.44
Sites with >50% BHFS measurements (3)	68	6.42	4.68	1.37
Sites with >50% equipment leaks (2)	64 (equipment leaks/on-site total)	5.14	4.41	1.17
Sites where BHFS measurements are expected to underreport high emitters (wellhead gas composition <82% CH_4)				
All sites (15)	28	18.2	21.4	0.85
Sites with ≥5% BHFS measurements (13)	35	10.9	19.6	0.56
Sites with ≥20% BHFS measurements (9)	45	6.10	12.5	0.49
Sites with >50% BHFS measurements (3)	69	2.27	5.68	0.40
Sites with >67% BHFS measurements (2)	75	0.42	1.52	0.28

BHFS, Bacharach Hi-Flow Sampler.

measurement total reported by the BHFS provides an estimate of the error in BHFS measurements made by Allen et al. [1].

Table 4 presents the results of this analysis, and shows that for the 13 sites with wellhead gas compositions <82% CH_4 and with at least 5% actual BHFS measurements (with an average of 35% of emission sources measured by BHFS; bottom half of Table 3), the actual measurement total of the BHFS is less than one-third of the expected total, and this appears consistent as sites with greater fractions of actual BHFS measurements are examined. For these sites, the emission rates for equipment leaks and pneumatics devices presented by [1] are approximately equal, so it is not possible to assign a larger error to one category or another. Additionally, the errors introduced by the sensor failure would be expected to vary from site to site depending on how many emitters were present with emission rates exceeding the sensor transition threshold ceiling. Nevertheless, for these 13 sites, the BHFS underreported emissions for equipment leaks and pneumatic devices on average by more than a factor of 3 (Table 4).

Although the magnitude of error due to BHFS sensor failure is not known for all the sites in [1], the tracer ratio measurements make clear that the BHFS measurements for sites with lower CH_4 content in the wellhead gas could be at least a factor of three too low. More precise estimates of errors in [1] are not possible because of the nature of the sensor failure. Unlike a simple calibration error, for which it might be possible to correct, when sensor transition failure occurs, it is not possible to know for any particular measurement if the failure has occurred, and if it has, what the resulting error was, since the reported emission rates could range from 20% to two orders of magnitude too low.

Implications

Sensor transition failure is clearly apparent in the BHFS measurements made in the UT study by Allen et al. [1], as evidenced by the rare occurrence of high emitters at sites with lower CH_4 (<91%) content in the wellhead gas. The occurrence of this sensor transition failure was corroborated by field tests of the UT BHFS during which it exhibited this sensor failure, as well as by tracer ratio measurements made by [1] at a subset of sites with lower wellhead gas CH_4 concentrations. At this subset of sites, the tracer ratio measurements indicate that the BHFS measurements were too low by at least a factor of three. Because BHFS measurements were the basis of 98% of the inventory developed by [1] using their own measurements (and 41% of their total compiled inventory), the inventory clearly underestimates CH_4 emissions from production sites. However, the extent of this error is difficult to estimate because the underreporting of emission rates due to BHFS sensor transition failure at any given site would vary depending on sampler performance and on how many high emitters were present at that site. Estimating this error is further complicated by the fact that the data set collected for pneumatic devices by [1] was an emitter data set; this might offset the effect of underreported high emitters in their pneumatic device emission factors. Finally, although real differences may exist in regional emission rates, the UT data set [1] should not be used to characterize them because the occurrence of sensor failure clearly varied between regions due to variations in wellhead CH_4 compositions, which may mask any actual regional differences that existed.

Although the performance of the BHFS may vary between instruments or with sensor age or calibration vintage, this analysis of the [1] data set shows that measurements made using a BHFS for NG streams with CH_4 content

Table 4. Estimation of underreporting in Allen et al. [1] BHFS measurements of CH_4 emission rates at sites with low CH_4 well head gas composition (<82%), using downwind tracer measurements (from Table 3).

Minimum percentage of on-site emissions reported by BHFS	Average percentage of on-site emissions reported by BHFS	No. of sites	Total emissions measured by tracer (scfm CH_4)	On-site emissions estimated by UT (excludes BHFS measurements) (scfm CH_4)	Expected BHFS measurement total (tracer – on-site estimates) (scfm CH_4)	Emissions reported by BHFS (scfm CH_4)	Ratio of reported BHFS to expected BHFS
≥5	35	13	19.63	7.09	12.54	3.81	0.30
≥20	45	9	12.50	3.34	9.16	2.76	0.30
>50	69	3	5.68	0.71	4.97	1.56	0.31
>67	75	2	1.52	0.11	1.42	0.31	0.22

BHFS, Bacharach Hi-Flow Sampler; UT, University of Texas.

up to 97% could lead to severe underreporting of NG leaks. That this failure can occur at such high CH_4 concentrations, which are close to the higher end of those found in transmission and distribution systems, indicates that past measurements in all segments of the NG supply chain could have been affected by this problem. Because the BHFS sensor transition failure phenomenon is not fully understood, it is not known how much this error may have affected past measurements of CH_4 emission rates. Two factors preclude this: first, the performance of any individual BHFS may vary, and second, once sensor transition failure occurs, there is no way to determine the magnitude of the measurement error in the absence of an independent flux or concentration measurement.

If BHFS sensor transition failure has occurred during industry monitoring at transmission, storage, and processing compressor stations where the BHFS is approved for leak measurements mandated by the USEPA Subpart W Greenhouse Gas Reporting Program (GHGRP) [23], then these errors could be larger than those observed at production sites. Leaks at transmission, storage, and processing compressor stations commonly exceed 0.4 scfm ($0.7 m^3 h^{-1}$) (the approximate threshold for BHFS sensor transition failure) and in some cases may range from 10 to over 100 scfm. Because the largest 10% of leaks typically account for 60–85% of the total leak rate at a given facility [9, 25], sensor transition failure in the BHFS could bias CH_4 emission inventories compiled by the USEPA GHGRP substantially low since the most significant leaks could be underreported. Additionally, leak measurements using the BHFS may be used to guide repair decisions at NG facilities, and underreporting of leaks could compromise safety if large leaks remain unrepaired as a result.

Finally, it is important to note that the BHFS sensor failure in the UT study [1] went undetected in spite of the clear artifact that it created in the emission rate trend as a function of wellhead gas CH_4 content and even though the authors' own secondary measurements made by the downwind tracer ratio technique confirmed the BHFS sensor failure. That such an obvious problem could escape notice in this high profile, landmark study highlights the need for increased vigilance in all aspects of quality assurance for all CH_4 emission rate measurement programs.

Acknowledgments

The author thanks Dave Allen (University of Texas at Austin) for making the UT BHFS available for field testing, and Adam Pacsi (University of Texas at Austin), Matt Harrison and Dave Maxwell (URS Corporation), and Tom Ferrara (Conestoga Rovers & Associates) for their assistance with the field testing of the BHFS. This paper was substantially improved by the comments of three anonymous reviewers.

Conflict of Interest

The author is the developer of high flow sampling technology (US Patent RE37, 403) and holds a license to use it for any purpose; however, he does not sell high flow samplers nor was he involved in the development of the Bacharach Hi-Flow Sampler.

References

1. Allen, D. T., V. M. Torres, J. Thomas, D. Sullivan, M. Harrison, A. Hendler, et al. 2013. Measurements of methane emissions at natural gas production sites in the United States. Proc. Natl. Acad. Sci. USA 110:17768–17773. doi: 10.1073/pnas.1304880110

2. Allen, D. T., A. Pacsi, D. Sullivan, D. Zavala-Araiza, M. Harrison, K. Keen, et al. 2014. Methane emissions from process equipment at natural gas production sites in the United States: pneumatic controllers. Environ. Sci. Technol. 49:633–640. doi: 10.1021/es5040156

3. Allen, D. T., D. W. Sullivan, and M. Harrison. 2015. Response to comment on "Methane emissions from process equipment at natural gas production sites in the United States: pneumatic controllers". Environ. Sci. Technol. 49:3983–3984. doi: 10.1021/acs.est.5b00941

4. Bacharach, Inc. 2010. Hi-Flow Sampler™ natural gas leak rate measurement. Instruction 0055-9017 Operation and Maintenance. Available at: http://www.bacharach-inc.com/PDF/Instructions/55-9017.pdf (accessed 20 July 2015).

5. Brandt, A. R., G. A. Heath, E. A. Kort, F. O'Sullivan, G. Pétron, S. M. Jordaan, et al. 2014. Methane leaks from North American natural gas systems. Science 343:733–735. doi: 10.1126/science.1247045

6. Eastern Research Group (ERG). 2011. City of Fort Worth natural gas air quality study final report, Fort Worth, TX. Available at: http://fortworthtexas.gov/uploadedFiles/Gas_Wells/AirQualityStudy_final.pdf (accessed 20 July 2015).

7. Fox Thermal Instruments, Inc. 2015. Fox gas flow meter model FT2A instruction manual. Available at: http://www.foxthermalinstruments.com/pdf/ft2a/FT2A_Manual.pdf (accessed 11 May 2015).

8. Howard, T., B. Lamb, W. L. Bamesberger, and P. Zimmerman. 1992. Measurement of hydrocarbon emission fluxes from refinery wastewater impoundments using atmospheric tracer techniques. J. Air Waste Manag. Assoc. 42:1337.

9. Howard, T., R. Kantamaneni, and G. Jones. 1999. Cost effective leak mitigation at natural gas compressor stations. Project No. RR-246-9526, Catalog No. L51802.

Pipeline Research Council International, Arlington, VA. Available at: http://prci.org (accessed 11 May 2015).

10. Howard, T., T. W. Ferrara, and A. Townsend-Small. 2015. Sensor transition failure in the high volume sampler: implications for methane emissions estimates from natural gas infrastructure. J. Air Waste Manag. Assoc. 65:856–862. doi: 10.1080/10962247.2015.1025925

11. Howard, T. 2015. Comment on "Methane emissions from process equipment at natural gas production sites in the United States: pneumatic controllers". Environ. Sci. Technol. 49:3981–3982. doi: 10.1021/acs.est.5b00507

12. Howarth, R. W. 2014. A bridge to nowhere: methane emissions and the greenhouse gas footprint of natural gas. Energy Sci. Eng. 2:47–60. doi: 10.1002/ese3.35

13. Howarth, R. W., R. Santoro, and A. Ingraffea. 2011. Methane and the greenhouse-gas footprint of natural gas from shale formations. Clim. Change 106:679–690. doi: 10.1007/s10584-011-0061-5

14. Karion, A., C. Sweeney, G. Pétron, G. Frost, R. M. Hardesty, J. Kofler, et al. 2013. Methane emissions estimate from airborne measurements over a western United States natural gas field. Geophys. Res. Lett. 40:4393–4397. doi: 10.1002/grl.50811

15. Lamb, B. K., S. L. Edburg, T. W. Ferrara, T. Howard, M. R. Harrison, C. E. Kolb, et al. 2015. Direct measurements show decreasing methane emissions from natural gas local distribution systems in the United States. Environ. Sci. Technol. 49:5161–5169. doi: 10.1021/es505116p

16. Lamb, B. K., J. B. McManus, J. H. Shorter, C. E. Kolb, B. Mosher, R. C. Harriss, et al. 1995. Development of atmospheric tracer methods to measure methane emissions from natural gas facilities and urban areas. Environ. Sci. Technol. 29:1468–1479.

17. Miller, S. M., S. C. Wofsy, A. M. Michalak, E. A. Kort, A. E. Andrews, S. C. Biraud, et al. 2013. Anthropogenic emissions of methane in the United States. Proc. Natl. Acad. Sci. USA 110:20018–20022. doi: 10.1073/pnas.1314392110

18. Modrak, M. T., M. S. Amin, J. Ibanez, C. Lehmann, B. Harris, D. Ranum, et al. 2012. Understanding direct emission measurement approaches for upstream oil and gas production operations. Proceedings of the Air & Waste Management Association 105th Annual Conference & Exhibition, San Antonio, TX. Available at: http://portal.awma.org/store/detail.aspx?id=411ACE12 (accessed 20 July 2015).

19. Myhre, G., D. Shindell, F.-M. Bréon, W. Collins, J. Fuglestvedt, J. Huang, et al. 2013. Anthropogenic and natural radiative forcing. Pp. 659–740 in T. F. Stocker,

D. Qin, G.-K. Plattner, M. Tignor, S. K. Allen, J. Boschung, A. Nauels, Y. Xia, V. Bex, P. M. Midgley, eds. Climate change 2013: the physical science basis. Contribution of Working Group I to the Fifth Assessment Report of the Intergovernmental Panel on Climate Change. Cambridge Univ. Press, New York, NY.

20. Pétron, G., G. Frost, B. R. Miller, A. I. Hirsch, S. A. Montzka, A. Karion, et al. 2012. Hydrocarbon emissions characterization in the Colorado Front Range: a pilot study. J. Geophys. Res. 117:D04304. doi: 10.1029/2011JD016360

21. Pétron, G., A. Karion, C. Sweeney, B. R. Miller, S. A. Montzka, G. J. Frost, et al. 2014. A new look at methane and nonmethane hydrocarbon emissions from oil and natural gas operations in the Colorado Denver-Julesburg Basin. J. Geophys. Res. 119:6386–6852. doi: 10.1002/2013JD021272

22. Prasino Group. 2013. Final report for determining bleed rates for pneumatic devices in British Columbia. Report to British Columbia Ministry of Environment, December 2013. Available at: http://www2.gov.bc.ca/gov/DownloadAsset?assetId=1F074ABD990D4EFB8AE555AEB3B8D771&filename=prasino_pneumatic_ghg_ef_final_report.pdf (accessed 11 May 2015).

23. United States Code of Federal Regulations. 2014. 40 CFR Part 98, subpart W. Available at: http://www.ecfr.gov/cgi-bin/text-idx?tpl=/ecfrbrowse/Title40/40cfr98_main_02.tpl (accessed 11 May 2015).

24. United States Environmental Protection Agency (USEPA). 1998. Inspection manual: federal equipment leak regulations for the chemical manufacturing industry. Volume III: petroleum refining industry regulations. EPA Office of Compliance: Chemical, Commercial Services and Municipal Division. EPA/305/B-98-011. Available at: http://www.epa.gov/compliance/resources/publications/assistance/sectors/insmanvol3.pdf (accessed 11 May 2015).

25. United States Environmental Protection Agency (USEPA). 2003. Directed inspection and maintenance at compressor stations. EPA Natural Gas Star Program. Available at: http://www.epa.gov/gasstar/documents/ll_dimcompstat.pdf (accessed 11 May 2015).

26. United States Environmental Protection Agency (USEPA). 2014. Inventory of U.S. greenhouse gas emissions and sinks: 1990–2012. EPA 430-R-14-003. Available at: http://www.epa.gov/climatechange/emissions/usinventoryreport.html (accessed 11 May 2015).

27. Wigley, T. M. L. 2011. Coal to gas: the influence of methane leakage. Clim. Change 108:601–608.

16

Li-ion batteries: basics, progress, and challenges

Da Deng

Department of Chemical Engineering and Materials Science, Wayne State University, Detroit, Michigan 48202

Keywords

Anode, cathode, electrolyte, Li-ion batteries, rechargeable, separator

Correspondence

Da Deng, Department of Chemical Engineering and Materials Science, Wayne State University, Detroit, MI 48202.
E-mail: da.deng@wayne.edu

Funding Information

The author appreciates Professor Simon Ng for comments on the manuscript and Wayne State University for support.

Abstract

Li-ion batteries are the powerhouse for the digital electronic revolution in this modern mobile society, exclusively used in mobile phones and laptop computers. The success of commercial Li-ion batteries in the 1990s was not an overnight achievement, but a result of intensive research and contribution by many great scientists and engineers. Then much efforts have been put to further improve the performance of Li-ion batteries, achieved certain significant progress. To meet the increasing demand for energy storage, particularly from increasingly popular electric vehicles, intensified research is required to develop next-generation Li-ion batteries with dramatically improved performances, including improved specific energy and volumetric energy density, cyclability, charging rate, stability, and safety. There are still notable challenges in the development of next-generation Li-ion batteries. New battery concepts have to be further developed to go beyond Li-ion batteries in the future. In this tutorial review, the focus is to introduce the basic concepts, highlight the recent progress, and discuss the challenges regarding Li-ion batteries. Brief discussion on popularly studied "beyond Li-ion" batteries is also provided.

Introduction

Li-ion batteries, as one of the most advanced rechargeable batteries, are attracting much attention in the past few decades. They are currently the dominant mobile power sources for portable electronic devices, exclusively used in cell phones and laptop computers [1]. Li-ion batteries are considered the powerhouse for the personal digital electronic revolution starting from about two decades ago, roughly at the same time when Li-ion batteries were commercialized. As one may has already noticed from his/her daily life, the increasing functionality of mobile electronics always demand for better Li-ion batteries. For example, to charge the cell phone with increasing functionalities less frequently as the current phone will improve quality of one's life. Another important expanding market for Li-ion batteries is electric and hybrid vehicles, which require next-generation Li-ion batteries with not only high power, high capacity, high charging rate, long life, but also dramatically improved safety performance and low cost. In the USA, Obama administration has set a very ambitious goal to have one million plug-in hybrid vehicles on the road by 2015. There are similar plans around the word in promotion of electric and hybrid vehicles as well. The Foreign Policy magazine even published an article entitled "The great battery race" to highlight the worldwide interest in Li-ion batteries [2].

The demand for Li-ion batteries increases rapidly, especially with the demand from electric-powered vehicles (Fig. 1). It is expected that nearly 100 GW hours of Li-ion batteries are required to meet the needs from consumer use and electric-powered vehicles with the later takes about 50% of Li-ion battery sale by 2018 [3]. Furthermore, Li-ion batteries will also be employed to buffer the intermittent and fluctuating green energy supply from renewable resources, such as solar and wind, to smooth the difference between energy supply and demand. For example, extra solar energy generated during the day time can be stored in Li-ion batteries that will supply energy at night when sun light is not available. Large-scale Li-ion batteries for grid application will require next-generation batteries to be produced at low cost.

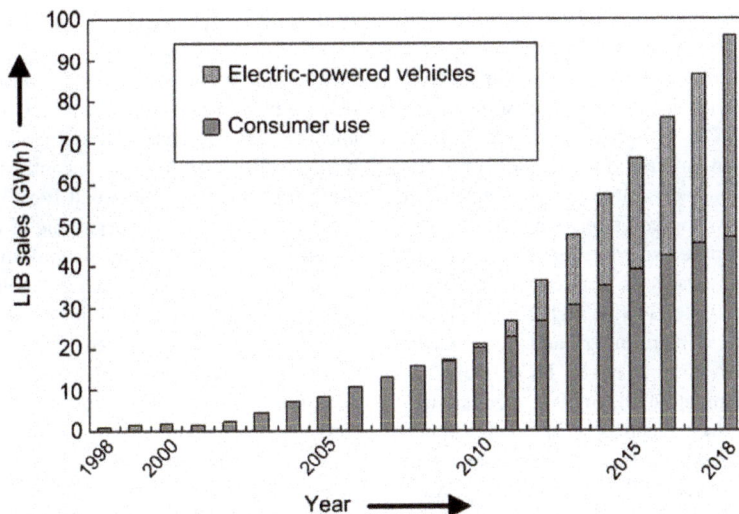

Figure 1. Demand for Li-ion batteries in two decades. Reproduced with permission [3].

Another important aspect of Li-ion batteries is related to battery safety. The recent fire on two Boeing 787 Dreamliner associated with Li-ion batteries once again highlights the critical importance of battery safety [4, 5]. This will trigger another wave of extensive research and development to enhance safety of Li-ion batteries, beyond pursuing high-energy density. In this tutorial review, I will try not to have a comprehensive coverage due to the limited scope, but instead I will highlight the basics, progress, and challenges regarding Li-ion batteries.

Li-ion batteries are highly advanced as compared to other commercial rechargeable batteries, in terms of gravimetric and volumetric energy. Figure 2 compares the energy densities of different commercial rechargeable batteries, which clearly shows the superiority of the Li-ion batteries as compared to other batteries [6]. Although lithium metal batteries have even higher theoretical energy densities than that of Li-ion batteries, their poor rechargeability and susceptibility to misuses leading to fire even explosion are known disadvantages. I anticipate that lithium metal batteries based on solid-state electrolytes with enhanced safety will be commercialized in the next decade. Recently, lithium-air and lithium-sulfur batteries regain wide interest, although the concepts have been proposed for a while. Promising progress has been achieved regarding Li-air and Li-sulfur batteries, but it may take another two decades to fully develop those technologies to achieve reliable performances that will be comparable to Li-ion batteries. It is expected that Li-ion batteries will still be dominant in rechargeable battery market, at least for the next decade, for advantages they offer. Li-ion batteries are design flexible. They can be formed into a wide variety

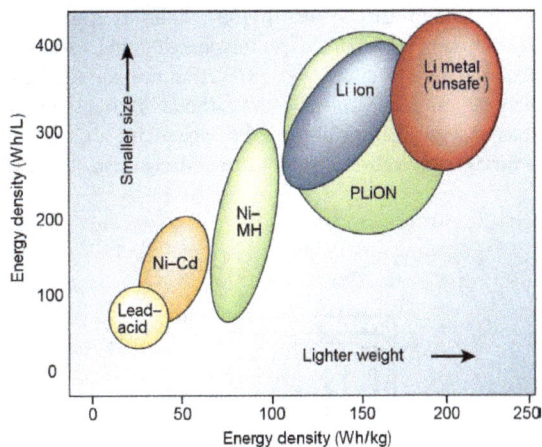

Figure 2. Comparison of energy densities and specific energy of different rechargeable batteries. Reproduced with permission [6].

of shapes and sizes, so as to efficiently fit the available space in the devices they power. Li-ion batteries do not suffer from the problem of memory effect, in contrast to Ni-Cd batteries. Li-ion batteries have voltages nearly three times the values of typical Ni-based batteries. The high single-cell voltage would reduce the number of cells required in a battery module or pack with a set output voltage and reduce the need for associated hardware, which can enhance reliability and weight savings of the battery module or pack due to parts reduction. The self-discharge rate is very low in Li-ion batteries – a typical figure is <5% per month which compares very favorably to 20–30% of Ni-based batteries.

It would be interesting to briefly review the history on the development of Li-ion batteries. The first rechargeable Li-ion batteries with cathode of layered TiS_2 and anode of metallic Li was reported by Whittingham while working at Exxon in 1976 [7]. Exxon subsequently tried to commercialize the Li-ion batteries, but was not successful due to the problems of Li dendrite formation and short circuit upon extensive cycling and safety concern [2]. Also in 1976, Besenhard proposed to reversibly intercalate Li^+ ions into graphite and oxides as anodes and cathodes, respectively [8, 9]. In 1981, Goodenough first proposed to use layered $LiCoO_2$ as high energy and high voltage cathode materials. Interestingly, layered $LiCoO_2$ did not attract much attention initially [10]. In 1983, Goodenough also identified manganese spinel as a low-cost cathode materials [11]. However, the lack of safe anode materials limited the application of layered oxide cathode of $LiMO_2$ (M = Ni, Co) in Li-ion batteries. It was discovered by Besenhard [8], Yazami [12], and Basu [13] that graphite, also with layered structure, could be a good candidate to reversibly store Li by intercalation/deintercalation in late 1970s and early 1980s. In 1987, Yohsino et al. [14]. filed a patent and built a prototype cell using carbonaceous anode and discharged $LiCoO_2$ as cathode (Fig. 3). Both carbon anode and LiCoO2 cathode are stable in air which is highly beneficial from the engineering and manufacturing perspectives. This battery design enabled the large-scale manufacturing of Li-ion batteries in the early 1990s.

It should be highlighted that Yohsion also carried out first safety test on Li-ion batteries to demonstrate their enhanced safety features without ignition by dropping iron lump on the battery cells, in contrast to that of metallic lithium batteries which caused fire [3]. Yohsino's success is widely considered the beginning of modern commercial Li-ion batteries. Eventually Sony, dominant maker of personal electronic devices such as Walkman at that time, commercialized Li-ion batteries in 1991. It was a tremendous success and supported the revolution of personal mobile electronics. To acknowledge their pioneering contribution to the development of Li-ion battery, Goodenough, Yazami, and Yoshino were awarded the 2012 IEEE Medal for Environmental and Safety Technologies. Japan is still leading the share of global Li-ion battery market dominating 57% global market in 2010. In the past two decades, there is some notable progress in development of Li-ion batteries, particularly the introduction of low-cost cathode of $LiFePO_4$ by Goodenough in 1996 and high capacity anode of C–Sn–Co by Sony in 2005. The recent development of high capacity anode based on nanostructured silicon (theoretical-specific capacity of 4200 mAh/g) is also worthy to be highlighted [15–17]. In 1990s, Dahn and colleagues pioneered the exploration of composites of C/Si obtained from pyrolysis of silicon-containing polymers as promising candidate as anode materials for Li-ion batteries [18–21]. In 2008, Cui and colleagues [15] reported potential application of Si nanowires for Li-ion batteries. Deng et al. [22–25] studied a number of tin-based negative electrode materials with different compositions and structures to achieve improved electrochemical performances. Deng et al. recently also studied iron and copper oxides based nanostructures that could be used as potential carbon alternatives with much higher capacity than that of graphite [26, 27]. There are a large number of excellent scientific papers published in the past few years, achieving impressive good progress. Future research should focus on full-cell evaluation and standardization of battery testing procedures.

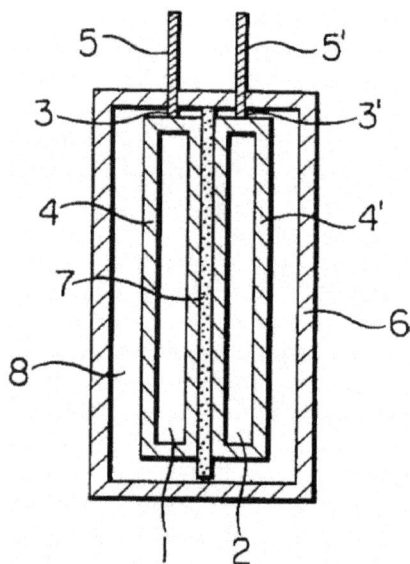

Figure 3. Illustration of first full cell of Carbon/LiCoO$_2$ coupled Li-ion battery patterned by Yohsino et al., with 1-positive electrode, 2-negative electrode, 3-current collecting rods, 4-SUS nets, 5-external electrode terminals, 6-case, 7-separator, 8-electrolyte. Reproduced with permission [14].

Basics of Li-Ion Batteries

A Li-ion battery is constructed by connected basic Li-ion cells in parallel (to increase current), in series (to increase voltage) or combined configurations. Multiple battery cells can be integrated into a module. Multiple modules can be intergrade into a battery pack. For example, the 85 kWh battery pack in a typical Tesla car contains 7104 cells. Typically, a basic Li-ion cell consists of a cathode (positive electrode) and an anode (negative electrode) which are contacted by an electrolyte containing lithium ions. The electrodes are isolated from each other by a separator, typically microporous polymer membrane, which allows the exchange of lithium ions between the two electrodes but not electrons. In addition to liquid electrolyte, polymer, gel, and ceramic electrolyte have also been explored for applications in Li-ion batteries. Figure 4 illustrates the basic operating principle of a typical Li-ion battery cell. The basic design of Li-ion cells today is still the same as those cells Sony commercialized two decades age, although various kinds of electrode materials, electrolyte, and separators have been explored.

The commercial cells are typically assembled in discharged state. The discharged cathode materials (e.g., $LiCoO_2$, $LiFePO_4$) and anode materials (e.g., carbon) are stable in atmosphere and can be easily handled in industrial practices. Yohsino made a significant contribution to the commercial production of Li-ion batteries by using discharged electrode materials in full cells for the first time. During charging process, the two electrodes are connected externally to an external electrical supply. The electrons are forced to be released at the cathode and move externally to the anode. Simultaneously the lithium ions move in the same direction, but internally, from cathode to anode via the electrolyte. In this way the external energy are electrochemically stored in the battery

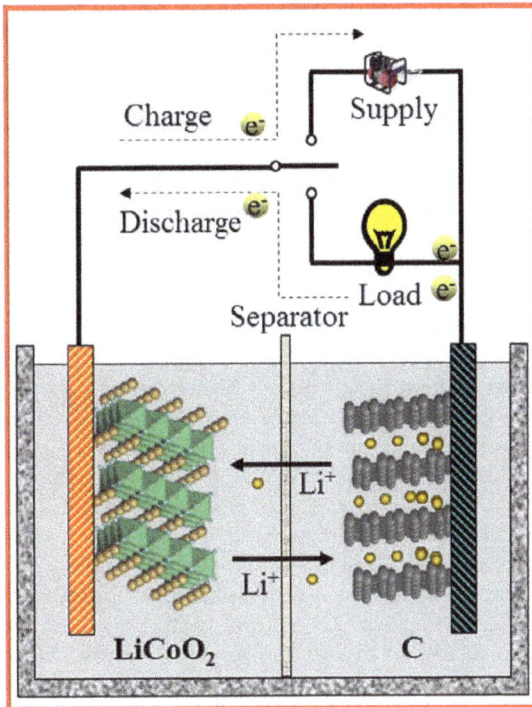

Figure 4. Illustration to show the basic components and operation principle of a Li-ion cell. Reproduced with permission [1].

in the form of chemical energy in the anode and cathode materials with different chemical potentials. The opposite occurs during discharging process: electrons move from anode to the cathode through the external load to do the work and Li ions move from anode to the cathode in the electrolyte. This is also known as "shuttle chair" mechanism, where the Li ions shuttle between the anode and cathodes during charge and discharge cycles.

Electrochemical reactions at the two electrodes released the stored chemical energy [1]. The total Gibbs free energy change due to the electrochemical reactions on the two electrodes is determined by the electrode materials selected. Given the overall electrochemical reaction and charges transferred, one can estimate the theoretical cell voltage ($\Delta E = -\Delta G/nF$). The performance of Li-ion batteries can be evaluated by a number of parameters, such as specific energy, volumetric energy, specific capacity, cyclability, safety, abuse tolerance, and the dis/charging rate. Specific energy (Wh/kg) measures the amount of energy that can be stored and released per unit mass of the battery. It can be obtained by multiplying the specific capacity (Ah/kg) with operating battery voltage (V). Specific capacity measures the amount of charge that can be reversibly stored per unit mass. It is closely related to number of electrons released from electrochemical reactions and the atomic weight of the host. Cyclability measures the reversibility of the Li-ion insertion and extraction processes, in terms of the number of charge and discharge cycles before the battery loses energy significantly or can no longer sustain function of the device it powers. Practically, the cycle life of Li-ion batteries is affected by depth of discharge (DOD) and state of charge (SOC), as well as operating temperature, in addition to the battery chemistry. Cycle life is enhanced with shallow DOD cycles and less SOC swing, and avoiding elevated temperature. Li dendrite formation on graphite anode can occur at low-temperature charge which should be avoided. Safety requirement is very high for Li-ion batteries with multiple cells. Battery management systems (BMS) are typically employed in battery cells/packs/modules to prevent any possible thermal runaway. For example, in the case of battery cell failure inside a battery pack, the BMS could detect and isolate the particular cell.

Abuse tolerance is a critical requirement for practical application of Li-ion batteries, especially in electric vehicles. Typically, mechanical, thermal, and electrical abuse evaluations are carried out on prototypes to evaluate abuse tolerance of the batteries. The mechanical abuse evaluation includes mechanical shock and drop, roll-over, nail penetration, and immersion in water tests. The thermal abuse evaluation includes radiant heat, thermal stability, overheat, and extreme cold tests. The electrical abuse evaluation includes short circuit, overcharge, over-discharge, and alternative current exposure tests. Those abuse tolerance tests are extremely important for their applications in electric vehicles, as electric vehicles are expected to compete with existing internal combustion engineer-powered vehicles that run well in rough conditions. The rate of charge or discharge measures how fast the battery can be charged and discharged, typically called C-rate. At 1 C, the battery is fully discharged releasing maximum capacity in 1 h. Common Li-ion batteries with carbonaceous anodes used in personal mobile devices take 1–4 h to return to the fully charged state. Li-ion batteries used in electric vehicles may take even longer, for example, overnight, to get fully charged, although it could be quickly charged to certain low SOC at high current with special charging devices. One of the active research directions in Li-ion battery community is to increase the rate performance so that the time consumed for charging a battery can be dramatically reduced, which is particularly crucial to the market acceptance of electric vehicles [28].

Theoretical capacity of active electrode materials can be estimated based on electrochemical reactions involved. For example, electrochemical reaction for the anode of graphite that can intercalate reversibly with lithium to form LiC_6, the reaction is

$$Li^+ + e^- + C_6 \leftrightarrow LiC_6 \tag{1}$$

The theoretical specific capacity (mAh/g) of anode graphite can be estimated as the following:

$$C_{specific} = xF/nM = 1 \times (96485\,C/mol)/6 \times (12\,g/mol)$$
$$= 372\,mAh/g$$

where x is number of electrons transferred in reaction (1), $F = 96485\,C/mol$ is Faraday's constant, n is number of moles of a chosen electroactive component that take place in the reaction, and M is the molecular weight of the same electroactive component.

The cathode reaction for $LiCoO_2$, with 0.5 as the practical number of electrons transferred, is

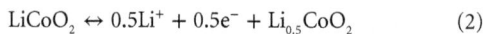

$$LiCoO_2 \leftrightarrow 0.5Li^+ + 0.5e^- + Li_{0.5}CoO_2 \tag{2}$$

The theoretical specific capacity can be estimated similarly

$$C_{specific} = xF/nM = 0.5 \times (96485\,C/mol)/1 \times (98\,g/mol)$$
$$= 137\,mAh/g$$

In practice, to evaluate specific capacity of a Li-ion battery cell, one not only has to take into consideration of the integration of cathode and anode materials but also other essential components, such as binders, conductive enhancers, separators, electrolyte, current collectors, case, tabs, as well as battery management systems (BMS). Therefore, the practical energy density is always less than that estimated based the battery chemistry.

Progress in Li-Ion Batteries

Since the commercialization of Li-ion batteries by Sony, Li-ion batteries have been attracting much attention world widely [6, 29–32]. The global production of Li-ion batteries continuously increase in the past two decades, especially with popularity of personal mobile electronics devices, such as mobile phones and personal computers, and electric vehicles. Although Li-ion batteries are highly successful commercially, there are still noticeable disadvantages. (1) The cost of Li-ion batteries based on per unit of energy stored ($/kWh) is still very high, although the price was decreasing over the past two decades. The Li-ion battery packs for electric vehicles could cost about $600/kWh, and it is anticipated that the cost could be reduced to about $200/kWh by 2020. In contrast, the average retail price of electricity to customers is about 0.1 $/kWh in 2014 according to the U.S. Energy Information Administration. (2) The performances of Li-ion batteries degrade at high temperature. (3) At the same time, it may not be safe if rapidly charged at low temperature. Therefore, protective circuits are typically used avoid overcharge and thermal runaway. However, the protective circuits could add weight burdens and decrease energy density of the whole batteries. (4) Another disadvantage is the possible venting and fire when crushed, which requires critical safety enhancement. The recent accidents of fires in Li-ion battery packs upon crushed by metal objects in the promising Tesla Model S cars highlights the importance of battery safety. Therefore, active research is continuing in all aspects of batteries, from anode, cathode, separator, electrolyte, safety, thermal control, packaging, to cell construction and battery management. The electrode materials selected are critical to the performances of Li-ion batteries, as they generally determine cell voltage, capacity, and cyclability. There are a number of potential alternative electrode materials to replace carbon-based anodes and $LiCoO_2$-based cathodes (Fig. 5). Composite alloys, Si, Sn-based materials and 3d-metal oxides materials have relatively higher specific capacities than graphite. However, they suffer from severe volume expansion during the process of lithiation, which causes the fracture of original structure upon lithium extraction, leading to bad electrical contact between particles and current collectors and poor cycling performance. Li metal has the issue of dendrite formation and is not safe as anode, which explained the failed commercialization of the Exxon's lithium ion batteries in the 70s. Given the high capacity of Li metal as anode, it should still be worthy for further exploration and research should focus on depressing the dendrite formation issues. In terms of cathode materials, a number of Co, Mn, Ni-based layered and spinel materials have been explored (mainly

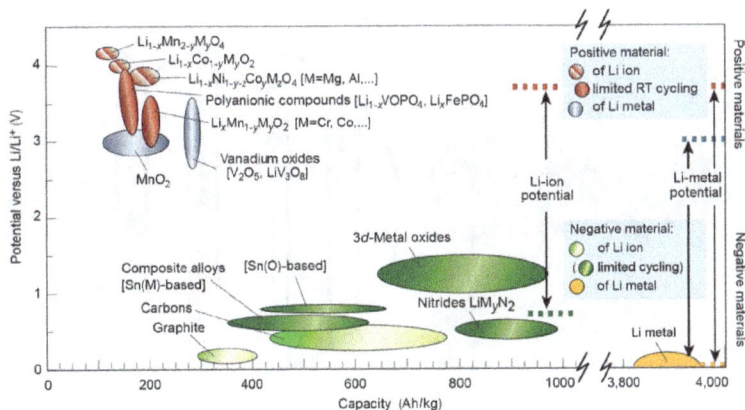

Figure 5. Voltage versus capacity for positive- and negative electrode materials presently used or under considerations for the next-generation of Li-ion batteries. Reproduced with permission [6].

concentrated in the left upper corner of Fig. 5). Recently, much effort is shifted to polyanionic components, such as $LiFePO_4$ and $LiMnPO_4$. The main requirements for cathode materials are high free energy of reaction with lithium, incorporating large quantities of lithium and insoluble in electrolytes. Although various promising electrode materials have been proposed, the slow lithium ion diffusivity, poor electronic conductivity and high cost are limiting their practical applications.

To address those issues associated with alternative electrode materials, recent efforts are to employ nanostructured materials for Li-ion batteries. It is generally believed that size and shape tunable properties of those lithium-active materials at nanoscale can offer additional parameters for further optimization of their electrochemical performances [29–31, 33]. The fascinating size-shape-related properties of nanomaterials bring new opportunities for a potential breakthrough in the development of next-generation Li-ion materials. Nanostructured electrode materials can offer various advantages not available in conventional bulk materials, and are believed to be the material-of-choice for the next-generation Li-ion batteries [1, 34–46]. However, nanomaterials bring along with new problems as will be discussed. In this tutorial review, I will try to provide an introduction on four aspects of Li-ion batteries, including cathode materials, anode materials, electrolyte, and separators, with emphasized on the progress achieved and challenges faced in those areas. Battery management and monitoring and optimization and cell integration are out of the scope of this review.

Cathode materials

There are a number of candidates that have been explored as cathode materials for Li-ion batteries. The cathode materials can be categorized based on voltage versus lithium. Typically: 2-Volt cathode materials are TiS_2 and MoS_2 with 2-D layered structure; 3-Volt cathode materials are MnO_2 and V_2O_5; 4-Volt cathode materials are $LiCoO_2$, $LiNiO_2$ with 2-D layered structure and 3-D spinel $LiMn_2O_4$ and olivine $LiFePO_4$; 5-Volt cathode materials are olivine $LiMnPO_4$, $LiCoPO_4$, and $Li_2MxMn_{4-x}O_8$ (M = Fe, Co) spinel 3-D structure. Generally, high cathode voltage is desirable as energy stored is proportional to the cell operating voltage. However, electrolyte stability has to be taken into consideration in selecting high voltage cathode materials.

Currently $LiCoO_2$ and $LiFePO_4$ are most widely used in commercial Li-ion batteries because of their good cycle life (>500 cycles). $LiCoO_2$ can be easily manufactured in large scale and is stable in air. Its practical capacity is ~140 mAh/g and the theoretical capacity is 274 mAh/g upon full charge. In addition to its low practical capacity, other noticeable disadvantages of the $LiCoO_2$ are their high material cost and the toxicity of cobalt. On the other hand, $LiFePO_4$-based cathode materials are attracting much attention in the past decade due to its low cost and low environmental impact. Compared to $LiCoO_2$, $LiFePO_4$ also offers a number of advantages, such as stability, excellent cycle life, and temperature tolerance (−20 to 70°C). However, $LiFePO_4$ has a problem of poor electronic and ionic conductivity at 10^{-10} S/cm and 10^{-8} cm²/sec, respectively, as well as relatively low capacity [47]. The other issue is one-dimensional channels for lithium ion diffusion which can easily be blocked by defects and impurities. Modeling and calculation reveal that the lowest Lithium ion migration energy is in the channels along [010] direction (Fig. 6) [48]. It also suggests that diffusion constant depends on particle size with diffusion in bulk being much slower than in nanoscale. Therefore, $LiFePO_4$ in nanoscale have been explored for and employed in Li-ion batteries.

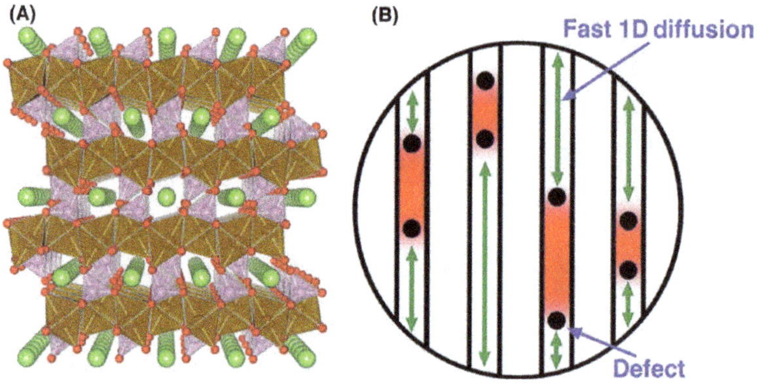

Figure 6. (A) Crystal structure of LiFePO$_4$ illustrating 1D Li$^+$ diffusion channels along the [010] direction; (B) Illustration of Li$^+$ diffusion impeded by immobile points defects in the channels. Reproduced with permission [48].

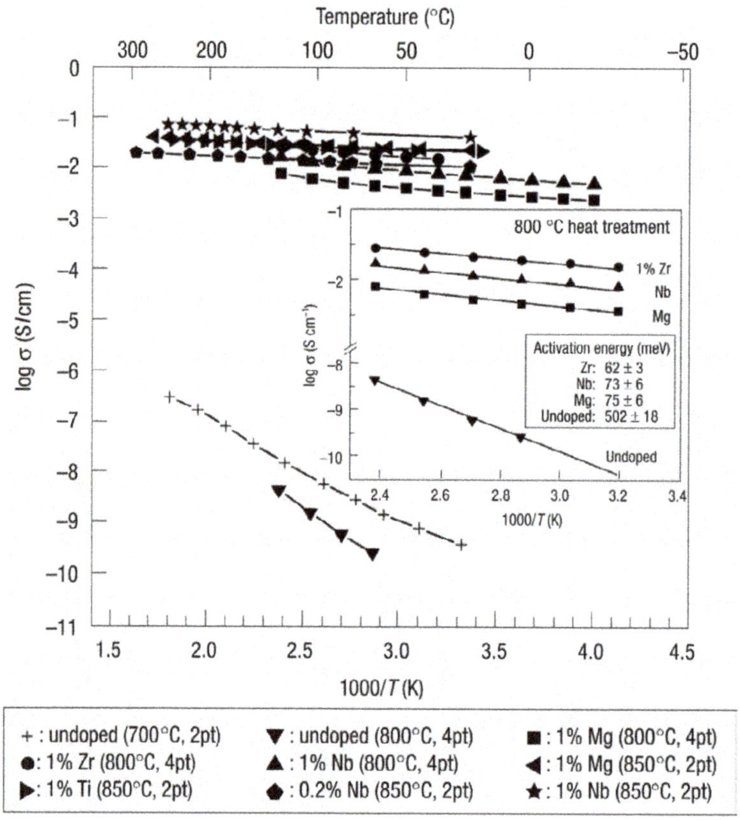

Figure 7. Electrical conductivity compared at different temperature with different cation doping for olivines of stoichiometry Li$_{1-x}$M$_x$FePO$_4$. versus undoped LiFePO$_4$. Reproduced with permission [47].

To enhance electronic and ionic conductivity of LiFePO$_4$, two common strategies are developed, namely doping by ions and coating by carbon [47, 49]. Chiang et al. reported that electronic conductivity of LiFePO$_4$ can be increased by a factor of 10^8 by cation doping (Fig. 7). The doped LiFePO$_4$ of stoichiometry Li$_{1-x}$M$_x$FePO$_4$ (M = Mg, Ti, Zr, Nb) could provide significant capacity with little polarization at rates as high as 6 A/g [47]. It was revealed by structural and electrochemical analysis that doping could reduce the lithium miscibility gap, increase phase transformation kinetics during cycling, and expand Li diffusion channels in the structure [49]. Combined doping and nanoscale size, it was possible to produce high power Li-ion batteries based on $LiFePO_4$. In fact, the start-up company A123 adopted the doping strategy to employ $LiFePO_4$ as cathode materials for making advanced Li-ion batteries with high power and was very successful initially and quickly established multiple manufacturing plants globally.

Recently, Kang and colleagues [50] reported that nanocrystalline $LiFe_{1-x}Sn_xPO_4$ ($0 \leq x \leq 0.07$) could be synthesized using Sn as dopant via an inorganic-based sol–gel method (Fig. 8). They noticed that under the synergetic effects between the charge compensation and the crystal distortion, the electrical conductivity first increases and then decreases with the increasing amount of Sn doping when tested at various rates. Their study suggested that a doping amount of about 3 mol% delivered the highest capacities at all rates attributed to the high electrical conductivity and the fast lithium ion diffusion velocity. It would be interesting to further understand the mechanism of change in specific capacity as a function of amount of Sn doped in $LiFePO_4$, and why the 3 mol% is the optimized value.

Improved rate performance of $LiFePO_4$ by Fe-site doping was observed [51]. When evaluated at 10 C rate, the capacity of $LiFe_{0.9}M_{0.1}PO_4$ for (M = Ni, Co, Mg) could be maintained at 80–90 mAh/g with 95% capacity retention after 100 cycles. Without doping, only 70% capacity can be retained. The improved rate performance and cyclability were attributed to the enhanced conductivity and good mobility of Li^+ ions in the doped electrode. In addition to Fe-site doping, Ceder reported an interesting strategy by designing $LiFePO_4$ off-stoichiometry (e.g., $LiFe_{1-2y}P_{1-y}O_{4-\delta}$) as cathode materials to achieve high lithium bulk mobility [28]. The procedure for materials preparation was impressively simple, suggesting the process could be easily carried out in manufacturing process. Precursors of Li_2CO_3, $FeC_2O_4 \cdot 2H_2O$ and $NH_4H_2PO_4$ in calculated amounts were ball-milled and heated to 350°C for 10 h and then heated at 600°C for 10 h under argon. The as-prepared nanoparticles are about 50 nm in size with poorly crystallized layer <5 nm thick (Fig. 9C). This two-stage heating was believed to provide the right conditions for construction of a fast ion-conducting surface phase of glassy lithium phosphate containing Fe^{3+}. The exceptionally high-rate performance (Fig. 9A and B) was attributed to the combination of nanoparticles with small size and the optimal coating with good lithium ion conductivity. It was estimated that a full battery discharge in 10–20 sec could be achieved. The very high-rate tests were carried out with 65 wt% carbon as conductive enhancer [28]. On the other hand, the ultrafast charging rate could require extremely high current up to 900 A for practically application, which is not recommended for safety consideration [52]. Similar to other fast charging batteries, more work is still needed to be done on addressing the issues of safety and heat generation.

In addition to doping, the other strategy of coating $LiFePO_4$ with carbon has also been demonstrated to improve electrochemical performance of $LiFePO_4$ as cathode for Li-ion batteries [53]. In fact, by a very simple procedure of dispersion of $LiFePO_4$ nanoparticles in a nanoporous carbon matrix, Guo and colleagues [54] demonstrated that $LiFePO_4$-based cathode could achieve both high power and high energy. The improved electrochemical performance could be attributed to the presence of conducting 3D nanonetwork in carbon matrix which enhances both Li ions and electrons to migrate and reach each $LiFePO_4$ particle achieving full utilization of the active materials. The nanoporous carbon matrix could serve as an electrolyte container for high-rate operation and be employed to support other active materials as well (Fig. 10)

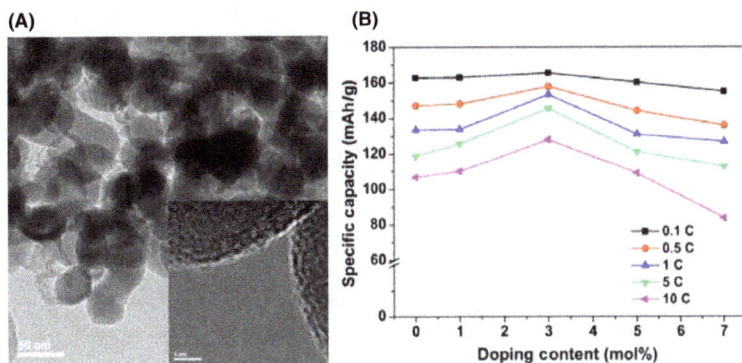

Figure 8. (A) The TEM image for the $LiFe_{1-x}Sn_xPO_4$ ($0 < x \leq 0.07$) nanoparticles and the inset shows the high-resolution TEM image. (B) The capability of the $LiFe_{1-x}Sn_xPO_4$ ($0 \leq x \leq 0.07$) samples as a function of the doping amount of tin and current rate (C-rate). Reproduced with permission [50].

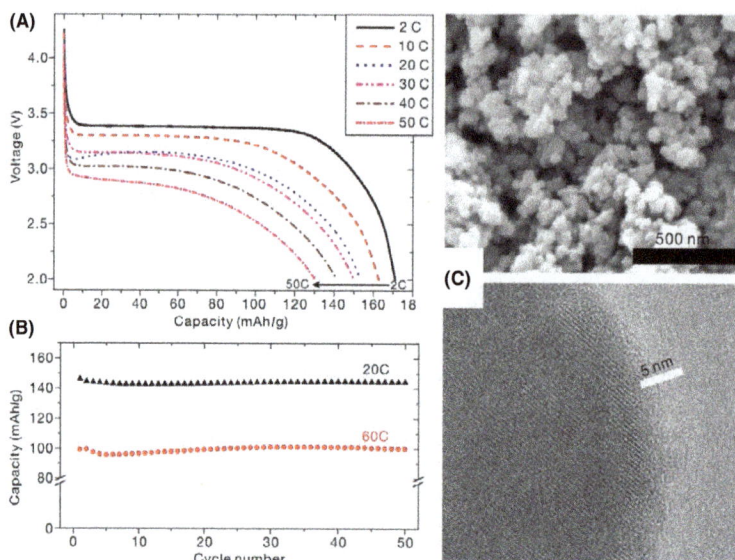

Figure 9. (A) Discharge rate capability and (B) capacity retention for $LiFe_{0.9}P_{0.95}O_{4-\delta}$ prepared by heating at 600°C under argon; (C) SEM (upper) and TEM (lower) images showing a particle size of ~50 nm and a poorly crystallized layer less than 5 nm thick on the edge of a particle, respectively. Reproduced with permission [28].

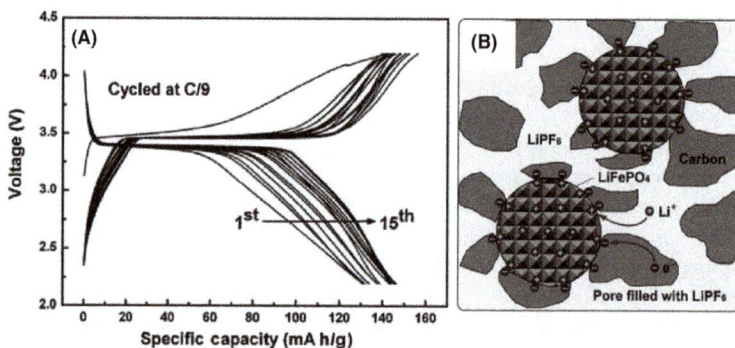

Figure 10. (A) First 15 cycles of galvanostatic charge/discharge profiles tested at a rate of C/9 of the composite of $LiFePO_4$ nanoparticles in a nanoporous carbon matrix; (B) Schematic illustration of the $LiFePO_4$@nanoporous carbon matrix for Li storage. Reproduced with permission [54].

[54]. It also interesting to note that the preparation procedure using sol–gel method followed by a solid-state reaction is very simple and can be used for industrial scale production in the future.

Amine et al. explored the coating of $LiFePO_4$ with carbon by vapor deposition process to improve the rate performance in Li-ion batteries. The carbon coating was achieved by decomposition of propylene balanced with N_2 in a preheated furnace at 700°C containing the olivine material [55]. The coating process is considered a low-cost coating technique, which could be suitable for mass production. Interestingly, few monolayers of carbon film formed on

the surface of olivine particles and the carbon also was deposited inside the pores of the particles. With just 3.4 wt% carbon, impressive cycling performance over 70 cycles could be achieved with no capacity fade at either room temperature or at 37°C (Fig. 11). It is particularly interesting to note that, without addition of carbon black as the conductivity enhancer typically for $LiFePO_4$, they could achieve specific capacity of 140 and 150 mAh/g when tested at room temperature and at 37°C, respectively. The improved performance was attributed to the network of carbon film on the surface of the olivine as well as the in the pores of the particles. The presence of carbon film dramatically

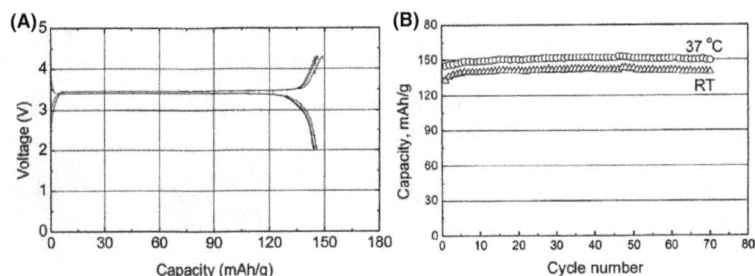

Figure 11. (A) The charge/discharge profiles of the carbon-coated olivine LiFePO$_4$; (B) Cycling performance of the carbon-coated olivine tested at room temperature and 37°C over 80 cycles of charge and discharge. Reproduced with permission [55].

improved the overall conductivity of the electrodes. It is believed the C-LiFePO$_4$ could be very promising to replace those toxic transition metal oxide-based cathodes.

In addition to carbon coating, other conductivity enhancers have been explored to improve the conductivity of LiFePO$_4$ as well. Chung and colleagues [56] explored the preparation of LiFePO$_4$ thin films with uniformly dispersed highly conductive silver to improve the conductivity of LiFePO$_4$ (Fig. 12). With a small fraction of dispersed silver at only 1.37 wt%, a superior electrochemical performance in terms of specific capacity, cyclability, and high charge–discharge rate has been achieved. The preparation

procedure for making this uniformly dispersed silver in LiFePO$_4$ thin films was remarkably simple. Pulsed laser deposition was employed to deposit LiFePO$_4$-Ag on a Si/ SiO$_2$/Ti/Pt substrate using 248 nm laser beam with the substrate kept at 600°C. The conductivity of the as-prepared LiFePO$_4$-Ag thin film is 1.29×10^{-3} S/cm, which is dramatically higher than that of bare LiFePO$_4$ film at 0.25×10^{-9} S/cm. The improved conductivity helped to achieve impressive electrochemical performances in terms of specific capacity, cyclability, and charging rate [56]. The thin film-based active materials deposited on Si substrate suggest that the Li-ion batteries eventually developed will

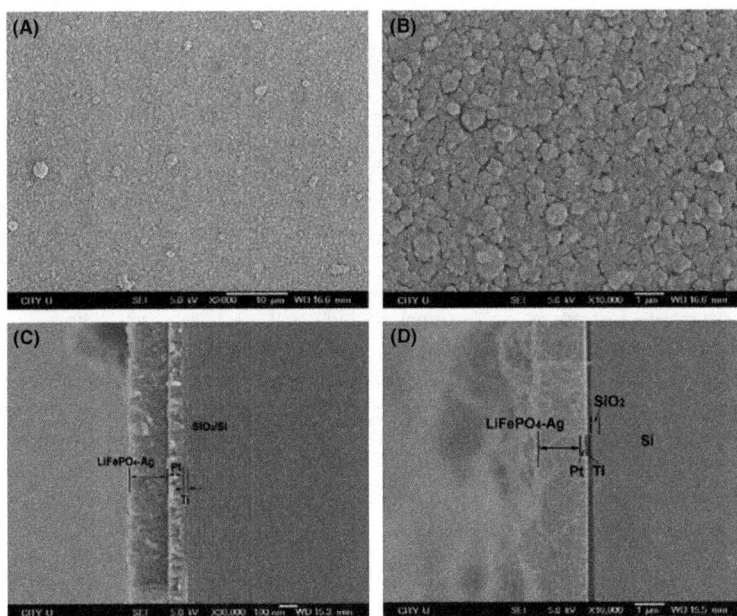

Figure 12. Planar view SEM images of the as-deposited LiFePO$_4$–Ag thin film at (A) low and (B) high magnifications. (C) Cross-sectional SEM image of the typical LiFePO$_4$-Ag film deposited on coated substrate. (D) Cross-sectional SEM image of the thick LiFePO$_4$–Ag thin film (1.4 μm) with a PLD deposition time of 4 h showing the film delaminated from substrate due to large stress of the thick film on the substrate. Reproduced with permission [56].

be for certain niche applications, such as microscale batteries, but not for mobile electronics or electric vehicles. On the other hand, thin film analysis is always helpful to reveal certain fundamental electrochemical mechanisms. The understanding and results obtained can be used to design better cathode materials beyond thin films.

Although $LiFePO_4$ has been demonstrated to show huge potential to replace traditional $LiCoO_2$ as safe cathode materials in Li-ion batteries, there are still challenges. The relatively moderate specific capacity or energy (by mass) of $LiFePO_4$ associated with its chemistry does not provide much space for further improvement. On the other hand, its rate performance or power can be further increased by the introduction of nanoscale structures. However, the introduction of nanoscale $LiFePO_4$ has two issues, namely the low volumetric density of the nanomaterials and the difficulty in preparation of nanoscale $LiFePO_4$ at low cost. The low packing density issue of nanoscale $LiFePO_4$ makes it difficult to produce batteries with high-energy density (by volume). Given that the interfacial behavior between $LiFePO_4$ with electrolyte plays critical roles in electrochemical performance, which is even more critical at nanoscale, it is crucial to gain better in-depth understanding of the interfacial phenomena involved. In situ investigation to monitor the interface during charge–discharge process may reveal the fundamental mechanisms involved in the interfacial electrochemical reactions. This kind of understanding will help to design better coating for $LiFePO_4$, and achieve better rate and cycling performance. Further study to address those challenges will enhance their electrochemical performance and generate big impact on energy storage.

Additional to layered compounds $LiMO_2$ (e.g., $LiCoO_2$) and olivine compounds $LiMPO_4$ (M = Fe, Mn, Ni, Co or combinations, e.g., $LiMn_{1-x}Fe_xPO_4$) discussed above, 5-Volt cathode materials of olivine $LiMnPO_4$, $LiCoPO_4$, and spinel $Li_2MxMn_{4-x}O_8$ (M = Fe, Co, Ni) are also being actively pursued. High voltage cathode materials could help to increase the energy density but demand for alternative stable electrolyte instead of conventional electrolyte. Another family of emerging polyanionic cathode is Li_2MSiO_4, (M = Mn, Fe, Co, Ni, e.g., Li_2MnSiO_4), which could offer much high capacity of 330 mAh/g. The obstacles to adopt those high-capacity Li_2MSiO_4 (Fig. 13) are their poor electronic conductivity, poor rate capability and fast capacity fading upon cycling [58].

Another notable progress was made recently by a team from the Argonne National Laboratory based on innovatively designed layered lithium nickel cobalt manganese oxide (with average composition of $Li[Ni_{0.68}Co_{0.18}Mn_{0.18}]O_2$) microparticles [59]. The cathode materials is unique in the way that the microparticles have concentration gradient, where the core is rich in Ni, and the outer layer is rich in Mn with decreasing Ni concentration and

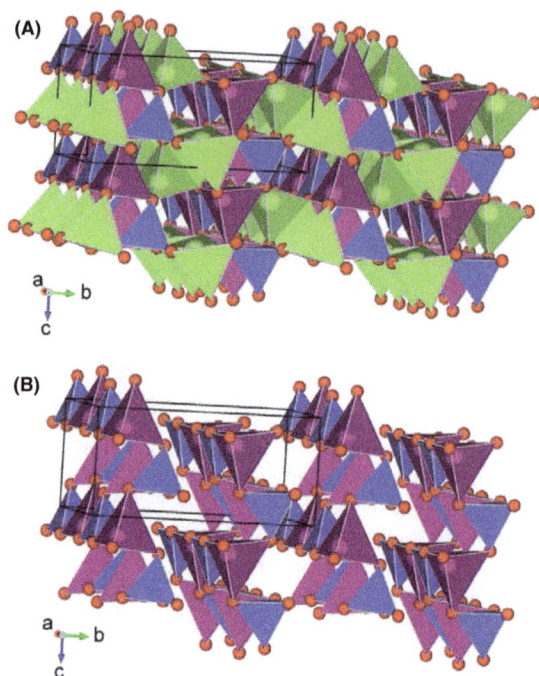

Figure 13. The crystal structure of (A) a typical form of Li_2MnSiO_4 (Pmnb) and (B) the hypothetical structure of the fully delithiated $MnSiO_4$ with SiO_4 shown in blue, LiO_4 in green, and MnO_4 in purple. Reproduced with permission [57].

increasing Mn and Co concentrations at the surface (Fig. 14). The bulk core of Ni-rich cathode provides high capacity. The concentration-gradient outer layer and the surface improve the thermal stability. The cathode materials demonstrated impressive high reversible capacity of 209 mAh/g and good safety characteristics. It should be noted that the materials preparation procedure based on coprecipitation in aqueous solution followed by calcination at 780°C is very simple and scalable, which is suitable for industrial scale production. This high-capacity concentration-gradient cathode material is promising to be produced on a large scale and used in next-generation commercialized Li-ion batteries. NCM cathodes are expected to play increasing roles in commercial Li-ion batteries.

Anode materials

Anode materials are extensively investigated and there is a bigger pool of candidates and materials. The electrochemical performances, including cyclability, charging rate, and energy density, of Li-ion batteries are significantly affected by anode materials selected. Since the first commercialization of carbonaceous anodes, carbon is still dominant in commercial Li-ion batteries today. Graphitic carbon with

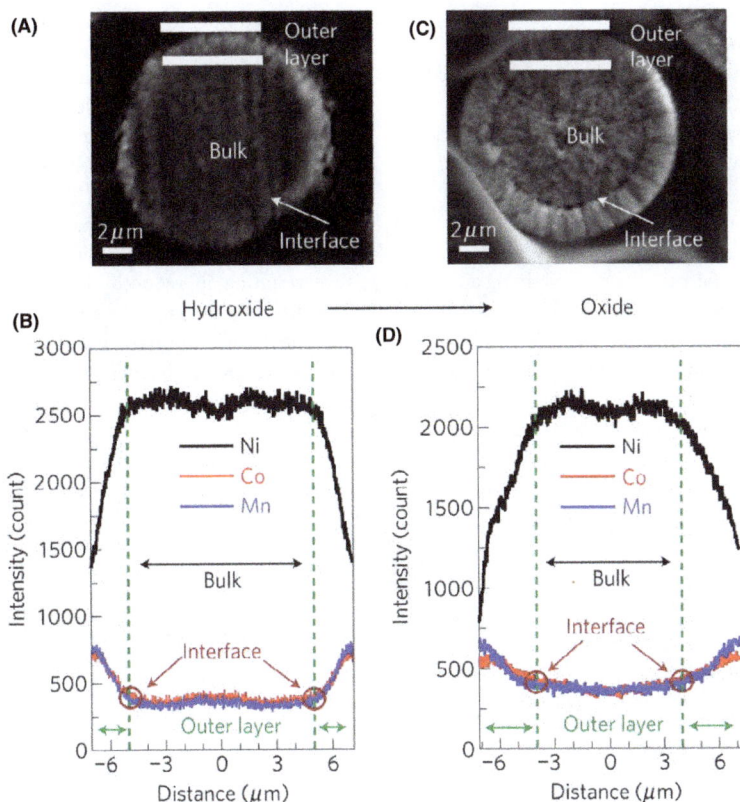

Figure 14. SEM images of (A) the precursor hydroxide obtained by coprecipitation, and (B) the final concentration-gradient lithiated NCM oxide; (C) and (D) are the corresponding electro-probe X-ray microanalysis line scans for (A) and (B), respectively. Reproduced with permission [59].

layered structure can facilitate the movement of lithium ions in and out of its lattice space with minimum irreversibly, resulting in an excellent cyclability [60]. However, the carbon anodes are soon approaching their theoretical maximum capacity of 372 mAh/g over the past two decades of development. Carbon alternatives with high-energy density and enhanced safety are required to meet the demands for increases in energy and power densities, especially to meet the demands from electric vehicles [61]. Fuji Film introduced tin composite oxide (TCO) as a carbon alternative in 1997 but was not very successful due to the poor cycling performance [62]. Sony Cooperation announced new-generation Li-ion batteries with the trade name of Nexelion in 2005. The anode of Nexelion is based on a carbon–tin–transition-metal composite (e.g., Sn–Co–C), and the compound is mainly amorphous or microcrystalline aggregates. Those efforts have rekindled another wave of interest in anode materials for Li-ion batteries [63].

Besides Sn, many other elements that are known to alloy with lithium, including silicon, are good candidates to replace carbon for lithium storage. These elements could alloy and de-alloy with lithium electrochemically at room temperature. However, the alloying/dealloying process during charging/discharging is accompanied by substantial variations in the specific volume of the material. The induced huge mechanical stress could lead to the destruction of the crystal structures and disintegrate the active materials and current collectors within a few cycles, or the so-called "pulverization" issue. The resulting poor cyclability has significantly limited their usability in practical situations. The engineering approach to solve the poor cyclability problem is to introduce composites. In such a composite material, one component (usually carbon) functions as a stress absorber whereas the other (such as silicon or tin) provides the boost in capacity. Through this approach a composite with capacity higher than carbon and cyclability better than Sn or Si can be achieved. A number of combinations involving carbon have been explored, among them Si/C [64] and SnO_2/C [65] have attracted much interest.

Carbons that are capable of reversible lithium ion storage can be classified as graphitic and nongraphitic (disordered) carbon. Graphitic carbons have a layered structure. Natural graphite, synthetic graphite, and pyrolytic graphite which consist of aggregates of graphite crystallites are commonly named graphite as well. The lithium insertion into graphite follows a stepwise occupation of the graphene interlayers at low concentrations of the lithium ions, known as stage formation [30]. Disordered carbons consist of carbon atoms that are arranged in a planar hexagonal network without an extended long-range order. There are amorphous domains cross-linking the crystalline graphitic flakes. Disordered carbon normally demonstrate high-specific capacity (x > 1 in $LixC_6$), but with issues of huge first-cycle irreversible capacity loss and capacity fading [66].

The excess charge consumed in the first cycle is generally attributed to the formation of a passivating solid-electrolyte interphase (SEI) on the carbon surface [30]. The SEI is caused by electrolyte decomposition at low potentials on fresh carbon surfaces. A poorly formed SEI

would continue to grow with time, leading to the increase in cell internal resistance and preventing full reversibility of lithium ion insertion into carbon. Subsequently the energy density of the cell decreases with the number of charging cycles. It is also known that some organic solvents would promote the insertion of lithium ions together with the solvent molecules. The solvated intercalation is accompanied by extreme expansion of the graphite matrix (~150%), which gradually deconstructs the graphite structure resulting in reduced charge storage capability [30]. The disordered carbons are normally synthesized by heat treatment of various carbon precursors under inert atmosphere. Recently, much effort is focused on disordered carbon, although the exact mechanism by which the high-specific capacity is achieved has not been fully understood [67–69]. Other carbon-based materials that have been extensively studied are buckminsterfullerene, carbon

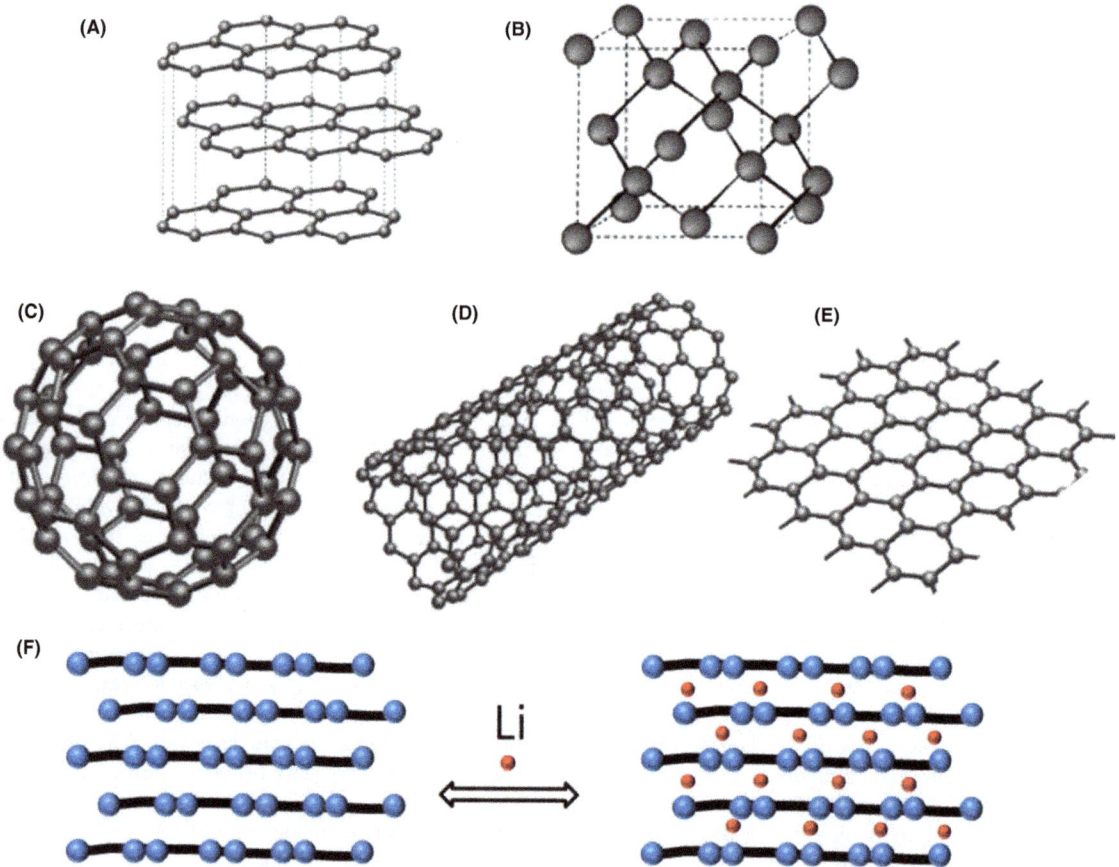

Figure 15. A family of carbon-based materials with different structure: (A) graphite with a stack of graphene layers, (B) diamond with carbon atoms arranged in a FCC structure, (C) buckminsterfullerene (C_{60}) with consisting of graphene balled into a sphere, (D) carbon nanotube with rolled-up cylinder of graphene, and (E) graphene of a single layer carbon, (F) the schematic of lithium intercalation and deintercalation between graphene layers in graphite. (A–E) were reproduced with permission [74].

nanotubes, and graphane (Fig. 15). Carbon nanotubes, in particular, can be a good lithium host on grounds of their excellent electronic conductivity and other properties associated with their linear dimensionality [70, 71]. However, current interest is focused on CNT- and graphene-based composites instead of pristine CNTs or graphene to achieve much higher capacity than that of pristine carbon [72, 73].

Another family of anode materials with high capacity is metal oxides, which have been widely studied since the first report by Tarascon's group [31]. Although metal oxides are generally poor in conductivity, properly tailored metal oxides at nanoscale have demonstrated promising characteristics. The reaction mechanism of lithiation and delithiation in metal oxides can be generally classified into three main types (Fig. 16): (1) the insertion/extraction, (2) the alloying/dealloying, and (3) the conversion mechanisms. The first mechanism is observed in different kinds of anode materials, including anatase TiO_2 [75]. In fact, most of cathode materials with layered or spinel structures also follow the insertion-extraction mechanism as discussed previously. Alloy reaction exists in some main-group elements, including Si, Sb, Ge, Bi and Sn, which can alloy with lithium forming Li_xM providing high-specific capacity. However, the lithium alloying and dealloying processes are typically associated with huge volumetric expansion (as high as 300%) and shrinkage, or pulverization, leading

to capacity fading upon cycling. The third mechanism of conversion is typically observed in transition metal oxides ($MxOy$, M = Mn, Fe, Co, Ni, Cu, etc.). Those conversion-type materials have relatively high theoretical capacities for they can incorporate more than one Li per metal. Metal grains and Li_2O will form during the process of lithiation. In many cases, the metals can reversibly alloy with lithium. Metal oxides also have the issue of pulverization, huge first-cycle irreversible capacity loss, as well as poor conductivity. To overcome those problems, much effort is focused on the preparation of nanoscale metal oxides and M_xO_y/carbon composites.

The electrochemical reactions of three types of reaction are listed.

1. Insertion/extraction mechanism:

$$MO_X + yLi^+ + ye^- \leftrightarrow Li_yMO_X$$

2. Li-alloy reaction mechanism:

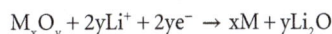

$$M_xO_y + 2yLi^+ + 2ye^- \rightarrow xM + yLi_2O$$

$$M + zLi^+ + ze^- \leftrightarrow Li_zM$$

3. Conversion reaction mechanism:

$$M_xO_y + 2yLi^+ + 2ye^- \leftrightarrow xM + yLi_2O$$

TiO_2-based anode materials have been extensively explored, although its practical capacity is not comparable to that of carbon anodes. In contrast to carbon where the

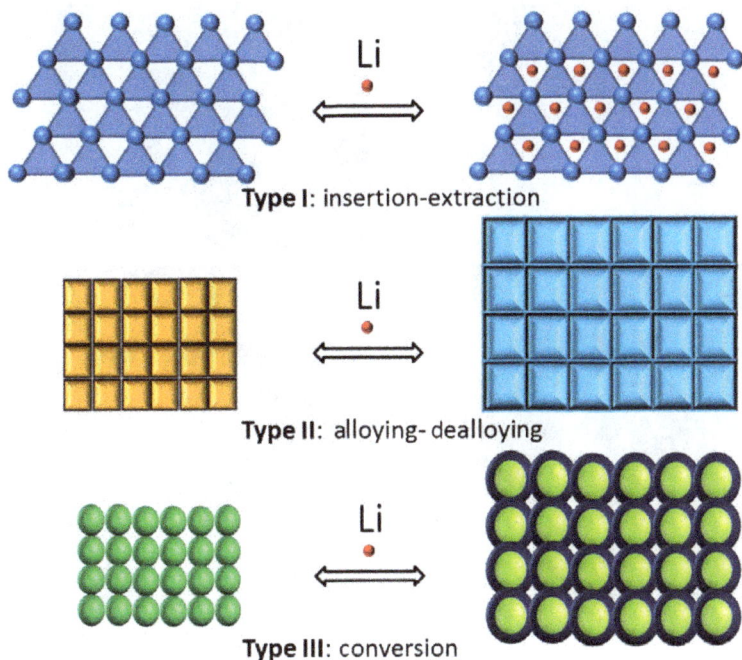

Figure 16. Schematic of different mechanisms of reversible lithium ion storage in metal oxides. The illustrate was drawn based on literature [76].

insertion of lithium into carbon takes place at a low voltage close to 0 V versus Li/Li$^+$ couple, the Li-ion insertion into TiO$_2$ is at about ~1.5 to 1.8 V versus Li/Li$^+$ couple. The electrochemical characteristics suggest that the potential risk of metal lithium deposition and dendrite formation, battery short circuit and thermal runaway in carbon-based batteries can be avoided if TiO$_2$ is used. Additionally, no surface electrolyte interphase (SEI) could form on TiO$_2$ in the potential window. In contrast, SEI formation on carbon causes the first-cycle irreversible capacity loss issue which would demand for additional supply of expensive cathode materials. Therefore, TiO$_2$-based anodes can find some niche applications where safety is the number one priority, for example, in aviation and aerospace applications. Among the four different phases of TiO$_2$, namely anatase, rutile, brookite and TiO$_2$(B), anatase TiO$_2$ is the most attracting one with theoretical specific capacity of 168 mAh/g based on the following electrochemical Li insertion-extraction reactions: TiO$_2$ + x(Li$^+$ + e$^-$) → Li$_x$TiO$_2$ where x = 0.5 for anatase TiO$_2$.

A facile gram-scale preparation of anatase TiO$_2$ high-order structures with subunits of tunable nanoparticle aggregates from one precursor for Li-ion batteries has been reported [75]. The nanoparticles were formed by basic building units aggregated controlled by calcination temperature (Fig. 17). Interestingly, the size of the basic building units of TiO$_2$ nanoparticles can significantly affect their electrochemical characteristics. When the crystallite size was at 17 nm, the anatase TiO$_2$ aggregates achieved an impressive high capacity 170 mAh/g, which is close to the theoretical value of 168 mAh/g. When charged at higher currents, good capacity retention was achieved indicating good rate performance. After 120 cycles, a reversible capacity of 160 mAh/g was still achieved at rate of ~C/3. However, when the crystallite size was at 33 nm, the amount of lithium insertion and extraction is very small. Electrochemical analysis revealed that the storage capacity was mainly in the form of pseudocapacity. Therefore, in the preparation of nanoparticles for Li-ion batteries, one has to control basic crystallite size to optimize their electrochemical performances.

Recently, much research focus of anode materials has been shifted to silicon-based anode. Silicon has the high theoretical capacity of 4200 mAh/g (lithiated to Li$_{4.4}$Si) or 3572 mAh/g (based on Li$_{3.75}$Si) which is about 10 times

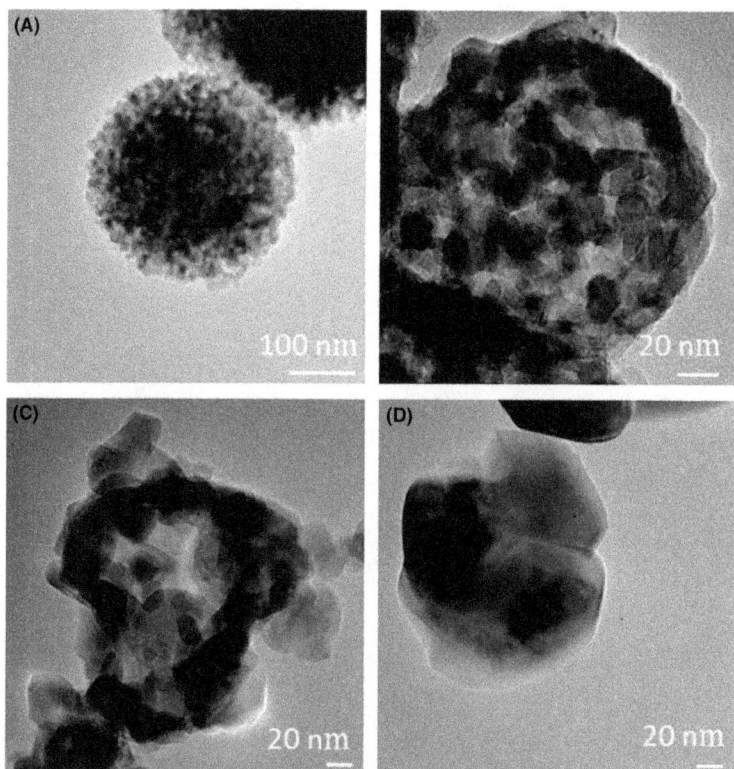

Figure 17. Anatase TiO$_2$ nanoparticle formed by nanoparticle aggregates with different unit size obtained from the same precursor by calcination at (A) 400, (B) 550, (C) 700 and (D) 900 °C characterized by TEM. Reproduced with permission [75].

of that graphite and almost four times of many metal oxides. However, silicon suffers two main drawbacks, namely poor conductivity and huge volume variation (400%) upon cycling, which makes it practically difficulty to use bulk silicon anode. The silicon nanostructured anode materials are very attractive. However, unlike metallic anodes, Si is a semiconductor and is more difficult to produce Si particles at nanoscale by simple cost-effective approaches on a large scale. Recently silicon nanowires have been reported to demonstrate promising reversible lithium storage properties [15]. Cui and colleagues [15] proposed and demonstrated that silicon nanowires were superior in lithium ion storage as compared to silicon thin film and particles (Fig. 18). The silicon nanowires could avoid the issue of pulverization and contact loss due to facile strain relaxation and efficient electron transport along each nanowire.

Bogart et al. [77] also demonstrated that silicon nanowires with carbon skin could enhance the cycling and rate performances of silicon nanowires in lithium storage. Recently, in another attempt, Ti@Si core–shell coaxial nanorods were proposed to further improve the electrochemical performances of Si nanorods (Fig. 19). As compared to pristine Si nanorods, the benefit of metallic core is that the axial resistance observed in solid Si nanorods could be dramatically reduced. The electrons released/acquired on electrochemical reactions of dealloying/alloying for LixSi could be transferred to the Ti foil current collectors easily via the metallic Ti core, in contrast to the solid semiconductor Si nanorod. The incorporation of metallic Ti core could dramatically reduce the Li diffusion distances from the radius of the solid Si nanorod to just the thickness of the shell of the core–shell Ti@Si nanorod. Additionally, the longitudinal direction charge transfer of the semiconductor solid Si nanorod with length of a few micrometers is reduced to thickness of the shell in the case of Ti@Si core–shell coaxial nanorods. Besides, the contact surface of Si with Ti increases, which could improve the connection between active materials and the collector. However, the capacity fading was still observed, suggesting further efforts should be put to optimize the core-shell materials. The concept should be extended to the preparation of other core–shell materials as well.

The author believes that silicon nanomaterials could potentially replace carbon anodes in the next 10 years,

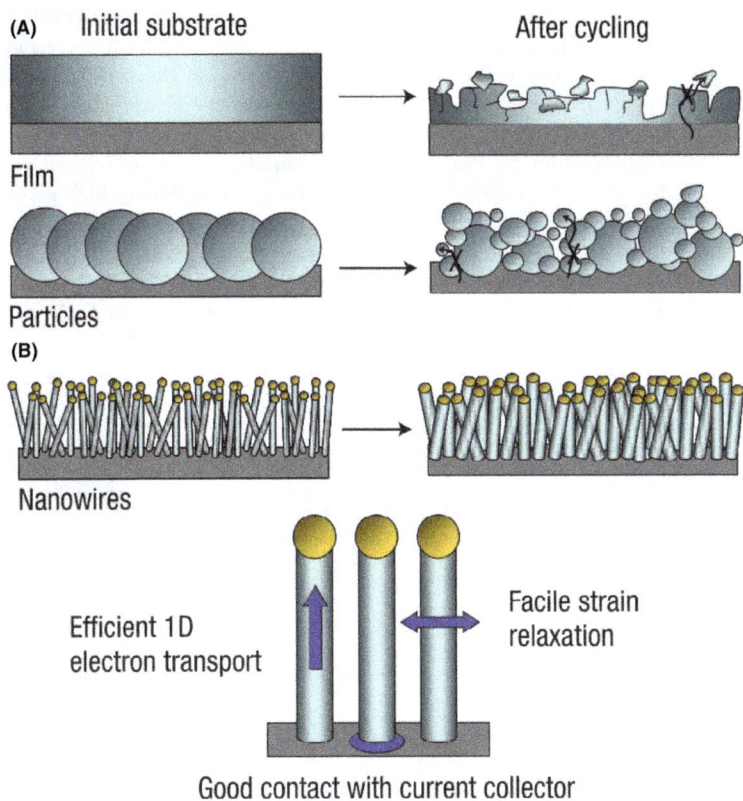

Figure 18. Schematic comparing the stability of (B) silicon nanowires with (A) thin film and particles upon repeated lithiation and delithiation. Reproduced with permission [15].

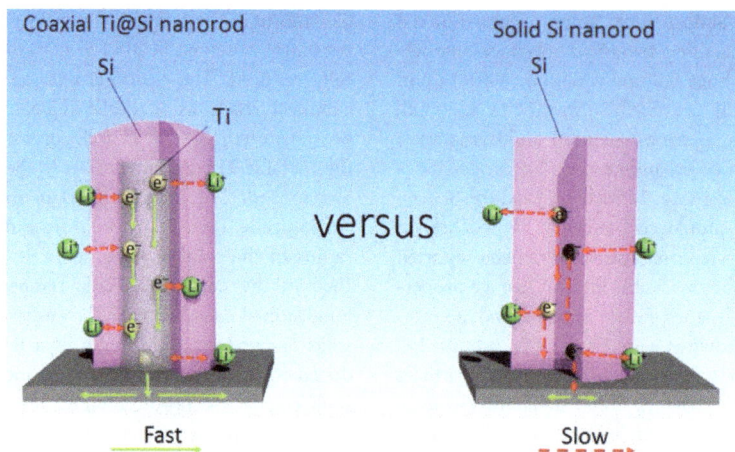

Figure 19. Illustration to compare (left) Ti@Si core–shell coaxial nanorod to that of (right) solid Si nanorod from the perspective of ion and electron transfer. Reproduced with permission [78].

provided that if the cost of production could be reduced significantly. Their application could be limited to devices that have no space restriction, for example, in smart energy grids, where high tap density is not the top priority. The production of nanostructured Si anodes must be scaled up to industrial level, which is also challenging. On the other hand, the volumetric density (mAh/cc) may be sacrificed in the adoption of silicon nanomaterials which have very low tap densities, typically in the order of 0.1 g/cc. Volumetric density (based on volume) has relatively less explored as compared to that of specific density (based on mass). Dahn and colleagues [79] has argued that although Si has the highest theoretical specific capacity (mAh/g), its energy density based on unit volume (Wh/cc) is just about the same as other materials (Fig. 20).

If one takes into consideration of low tap density of Si nanomaterials as compared to bulk Si, the future Si nanomaterials based Li-ion batteries will have even lower energy density. Therefore, a tradeoff between specific capacity and energy density should be considered, which is true for many other kinds of nanomaterials based electrodes. Similar to the problem faced in the preparation of silicon nanostructures, the fabrication of Si/C nanocomposites requires extensive studies as well. Mostly likely, the commercial Si-based electrode will be a composite containing a significantly high percentage of carbon and a small percentage of Si.

As one of the intensively studied anode candidates, SnO_2-based anodes are promising to replace carbon anodes. In its first cycle of lithiation, SnO_2 is converted to Sn, and

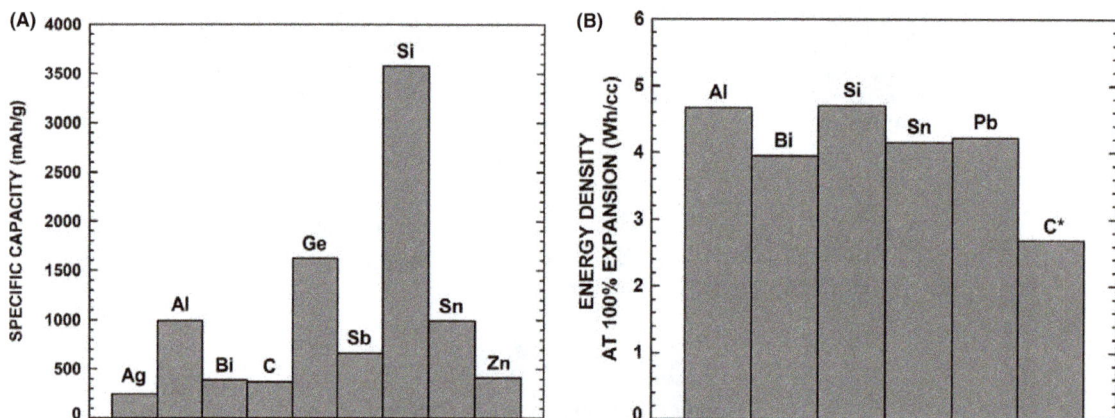

Figure 20. (A) Specific capacity (mAh/g) of a number of materials; (B) the volumetric energy density of the materials at a 100% volume expansion. Reproduced with permission [79].

subsequently, the Sn phase could store and release Li ions according to the Li-Sn alloying and dealloying reactions. SnO_2 has a theoretical capacity of 781 mAh/g, which is about two times higher than graphite. However, its poor cyclability is not acceptable from the perspective of cycling performance. Nanostructured SnO_2 has been proposed as a possible solution [25, 72, 80–84]. One dimensional (1-D) SnO_2 nanotubes could be prepared by template-assisted method using an anodic alumina oxide membranes with 1-D tunnels as the template [83]. In another study, polycrystalline hollow SnO_2 spheres could be prepared by deposited the precursor on hollow carbon nanospheres as the template and subsequent calcination in air at 800°C could remove the hollow carbon spheres and compacted the SnO_2 nanoparticles into hollow spheres. However, capacity fading was generally observed in SnO_2-based anodes [81].

Another interesting SnO_2-based anode material is electrospray deposited thin films of particles with unique porous spherical multideck-cage morphology (Fig. 21). The reversible capacity was reported to be as high as 1158 mAh/g, which is even higher than theoretical value. The improved electrochemical performance was attributed to the unique structure and the presence of Li_2O and CuO phases in the composite film [85]. Interestingly, even higher capacity of 2050 mAh/g with excellent capacity retention (97% upon 60 cycles) has also been reported for Sn-Si/C anode materials [86]. Additional effort should put to reveal the unusually high capacity observed. Besides metallic Sn- or SnO_2-based anodes, other Sn-based compounds, including tin phosphate [87, 88], SnO_2 filled mesoporous tin phosphate [87], teardrop-shaped SnP, [89] $AlPO_4$-coated SnO_2 [90], and Zn_2SnO_4 [91] have been reported, achieving certain degree of success in terms electrochemical characteristics.

Iron oxides have been considered as promising carbon alternative anode materials for high-capacity Li-ion batteries. It has advantages of low cost, nontoxicity, and environmental benignity. Iron is the fourth abundant element in the earth crust. It is abundant and has much lower cost than other 3d metal oxides, such as Co and Ni. However, similar to other metal oxides, iron oxides also suffer from poor conductivity and large volume change during the charge/discharge processes, which limits their practical use in commercial Li-ion batteries. Again, one popular strategy is to prepare iron oxide materials in nanoscale with tailored morphologies and synthesize iron oxide/carbon composites to accommodate the volume changes as well as improve conductivity. Various iron oxides with different nanostructures have been synthesized, including nanoparticles [92], nanowires [93–95], nanobelts [94, 96], nanorods [97–99], nanotubes [100, 101], nanoflakes [102], nanodisks [103] and nanorings [104], nanocubes [105–107], nanospheres [108–111], nanospindles [108, 112], nanourchins [113], and nanoflowers [103, 108, 114–119]. Hollow structured iron oxides are of great interest as the space could, ideally, accommodate the volume expansion during Li insertion [108, 115–119].

Popular routes to prepare iron metal oxides are template-assisted, hydrothermal, sol–gel and thermal oxidizing methods. Wen et al. [94] reported the preparation of vertically aligned iron oxide nanobelt and nanowire arrays by direct thermal oxidation of iron substrates under the flow of O_2. Nanobelts were produced in the low-temperature region (~700°C) and cylindrical nanowires tens of nanometers thick were formed at relatively higher temperatures (~800°C). Both nanobelts and nanowires are mostly bicrystallites with a length of tens of micrometers which grow uniquely along the [110] direction [94]. In another study, nanoflakes of α-Fe_2O_3 on Cu foil were achieved by thermal treatment method. The as-prepared α-Fe_2O_3 nanoflakes exhibited a stable capacity of 680 mAh/g, with no noticeable capacity fading up to 80 cycles when cycled in the voltage range 0.005–3.0 V at a rate of 65 mA/g [102]. Xu et al. reported the preparation of spindle-like porous α-Fe_2O_3 from an iron-based organic framework template, MIL-88-Fe. MIL-88-Fe was prepared by a modified solvothermal method using $FeCl_3 \cdot 6H_2O$ and 1,4-benzenedicarboxylate (BDC) as starting materials. The as-prepared spindle-like porous α-Fe_2O_3 exhibited very high charge capacity of 911 mAh/g after 50 cycles at current rate of 0.2C (1C = 1000 mA/g). Meanwhile, the capacity of commercially available bulk Fe_2O_3 (<5 μm; Sigma-Aldrich, Missouri, USA) quickly faded to <630 mAh/g. The spindle-like porous α-Fe_2O_3 also show a superior rate performance as compared to bulk Fe_2O_3 (Fig. 22) [112].

Chen et al. [103] reported a top-down approach to fabricate uniform single-crystal α-Fe_2O_3 nanodicks and melon-like α-Fe_2O_3 microparticles through selective oxalic acid etching, using phosphate ions as capping agent to

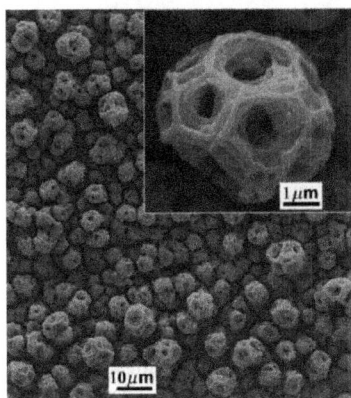

Figure 21. SEM images of the as-deposited film of nanostructured SnO_2 on copper current collector. Reproduced with permission [85].

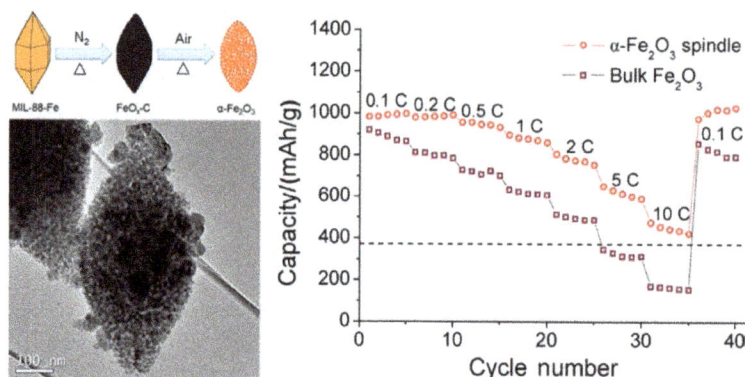

Figure 22. Schematic of the formation and TEM image of spindle-like porous α-Fe$_2$O$_3$ and it electrochemical performances compared to bulk Fe$_2$O$_3$. Reproduced with permission [112].

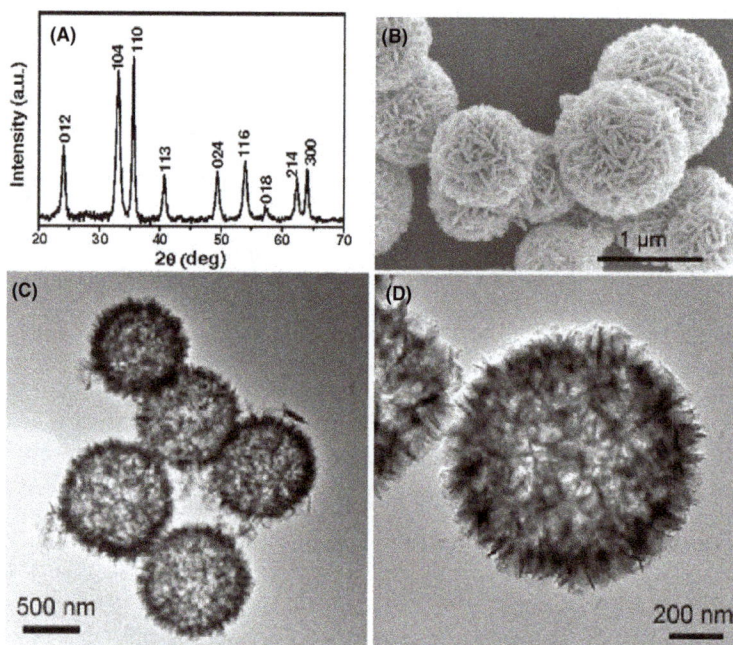

Figure 23. (A) XRD, (B) SEM, and (C, D) TEM images of Fe$_2$O$_3$ hollow spheres for Li-ion batteries. Reproduced with permission [111].

control the etching along [001] direction. The capacities of 662 and 341 mAh/g were retained after 100 cycles, for porous and solid melon-like α-Fe$_2$O$_3$ microparticles, respectively. Wang et al. reported the preparation of hollow α-Fe$_2$O$_3$ sphere with sheet-like subunits with quasiemulsion-templated methods (Fig. 23). The capacity of as-prepared hollow spheres around 1 μm faded very slowly from ~900 to 710 mAh/g after being charge/discharged from 2nd to 100th cycle, at testing rate of 200 mA/g. In contrast, the capacity of α-Fe$_2$O$_3$ microparticles faded from ~900 to 340 mAh/g, which was less

than one half of the capacity retained of hollow spheres [111]. Other hollow α-Fe$_2$O$_3$ has also been reported demonstrating advantages of hollow structures [120]. Beside spherical hollow materials, other structured Fe$_2$O$_3$ anode materials have been reported. For example, cocoon-like Fe$_2$O$_3$ mesoparticles of nanoparticle aggregates have also been reported [120]. Microscale single crystals of hematite with nearly 100% exposed {104} facets have been synthesized with a fast hydrothermal method [121].

Addition to nanostructures, carbon coating was used to improve the electrochemical performances of iron oxides. It is generally believed that the carbon coating layer can enhance the conductivity of each single units as well as improve the better electrical contact between each single units. The carbon layer can also act as a buffer layer to alleviate the stress caused by huge volume expansion and prevent the structure from collapsing or improved cyclability. Additionally, given that nanomaterials are more active than bulk materials, the high contact area between electrode and electrolyte may lead to more

significant side reactions. The carbon-coated layer can act as a protective layer to prevent the direct contact between the active nanomaterials and the electrolyte. Wang et al. [122] reported a novel hierarchical nanostructure composed of carbon-coated α-Fe_2O_3 hollow nanohorns grafted on CNT backbones by bottom-up assembly of β-FeOOH nanospindles on CNTs and subsequent in-situ transformation and further modification with carbon nanocoating (Fig. 24). The specific capacity of carbon-coated CNT@

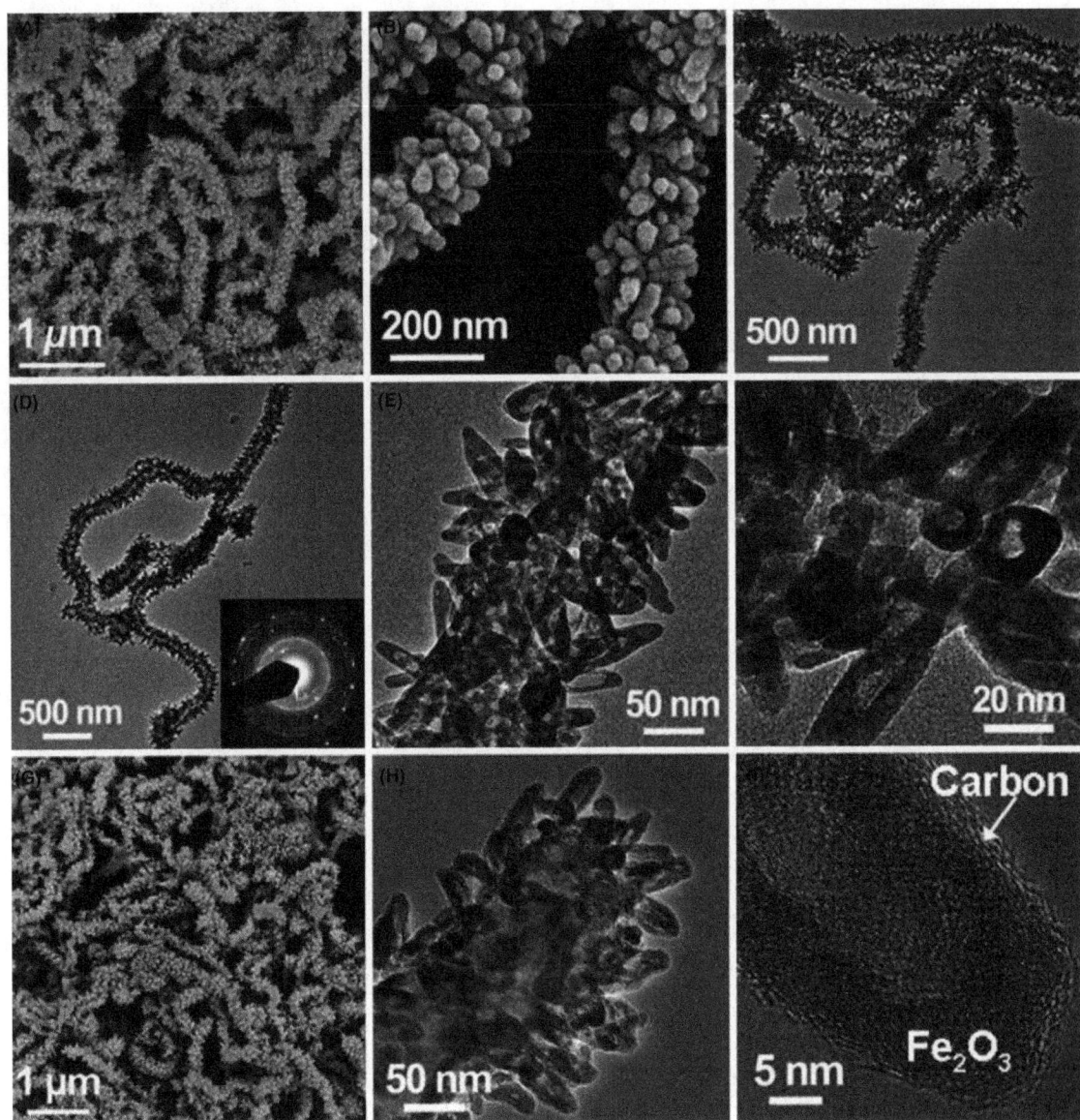

Figure 24. CNT@Fe_2O_3 hierarchical structures characterized by (A and B) FESEM and (C and D) TEM at different magnification; (E and F) TEM images revealing the formation of hollow nanohorns on CNT; (G) FESEM image and (H) TEM image of carbon-coated CNT@ Fe_2O_3 hierarchical structures; (I) HRTEM image of uniform carbon coating on α- Fe_2O_3 hollow nanohorns. Reproduced with permission [122].

Fe_2O_3 hollow nanohorns gradually rose from 660 to 820 mAh/g from 2nd to 100th cycle at current density of 500 mA/g, while the capacity of uncoated $CNT@Fe_2O_3$ slowly faded from 1200 to around 500 mAh/g, from 1st to 100th cycle. Both materials show superior performance compared to α-Fe_2O_3 nanoparticles, which had a capacity of 300 mAh/g after 100 cycles. The results evidenced that properly tailored structure with designed coating could further improve their electrochemical performances.

Another family of iron oxide, magnetite Fe_3O_4, which has a theoretical capacity of 927 mAh/g, has been attracting much attention due to its low cost, abundance, environmental friendliness, and especially the high electrical conductivity at room temperature of about 2.5×10^2 S/cm [123]. High electrical conductivity is rarely observed in other metal oxides explored for application in LIBs (e.g., α-Fe_2O_3 has an electrical conductivity of $\sim 10^{-4}$ S/cm which is six order of magnitude or $\times 10^{-6}$ lower than electrical conductivity of magnetite). High conductivity is highly desirable for electrodes in LIBs to facilitate charge transfer [124, 125]. Carbon-coated magnetite nanospindles prepared by in situ partial reduction in hematite nanospindles with carbon shells was reported [126]. The specific capacity of Fe_3O_4 spindle/carbon composites remained 530 mAh/g after 80 cycles, while the specific capacities of bare α-Fe_2O_3 spindles and the commercial Fe_3O_4 particles faded to 105 and around 152 mAh/g, respectively. In another study, Yang et al. reported a facile, scalable emulsion polymerization technique for synthesizing Fe_3O_4

nanoparticles around 20 nm embedded in a porous carbon matrix [127]. The as-prepared materials had specific capacities around 600 and 450 mAh/g with little fading after being discharged/charged at current rates of 1C and 0.2C, respectively. In another interesting study, Wang et al. [128] successfully synthesized $Fe_3O_4@C$ core–shell nanorings by a synchronous reduction and carbon deposition process from the reduction in the Fe_2O_3 nanoring precursor with the assistance of acetylene (Fig. 25). The specific capacity of $Fe_3O_4@C$ core–shell nanorings was about 900 mAh/g after 160 cycles at charge rate of 200 mA/g. It was proposed that the shorter pathways for diffusion, nanospace in the structures and conductive carbon shells all contributed to the superior electrochemical performances.

Carbon-decorated Fe_3O_4 also demonstrated interesting electrochemical performance, even in the absence of carbon black additives, PVDF binders [129]. It is particularly interesting to highlight that a new procedure in making electrode using centrifugation-assisted deposition (CAD) was demonstrated, where the active materials of Fe_3O_4/C were directly fabricated on Cu disk which can be used as an electrode without any further process. In contrast, traditional electrode preparation involves multiple steps of slurry preparation, coating, drying in a vacuum oven and additives (e.g., PVDF binder, carbon black conductivity enhancer and NMP solvent) are used. The as-prepared additive-free porous Fe_3O_4/C-based electrode exhibited superior electrochemical performance. Excellent cycling

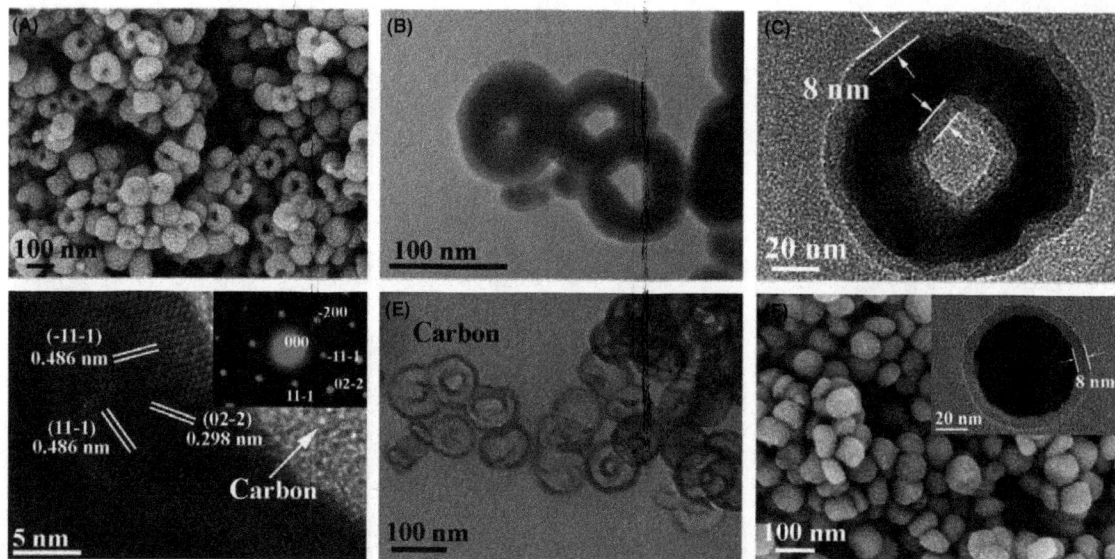

Figure 25. Characterization of $Fe_3O_4@C$ core–shell nanorings: (A and B) SEM and TEM images, (C) high magnification TEM image and (D) HRTEM image and (inset) SAED patterns; (E) TEM image of the carbon shell; (F) SEM and (inset) TEM images of erythrocyte-like $Fe_3O_4@C$ nanoparticle. Reproduced with permission [128].

performance was achieved with a specific capacity at ~800 mAh/g for more than 200 cycles. Impressive rate performance was accomplished when tested under different currents: a specific capacity of 730 mAh/g could be maintained at current of 1500 mA/g (Fig. 26). It is interesting to highlight that new techniques or procedures in the preparation of electrodes and their effects on materials performance are relatively less studied as compared to the large number of publications on materials designs. Future studies should try to define the optimized electrode preparation procedure for individual electrode materials.

Electrolyte

Electrolyte must be carefully chosen to withstand the redox environment at both cathode and anode sides and the voltage range involved without decomposition or

degradation. Additionally, electrolyte should be inert and stable in an acceptable temperature range. In commercial Li-ion batteries, typically a liquid electrolyte is a solution of lithium salts in organic solvents. However, the existing organic liquid electrolyte can potentially catch fires under conditions of thermal runaway or short circuit due to volatile and flammable nature of the solvents which are highly toxic. Ideally, the electrolyte should also be environmentally benign and can be produced at low cost in the future. Polar aprotic organic solvents, such as carbonate solvents with high dialectic constant, are selected to solvate lithium salts at a high concentration (1 M typically). On the other hand, solvents with low viscosity and low melting point are required to meet the requirement for high ionic mobility in the operating temperature range. Various organic solvents have been explored, including dimethyl carbonate, diethyl carbonate, ethyl methyl carbonate, propylene carbonate, ethylene carbonate, diethoxyethane,

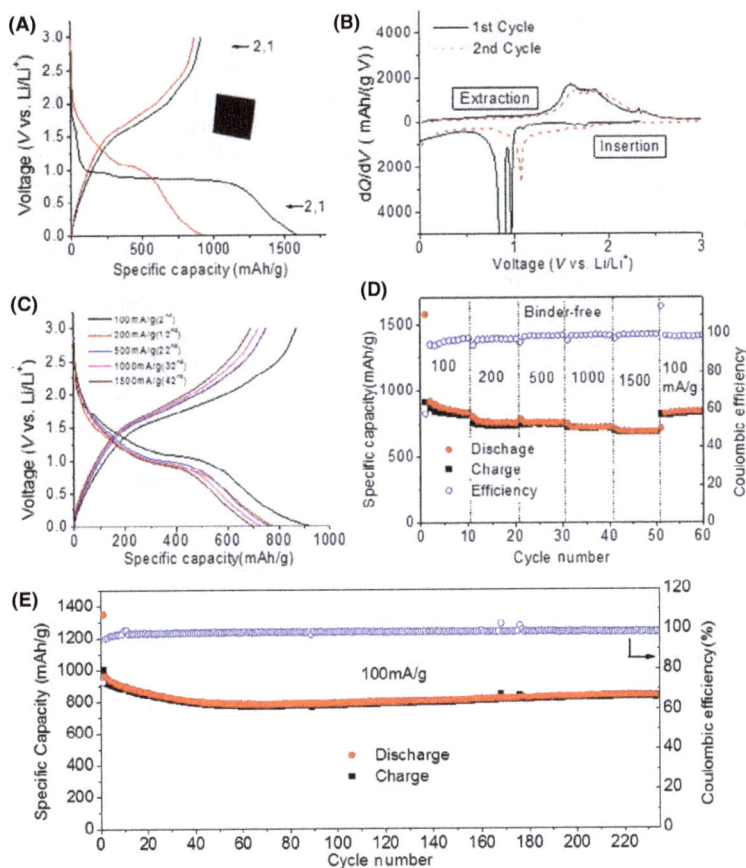

Figure 26. Electrochemical measurement of additive-free carbon decorated Fe_3O_4 fabricated directly on a copper current collector by centrifugation assisted electrode preparation: (A) first two cycle charge–discharge profiles; the inset of (A) shows the optical image of black additive-free Fe_3O_4 on a copper disk as a ready electrode; (B) differential capacity profiles for first two cycles; (C) charge–discharge voltage profiles and (D) rate performances and Coulombic efficiency at different currents from 100 to 1500 mA/g; (E) cycling performance at rate of 100 mA/g. Reproduced with permission [129].

dioxolane, γ-butyrolactone, and tetrahydrofuran. Heteroatom-containing organic solvents have also been explored. Various lithium salts have been explored, including $LiPF_6$, $LiBF_4$, $LiAsF_6$, $LiClO_4$, and $LiCF_3SO_3$. It should be noticed that anions are selected to avoid being oxidized on the charged surface of cathodes, which rules out those simple anions of Cl^-, Br^-, and I^-. $LiPF_6$ is a particular outstanding lithium salt from the perspective of safety, conductivity and the balance between ionic mobility and dissociation constant. However, $LiPF_6$ can react with water to form highly corrosive HF. Therefore, moisture must be minimized in handling of $LiPF_6$ electrolyte. In fact, the success of first commercial Li-ion batteries could be ascribed to the industrial scale availability of high-purity $LiPF_6$ with minimal amount of water.

The solvents are typically formulated and mixed to address the requirements on viscosity, conductivity and stability and to match with the lithium salts selected. For example, high dielectric solvents with high viscosity are typically mixed and balanced with solvents with low viscosity to achieve a liquid state electrolyte within a required temperature window. The commonly used electrolyte is 1M $LiPF_6$ in a 50:50 w/w mixture of ethylene carbonate (EC, melting point of 34°C and boiling point of 260.7°C) and diethyl carbonate (DEC, melting point of 2–4°C and boiling point of 90°C) or dimethyl carbonate (DMC, melting point of 2°C and boiling point of 90°C) (Fig. 27). Generally, the EC can bind Li^+ ions more strongly than DEC or DMC [130]. The formulated electrolyte offer reasonably good stability over a wide potential range. Additives to further enhance the electrolyte stability and improve the formation of good solid-electrolyte interphase (SEI) are also added in many cases. The various formulated additives in terms of chemical compositions and percentages used by different companies in making Li-ion

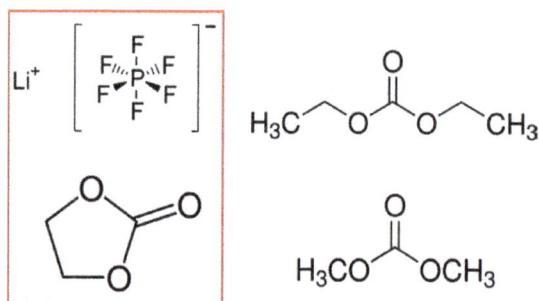

Figure 27. Chemcial structures of common components in commercial electrolyte: $LiPF_6$ as the lithium salt and ethylene carbonate as the solvent are present in almost all commercial electrolyte; whereas diethyl carbonate or dimethyl carbonate are also present to reduce viscosity to promote ion transfer.

batteries could be considered business secrets. Generally, SEI formed in the first cycle is essential to isolate the electrode and electrolyte to avoid further electrolyte decomposition upon cycling. It is generally believed that the SEI is electrically insulating but ionically conductive. Although the chemical composition of SEI is relatively well characterized, the formation mechanism of SEI is still poorly understood and various SEI formation models have been proposed. SEI formation is a hot topic in battery community and the knowledge obtain could help to design better Li-ion batteries with improve abuse tolerance capability. A very informative review by Xu is an excellent source of information regarding liquid electrolyte in Li-ion batteries [131].

Other types of electrolyte have also been developed and proposed for Li-ion batteries, such as polymer, gel and ceramic electrolyte. Polymer electrolytes are solvent-free using high molecular weight-based polymers with dissolved lithium salts [132]. One should be aware that polymer electrolyte is not considered polyelectrolyte. The obvious advantages of polymer electrolyte over liquid electrolyte are (1) improved safety properties due to low volatility, (2) design flexibility, and (3) potential to eliminate separators. Arguably, polymer electrolyte could be more conveniently processed as compared to that of liquid electrolyte. Simplified processes could reduce the cost significantly. Similar to other electrolyte, polymer electrolyte must be stable under the operating conditions of Li-ion batteries from electrochemical, thermal and mechanical perspectives. One of the widely studied polymer is poly(ethylene oxide) which has been coupled with various lithium salts, such as $LiCF_3SO_3$ and $LiClO_4$. The ion conduction in poly(ethylene oxide) mainly occurs at amorphous phases. The ions can be transported by the semirandom motion of short polymer segments. In order to maintain a good mechanical stability, ionic conductivity will be sacrificed. The conductivity is typically about 10^{-8} S/cm, which is significantly less than that of liquid electrolyte. Li-ion batteries based on polymer electrolyte are design flexible and can be fabricated as cylindrical, coin, prismatic, flat cells and other configurations. There are good reviews on polymer electrolyte in literature [133, 134].

Another type of electrolyte is based gels in which both lithium salts and polar solvents are dissolved and added into inactive networks of high-molecular-weight polymers. $LiPF_6$ and carbonate solvents are typically used similar to those in liquid electrolyte discussed above. The liquid phases are fully absorbed within the polymers which can avoid the leakage issue in contrast to that of pristine liquid electrolyte. Meanwhile, the ionic conductivity of gel electrolyte could be dramatically increased as compared to that of polymer electrolyte. There are a number of polymers explored as the hosts, including polyacrylonitrile, polyvinyl chloride, polyvinylidene fluoride and Poly(methyl methacrylate). In the preparation of gel electrolyte, one can simply increase the viscosity of liquid electrolyte by adding soluble polymers. Alternatively, one can soak the microporous polymer matrix into the electrolyte.

Recently, ceramic electrolyte is also re-attracting much attention. Ceramic electrolyte has long being explored for fuel cells, and their application in Li-ion batteries is attracting increasing interest. The obvious advantage to use ceramic electrolyte is safety, for example, no more flammable organic solvents needed. Those batteries with ceramic electrolyte can find applications in high-temperature environment, including handheld orthopedic tools and other batteries powered medical devices that need to be sterilized in autoclaves under high temperature and high pressure conditions. The batteries autoclaved should be able to withstand for at least 130°C and are impermeable up to 30 psi in heated water, and importantly, deterioration in performance should be minimum after sterilization. Another interesting advantage to use ceramic electrolyte for high-temperature applications is that the ionic conductivity of ceramic electrolyte increases with increasing temperature. This is because the creation and movement of ionic point defects, which determines the ionic conductivity, requires energy. One area of intensive research is to achieve ceramic electrolyte with reasonably high conductivity at room temperature. Various sulfides, oxides and phosphates have been explored. The author would anticipate that ceramic electrolyte will eventually be used in next-generation Li-ion batteries in electric vehicles, mainly for its excellent safety performance which can enhance the customer confidence and acceptance. Fergus provided a good review on various ceramic electrolyte explored for Li-ion batteries [135].

Separators

Separators are essential components of Li-ion batteries. In fact, separators are commonly used in most electrochemical systems with liquid electrolyte, including fuel cells, capacitors and various kinds of batteries based on different chemistry. The separator in a Li-ion battery plays the critical roles to avoid direct physical contact between the cathode and anode, and prevents short circuit to occur. At the same time, the separator allows lithium ions in the electrolyte to pass through it. The separators must be chemically stable and inert in contact with both electrolyte and electrodes. At the same time, it is required to be mechanically robust to withstand the tension and puncture by electrode materials and the pore size should be less than 1 μm. Although various separators, including microporous polymer membranes, nonwoven fabric mats and inorganic membranes have been explored, the

microporous polyolefin materials based polymer membranes are dominantly used in commercial Li-ion batteries with liquid electrolyte.

The microporous polymer membranes could be made very thin (typically about ~25 μm) and highly porous (typically 40%) to reduce the resistance and enhance ionic conductivity. At the same time, the polymer membranes could still be mechanically robust. Other parameters that have to be considered in the selection of microporous polymer membranes are low yield or shrinkage, permeability, wettability and cost. Another interesting advantage to use microporous polymer membrane as the separator is that, with properly designed multilayer composites, the separator can shut the battery in the case of short circuit or thermal runaway, functioning similar to a thermal fuse. It is required to have at least two functional parts in the separator: one part that will melt to close the pores and the other part provides mechanical strength to keep isolating the anode and cathode. One typical example is the Celgard® (North Carolina, USA) microporous separator made of both polyethylene (PE) and polypropylene (PP), in the form of trilayer of PP-PE-PP. The melting points of PE and PP are 135 and 165°C, respectively. In the case of over-temperature approaching that of melting point of PE, the porosity of the membrane could be closed by PE, preventing further reactions. So for commercial Li-ion batteries, the shutdown temperature is about 130°C.

For the development of future Li-ion batteries for high-temperature applications, inorganic membranes as separators are highly attractive. The all-solid Li-ion batteries should also be further investigated to meet those niche markets of high-temperature applications. Another parameter that determines the commercial success of a separator is cost. The cost of the existing polymer separator in a Li-ion battery could be as high as one-fifth of the total cost of the battery. Therefore, intensified research on the development of highly improved separators at reasonably low price for Li-ion batteries is required. Zhang wrote a good review article on the separators of liquid electrolyte Li-ion batteries [136].

Challenges in Li-Ion Batteries

Li-ion batteries have been commercialized for about two decades. The technology is considered relatively mature based on the current battery chemistry. Li-ion batteries have been dominantly used in mobile electronic devices, including cell phones and laptop computers, and are starting to play increasing role in electric vehicles. Li-ion batteries will also be considered in sustainable energy grids to store sustainable energy generated from renewable sources. The increasing demand for energy storage requires further improvements in the existing Li-ion batteries and the development of next-generation Li-ion batteries, in particularly, to reduce the cost of Li-ion batteries. It is still colossally challenging to develop new battery chemistry to replace the existing Li-ion battery technology.

In order to increase energy density Li-ion batteries, it is desirable to find electrode couples with both high-specific capacities and high operating cell voltage. As discussed previous, there are a large number of anode candidates that could dramatically increase the specific capacities, in particularly, with highly attractive Si- and Sn-based anodes. It is still challenging to prepare Si nanomaterials on a large scale with low cost. Sn-based anodes suffer from the issue of poor cycling performance due to pulverization. Therefore, one of the possible future anodes could be Si-Sn-based composites [137]. In contrast to that of anode candidates, the cell capacity is mainly limited by the low capacity of cathode candidates. The existing cathode material of $LiCoO_2$ is expensive and highly toxic. The increasingly popular $LiFePO_4$ has a low capacity. The facilely prepared Ni-Co-Mn-based cathodes developed by Argonne National Laboratory are highly attractive, especially from industrial prospective. However, the specific capacity is still considered moderate, and both Co and Ni are expensive and toxic. Future cathode materials should try to avoid the use of either Co or Ni, or other toxic elements, from environmental perspectives. Additionally, the ideal cathode should be able to reversibly insert/extract multiple electrons per 3d metal. The future low-cost cathode materials could be Mn- and/or Fe-based. The issue of intrinsically low conductivity should be creatively addressed, most likely by nanotechnology and nanocomposites. There is relatively no much room to increase the operating cell voltages with the current known cathode candidates under exploration. Composite cathodes with two or three 3d metals and polyanions are highly promising. New cathode chemistry may be developed in the next 10 years. The new cathode chemistry may require for new electrolytes as well.

The safety concern is another challenge that needs to be properly addressed. The recent news on fires of Li-ion batteries, involving the Boeing 787 passenger aircrafts, Tesla Model S cars, highlights the importance of battery safety. To ensure the wide acceptance of electric vehicles and expanded the market of Li-ion battery powered vehicles, automakers should invest significantly on the battery management systems to enhance safety of the huge battery packs in vehicles. Alternatively, nonflammable Li-ion batteries should be developed, including those Li-ion batteries based on aqueous electrolyte or ceramic electrolyte, and all-solid-state batteries. Next-generation Li-ion batteries, most likely, will be using high voltage (5 V) cathodes and high capacity anodes (such as Si- or Sn-based). Therefore, intensive research is required to gain better understanding

about those electrode materials in terms of stability and interaction with electrolyte. Instead of intensively pursuing of high-energy density, there should be increasing emphasis on battery safety as well. Standardized battery safety testing procedures should be widely employed.

It is still challenging to develop electrode materials with low carbon footprint, or the so-called "green batteries". Ideally, future Li-ion batteries should use biologically derived organic or inorganic electrodes, using aqueous electrolyte. Carbon and silicon can be derived from biomasses. The recent attempts to explore virus-assisted synthesis of electrode materials for Li-ion batteries attracted much enthusiasm [138]. It will be interesting to explore large-scale synthesis at room temperature using biological templates, including genetically modified virus. Organic electrodes that will not be easily dissolved by electrolyte can be further developed for sustainable Li-ion batteries [139]. Therefore, one area of future research could be focusing on "sustainable" and "green" Li-ion batteries.

Additionally, future batteries research should keep in mind the life cycle assessment (LCA) of the technologies to evaluate whether the batteries are truly green or not. The existing materials involved in commercial Li-ion batteries (typically, $LiCoO_2$, $LiMn_2O_4$, $LiNiO_2$, $LiFePO_4$ as cathodes; Graphite, $Li_4Ti_5O_{12}$ as anodes; Ethylene carbonate, diethyl carbonate, $LiPF_6$, $LiBF_4$, $LiClO_4$ as electrolytes/salts; Polypropylene, polyethylene as separator) still require further studies from the perspectives of LCA. Little is known about the environmental impacts of the production, use and disposal of next-generation Li-ion batteries [140]. Mass production of Li-ion batteries for electric vehicles will result in large volumes of contaminated waste and places. It could reduce agricultural productivity near mine sites, decrease air quality near processing facilities, and increase energy cost near factories. Furthermore, additional fossil fuel use may be required to meet factory demand in production of Li-ion batteries. LCA may provide the full scenarios of their environmental aspects and potential environmental impacts of throughout their life cycle from raw minerals acquisition through production, use, end-of-life treatment, recycling, and final disposal. Therefore, LCA is an important tool for the development of truly "green" Li-ion batteries in the future, which can contribute to future sustainability. Another important area that should attract deserved attention is techno-economic analysis (TEA) of Li-ion batteries. TEA will help to determine whether a new Li-ion battery system developed is technically and economically feasible for large-scale production. The combination of LCA and TEA studies will help to develop next-generation Li-ion batteries achieving optimized social, environmental and economic impacts.

There are increasing worldwide interests in preparation of "nano" materials to increase the storage capacity for next-generation Li-ion batteries. The development of nanomaterials could offer tremendous opportunities, as it is known that electrochemical properties of nanomaterials are size and shape dependent. "Nano" has been attracting much attention in the battery community in the past two decades. However, the challenging issue rarely mentioned is that the energy density (by volume) of "nano" materials is very low due to the low volumetric density of nanomaterials. This critical issue of low tap density is most vividly illustrated by the following example: the tap density of intensely pursued graphene is 0.03 g/cc, as compared to that of MCMB graphite power at 1.3 g/cc, or about 40 time difference by volume for a given weight. Other issues associated with electrode materials at nanoscale are poor electrical properties of the electrode due to interparticle resistance and low Coulombic efficiency attributed to large surface area induced side reactions between the electrode and electrolyte. The author would like to suggest the use of microstructures of nanoparticles as the building units, instead of random "nano" particles, to simultaneously overcome the issues discussed above, in particular, achieving electrode materials with high packing density, while preserving certain beneficial features of nanomaterials. We have been working on the preparation of micromaterials of nanoparticle aggregates to increase volumetric density for a while [27, 91, 120, 121, 129, 141]. For example, the microscale cubes of Zn_2SnO_4 & Sn@C of nanoparticle aggregates have a tap density of 0.98 g/cc which is nearly eight times higher than that of commercial P25 TiO_2 nanoparticles [91]. Additionally, it seems that the electrochemical performances of micromaterials are comparable to that of nanomaterials. In many cases, the devices powered by Li-ion batteries have restricted space to hold the batteries, such as in cell phone and electric vehicles. In other words, electrode materials with high tap density are required in the future. Therefore, the author would dare to anticipate that further Li-ion batteries will be based on micromaterials with features of nanomaterials, from the perspective of volumetric density. Future battery research should enable the preparation of next-generation batteries with smaller size but better performances as compared to the existing ones (Fig. 28).

In addition to specific technical challenges discussed above, the other less disclosed challenge is the limitation of lithium reserves to meet future 100 million 40 kWh Li-based batteries for electric vehicles annually. The limited or inaccessible lithium resources could make it very challenging to reduce the cost of Li-ion batteries in the future. Of course, Li-ion batteries should be recycled as much as we can. We still have to intensify the research on

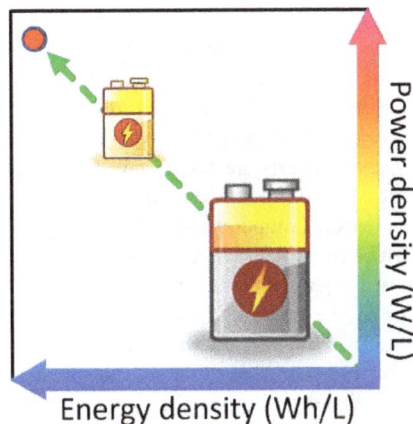

Figure 28. The illustration to demonstration that future Li-ion batteries should be light and small without any compromise on energy and power.

alternative Li-ion batteries or beyond Li-ion batteries, including Na-ion, Mg-ion, Ca-ion, Al-ion, and F-ion batteries. Learning for the experience on the development of Li-ion batteries, with access to increasing powerful computational tools, future development of batteries beyond Li-ion batteries must leverage the simulation and modeling power of modern computers to guide and accelerate electrode discovery and synthesis. Therefore, the time required for the development of those advanced batteries beyond Li-ion batteries will be significantly shorter than that used in the development of Li-ion batteries.

Beyond Li-Ion Batteries

In line with the aim to develop batteries with even higher energy, Li-air ($Li-O_2$) and Li-sulfur (Li-S), achieving impressively high-energy density theoretically, have again attracted much attention recently. Although Zn-air and high-temperature Na-sulfur have been commercialized for a while, the development of rechargeable $Li-O_2$ and Li-S batteries still have technical barriers. On the anode side, the critical understanding of and control over Li-electrolyte interface must be thoroughly studied. In fact, Exxon had tried to commercialize the Li metal-based batteries more than 40 years ago without a success, which was partially attributed to the poor Li-electrolyte interface and dendrites. Then Li-ion batteries were developed later to avoid the Li-electrolyte interface by completely taking out the lithium metal. We have to revisit the issue of Li-electrolyte interface in the development of $Li-O_2$ or -S batteries now. On the cathode side, the designing oxygen electrodes, which has been "a nightmare for fuel cells" [142], for Li-air batteries is colossally difficult. The good news is that the knowledge accumulated over the past few decades on electrodes for fuel cells can be used as a good source of references for development of Li-O$_2$ batteries. Another notable issue is the high overpotential and low electrical efficiency observed, which make Li-O$_2$ or –S less attractive from the perspective of energy efficiency and sustainability. Currently, many efforts are focused on the development of efficient catalysts for the cathodes, but mainly based on trial and error. Computer assisted modeling and simulation must be employed to predict and design the catalysts with exposed highly active facets, similar to those have been done in the field of fuel cells, for those catalysts-based cathodes. However, the solid phases formed in the Li-O$_2$/-S batteries are different from that of fuel cells at either gas or liquid phases. In other words, the challenges faced in the development of Li-O$_2$ batteries are even more difficult to overcome than that in fuel cells. Another issue is associated with the volatile electrolyte, which can be evaporated into gas phases leading to death of the battery. For Li-S batteries, the formation of soluble polysulfide (LiS$_x$) species is another critical problem, in addition to its poor conductivity. Confining sulfur into voids of mesoporous nanostructures is a popular strategy to overcome the solubility of LiS$_x$. It is believed that rechargeable Li-O$_2$ or Li-S batteries will eventually be commercialized in the future, which will require intensified studies. The good news is that increasing number of research groups and battery companies are starting to investigate Li-O$_2$ or Li-S batteries, which should help to accelerate their development. However, due to the problems with Li-O$_2$ or Li-S batteries, Li-ion batteries will not be replaced by them in the next decade.

Sodium-ion batteries (SIBs) are outstanding candidates that could potentially replace Li-ion batteries. With respect to large-scale stationary energy storage systems for energy grids in sustainable energy networks of wind and solar energy, low-cost SIBs are expected to be produced at lower cost than that of Li-ion batteries in the future [143–146]. Mass produced SIBs with comparable performances with those existing rechargeable batteries can significantly impact our environment and society [147]. Additionally, due to the lower standard half-reaction potential of sodium (2.714 V for Na/Na$^+$) as compared to that of lithium (3.045 V for Li/Li$^+$), electrolyte degradation will be reduced and safe electrolyte with low decomposition potential can be used in SIBs to enhance safety [148]. In other words, not only low cost, SIBs will be good candidates as safe rechargeable batteries. The performances of the highly attractive reemerging SIBs will be strongly dependent on electrode materials developed [149–152]. A number of positive electrode materials have been explored [147]. For example, cathode materials: carbon-coated NaVPO$_4$F could deliver a reversible capacity of 98 mAh/g; [153] Na$_{0.44}$MnO$_2$ particles delivered

65 mAh/g; [154] Na$_x$Co[F\e(CN)$_6$]$_{0.90}$2.9H$_2$O delivered 135 mAh/g; [155] layered Na$_{0.71}$CoO$_2$ delivered 80 mAh/g; [156] KFe$_{(II)}$Fe$_{(III)}$(CN)$_6$ delivered 100 mAh/g; [157] and Na$_2$FePO$_4$F exhibited high voltage of 3.5 V [144]. In terms of anode materials, the widely used graphite in Li-ion batteries, however, cannot reversibly store significant amount of Na, typically <5 mAh/g due to lack of Na intercalation [147, 151, 158, 159]. On the other hand, disordered nongraphitic carbons can reversibly store more Na as compared to graphite [147, 151]. Carbon nanowires prepared by pyrolyzation of polyaniline demonstrated impressive performance in Na storage [160]. Nanocellular carbon foams delivered 152 mAh/g [161]. Furthermore, metals, metal oxides, and phosphorus have been investigated as anode as well [162–171]. Interestingly, for both Na- and Li-active materials of Sn@C nanospheres, it would be broken in SIBs but well preserved in Li-ion batteries tests [172]. Therefore, in the selection of materials for SIBs, the knowledge learned from Li-ion batteries may not always be applicable.

There are many other promising battery technologies under development. The new battery technologies may not be necessary to directly compete with Li-ion batteries in market share due to their unique characteristics in terms of performance, cost and size. Redox flow batteries using solutions of redox complexes or ions, or so-called catholyte and anolyte, are good options for stationary applications, such as in energy grids based on sustainable sources. Mg-ion batteries have the advantages of no dendrite formation and therefore enhanced safety, and high-energy density comparable to Li-ion batteries but at significantly lower cost. Therefore, Mg-ion batteries are highly attractive for electric vehicles. Toyota is leading the industrial efforts to commercialize Mg-ion batteries which are expected to power electric vehicles by 2020. Al-ion batteries, with three-electron transfer, could achieve three times higher energy density than that of Li-ion batteries. However, the use of expensive ionic liquids will limit Al-ion batteries to mainly special applications. Rechargeable Zn-air batteries may be more promising as compared to Li-air batteries. Primary Zn-air batteries were invented more than 100 years ago. The recent efforts are mainly focusing on the development of rechargeable Zn-air batteries. Rechargeable Zn-air batteries can find applications in electric vehicles.

Summary

In summary, Li-ion batteries have affected almost everyone in the world. The success of commercial Li-ion batteries was a result of intensive research and contribution by many great scientists over few decades. Recently, much effort was put into further improvement on the performances of Li-ion batteries, achieving certain level of success. There are still notable issues. Intensified research is required to achieve next-generation Li-ion batteries. Private companies are investing significantly into the research and development of Li-ion batteries which could lead to incrementally advanced products achieving significant direct impacts to our society. Academic sections could make contributions by generating out-of-the-box ideas and concepts, especially those beyond Li-ion batteries that could be commercialized in the next few decades. Battery safety and sustainable batteries should receive their deserved emphasis and attention in the future. New battery concepts have to be further developed to go beyond Li-ion batteries in the future. LCA and TEA are good tools for evaluation next-generation Li-ion batteries and beyond Li-ion batteries.

Acknowledgment

The author appreciates Professor Simon Ng for comments on the manuscript and Wayne State University for support.

Conflict of Interest

None declared.

References

1. Deng, D., M. G. Kim, J. Y. Lee, and J. Cho. 2009. Green energy storage materials: Nanostructured TiO_2 and Sn-based anodes for lithium-ion batteries. Energ. Environ. Sci. 2:818–837.
2. Levine, S.. 2010. The Great Battery Race. Foreign Policy 182:88–95.
3. Yoshino, A. 2012. The Birth of the Lithium-Ion Battery. Angew. Chem. Int. Edit. 51:5798–5800.
4. New York Times 2013, 162, B5–B5.
5. Joan Lowy 2013. NTSB: Boeing 787 battery shows short-circuiting. The Associated Press.
6. Tarascon, J. M., and M. Armand. 2001. Issues and challenges facing rechargeable lithium batteries. Nature 414:359–367.
7. Whittingham, M. S. 1976. Electrical Energy Storage and Intercalation Chemistry. Science 192:1126–1127.
8. Besenhard, J. O., and G. Eichinger. 1976. High energy density lithium cells: Part I. Electrolytes and anodes. J. Electroanal. Chem. Interfacial Electrochem. 68:1–18.
9. Eichinger, G., and J. O. Besenhard. 1976. High energy density lithium cells: Part II. Cathodes and complete cells. J. Electroanal. Chem. Interfacial Electrochem. 72:1–31.
10. Mizushima, K., P. C. Jones, P. J. Wiseman, and J. B. Goodenough. 1981. LixCoO$_2$ (0<x≤1): A new cathode material for batteries of high energy density. Solid State Ionics 3–4:171–174.
11. Thackeray, M. M., W. I. F. David, P. G. Bruce, and J. B. Goodenough. 1983. Lithium insertion into manganese spinels. Mater. Res. Bull. 18:461–472.
12. Yazami, R., and P. Touzain. 1983. A reversible graphite-lithium negative electrode for electrochemical generators. J. Power Sources 9:365–371.
13. Basu, S., C. Zeller, P. J. Flanders, C. D. Fuerst, W. D. Johnson, and J. E. Fischer. 1979. Synthesis and properties of lithium-graphite intercalation compounds. Mat. Sci. Eng. 38:275–283.
14. Yoshino, A., K. Sanechika, and T. Nakajima. 1987. USP4,668,595.
15. Chan, C. K., H. Peng, G. Liu, K. McIlwrath, X. F. Zhang, R. A. Huggins, et al. 2008. High-performance lithium battery anodes using silicon nanowires. Nat. Nanotechnol. 3:31–35.
16. Bates, J. B., N. J. Dudney, B. Neudecker, A. Ueda, and C. D. Evans. 2000. Thin-film lithium and lithium-ion batteries. Solid State Ionics 135:33–45.
17. Li, H., X. J. Huang, L. Q. Chen, Z. G. Wu, and Y. Liang. 1999. A high capacity nano-Si composite anode material for lithium rechargeable batteries. Electrochem. Solid State Lett. 2:547–549.
18. Xing, W. B., A. M. Wilson, G. Zank, and J. R. Dahn. 1997. Pyrolysed pitch-polysilane blends for use as anode materials in lithium ion batteries. Solid State Ionics 93:239–244.
19. Xing, W. B., A. M. Wilson, K. Eguchi, G. Zank, and J. R. Dahn. 1997. Pyrolyzed polysiloxanes for use as anode materials in lithium-ion batteries. J. Electrochem. Soc. 144:2410–2416.
20. Wilson, A. M., G. Zank, K. Eguchi, W. Xing, and J. R. Dahn. 1997. Pyrolysed silicon-containing polymers as high capacity anodes for lithium-ion batteries. J. Power Sources 68:195–200.
21. Xue, J. S., K. Myrtle, and J. R. Dahn. 1995. An epoxy-silane approach to prepare anode materials for rechargeable lithium ion batteries. J. Electrochem. Soc. 142:2927–2935.
22. Deng, D., and J. Y. Lee. 2010. Direct fabrication of double-rough chestnut-like multifunctional Sn@C composites on copper foil: lotus effect and lithium ion storage properties. J. Mater. Chem. 20:8045–8049.
23. Deng, D., and J. Y. Lee. 2009. Reversible Storage of Lithium in a Rambutan-Like Tin-Carbon Electrode. Angew. Chem. Int. Ed. 48:1660–1663.
24. Lou, X. W., D. Deng, J. Y. Lee, and L. A. Archer. 2008. Preparation of SnO$_2$/Carbon Composite Hollow Spheres and Their Lithium Storage Properties. Chem. Mater. 20:6562–6566.
25. Deng, D., and J. Y. Lee. 2008. Hollow core-shell mesospheres of crystalline SnO$_2$ nanoparticle aggregates

for high capacity Li$^+$ ion storage. Chem. Mater. 20:1841–1846.

26. Zhu, J., K. S. Ng, and D. Deng. 2014. Hollow Cocoon-Like Hematite Mesoparticles of Nanoparticle Aggregates: Structural Evolution and Superior Performances in Lithium Ion Batteries. ACS Appl. Mat. Interfac. 6:2996–3001.

27. Deng, D., and J. Y. Lee. 2014. Meso-oblate Spheroids of Thermal-Stabile Linker-Free Aggregates with Size-Tunable Subunits for Reversible Lithium Storage. ACS Appl. Mat. Interfac. 6:1173–1179.

28. Kang, B., and G. Ceder. 2009. Battery materials for ultrafast charging and discharging. Nature 458:190–193.

29. Arico, A. S., P. Bruce, B. Scrosati, J. M. Tarascon, and W. Van Schalkwijk. 2005. Nanostructured materials for advanced energy conversion and storage devices. Nat. Mater. 4:366–377.

30. Winter, M., J. O. Besenhard, M. E. Spahr, and P. Novak. 1998. Insertion electrode materials for rechargeable lithium batteries. Adv. Mater. 10:725–763.

31. Poizot, P., S. Laruelle, S. Grugeon, L. Dupont, and J. M. Tarascon. 2000. Nano-sized transition-metaloxides as negative-electrode materials for lithium-ion batteries. Nature 407:496–499.

32. Winter, M., and R. J. Brodd. 2004. What are batteries, fuel cells, and supercapacitors? Chem. Rev. 104:4245–4269.

33. Nam, K. T., D. W. Kim, P. J. Yoo, C. Y. Chiang, N. Meethong, P. T. Hammond, et al. 2006. Virus-enabled synthesis and assembly of nanowires for lithium ion battery electrodes. Science 312:885–888.

34. Yi, T. F., X. Y. Li, H. P. Liu, J. Shu, Y. R. Zhu, and R. S. Zhu. 2012. Recent developments in the doping and surface modification of LiFePO$_4$ as cathode material for power lithium ion battery. Ionics 18:529–539.

35. Xu, B., D. N. Qian, Z. Y. Wang, and Y. S. L. Meng. 2012. Recent progress in cathode materials research for advanced lithium ion batteries. Mat. Sci. Eng. R 73:51–65.

36. Whittingham, M. S. 2012. History, Evolution, and Future Status of Energy Storage. Proc. IEEE 100:1518–1534.

37. Wen, J. W., Y. Yu, and C. H. Chen. 2012. A Review on Lithium-Ion Batteries Safety Issues: Existing Problems and Possible Solutions. Mater Exp. 2:197–212.

38. Qu, C. Q., Y. J. Wei, and T. Jiang. 2012. Research Progress in Li$_3$V$_2$(PO$_4$)(3) as Polyanion-type Cathode Materials for Lithium-ion Batteries. J. Inorg. Mater 27:561–567.

39. Oudenhoven, J. F. M., R. J. M. Vullers, and R. Schaijk. 2012. A review of the present situation and future developments of micro-batteries for wireless autonomous sensor systems. Int. J. Energ. Res. 36:1139–1150.

40. Du, N., H. Zhang, and D. R. Yang. 2012. One-dimensional hybrid nanostructures: synthesis via layer-by-layer assembly and applications. Nanoscale 4:5517–5526.

41. de las Casas, C., and W. Z. Li. 2012. A review of application of carbon nanotubes for lithium ion battery anode material. J. Power Sources 208:74–85.

42. Chu, D. B., J. Li, X. M. Yuan, Z. L. Li, X. Wei, and Y. Wan. 2012. Tin-Based Alloy Anode Materials for Lithium Ion Batteries. Prog. Chem. 24:1466–1476.

43. Zhang, W. J. 2011. Structure and performance of LiFePO$_4$ cathode materials: A review. J. Power Sources 196:2962–2970.

44. Marom, R., S. F. Amalraj, N. Leifer, D. Jacob, and D. Aurbach. 2011. A review of advanced and practical lithium battery materials. J. Mater. Chem. 21:9938–9954.

45. Yi, T. F., Y. R. Zhu, X. D. Zhu, J. Shu, C. B. Yue, and A. N. Zhou. 2009. A review of recent developments in the surface modification of LiMn$_2$O$_4$ as cathode material of power lithium-ion battery. Ionics 15:779–784.

46. Yi, T. F., C. B. Yue, Y. R. Zhu, R. S. Zhu, and X. G. Hu. 2009. A Review of Research on Cathode Materials for Power Lithium Ion Batteries. Rare Metal. Mat. Eng. 38:1687–1692.

47. Chung, S. Y., J. T. Bloking, and Y. M. Chiang. 2002. Electronically conductive phospho-olivines as lithium storage electrodes. Nat. Mater. 1:123–128.

48. Malik, R., D. Burch, M. Bazant, and G. Ceder. 2010. Particle Size Dependence of the Ionic Diffusivity. Nano Lett. 10:4123–4127.

49. Meethong, N., Y. H. Kao, S. A. Speakman, and Y. M. Chiang. 2009. Aliovalent Substitutions in Olivine Lithium Iron Phosphate and Impact on Structure and Properties. Adv. Funct. Mater. 19:1060–1070.

50. Ma, J., B. H. Li, H. D. Du, C. J. Xu, and F. Y. Kang. 2011. Effects of tin doping on physicochemical and electrochemical performances of LiFe$_{1-x}$Sn$_x$PO$_4$/C $(0 <= x <= 0.07)$ composite cathode materials. Electrochim. Acta 56:7385–7391.

51. Wang, D. Y., H. Li, S. Q. Shi, X. J. Huang, and L. Q. Chen. 2005. Improving the rate performance of LiFePO$_4$ by Fe-site doping. Electrochim. Acta 50:2955–2958.

52. Zaghib, K., J. B. Goodenough, A. Mauger, and C. Julien. 2009. Unsupported claims of ultrafast charging of LiFePO$_4$ Li-ion batteries. J. Power Sources 194:1021–1023.

53. Wang, J. J., and X. L. Sun. 2012. Understanding and recent development of carbon coating on LiFePO$_4$

cathode materials for lithium-ion batteries. Energ. Environ. Sci. 5:5163–5185.

54. Wu, X.-L., L.-Y. Jiang, F.-F. Cao, Y.-G. Guo, and L.-J. Wan. 2009. LiFePO$_4$ Nanoparticles Embedded in a Nanoporous Carbon Matrix: Superior Cathode Material for Electrochemical Energy-Storage Devices. Adv. Mater. 21:2710–2714.

55. Belharouak, I., C. Johnson, and K. Amine. 2005. Synthesis and electrochemical analysis of vapor-deposited carbon-coated LiFePO$_4$. Electrochem. Commun. 7:983–988.

56. Lu, Z. G., H. Cheng, M. F. Lo, and C. Y. Chung. 2007. Pulsed laser deposition and electrochemical characterization of LiFePO$_4$-Ag composite thin films. Adv. Funct. Mater. 17:3885–3896.

57. Gummow, R. J., N. Sharma, V. K. Peterson, and Y. He. 2012. Crystal chemistry of the Pmnb polymorph of Li$_2$MnSiO$_4$. J. Solid State Chem. 188:32–37.

58. Gummow, R. J., and Y. He. 2014. Recent progress in the development of Li$_2$MnSiO$_4$ cathode materials. J. Power Sources 253:315–331.

59. Sun, Y.-K., S.-T. Myung, B.-C. Park, J. Prakash, I. Belharouak, and K. Amine. 2009. High-energy cathode material for long-life and safe lithium batteries. Nat. Mater. 8:320–324.

60. Megahed, S., and B. Scrosati. 1994. Lithium-ion rechargeable batteries. J. Power Sources 51:79–104.

61. Deng, D., and J. Y. Lee. 2013. Meso-oblate Spheroids of Thermal-Stabile Linker-Free Aggregates with Size-Tunable Subunits for Reversible Lithium Storage. ACS Appl. Mat. Interfac. 6:1173–1179.

62. Idota, Y., T. Kubota, A. Matsufuji, Y. Maekawa, and T. Miyasaka. 1997. Tin-based amorphous oxide: A high-capacity lithium-ion-storage material. Science 276:1395–1397.

63. Inoue, H. 2006. International Meeting on Lithium Batteries abstr #228, Biarritz, France.

64. Dimov, N., Y. Xia, and M. Yoshio. 2007. Practical silicon-based composite anodes for lithium-ion batteries: Fundamental and technological features. J. Power Sources 171:886–893.

65. Winter, M., and J. O. Besenhard. 1999. Electrochemical lithiation of tin and tin-based intermetallics and composites. Electrochim. Acta 45:31–50.

66. Bonino, F., S. Brutti, P. Reale, B. Scrosati, L. Gherghel, J. Wu, et al. 2005. A disordered carbon as a novel anode material in lithium-ion cells. Adv. Mater. 17:743–746.

67. Zheng, T., J. S. Xue, and J. R. Dahn. 1996. Lithium insertion in hydrogen-containing carbonaceous materials. Chem. Mater. 8:389–393.

68. Ebert, L. B. 1996. The interrelationship of hydrogen-containing carbon and lithium. Carbon 34:671–672.

69. Dahn, J. R., T. Zheng, Y. H. Liu, and J. S. Xue. 1995. Mechanisms for lithium insertion in carbonaceous materials. Science 270:590–593.

70. Baughman, R. H., A. A. Zakhidov, and W. A. de Heer. 2002. Carbon nanotubes-the route toward applications. Science 297:787–792.

71. Che, G. L., B. B. Lakshmi, E. R. Fisher, and C. R. Martin. 1998. Carbon nanotubule membranes for electrochemical energy storage and production. Nature 393:346–349.

72. Wang, Y., H. C. Zeng, and J. Y. Lee. 2006. Highly reversible lithium storage in porous SnO$_2$ nanotubes with coaxially grown carbon nanotube overlayers. Adv. Mater. 18:645–649.

73. Kumar, T. P., R. Ramesh, Y. Y. Lin, and G. T. K. Fey. 2004. Tin-filled carbon nanotubes as insertion anode materials for lithium-ion batteries. Electrochem. Commun. 6:520–525.

74. Scarselli, M., P. Castrucci, and M. D. Crescenzi. 2012. Electronic and optoelectronic nano-devices based on carbon nanotubes. J. Phys. 24:313202.

75. Charette, K., J. Zhu, S. O. Salley, K. Y. S. Ng, and D. Deng. 2014. Gram-scale synthesis of high-temperature (900°C) stable anatase TiO$_2$ nanostructures assembled by tunable building subunits for safer lithium ion batteries. RSC Adv. 4:2557–2562.

76. Liu, F., S. Song, D. Xue, and H. Zhang. 2012. Selective crystallization with preferred lithium-ion storage capability of inorganic materials. Nanoscale Res. Lett. 7:149.

77. Bogart, T. D., D. Oka, X. Lu, M. Gu, C. Wang, and B. A. Korgel. 2013. Lithium Ion Battery Peformance of Silicon Nanowires with Carbon Skin. ACS Nano 8:915–922.

78. Meng, X., and D. Deng. 2015. Core–Shell Ti@Si Coaxial Nanorod Arrays Formed Directly on Current Collectors for Lithium-Ion Batteries. ACS Appl. Mat. Interfac. 7:6867–6874.

79. Obrovac, M. N., L. Christensen, D. B. Le, and J. R. Dahn. 2007. Alloy Design for Lithium-Ion Battery Anodes. J. Electrochem. Soc. 154:A849–A855.

80. Lou, X. W., C. L. Yuan, and L. A. Archer. 2007. Double-walled SnO$_2$ nano-cocoons with movable magnetic cores. Adv. Mater. 19:3328–3332.

81. Wang, Y., F. B. Su, J. Y. Lee, and X. S. Zhao. 2006. Crystalline carbon hollow spheres, crystalline carbon-SnO$_2$ hollow spheres, and crystalline SnO$_2$ hollow spheres: Synthesis and performance in reversible Li-ion storage. Chem. Mater. 18:1347–1353.

82. Lou, X. W., Y. Wang, C. L. Yuan, J. Y. Lee, and L. A. Archer. 2006. Template-free synthesis of SnO$_2$ hollow nanostructures with high lithium storage capacity. Adv. Mater. 18:2325–2329.

83. Wang, Y., J. Y. Lee, and H. C. Zeng. 2005. Polycrystalline SnO$_2$ nanotubes prepared via infiltration

casting of nanocrystallites and their electrochemical application. Chem. Mater. 17:3899–3903.

84. Wang, Y., J. Y. Lee, and T. C. Deivaraj. 2004. Controlled synthesis of V-shaped SnO_2 nanorods. J. Phys. Chem. B 108:13589–13593.

85. Yu, Y., C. H. Chen, and Y. Shi. 2007. A tin-based amorphous oxide composite with a porous, spherical, multideck-cage morphology as a highly reversible anode material for lithium-ion batteries. Adv. Mater. 19:993–997.

86. Kwon, Y., and J. Cho. 2008. High capacity carbon-coated $Si_{70}Sn_{30}$ nanoalloys for lithium battery anode material. Chem. Commun. 9:1109–1111.

87. Kim, J. Y., and J. Cho. 2006. SnO_2 filled mesoporous tin phosphate - High capacity negative electrode for lithium secondary battery. Electrochem. Solid-State Lett. 9:A373–A375.

88. Kim, E., D. Son, T. G. Kim, J. Cho, B. Park, K. S. Ryu, et al. 2004. A mesoporous/crystalline composite material containing tin phosphate for use as the anode in lithium-ion batteries. Angew. Chem. Int. Ed. 43:5987–5990.

89. Kim, Y., H. Hwang, C. S. Yoon, M. G. Kim, and J. Cho. 2007. Reversible lithium intercalation in teardrop-shaped ultrafine SnP0.94 particles. Adv. Mater. 19:92–96.

90. Kim, T. J., D. Son, J. Cho, B. Park, and H. Yang. 2004. Enhanced electrochemical properties of SnO_2 anode by $AlPO_4$ coating. Electrochim. Acta 49:4405–4410.

91. Addu, S. K., J. Zhu, K. Y. S. Ng, and D. Deng. 2014. A Family of Mesocubes. Chem. Mater. 26:4472–4485.

92. Bronstein, L. M., X. Huang, J. Retrum, A. Schmucker, M. Pink, B. D. Stein, et al. 2007. Influence of Iron Oleate Complex Structure on Iron Oxide Nanoparticle Formation. Chem. Mater. 19:3624–3632.

93. Xiong, Y., Z. Li, X. Li, B. Hu, and Y. Xie. 2004. Thermally stable hematite hollow nanowires. Inorg. Chem. 43:6540–6542.

94. Wen, X., S. Wang, Y. Ding, Z. L. Wang, and S. Yang. 2005. Controlled growth of large-area, uniform, vertically aligned arrays of α-Fe_2O_3 nanobelts and nanowires. J. Phys. Chem. B 109:215–220.

95. Ling, Y., G. Wang, D. A. Wheeler, J. Z. Zhang, and Y. Li. 2011. Sn-Doped Hematite Nanostructures for Photoelectrochemical Water Splitting. Nano Lett. 11:2119–2125.

96. Yuan, L., Q. Jiang, J. Wang, and G. Zhou. 2012. The growth of hematite nanobelts and nanowires—tune the shape via oxygen gas pressure. J. Mater. Res. 27:1014–1021.

97. Zhou, W., K. Tang, S. Zeng, and Y. Qi. 2008. Room temperature synthesis of rod-like $FeC_2O_4 \cdot 2H_2O$ and its transition to maghemite, magnetite and hematite

nanorods through controlled thermal decomposition. Nanotechnology 19:065602.

98. Ganguli, A. K., and T. Ahmad. 2007. Nanorods of Iron Oxalate Synthesized Using Reverse Micelles: Facile Route for-Fe_2O_3 and Fe_3O_4 Nanoparticles. J. Nanosci. Nanotechnol. 7:2029–2035.

99. Cho, W., S. Park, and M. Oh. 2011. Coordination polymer nanorods of Fe-MIL-88B and their utilization for selective preparation of hematite and magnetite nanorods. Chem. Commun. 47:4138–4140.

100. Chen, J., L. Xu, W. Li, and X. Gou. 2005. α-Fe_2O_3 Nanotubes in Gas Sensor and Lithium-Ion Battery Applications. Adv. Mater. 17:582–586.

101. Kang, N., J. H. Park, J. Choi, J. Jin, J. Chun, I. G. Jung, et al. 2012. Nanoparticulate Iron Oxide Tubes from Microporous Organic Nanotubes as Stable Anode Materials for Lithium Ion Batteries. Angew. Chem. Int. Ed. 51:6626–6630.

102. Reddy, M., T. Yu, C. H. Sow, Z. X. Shen, C. T. Lim, G. Subba Rao, et al. 2007. α-Fe_2O_3 Nanoflakes as an Anode Material for Li-Ion Batteries. Adv. Funct. Mater. 17:2792–2799.

103. Chen, J. S., T. Zhu, X. H. Yang, H. G. Yang, and X. W. Lou. 2010. Top-Down Fabrication of α-Fe_2O_3 Single-Crystal Nanodiscs and Microparticles with Tunable Porosity for Largely Improved Lithium Storage Properties. J. Am. Chem. Soc. 132:13162–13164.

104. Hu, X., J. C. Yu, J. Gong, Q. Li, and G. Li. 2007. α-Fe_2O_3 Nanorings Prepared by a Microwave-Assisted Hydrothermal Process and Their Sensing Properties. Adv. Mater. 19:2324–2329.

105. Xiong, S., J. Xu, D. Chen, R. Wang, X. Hu, G. Shen, et al. 2011. Controlled synthesis of monodispersed hematite microcubes and their properties. CrystEngComm 13:7114–7120.

106. Liang, X., X. Wang, J. Zhuang, Y. Chen, D. Wang, and Y. Li. 2006. Synthesis of nearly monodisperse iron oxide and oxyhydroxide nanocrystals. Adv. Funct. Mater. 16:1805–1813.

107. Cao, H., G. Wang, J. H. Warner, and A. A. Watt. 2008. Amino-acid-assisted synthesis and size-dependent magnetic behaviors of hematite nanocubes. Appl. Phys. Lett. 92:013110–013113.

108. Zeng, S., K. Tang, T. Li, Z. Liang, D. Wang, Y. Wang, et al. 2007. Hematite hollow spindles and microspheres: selective synthesis, growth mechanisms, and application in lithium ion battery and water treatment. J. Phys. Chem. C 111:10217–10225.

109. Jia, X. H., and H. J. Song. 2012. Facile synthesis of monodispersed α-Fe_2O_3 microspheres through template-free hydrothermal route. J. Nanopart. Res. 14:1–8.

110. Cao, S.-W., and Y.-J. Zhu. 2008. Hierarchically Nanostructured α-Fe_2O_3 Hollow Spheres: Preparation,

Growth Mechanism, Photocatalytic Property, and Application in Water Treatment. J. Phys. Chem. C 112:6253–6257.

111. Wang, B., J. S. Chen, H. B. Wu, Z. Wang, and X. W. Lou. 2011. Quasiemulsion-Templated Formation of α-Fe_2O_3 Hollow Spheres with Enhanced Lithium Storage Properties. J. Am. Chem. Soc. 133:17146–17148.

112. Xu, X., R. Cao, S. Jeong, and J. Cho. 2012. Spindle-like Mesoporous α-Fe_2O_3 Anode Material Prepared from MOF Template for High-Rate Lithium Batteries. Nano Lett. 12:4988–4991.

113. Du, D., and M. Cao. 2008. Ligand-assisted hydrothermal synthesis of hollow Fe_2O_3 urchin-like microstructures and their magnetic properties. J. Phys. Chem. C 112:10754–10758.

114. Zeng, S., K. Tang, T. Li, Z. Liang, D. Wang, Y. Wang, et al. 2008. Facile route for the fabrication of porous hematite nanoflowers: its synthesis, growth mechanism, application in the lithium ion battery, and magnetic and photocatalytic properties. J. Phys. Chem. C 112:4836–4843.

115. Wu, Z., K. Yu, S. Zhang, and Y. Xie. 2008. Hematite hollow spheres with a mesoporous shell: controlled synthesis and applications in gas sensor and lithium ion batteries. J. Phys. Chem. C 112:11307–11313.

116. Kim, H.-J., K.-I. Choi, A. Pan, I.-D. Kim, H.-R. Kim, K.-M. Kim, et al. 2011. Template-free solvothermal synthesis of hollow hematite spheres and their applications in gas sensors and Li-ion batteries. J. Mater. Chem. 21:6549–6555.

117. Song, H.-J., N. Li, and X.-Q. Shen. 2011. Template-free synthesis of hollow α-Fe_2O_3 microspheres. Appl. Phys. A Mater. Sci. Process. 102:559–563.

118. Hu, C.-Y., Y.-J. Xu, S.-W. Duo, W.-K. Li, J.-H. Xiang, M.-S. Li, et al. 2010. Preparation of Inorganic Hollow Spheres Based on Different Methods. J. Chin. Chem. Soc. 57:1091.

119. Liu, J., Y. Li, H. Fan, Z. Zhu, J. Jiang, R. Ding, et al. 2009. Iron oxide-based nanotube arrays derived from sacrificial template-accelerated hydrolysis: large-area design and reversible lithium storage. Chem. Mater. 22:212–217.

120. Zhu, J., K. Y. S. Ng, and D. Deng. 2014. Hollow Cocoon-Like Hematite Mesoparticles of Nanoparticle Aggregates: Structural Evolution and Superior Performances in Lithium Ion Batteries. ACS Appl. Mat. Interfac. 6:2996–3001.

121. Zhu, J., K. Y. S. Ng, and D. Deng. 2014. Micro Single Crystals of Hematite with Nearly 100% Exposed {104} Facets: Preferred Etching and Lithium Storage. Cryst. Growth Des. 14:2811–2817.

122. Wang, Z., D. Luan, S. Madhavi, Y. Hu, and X. W. Lou. 2012. Assembling carbon-coated [small

alpha]-Fe_2O_3 hollow nanohorns on the CNT backbone for superior lithium storage capability. Energ. Environ. Sci. 5:5252–5256.

123. Tsuda, N. 2000. Electronic conduction in oxides, Vol. 94. Springer, 207.

124. Yuan, S. M., J. X. Li, L. T. Yang, L. W. Su, L. Liu, Z. Zhou, et al. 2011. Preparation and Lithium Storage Performances of Mesoporous Fe_3O_4@C Microcapsules. Mater. Interfaces 3:705–709.

125. Chen, Y., H. Xia, L. Lu, and J. Xue. 2012. Synthesis of porous hollow Fe_3O_4 beads and their applications in lithium ion batteries. J. Mater. Chem. 22:5006–5012.

126. Zhang, W. M., X. L. Wu, J. S. Hu, Y. G. Guo, and L. J. Wan. 2008. Carbon Coated Fe_3O_4 Nanospindles as a Superior Anode Material for Lithium-Ion Batteries. Adv. Funct. Mater. 18:3941–3946.

127. Yang, Z., J. Shen, and L. A. Archer. 2011. An in situ method of creating metal oxide-carbon composites and their application as anode materials for lithium-ion batteries. J. Mater. Chem. 21:11092–11097.

128. Wang, L., J. Liang, Y. Zhu, T. Mei, X. Zhang, Q. Yang, et al. 2013. Synthesis of Fe_3O_4@ C core–shell nanorings and their enhanced electrochemical performance for lithium-ion batteries. Nanoscale 5:3627–3631.

129. Zhu, J., K. Y. S. Ng, and D. Deng. 2014. Porous olive-like carbon decorated Fe_3O_4 based additive-free electrodes for highly reversible lithium storage. J. Mater. Chem. A 2:16008–16014.

130. Yang, L., A. Xiao, and B. L. Lucht. 2010. Investigation of solvation in lithium ion battery electrolytes by NMR spectroscopy. J. Mol. Liq. 154:131–133.

131. Xu, K. 2004. Nonaqueous Liquid Electrolytes for Lithium-Based Rechargeable Batteries. Chem. Rev. 104:4303–4418.

132. Croce, F., G. B. Appetecchi, L. Persi, and B. Scrosati. 1998. Nanocomposite polymer electrolytes for lithium batteries. Nature 394:456–458.

133. Dias, F. B., L. Plomp, and J. B. J. Veldhuis. 2000. Trends in polymer electrolytes for secondary lithium batteries. J. Power Sources 88:169–191.

134. Meyer, W. H. 1998. Polymer Electrolytes for Lithium-Ion Batteries. Adv. Mater. 10:439–448.

135. Fergus, J. W. 2010. Ceramic and polymeric solid electrolytes for lithium-ion batteries. J. Power Sources 195:4554–4569.

136. Zhang, S. S. 2007. A review on the separators of liquid electrolyte Li-ion batteries. J. Power Sources 164:351–364.

137. Wang, X., Z. Wen, Y. Liu, and X. Wu. 2009. A novel composite containing nanosized silicon and tin as anode material for lithium ion batteries. Electrochim. Acta 54:4662–4667.

138. Lee, Y. J., H. Yi, W.-J. Kim, K. Kang, D. S. Yun, M. S. Strano, et al. 2009. Fabricating Genetically

Engineered High-Power Lithium-Ion Batteries Using Multiple Virus Genes. Science 324:1051–1055.

139. Chen, H., M. Armand, G. Demailly, F. Dolhem, P. Poizot, and J.-M. Tarascon. 2008. From Biomass to a Renewable $Li_xC_6O_6$ Organic Electrode for Sustainable Li-Ion Batteries. Chemsuschem 1:348–355.

140. Zackrisson, M., L. Avellán, and J. Orlenius. 2010. Life cycle assessment of lithium-ion batteries for plug-in hybrid electric vehicles – Critical issues. J. Clean. Prod. 18:1519–1529.

141. Deng, D., and J. Y. Lee. 2011. Linker-free 3D assembly of nanocrystals with tunable unit size for reversible lithium ion storage. Nanotechnology 22:355401.

142. Tarascon, J.-M. 2010. Key challenges in future Li-battery research, Philos. Trans. A. Math. Phys. Eng. Sci. 368:3227–3241.

143. Yabuuchi, N., M. Kajiyama, J. Iwatate, H. Nishikawa, S. Hitomi, R. Okuyama, et al. 2012. P2-type $Nax[Fe_{1/2}Mn_{1/2}]O_2$ made from earth-abundant elements for rechargeable Na batteries. Nat. Mater. 11:512–517.

144. Ellis, B. L., W. R. M. Makahnouk, Y. Makimura, K. Toghill, and L. F. Nazar. 2007. A multifunctional 3.5 V iron-based phosphate cathode for rechargeable batteries. Nat. Mater. 6:749–753.

145. Hayashi, A., K. Noi, A. Sakuda, and M. Tatsumisago. 2012. Superionic glass-ceramic electrolytes for room-temperature rechargeable sodium batteries. Nat. Commun. 3:856.

146. Xiong, H., M. D. Slater, M. Balasubramanian, C. S. Johnson, and T. Rajh. 2011. Amorphous TiO_2 Nanotube Anode for Rechargeable Sodium Ion Batteries. J Phys Chem Lett 2:2560–2565.

147. Slater, M. D., D. Kim, E. Lee, and C. S. Johnson. 2012. Sodium-Ion Batteries. Adv. Funct. Mater. 23:947–958.

148. Stevens, D. A., and J. R. Dahn. 2000. High Capacity Anode Materials for Rechargeable Sodium-Ion Batteries. J. Electrochem. Soc. 147:1271–1273.

149. Lu, X. C., G. G. Xia, J. P. Lemmon, and Z. G. Yang. 2010. Advanced materials for sodium-beta alumina batteries: Status, challenges and perspectives. J. Power Sources 195:2431–2442.

150. Pan, H. L., Y. S. Hu, and L. Q. Chen. 2013. Room-temperature stationary sodium-ion batteries for large-scale electric energy storage. Energ. Environ. Sci. 6:2338–2360.

151. Palomares, V., P. Serras, I. Villaluenga, K. B. Hueso, J. Carretero-Gonzalez, and T. Rojo. 2012. Na-ion batteries, recent advances and present challenges to become low cost energy storage systems. Energ. Environ. Sci. 5:5884–5901.

152. Kim, S. W., D. H. Seo, X. H. Ma, G. Ceder, and K. Kang. 2012. Electrode Materials for Rechargeable Sodium-Ion Batteries: Potential Alternatives to Current Lithium-Ion Batteries. Adv. Energ. Mater. 2:710–721.

153. Lu, Y., S. Zhang, Y. Li, L. G. Xue, G. J. Xu, and X. W. Zhang. 2014. Preparation and characterization of carbon-coated $NaVPO_4F$ as cathode material for rechargeable sodium-ion batteries. J. Power Sources 247:770–777.

154. Ruffo, R., R. Fathi, D. J. Kim, Y. H. Jung, C. M. Mari, and D. K. Kim. 2013. Impedance analysis of Na 0.44 MnO_2 positive electrode for reversible sodium batteries in organic electrolyte. Electrochim. Acta 108:575–582.

155. Takachi, M., T. Matsuda, and Y. Moritomo. 2013. Cobalt Hexacyanoferrate as Cathode Material for Na^+ Secondary Battery. Appl. Phys. Express 6:2.

156. D'Arienzo, M., R. Ruffo, R. Scotti, F. Morazzoni, C. M. Maria, and S. Polizzi. 2012. Layered $Na0.71CoO_2$: a powerful candidate for viable and high performance Na-batteries. Phys. Chem. Chem. Phys. 14:5945–5952.

157. Lu, Y. H., L. Wang, J. G. Cheng, and J. B. Goodenough. 2012. Prussian blue: a new framework of electrode materials for sodium batteries. Chem. Commun. 48:6544–6546.

158. Ge, P., and M. Fouletier. 1988. Electrochemical intercalation of sodium in graphite. Solid State Ionics 28–30(Pt 2):1172–1175.

159. Wenzel, S., T. Hara, J. Janek, and P. Adelhelm. 2011. Room-temperature sodium-ion batteries: Improving the rate capability of carbon anode materials by templating strategies. Energ. Environ. Sci. 4:3342–3345.

160. Cao, Y. L., L. F. Xiao, M. L. Sushko, W. Wang, B. Schwenzer, J. Xiao, et al. 2012. Sodium Ion Insertion in Hollow Carbon Nanowires for Battery Applications. Nano Lett. 12:3783–3787.

161. Shao, Y. Y., J. Xiao, W. Wang, M. Engelhard, X. L. Chen, Z. M. Nie, et al. 2013. Surface-Driven Sodium Ion Energy Storage in Nanocellular Carbon Foams. Nano Lett. 13:3909–3914.

162. Qian, J. F., X. Y. Wu, Y. L. Cao, X. P. Ai, and H. X. Yang. 2013. High Capacity and Rate Capability of Amorphous Phosphorus for Sodium Ion Batteries. Angew. Chem. Int. Edit. 52:4633–4636.

163. Kim, Y., Y. Park, A. Choi, N. S. Choi, J. Kim, J. Lee, et al. 2013. An Amorphous Red Phosphorus/Carbon Composite as a Promising Anode Material for Sodium Ion Batteries. Adv. Mater. 25:3045–3049.

164. Hariharan, S., K. Saravanan, V. Ramar, and P. Balaya. 2013. A rationally designed dual role anode material for lithium-ion and sodium-ion batteries: case study of eco-friendly Fe_3O_4. Phys. Chem. Chem. Phys. 15:2945–2953.

165. Bi, Z. H., M. P. Paranthaman, P. A. Menchhofer, R. R. Dehoff, C. A. Bridges, M. F. Chi, et al. 2013.

Self-organized amorphous TiO_2 nanotube arrays on porous Ti foam for rechargeable lithium and sodium ion batteries. J. Power Sources 222:461–466.

166. Zhu, H. L., Z. Jia, Y. C. Chen, N. Weadock, J. Y. Wan, O. Vaaland, et al. 2013. Tin Anode for Sodium-Ion Batteries Using Natural Wood Fiber as a Mechanical Buffer and Electrolyte Reservoir. Nano Lett. 13:3093–3100.

167. Liu, Y. H., Y. H. Xu, Y. J. Zhu, J. N. Culver, C. A. Lundgren, K. Xu, et al. 2013. Tin-coated viral nanoforests as sodium-ion battery anodes. ACS Nano 7:3627–3634.

168. Datta, M. K., R. Epur, P. Saha, K. Kadakia, S. K. Park, and P. N. Kuma. 2013. Tin and graphite based nanocomposites: Potential anode for sodium ion batteries. J. Power Sources 225:316–322.

169. Su, D. W., H. J. Ahn, and G. X. Wang. 2013. SnO_2@ graphene nanocomposites as anode materials for Na-ion batteries with superior electrochemical performance. Chem. Commun. 49:3131–3133.

170. Xu, Y. H., Y. J. Zhu, Y. H. Liu, and C. S. Wang. 2013. Electrochemical Performance of Porous Carbon/Tin Composite Anodes for Sodium-Ion and Lithium-Ion Batteries. Adv. Energ. Mater. 3:128–133.

171. Huang, J. P., D. D. Yuan, H. Z. Zhang, Y. L. Cao, G. R. Li, H. X. Yang, et al. 2013. Electrochemical sodium storage of TiO_2(B) nanotubes for sodium ion batteries. RSC Adv. 3:12593–12597.

172. Chen, W., and D. Deng. 2015. Deflated Carbon Nanospheres Encapsulating Tin Decorated on Layered 3-D Carbon Structures for Low-Cost Sodium Ion Batteries. ACS Sustain. Chem. Eng. 3:63–70.

17

A three-dimensional multiphysics modeling of thin-film amorphous silicon solar cells

Ahmadreza Ghahremani & Aly E. Fathy

Electrical Engineering Department, University of Tennessee Knoxville, Knoxville, Tennessee

Keywords

3D multiphysics modeling, thin-film amorphous silicon, nanoparticles, plasmonic, solar cells

Correspondence

Ahmadreza Ghahremani, Electrical Engineering Department, University of Tennessee Knoxville, Min H. Kao Building, Suite 609 1520 Middle Drive, Knoxville, TN 37996-2250, USA.

E-mail: aghahrem@utk.edu

Funding Information

This work was supported by the grant from the National Science Foundation of USA (Grant No. NSF EPS-1004083).

Abstract

A 3D multiphysics simulation toolbox for thin-film amorphous silicon solar cells has been developed. The simulation is rigorous and is based on developing three modules: first to analyze light propagation using electromagnetic techniques, second to account for charge generation and transportation based on the physics of the semiconductor device, and third including electrode modeling by applying electrostatic techniques. Published results of a P-I-N thin-film amorphous silicon solar cell fabricated on a thick glass and experimentally evaluated was used as a vehicle to validate our 3D multiphysics toolbox and demonstrate its capabilities. The toolbox utilizes COMSOL for solving the partial differential equations describing the three modules, and MATLAB to input data, control the solver, and provides the coupling between the three modules. The developed toolbox was used to investigate both the effect of embedding Metallic Nanoparticles (MNPs) and the impact of defects on the external quantum efficiency. The simulator, besides being rigorous, is suitable to model various types of solar cells (organic, inorganic, thick film, thin film, heterojunction, or plasmonic) as well.

Introduction

Currently thin-film amorphous silicon solar cell has relatively poor efficiency due to low photocurrent generation. Although efficiency can be significantly increased upon enhancing light management and absorption mechanisms for absorber layers [1]. Fortunately, there are different methods to achieve such enhancement. For instance, applying a randomly textured layer inside the device is a standard approach to gain more effective scattered rays inside the device [2–5]. Additionally, introducing a periodic structure as a reflector to enable the increase of light path inside the absorber [6–8]; thus enhancing light absorption. Alternatively, scientists have tried to utilize metallic nanoparticles (MNPs) to improve the cell efficiency upon creating a high-intensity localized fields around these MNPs [9–11], but unfortunately not much success has been reported yet. To investigate such effects, several groups carried out many studies on solar cells performance enhancement using 1D and 2D analysis [12–16]. However, for accurate modeling, rigorous 3D

analysis [17–19] is indeed required to resolve many associated real practical fabrication problems. For example, only 3D multiphysics tools can be used to analyze the effects of placing spherical MNPs, also accounting for the random polarization of rays (incident waves), and considering their different multipaths behaviors after being scattered from these randomly shaped metallic nanoparticles. Table 1 shows different commercial CAD tools, their features and compares their capabilities and analysis methods to address solar cells multiphysics problems. However, solving such coupled 3D multiphysics problem (solar cell) is very challenging as it includes: light propagation—an optics problem, energy absorption—a semiconductor problem, and light conversion to photocurrent—an electrostatic problem. These three distinct physical fields need to be understood very well and accurately modeled by utilizing a fast and accurate simulator and account for their coupling. In other words, we need to first accurately predict the propagation of light within the different regions like dielectrics (passive region), semiconductors (active region) plasmonic nanoparticles (parasitic region), and effects of metals as reflectors/scatterers. Second, we need to solve the physics of the semiconductor devices so that we estimate the number of free electrons and holes. Finally, the third step is applying electrostatic formulas to calculate the photocurrent that is collected by the two electrodes placed at both sides of the device.

In this study, a 3D multiphysics toolbox has been developed to accurately model various types of solar cells. The toolbox main features are given in Table 1, and its novelty is the 3D Model of plasmon nanoparticles embedded inside semiconductors and the analysis of the effect of defects created by the plasmon layer. The main analysis concepts are explained in section Background, equations for the 3D problem are explained in detail in Appendix A1, supplemented by a description of the simulator initialization given in Appendix A2 and a list of some major parameters required for electrical characteristic of solar cells is derived in Appendix A3. Finally, the results of this study are presented for solar cells-no plasmon, simulation of solar cell-no plasmon with 3D gratings, electromagnetic field analysis in the presence of MNPs, and finally the effect of utilizing MNPs in sections (Simulated and Measured Results-no Plasmon, Analysis of Simulation Results of Solar Cell-No Plasmon With 3-D Gratings, Electromagnetic Field Analysis of MNPs, and Simulated and Measured Results-With Plasmon), respectively. Our conclusion is given in section Conclusion.

Background

To numerically analyze a solar cell structure using finite element of finite difference methods, first the 3D structure is drawn and meshed using a nonuniform grid. Definitely, the accuracy and speed of simulation are completely dependent on the selected mesh density. For instance, the critical regions like electrodes, semiconductor junctions, plasmon, and sharp or tiny structures should be generated using very dense mesh ($\sim\lambda/50$). Next, initial conditions for all variables and boundary conditions are set, respectively, to numerically solve the partial differential equations (i.e., Maxwell's equations, and the physics of the device transport equations). Given that the solution of the nonlinear PDEs system is sensitive to these initial and boundary conditions. For example, initializing the intensity of the impinging light (based on the latest published data for the Sun spectrum), setting up correctly all electric parameters including the electro-optical material properties (like complex refractive indices for semiconductors), and using a fast PDE solver are essential ingredients to get an accurate realistic results.

In general, electromagnetic modeling is carried out by numerically solving Maxwell's equations using one of the available techniques [20–24], like a finite element method that has been applied here. Meanwhile, Drude–Lorentz dispersion model [25] has been utilized here to model metallic nanoparticles. Subsequently, the result of this EM simulation is used to calculate the light power intensity distribution inside the semiconductor region (active layer). Then, the

Table 1. Introduce some commercial CAD Tools were used to solve solar cells' model.

Reference	Tools	Analysis Method	Type of modeling	Validation	Problem solved
[17]	HFSS	FEM	Optical model	Validated by another CAD tool (ASA)	3D modeling of Single junction thin-film silicon Solar cells with flat surface and 1D, and 2D modeling of trapezoid-shaped surface
[15]	SCAPS	Numerical methods (Newton)	Semiconductor and Electrostatic		1D modeling of Polycrystalline thin-film solar cells, The problem of transients of current to time dependent bias conditions is solved
[16]	D-AMPS-1D (1D simulation program Analysis of Microelectronic and Photonic Devices)	Finite differences (Newton Raphson iteration method)	Semiconductor and Electrostatic	Validated by measurement	1D modeling of structures such as homojunctions, heterojunctions, multijunctions, and Ge photovoltaic devices
[18]	COMSOL	FEM	Multiphysics	Validated by measurement	3D modeling of plasmonic solar cells; existing metallic nanoparticles at front of P-I-N
[19]	FDTD Solution TCAD Sentaurus	FDTD FEM	Multiphysics		Thin-film amorphous silicon; existing metallic nanoparticles at back of P-I-N
Our approach	COMSOL+MATLAB	FEM	Multiphysics	Validated by measurement	Thin-film amorphous silicon; existing metallic nanoparticles inside P-I-N and 3D modeling of trapezoid-shaped surface

number of electron-hole pairs is estimated as a function of the previously calculated light intensity using chemical absorption data [26]. In our analysis, the free charge carriers' recombination due to different factors (free carrier life time, energy trap level, and density of carrier concentrations) are considered, as they could cause significant impact in predicting the final results. Keep in mind that the recombination rate is a strong function of both the doping concentration and density of defects inside the semiconductor. As a final step, the number of electron-hole pairs captured by the two electrodes at both sides of the active region is calculated using Poisson's Electrostatics Equation. Meanwhile, other major parameters like external quantum efficiency, solar cell efficiency, J–V curve, and fill factor can be extracted as well as by-products [27–29]. A step-by-step illustration of the above calculations are given in the Appendix A1.

Simulated and Measured Results-no Plasmon

For model validation, the simulation results for a thin-film P-I-N solar cell model were compared to previously

published measured results [10]. However, some material properties were not cited on this publication and their values were assumed here based on well-known references like [13] and [30–34].

The structure of the modeled solar cell shown in Figure 1A consists of five stacked layers: a Glass (for protection) on top, a SnO:F TCO (transparent conductive oxide), followed by absorber layers (a-Si:H) as a P-I-N structure, then a TCO on the back, and all on top of a back reflector. The Amorphous silicon has 260-nm-thick intrinsic layer, the front electrode has a 75 nm-thick (SnO:F) layer, and the back electrode (AZO) is 75-nm-thick layer on top of the reflector. The N+ and P+ layers are modeled here too as 40-nm-thick layer each. Figure 1B shows a comparison between the simulated and measured EQE results, where only a slight discrepancy is seen.

Given that most of the material parameters were obtained from well-known references (listed in Table 2) but their

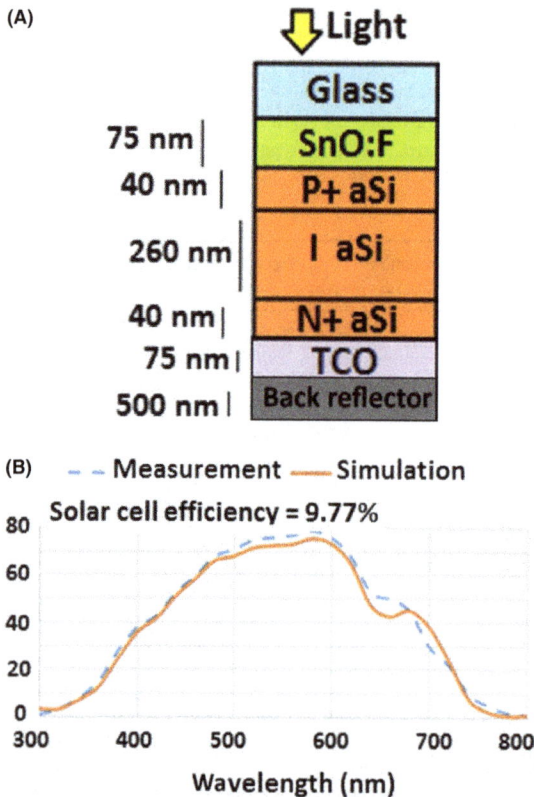

Figure 1. (A) Schematic of the P-I-N device; (B) EQE curve (comparison between measurement [10], and simulation result).

Table 2. The utilized value of each parameter for the utilized validation example.

Parameter – Name	Value
T (temperature)	300 [K]
Ni -Ref [34]	0.949×10^6 [cm^{-3}]
Doping (N+)	1×10^{20} [cm^{-3}]
Doping (P+)	10^{21} [cm^{-3}]
Thickness (N+, a-Si)-Ref [10]	40 [nm]
Thickness (intrinsic, a-Si)- Ref [10]	260 [nm]
Thickness (P+, a-Si)- Ref [10]	40 [nm]
Thickness (AZO) -(TCO in front)- Ref [10]	75 [nm]
Thickness (glass)- Ref [10]	200 [μm]
Thickness (Air)- Ref [10]	20 [μm]
Thickness -(TCO BACK)- Ref [10]	75 [nm](AZO)
Thickness (Silver) Reflector back- Ref [10]	500 [nm]
Electron mobility, a-Si -intrinsic-Ref [13]	20 [cm^2/(V s)]
Hole mobility, a-Si -intrinsic-Ref [13]	2 [cm^2/(V s)]
Electron mobility, a-Si N+ -Ref [13]	20 [cm^2/(V s)]
Hole mobility, a-Si N+ -Ref [13]	2 [cm^2/(V s)]
Electron mobility, a-Si P+ -Ref [13]	20 [cm^2/(V s)]
Hole mobility, a-Si P+ -Ref [13]	2 [cm^2/(V s)]
Electron life time, a-Si -intrinsic-Ref [33,34]	20 [ns]
Hole life time, a-Si -intrinsic-Ref [33,34]	20 [ns]
Electron life time, a-Si N+ -Ref [33,34]	0.0001 [ns]
Hole life time, a-Si N+ -Ref [33,34]	10 [ns]
Electron life time, a-Si P+ -Ref [21, 22]	10 [ns]
Hole life time, a-Si P+ -Ref [21, 22]	0.0001 [ns]
Density of state valence band, a-Si -Ref [31]	2.5×10^{20} [cm^{-3}]
Density of state conduction band, a-Si -Ref [31]	2.5×10^{20} [cm^{-3}]
Difference between defect level and intrinsic level N+, P+ -Ref [27]	0.7
Difference between defect level and intrinsic level intrinsic-Ref [27]	0.3
Energy band gap a-Si-Ref [34]	1.74
Diameter silver NPs-Ref [10]	20 [nm]
Affinity, a-Si (electro affinity)-Ref [13]	4.00 eV
Incident light angle	0 [deg]

real values could be within a specified uncertainty range; also the physical layers dimensions were guessed based on the intended design but could deviate slightly from the fabricated ones. In addition, the dopant distribution and the density of the traps (recombination) were not listed in the experimental description and were assumed as well. Considering the approximations, the simulated results are still indicating a similar behavior to the experimentally demonstrated ones. The solar efficiency and fill factor FF were used as a base line for comparison and their values were 9.77%, and 74%, respectively, and the maximum EQE occurs at 580 nm. The calculated current density as a function of voltage is shown in Figure 2. Additionally, the measured and simulated short circuit current J_{sc} and the open circuit voltage V_{oc} were listed too.

Additional parameters including the reflection and absorbance coefficient were calculated for the whole solar cell as a function of wavelength and are shown in Figure 3. The transmission coefficient, however, is almost zero.

When comparing Figures 3 and 1B, the level of light absorbance coefficient in some regions of the spectrum is higher than what is indicated by the EQE curve. This difference demonstrates that the energy was not totally converted to current and this drop is due to internal device losses. Figure 4 represents the extinction coefficient of amorphous silicon calculated based on [35], and it can be observed that the absorbance of light inside the amorphous silicon increases at higher frequencies. For this reason in the UV region most of the impinging solar energy on the cell is absorbed at the top layer of the semiconductor (P+ region) before approaching the depletion region. It turns out that the probability of creating free electron-hole pairs outside the depletion region increases; thus unfortunately increasing the probability of free carrier recombination immediately and this would be an energy loss. If the light could propagate for a longer path inside the semiconductor, it can generate free electron-hole pairs inside the depletion region, subsequently the electrostatic electric field (created based on majority carriers in the depletion region) could cause charge separation (electrons and holes) creating electricity. In other words, the presence of material defects inside the semiconductors can elevate the recombination probability of free carriers as well and reduce the overall conversion efficiency.

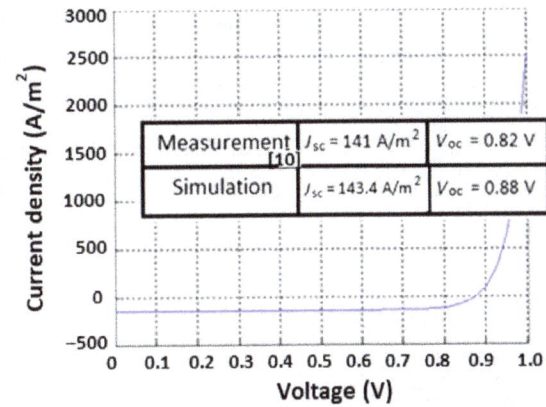

Figure 2. J–V curve (simulation), and a comparison between simulation and measured data [10] is shown in the embedded Table.

Analysis of Simulation Results of Solar Cell-no Plasmon with 3D Gratings

Surface roughness of solar cell layers can affect the solar cell performance. Where most of the time, the surface roughness is created by the morphology of the transparent conductive oxide (TCO) located at the top and bottom of the semiconductor layers. Use of large grains in the TCO film too can increase surface roughness [36–38]. At the same time, the size of the grains is correlated to the thickness of the thin-film TCO layer, that is, a thinner film would have less surface roughness. For instance, the thickness of the thin-film TCO considered for this model here is 75 nm, and its surface roughness is estimated to be <10 nm. To model the surface roughness effect, in general, a 3D device model can be used where a trapezoidal grating can be assumed. A periodic structure of a trapezoidal shape (like that of [17]) with a 30° degree slope in each side, and a 10 nm height (with periodicity of 200 nm) is assumed and implemented to model the 3D gratings of the solar cells (see Fig. 5A). Figure 5B shows that the results of the simulation for the EQE for both cases (flat and with gratings) are very similar for the thin-layered case. After analyzing the simulation's raw data, it was clear that the surface roughness at low frequencies (IR region) is very small compared

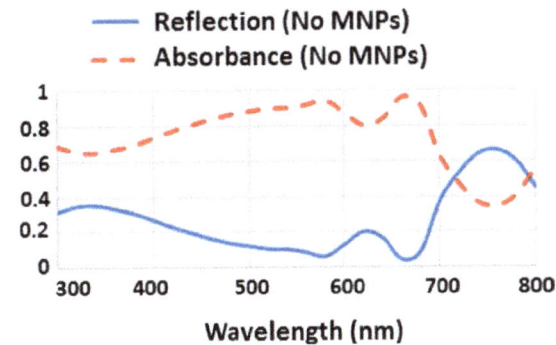

Figure 3. Simulation results for solar cells without MNPs; the blue solid line represents reflection coefficient; and red dash line shows the absorbance coefficient versus wavelength.

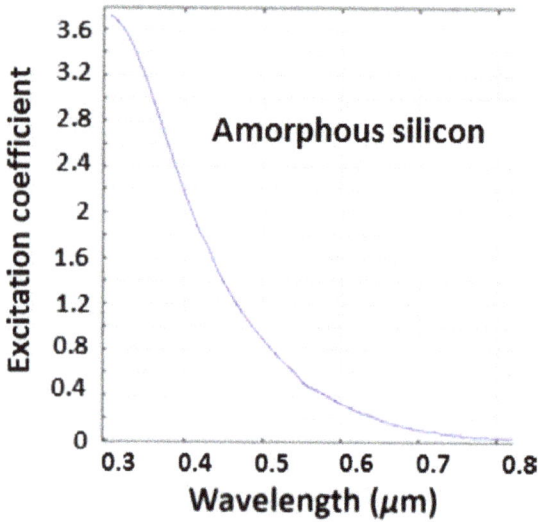

Figure 4. Extinction coefficient of amorphous silicon.

to the wavelength of solar rays. So, at these frequencies, it is a good approximation to assume that it behaves like a flat surface (from optics point of view). On the other hand, at higher frequencies (UV region) the gratings cannot be neglected, because its size is close to a wavelength. The

gratings lead to ray scattering and it could cause substantial increase in light path length inside the absorber (amorphous silicon), and it is anticipated that longer path would improve solar cell efficiency. Although, no efficiency improvement has been observed in the UV region, because the absorption of the amorphous silicon in this region is very high (see Fig. 4), and the penetration depth (see Appendix A1, equations 3–4) is rather very small compared to the path length of the ray inside the absorber. Hence, the effect of the gratings could only be observed weakly over a narrow bandwidth (510–620 nm). Apart from slight ripples seen in Figure 5B, the assumption of a flat surface for very thin layers has been validated as the simulation results are very similar to the measurements of [10], and for simplification it will be implemented in our analysis of the effect of MNPs.

Electromagnetic Field Analysis of MNPs

One way to prove that our electromagnetic 3D model for a MNP works properly, was to validate our simulation results with an analytical method. Here, we use the well-known conic problem of scattered fields from a metallic sphere in a vacuum space (shown in Ref [39]) to validate our models for MNPs. If an electric field of a uniform plane wave is polarized in the x direction, and is traveling along the z-axis, then the monostatic radar cross section can be expressed by

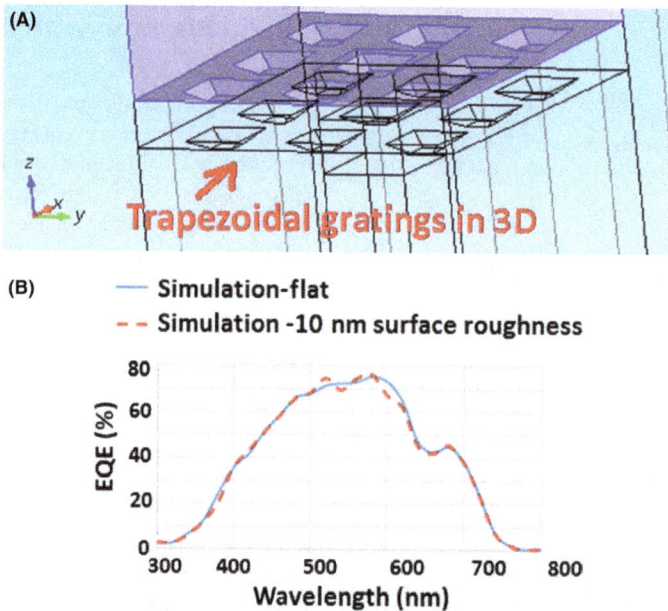

Figure 5. (A) schematic of the gratings in 3D inside the TCO; (B) EQE curve (comparison between simulation flat, and simulation with gratings (surface roughness = 10 nm).

$$\sigma_{3-D}\,(\text{monostatic}) = \lim_{r \to \infty}\left[4\pi r^2\frac{|E^s|^2}{|E^i|^2}\right]$$

$$= \frac{\lambda^2}{4\pi}\left|\sum_{n=1}^{\infty}\frac{(-1)^n(2n+1)}{H'^{(2)}(\beta a)\,H^{(2)}(\beta a)}\right|^2 \quad (1)$$

where r, E^s, E^i, λ, β, a, and $H^{(2)}$ are represent the range, scattered electric field, incident electric field, wavelength, wave propagation constant, radius of the metallic sphere, and spherical Hankel function, respectively. A plot of equation 1 as a function of the radius of the sphere is shown in Figure 6.

The results can be divided into three regions; the *Rayleigh*, the *Mie* (or *resonance*), and the *optical* region. The Rayleigh region represents the part of the curve for small radii values ($a < 0.1\lambda$). Hence, in the Rayleigh region, equation 1 can be reduced to

$$\sigma_{3-D}\,(\text{monostatic}) \cong \frac{9\lambda^2}{4\pi}(\beta a)^6 \quad (2)$$

Hence, we compared our 3D model for a sphere in vacuum and the analytical method and results are shown in Figure 7. A good accuracy is seen for the long wavelengths region. Some slight deviation is, however, seen for longer wavelength but still adequate for our modeling efforts here. To improve our model at long wavelength region, finer mesh can still be used but will significantly reduce computation speed.

Simulated and Measured Results-with Plasmon

Our next step was to analyze the effect of adding silver nanoparticles in each layer; one layer at a time to determine the optimum location and position that would

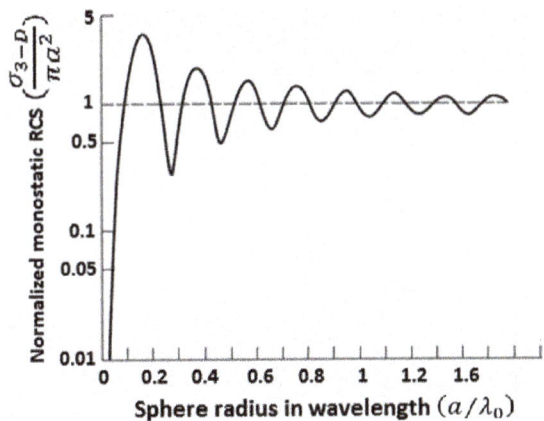

Figure 7. Normalized monostatic RCS for a single silver nanoparticle (radius = 10 nm) inside vacuum; solid blue line represents simulation result and red dash line shows the analytical method.

enhance the solar cell efficiency. In our model, silver nanoparticles are modeled as spheres with 10 nm radius [10] and are arranged in a random 2D array (in xy-plane) with a maximum center to center spacing of 50 nm. Initially, these silver NPs were placed 2 nm above the absorber layer along the SnO:F P-type A-Si interface as shown in Figure 8A as suggested by [10]. Although this resulted in a pronounced drop in solar cell efficiency

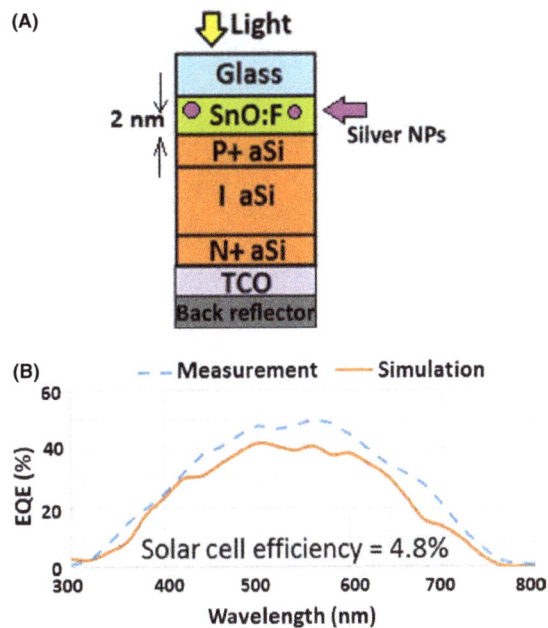

Figure 8. (A) Schematic of a PIN device; (B) EQE curve using MNPs embedded inside SnO:F.; blue dash line represents measurement [10]; orange solid line shows simulation.

Figure 6. Normalized monostatic radar cross section for a conducting sphere as a function of sphere radius [40].

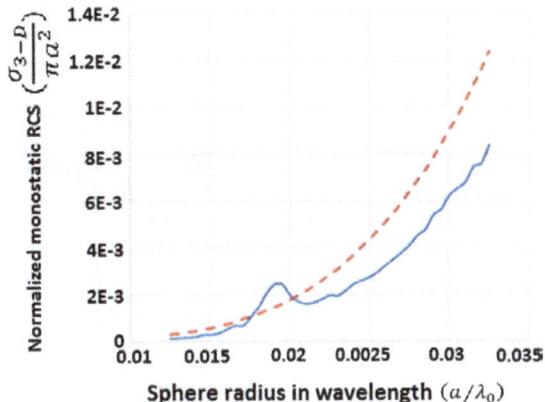

(A)

Particles	J_{sc} (mA/cm^2)	V_{oc} (V)	FF	Efficiency (%)
Measurement				
Ag	9.06	0.806	72.7	5.32
No particles	14.10	0.823	73.2	8.51
Simulation				
Ag	7.2	0.87	73	4.8
No particles	14.34	0.878	74	9.77

(B)

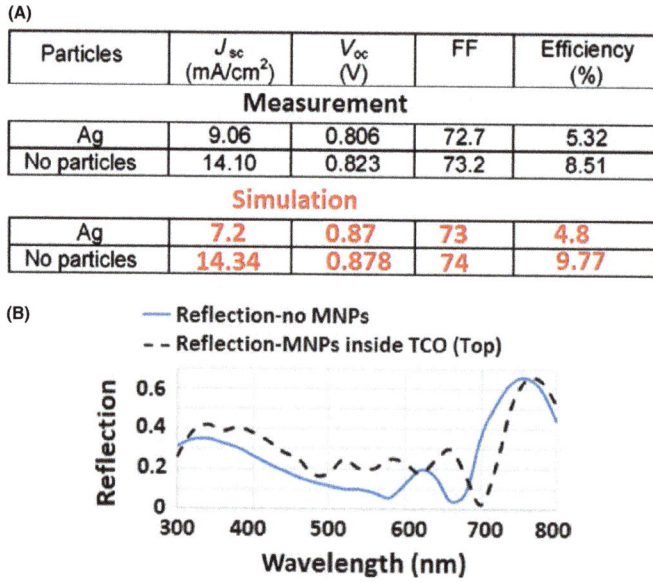

Figure 9. (A) A comparison between the measurements of [10] and simulation data for the electrical outputs; (B) Simulation results for the Reflection coefficient; solid blue line represents reflection from the solar cell without MNPs; dash black line shows reflection from the solar cell with MNPs embedded inside the TCO layer on top of the PIN device.

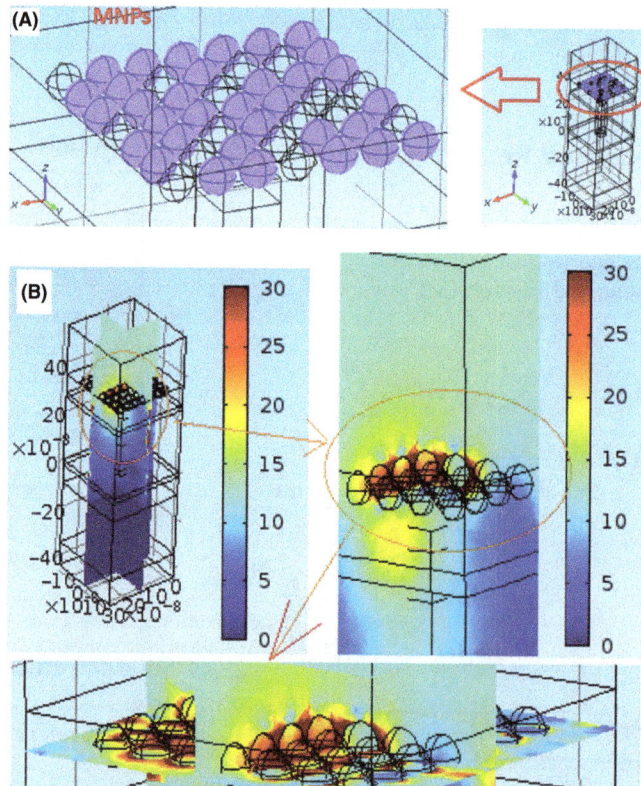

Figure 10. (A) schematic of MNPs are shown as blue spheres embedded inside TCO (B) the intensity of power flow of the light (wavelength@720 nm).

as noticed in Figure 8B, which is also consistent with the observations of Ref. [10]; where the efficiency has dropped to only 4.8%. Figure 9A shows a comparison between the simulated and measured electrical outputs of a solar cell with and without MNPs. The simulated reflection coefficient for these cases are also shown in Figure 9B.

Next step, the optical behavior of the solar cell excited by a polarized plane wave is analyzed in 3D. Figure 10A shows a schematic of a random 2D array of MNPs in xy-plane. Silver nanoparticles are embedded inside TCO (on top of the PIN device). An x-polarized electromagnetic plane wave (wavelength@720 nm) is propagating in the z direction toward the structure (assuming normal incidence). The power intensity of the propagating plane wave is dropped just after passing through the MNPs (see Figure 10B). Two MNPs (next to each other) create strong localized fields (see Figure 10B) as expected.

The above test was repeated, but this time with these silver nanoparticles embedded in the absorber layer (t = 50 nm under P+ layer) as seen in Figure 11A. However, it is believed that when MNPs are integrated into the amorphous silicon, they will cause an increase in the amount of defects inside the amorphous silicon structure. Unfortunately, it turns out that if this degradation phenomenon is not accounted for, then the simulation result is totally very optimistic and unrealistic. In summary, this misleading simulation result is shown in Figure 11B and indicates that significant efficiency improvement can be achieved, and as a matter of fact initially has led to a considerable confusion and a myth that adding MNPs should automatically lead to efficiency enhancement; but this has not been experimentally materialized yet due to significant fabrication defects. Practically, we needed to revise our model too to account for the impact of defects on performance, hence we modified the device physics equations as suggested by [33] to include these defects effects as well.

The modified model results after including the defects, gives results that are now in a fairly good agreement with measured data (Fig. 12). The results indicate a rather significant drop in efficiency which is consistent again with the observation of [10]. For t = 50 nm, the efficiency, for example, has dropped to only 3.5% in contradiction to the common belief that efficiency should have been enhanced. The pronounced efficiency drop was even worsening upon moving the MNPs from the top of the intrinsic layer to its bottom, where the solar cell efficiency dropped from 3.5% to even 1.89%. A plausible reason to explain this phenomenon is that; by varying the position of the MNPs inside the absorber and moving further down toward the bottom of the structure, the

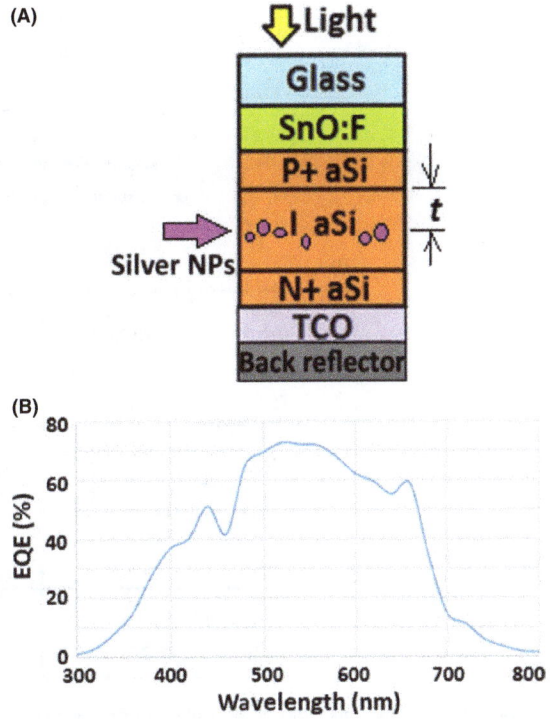

Figure 11. (A) Schematic of the PIN device; (B) EQE curve (simulation), using MNPs embedded inside intrinsic layer (t = 50 nm) without considering of defects.

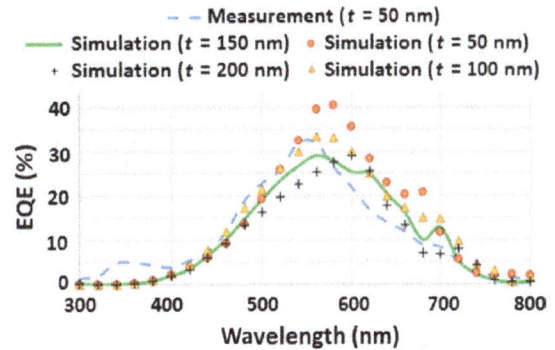

Figure 12. EQE curve in case of existing MNPs embedded inside intrinsic layer (a-Si) blue dash line measurement (t = 50 nm) [10]; other curves represents simulation results (after considering of defects).

intensity of the localized field created around the MNPs decreases exponentially, because most part of the solar energy in the UV region has been absorbed significantly after passing the top layers.

Hence, fields would become very weak to excite the MNPs. For this reason the generation rate compared to the recombination rate in the intrinsic layer goes down

(A)

(B)

t_{top} (nm)	J_{sc} (mA/cm^2)	V_{oc} (V)	FF	Efficiency (%)
Measurement				
50	6.38	0.882	55.3	3.11
100	4.18	0.899	57.0	2.14
150	3.96	0.881	56.5	1.97
200	3.14	0.876	53.7	1.48
Simulation				
50	6.4	0.67	78	3.5
100	4.3	0.64	81	2.34
150	4.1	0.61	77	2.02
200	3.8	0.59	80	1.89

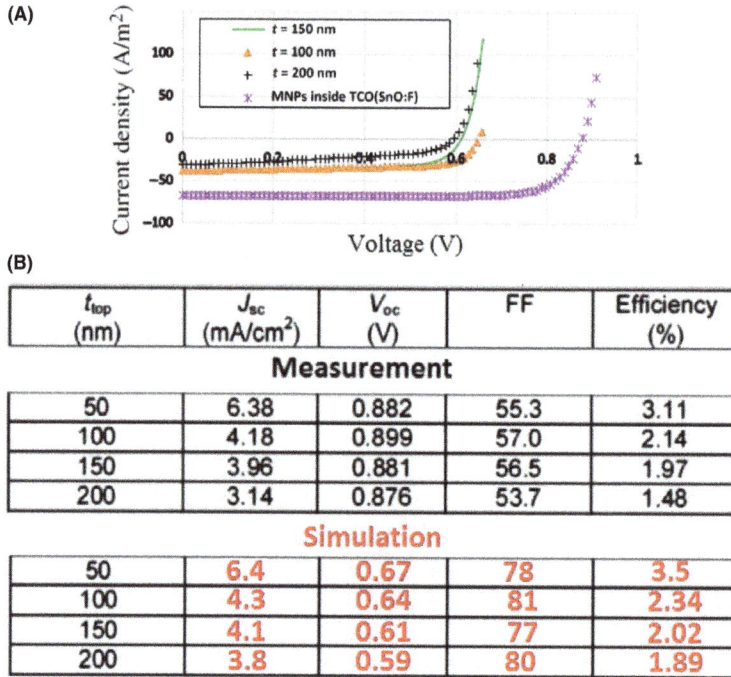

Figure 13. (A) J–V curve of the solar cell; for existing MNPs inside the structure; (B) comparison electrical outputs from measurement [9], and simulation.

upon moving the MNPs toward the bottom. The simulation result for the JV curves, with existing MNPs in different locations inside the solar cells are shown in Figure 13A. Figure 13B shows the electrical outputs for the measured [10] and simulated data. Finally, the silver NPs were relocated to the back (embedded inside the TCO), and that represented the most suitable location as seen in Figure 14A. The maximum solar efficiency now is 9.8% which is very close to the solar cell case with no NPs (see Fig. 14B). Again this observation is consistent with [10], but definitely much lower than our expectation of enhancing the efficiency upon adding MNPs.

Conclusion

In this study, an effective rigorous 3D multiphysics modeling of solar cells was presented. Our developed simulator is a real multiphysics modeling toolbox that is comprised of three coupled modules: Optics, carrier transport in semiconductors, and Electrostatic. To solve their associated nonlinear partial differential equations (PDEs) in 3D, we used two commercial tools (COMSOL, and MATLAB). One of the main reasons to carry out this 3D simulation is to accurately predict the electric field distribution due to the light scattering of the 3D plasmonic particles excited by the randomly polarized sun light. Our 3D tool has

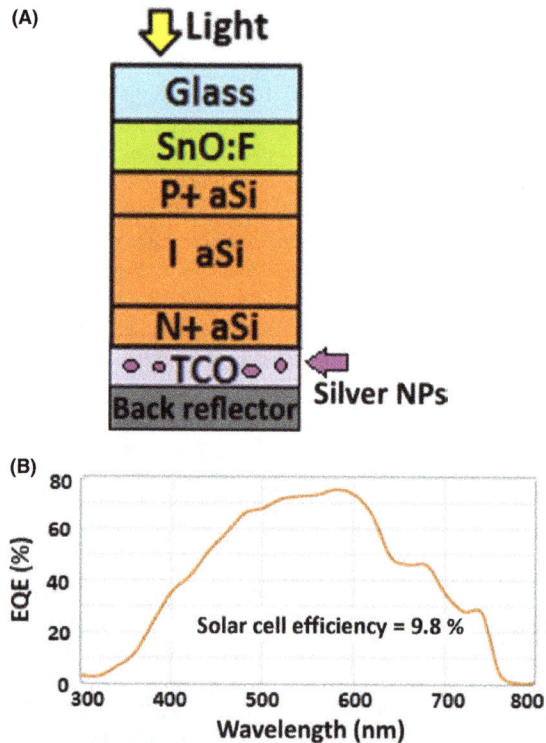

(A)

(B)

Figure 14. (A) Schematic of the PIN device; (B) EQE curve (simulation), using MNPs embedded inside TCO (back).

been validated by comparing its results with published measured ones.

The comparison between both measured and simulated results indicates a very good agreement even though some material parameters were assumed like the dopant distribution, and the density of traps (recombination), in addition to some assumed layers' thicknesses. The developed toolbox has been used to observe some major phenomena like strong localized fields around the MNPs and relate intense recombination to defects. The presence of MNPs might be helpful to enhance light absorption, although it would not guarantee improving solar cells' efficiency. Our simulation results demonstrated too that the trap density caused by defects around MNPs is the main reason for efficiency degradation that was seen experimentally. Here, we have verified that embedding MNPs randomly can lead to a significant external quantum efficiency degradation which is consistent with various experimental observations. Meanwhile, relocating the MNPs to the TCO layer close to the bottom electrode could provide some efficiency enhancement even though not pronounced. So, work is still under way to simulate other cases using our 3D toolbox to get significant efficiency improvement and to address the defect degradation issue, preliminary results are encouraging but would require more validation at this point.

Acknowledgments

This work was supported by the grant from the National Science Foundation of USA (Grant No. NSF EPS-1004083).

Conflict of Interest

None declared.

References

1. Zeman, M. 2006. Advanced amorphous silicon solar cell technology. Pp. 173–236 *in* J. Poortmans, V. Archipov, eds. Thin film solar cells: fabrication, characterization and applications. Wiley, Chichester. ISBN: 978-0-470-09126-5.

2. Deckman, H. W., C. B. Roxlo, and E. Yablonovitch. 1983. Maximum statisticalincrease of optical absorption in textured semiconductor films. Opt. Lett. 8:491–493.

3. Isabella, O., F. Moll, J. Krč, and M. Zeman. 2010. Modulated surface textures using zinc-oxide films for solar cells application. Phys. Status Solidi A 207:642–646.

4. Kambe, M., A. Takahashi, N. Taneda, K. Masumo, T. Oyama, and K. Sato. 2008. Fabrication of a-Si:H solar cells on high haze SnO2:F thin films. p. 1, *in* Photovoltaic Specialist Conference, San Diego, USA. 33rd IEEE (doi: 10.1109/PVSC.2008.4922507)

5. Borri, C., and M. Paggi. 2015. Topological characterization of antireflective and hydrophobic rough surfaces: are random process theory and fractal modeling applicable? J. Phys. D: Appl. Phys. 48:045301.

6. Haug, F.-J., T. Söderström, O. Cubero, V. Terrazzoni-Daudrix, X. Niquille, S. Perregeaux, et al. 2008. Periodic textures for enhanced current in thin film silicon solar cells. Mater. Res. Soc. Symp. Proc. 1101:KK13–KK. doi: 10.1557/PROC-1101-KK13-01.

7. Zeman, M., O. Isabella, K. Jäger, R. Santbergen, S. Solntsev, M. Topic, and J. Krc. 2012. Advanced Light management approaches for thin-film silicon solar cells. Energy Procedia 15:189–199.

8. Pahud, C., O. Isabella, A. Naqvi, F.-J. Haug, M. Zeman, H. P. Herzig, et al. 2013. Plasmonic silicon solar cells: impact of material quality and geometry. Opt. Express 21:A786–A797. doi: 10.1364/OE.21.00A786.

9. Toroghi, S., L. Chatdanai, and P. G. Kik. 2013. Cascaded plasmon resonances multi-material nanoparticle trimers for extreme field enhancement. Proc. SPIE 8809, Plasmonics: Metallic Nanostructures and Their Optical Properties XI, 88091M. doi: 10.1117/12.2024709.

10. Santbergen, R., A. H. M. Smets, and M. Zeman. 2011. Silver nanoparticles for plasmonic light trapping in A-Si:H solar cells. Pp 000673–000678 *in* Photovoltaic Specialists Conference (PVSC), 37th IEEE, Seattle, WA.

11. Stuart, H. R., and D. G. Hall. 1998. Island size effects in nanoparticle-enhanced photodetectors. Appl. Phys. Lett. 73:3815.

12. Malm, U., and M. Edoff. 2009. 2D device modelling and finite element simulations for thin-film solar cells. Sol. Energy Mater. Sol. Cells 93:1066–1069.

13. Ihalane, E., M. Meddah, A. Elfanaoui, L. Boulkaddat, E. El Hamri, X. Portier, et al. 2011. Numerical simulation of photocurrent in a solar cell based amorphous silicon. Moroccan J. Condens. Matter. 13:83–87.

14. Basore, P. A. 1990. Numerical modeling of textured silicon solar cells using PC-1D. IEEE Trans. Electron Devices 37:337–343.

15. Degrave, S., M. Burgelman, and P. Nolle. Modelling of polycrystalline thin film solar cells: new features in scaps version 2.3. 3rd World Conference on Pholovolroic Lnirergv Conversion Mov 11–18, 2003 Osaka, Japan.

16. Barrera, M., F. Rubinelli, I. Rey-Stolle, and J. Pla. 2012. Numerical simulation of Gesolar cells using D-AMPS-1Dcode. Phys. B 407:3282–3284.

17. Isabella, O., S. Solntsev, and M. Zeman. 2013. 3-D optical modeling of thin-film silicon solar cells on diffraction gratings. Prog. Photovoltaics Res. Appl. 21:94108.

18. Li, X., N. P. Hylton, V. Giannini, K.-H. Lee, N. J. Ekins Daukes, and S. A. Maier. 2011. Bridging electromagnetic and carrier transport calculations for three-dimensional modelling of plasmonic solar cells. Opt. Express 19:A888–A896.

19. Deceglie, M. G., V. E. Ferry, A. P. Alivisatos, and H. A. Atwater. 2012. Design of nanostructured solar cells using coupled optical and electrical modeling. Nano Lett. 12:2894–2900. doi: 10.1021/nl300483y.

20. Yee, K. S. 1966. Numerical solution of initial boundary value problems involving Maxwell's equations. IEEE Trans. Antennas Propag. 14:302–307.

21. Weiland, T. 1977. A discretization method for the solution of Maxwell's equations for six-component fields. Int. J. Electron. Commun. AEU 31:116–120.

22. Moharam, M. G., and T. K. Gaylord. 1981. Rigorous coupled-wave analysis of planar-grating diffraction. J. Opt. Soc. Am. 71:811–818.

23. Gibson, W. C. 2008. The method of moments in electromagnetics. Chapman & Hall/CRC, Boca Raton. ISBN 978-1-4200-6145-1.

24. Jin, J.-M. 2002. The finite element method in electromagnetics, 2nd ed. John Wiley & Sons, New York. ISBN 978-0-471-43818-2.

25. Quinn, J. J., and K.-S. Yi. 2009. Solid state physics: principles and modern applications. Springer-Verlag, Berlin Heidelberg.

26. Dewan, R., S. Fischer, V. B. Meyer-Rochow, Y. Özdemir, S. Hamraz, and D. Knipp. 2012. Studying nanostructured nipple arrays of moth eye facets helps to design better thin film solar cells. Bioinspir. Biomim. 7:016003.

27. SZE. 1981. Physics of semiconductor devices, 2nd ed. Wiley-Interscience, New York.

28. Neamen, D. A. 2003. Semiconductor physics and devices: basic principles, 3rd ed. McGraw Hill, New York.

29. Madelung, O. 2004. Semiconductors: data handbook. Springer-Verlag Berlin Heidelberg.

30. Deng, X., and E. A. Schiff. 2005. Amorphous silicon–based solar cells, chapter 12. John Wiley & Sons, Ltd, Chichester, UK.

31. Bakr, N. A., A. M. Funde, V. S. Waman, M. M. Kamble, R. R. Hawaldar, D. P. Amalnerkar, et al. 2010. Determination of the optical parameters of a-Si:H thin films deposited by hot wire–chemical vapor deposition technique using transmission spectrum only. Pramana J Physics. 76:519–531.

32. Alamo, J. A., and R. M. Swanson. 1987. Modeling of minority-carrier transport in heavily doped silicon emitters. Solid-State Electron. 30:1127–1136.

33. Piprek, J. 2003. Semiconductor optoelectronic devices introduction to physics and simulation. Academic Press; 1 edition (January 21, 2003). ISBN-13: 978-0125571906.

34. Sakata, I., and Y. Hayashi. 1985. Theoretical analysis of trapping and recombination of photo generated carriers in amorphous silicon solar cells. Appl. Phys. A 37:153–164.

35. Pierce, D. T., and W. E. Spicer. 1972. Electronic structure of amorphous SI from photoemission and optical studies. Pierce Spicer Phys. Rev. B. 5:3017.

36. Gracia, M., F. Rojas, and G. Gordillo. Morphological and optical characterization of SnO_2:F thin films deposited by spray pyrolysis. 20th European Photovoltaic Solar Energy Conference, 6–10 June 2005, Barcelona, Spain.

37. Chaaya, A. A., R. Viter, M. Bechelany, Z. Alute, D. Erts, A. Zalesskaya, et al. 2013. Evolution of microstructure and related optical properties of ZnO grown by atomic layer deposition. Beilstein J. Nanotechnol. 4:690–698. doi: 10.3762/bjnano.4.78.

38. Kumar, V., N. Singh, R. M. Mehra, A. Kapoor, L. P. Purohit, and H. C. Swart. 2013. Role of film thickness on the properties of ZnO thin films grown by sol-gel method. ELSEVIER Thin Solid Films 539:161–165.

39. Balanis, C. A. 1989. Advanced engineering electromagnetics. Wiley, Hoboken, New Jersey, USA.

40. Aden, A. L. 1951. Scattering from spheres with sizes comparable to the wavelength. J. Appl. Phys. 22:601.

41. http://www.astm.org/Standards/G159.htm.

42. Hall, R. N. 1952. Electron-hole recombination in silicon. Phys. Rev. 87:387.

43. Shockley, W., and W. T. Read. 1952. Statistics of the recombinations of electrons and holes. Phys. Rev. 87:835–842.

44. Johnson, P. B., and R. W. Christy. 1963. Optical constants of the noble metals. Phys. Rev. Lett. 11:541.

45. OptiFDTD: Technical Background and Tutorials.

46. Terrestrial photovoltaics measurement procedures, NASA Tech. Memo. TM 73702, National Aeronautics and Space Administration, Cleveland, OH, 1977.

Appendix A1

Solving Solar Cell Problem in 3D

If α and $I_f(x, y, z)$ are the absorption coefficient and the intensity of the photons at position (x, y, z), respectively (Fig. 15), then the relationship for the optical absorption for the differential length (Dx, Dy, Dz) is given bellow.

$$\vec{I}_f(x,y,z) = I_{f_x}(x,y,z)\,\hat{x} + I_{f_y}(x,y,z)\,\hat{y} + I_{f_z}(x,y,z)\,\hat{z} \quad (3)$$

$$\frac{dI_{f_x}(x,y,z)}{\partial x}\hat{x} + \frac{\partial I_{f_y}(x,y,z)}{\partial y}\hat{y} + \frac{\partial I_{f_z}(x,y,z)}{\partial z}\hat{z} = -\alpha \vec{I}_f(x,y,z) \quad (4)$$

This equation clearly shows that the intensity of a photon decreases exponentially with the propagation distance through the semiconductor material due to the absorption [25,27–29]. Meanwhile, the number of electron-hole pairs generated by the light is:
Generation rate [28]:

$$G(x,y,z) = \frac{\alpha\left(\sqrt{I_{f_x}(x,y,z)^2 + I_{f_y}(x,y,z)^2 + I_{f_z}(x,y,z)^2}\right)}{hf} \quad (5)$$

where h is plank constant, and f is frequency of the wave.

Subsequently, to calculate the free electron density and free hole density, Poisson's Equation is used:

$$\vec{\nabla}\cdot\vec{\nabla}\varphi = -\frac{q}{\varepsilon}(p-n+c) \quad \text{PDE: } \frac{\partial^2\varphi}{\partial x^2} + \frac{\partial^2\varphi}{\partial y^2} + \frac{\partial^2\varphi}{\partial z^2} \quad (6)$$

$$= -\frac{q}{\varepsilon}(p-n+c)$$

Shockley Read Hall recombination rate (7–11) [42,43]:

$$\Delta E_t = E_t - E_i \quad (7)$$

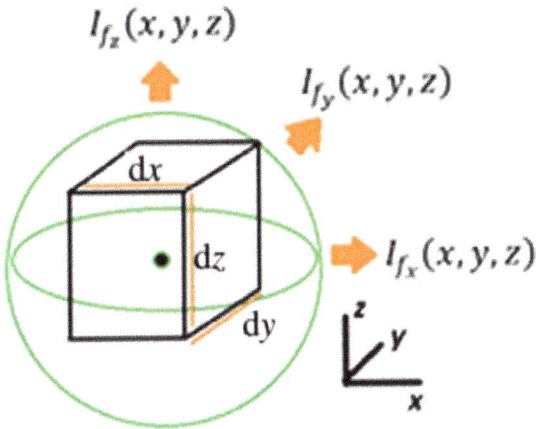

$$I_{f_z}(x,y,z)$$

$$I_{f_y}(x,y,z)$$

$$I_{f_x}(x,y,z)$$

Figure 15. Photon intensity in a 3D cell.

$$n_1 = \gamma_n \sqrt{N_c N_v}\,\exp\left(-\frac{E_g - \Delta E_g}{2V_{th}}\right)\exp\left(\frac{\Delta E_t}{V_{th}}\right) \quad (8)$$

$$p_1 = \gamma_p \sqrt{N_c N_v}\,\exp\left(-\frac{E_g - \Delta E_g}{2V_{th}}\right)\exp\left(\frac{\Delta E_t}{V_{th}}\right) \quad (9)$$

$$n_i = \gamma_n \gamma_p \sqrt{N_c N_v}\,\exp\left(-\frac{E_g - \Delta E_g}{2V_{th}}\right) \quad (10)$$

$$R(x,y,z) = R_n(x,y,z) = R_p(x,y,z) = \frac{np - n_i^2}{\tau_p(n_1 + n)\tau_n(p_1 + p)} \quad (11)$$

$$\text{Total rate: } R_t(x,y,z) = G(x,y,z) - R(x,y,z) \quad (12)$$

Transport of Diluted species (Electron)

$$\text{PDE: } \vec{\nabla}\cdot\vec{J}_n - q\frac{\partial n}{\partial t} - q\frac{\partial \rho_n}{\partial t} = qR_t(x,y,z)$$

$$\frac{\partial J_{nx}}{\partial x} + \frac{\partial J_{ny}}{\partial y} + \frac{\partial J_{nz}}{\partial z} - q\frac{\partial n}{\partial t} - q\frac{\partial \rho_n}{\partial t} = qR_t(x,y,z) \quad (13)$$

$$\text{PDE: } \vec{J}_n = \vec{J}_{nx} + \vec{J}_{ny} + \vec{J}_{nz}$$

$$\vec{J}_n = \left(-qn\mu_n\frac{\partial\varphi}{\partial x} + qD_{nx}\frac{\partial n}{\partial x}\right)\hat{x} + \left(-qn\mu_n\frac{\partial\varphi}{\partial y} + qD_{ny}\frac{\partial n}{\partial y}\right)\hat{y} \quad (14)$$

$$+ \left(-qn\mu_n\frac{\partial\varphi}{\partial z} + qD_{nz}\frac{\partial n}{\partial z}\right)\hat{z}$$

$$\text{Space charge density (Electron): } \rho_n = -\frac{q}{\varepsilon}n \quad (15)$$

where φ (Electric Potential), p (Hole concentration), and n (Electron concentration) are variables, q (Electron charge), ε (Optical property of the material), c (Initial value for carrier concentration), n_i (intrinsic concentration), τ_n (Electron life time), τ_p (Hole life time), ϕ_0 (incident photon flux), α, (absorption coefficient of material), μ_n (Electron mobility), μ_p (Hole mobility), and D_n (Electron Diffusivity) are constants.

γ_n and γ_p are the electron and hole degeneracy factors, N_c and N_v are the effective densities of states for the conduction and valence band, E_g is the band gap, ΔE_g the band gap narrowing and E_t is the trap energy level. Energy difference between the defect level and the intrinsic level is ΔE_t.

The same equations can be derived for holes.

After coupling equations (13), (14), and (15), there will be only two nonlinear PDEs (16), (17).

$$-n\mu_n\left(-\frac{q}{\varepsilon}(p-n+c)\right) + D_{nx}\frac{\partial^2 n}{\partial x^2} + D_{ny}\frac{\partial^2 n}{\partial y^2} + D_{nz}\frac{\partial^2 n}{\partial z^2} \quad (16)$$

$$+ \left(-1 + \frac{q}{\varepsilon}\right)\frac{\partial n}{\partial t} = R_t(x,y,z)$$

$$p\mu_p\left(-\frac{q}{\varepsilon}\left(p-n+c\right)\right)-D_{px}\frac{\partial^2 p}{\partial x^2}-D_{py}\frac{\partial^2 p}{\partial y^2}-D_{pz}\frac{\partial^2 p}{\partial z^2} \quad (17)$$

$$+\left(1+\frac{q}{\varepsilon}\right)\frac{\partial p}{\partial t}=-R_t\left(x,y,z\right)$$

Gaussian distribution (for both acceptor and donor) is considered as a dopant distribution. The junction depths

in x, y, and z direction are set at 2 µm, 2 µm, and 300 [nm], respectively.

Figure A1 is shown in below that could be a good representation of our 3D model.

The flowchart clearly shows that the proposed solution is a staggered one, and not a fully coupled one. First, the optical problem is solved. Then the two other physics

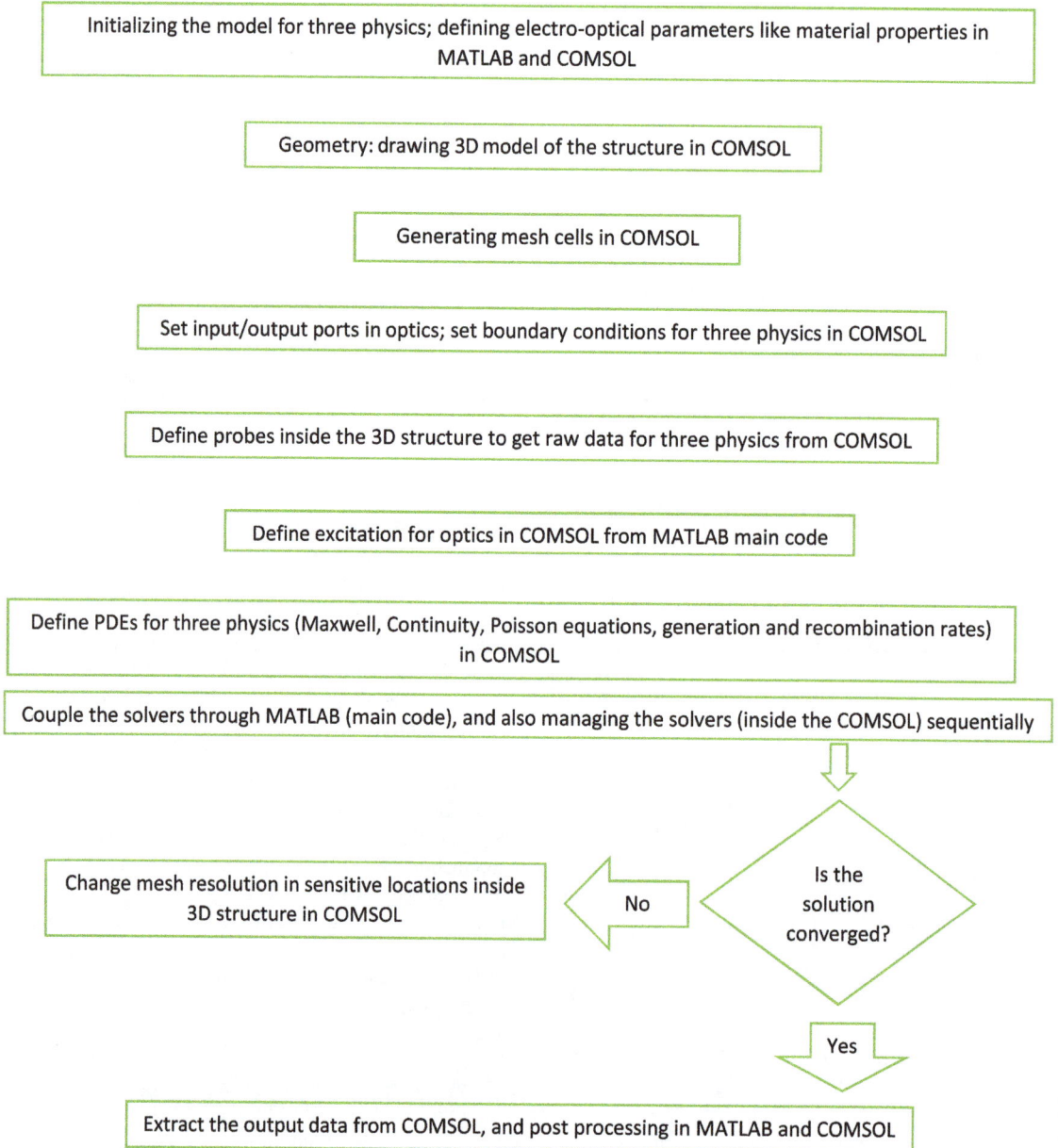

Initializing the model for three physics; defining electro-optical parameters like material properties in MATLAB and COMSOL

Geometry: drawing 3D model of the structure in COMSOL

Generating mesh cells in COMSOL

Set input/output ports in optics; set boundary conditions for three physics in COMSOL

Define probes inside the 3D structure to get raw data for three physics from COMSOL

Define excitation for optics in COMSOL from MATLAB main code

Define PDEs for three physics (Maxwell, Continuity, Poisson equations, generation and recombination rates) in COMSOL

Couple the solvers through MATLAB (main code), and also managing the solvers (inside the COMSOL) sequentially

Change mesh resolution in sensitive locations inside 3D structure in COMSOL ← No ← Is the solution converged?

Yes

Extract the output data from COMSOL, and post processing in MATLAB and COMSOL

Figure A1. The flow diagram of the 3D model.

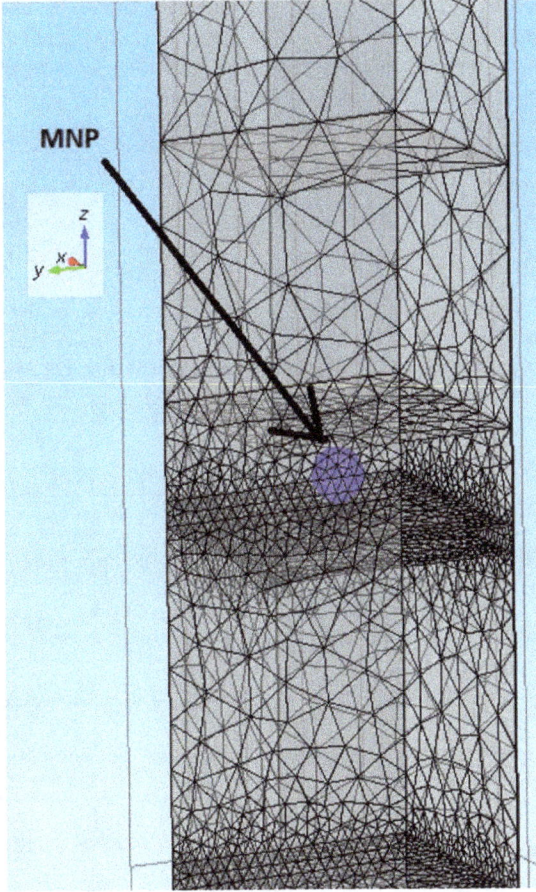

Figure 16. Tetrahedral Mesh Generation of the 3D device for one unit cell.

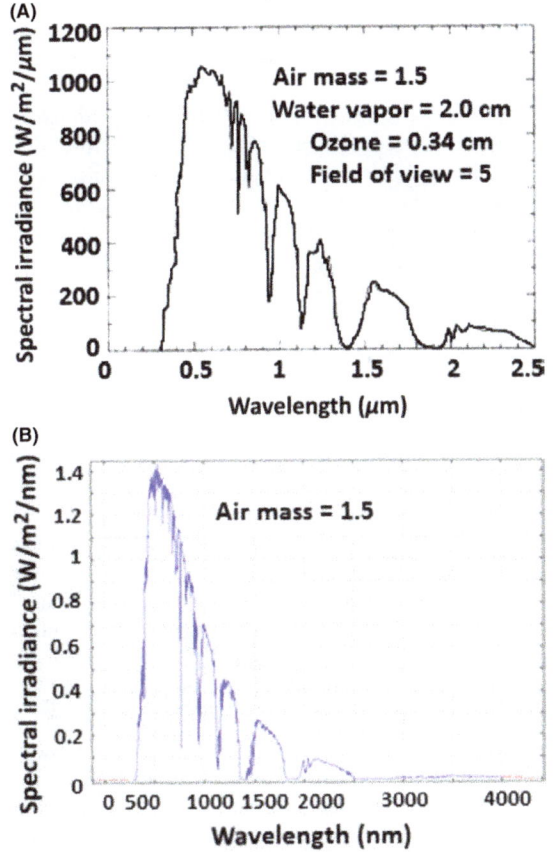

Figure 17. (A) Standard solar spectrum generated with the BRITE Monte Carlo Solar Insolation Model using the atmospheric conditions specified in [41]; (B) the utilized update of the spectral irradiance—downloaded from [26]..

(the semiconductor device and the electrostatic) are solved sequentially.

In this analysis, the PDEs are solved in steady state. So the transient solutions are not considered. It means that $((\partial n/\partial t) = 0$ or $(\partial p/\partial t) = 0)$. The excess minority carrier hole concentration increases with time with the time constant (which is the excess minority carrier lifetime). The excess carrier concentration reaches a steady-state value as time goes to infinity, even though a steady-state generation of excess electrons and hole exists [28]. The generation rate G a function of wavelength and time. However, here we are solving only for steady state and ignore any time dependence. Wavelength dependence is often included by summing the result obtained at one wavelength over all wavelengths of interest.

The RF interfaces formulate and solve the differential form of Maxwell's equations together with the initial and boundary conditions. The equations are solved using the finite element method. The steps are defined from MATLAB, and it can control COMSOL through a live link.

1-Define the geometry 2-Select materials 3-Select a suitable RF interface 4-Define boundary and initial conditions 5-Define finite element mesh 6-Define PDEs 7-Visualize the results.

To minimize the amount of physical memory (RAM) for computation and speed up computation we can use periodic boundary conditions along x, and y axis, which means that we solve equations for only one unit cell. Tetrahedral has been chosen as a type of mesh. Sharp edges or small particles like MNPs should be generated with high mesh density. Also the mesh size physically placed close to the critical edges (p and n type region) should be considered fine to make sure the PDEs solver is converged properly (Fig. 16).

Appendix A2

Initialization of the Simulator

In order to solve these partial differential equations, all parameters are needed to be initialized. The electro-optical properties like dielectric constant, refractive indices (real part and imaginary part), electron/hole mobility and the other electronic properties have been extracted from [13], [30–34]. In case of using MNPs or a plasmon layer the dielectric constant value has to be calculated for each optical wavelength of interest. For gold or silver NPs, optical constants were measured and given by Drude [44]. Drude-Lorentz dispersion model is a well-known model and is given by

$$\varepsilon_r = \varepsilon_\infty + \sum_{j=1}^{M} \frac{f_j \omega_p^2}{\omega_{oj}^2 - \omega^2 + i\Gamma_j\omega}, \sigma = 0 \tag{18}$$

Additional references can be used like [45,46] to double check materials properties.

Generally, the spectrum of incident light and intensity of impinging on the device is the most important initializing step to get accurate results from the simulation. Because some of the simulation outputs (like solar cell efficiency, and short circuit current) are directly related to the impinging light intensity as a function of wavelength, we needed then to utilize very accurate data. Hence, we utilized the spectral irradiance chart that has been measured several times and published by NASA [41] (Fig. 17A). Although, we need to keep in mind that the spectral irradiance depends on different factors like: the height from the sea level, water vapor, and air pollution. It is time dependent as well. For simplicity, we assume as a constant for the time being. Figure 17B shows the latest update that was found from [26], and utilized in our simulator.

Appendix A3

Electrical Outputs

Our model can be used to calculate many parameters like: External Quantum Efficiency (EQE), Solar Cell Efficiency, open circuit voltage, and short circuit [27,28].

The external quantum efficiency (EQE) is defined as:

$$\text{EQE (wavelength)} = \frac{\text{Number of electrons is collected by electrode}}{\text{Number of photons light source}} \tag{19}$$

The short-circuit current density can be computed as:

$$J_{sc} = \text{Charge of electron} \times \text{number of electrons collected by electrode} \tag{20}$$

Solar Cell Efficiency is: $\text{Efficiency} = \dfrac{J_{sc}V_{oc}\text{FF}}{P_{in}}$ (21)

FF is the filling factor, which can be derived from J–V curve of the device.

$$\text{FF} = \frac{I_m V_m}{I_{sc} V_{oc}} \tag{22}$$

18

Ozone pretreatment of humid wheat straw for biofuel production

Ali Kamel H. Al jibouri, Ginette Turcotte, Jiangning Wu & Chil-Hung Cheng

Chemical Engineering Department, Ryerson University, 350 Victoria Street, Toronto, Ontario M5B 2K3, Canada

Keywords

Biofuel, delignification, enzymatic hydrolysis, ozonolysis, wheat straw

Correspondence

Ginette Turcotte, Chemical Engineering Department, Ryerson University, 350 Victoria Street, Toronto, Ontario M5B 2K3, Canada.

E-mail: gturcott@ryerson.ca

Funding Information

The authors gratefully acknowledge the support of Natural Sciences and Engineering Research Council of Canada (NSERC), Ryerson University, and Ontario Graduate Scholarship (OGS).

Abstract

In an attempt to maximize the amount of ozone reacting with lignin inside humid wheat straw, some of the ozone-reactive lignin degradation products were washed away before a second ozonolysis delignification stage. The total contact time for the two stages was kept the same as that for a one-stage process for comparison. A significant decrease in the Acid Insoluble Lignin (AIL) content of the straw resulted: from 13.04 wt. % (after a 30-min one-stage ozonolysis) to 9.34 wt. % (after a 30-min two-stage ozonolysis, separated by a washing step). This significant improvement was accompanied by an increase in released fermentable sugars from an enzymatic hydrolysis. The yield increased from 60% theoretical sugars to 80%. A further improvement in AIL (down to 7.36 wt. %) and released sugars (up to 90% theoretical) occurred when the moisture content (MC) of the straw entering the second stage was adjusted to the optimum value of the straw entering first stage (45 wt. %, predicted from an experimental design). The authors believe this is the first time results are published for the introduction of a two-stage process separated by a washing step.

Introduction

Securing energy and reducing greenhouse gas emissions are some of the global major concerns of these days. The conversion of abundant lignocellulosic biomass to biofuels as transportation fuels could be the answer for these concerns [1]. Biofuels are considered cleaner-burning fuels because they do not add net CO_2 to the atmosphere and they have the potential to cut greenhouse gas emissions by 86% [2]. Lignocellulosic biomass consists mainly of cellulose, hemicellulose, lignin, and pectin [3,4]. Agricultural residues, such as wheat straw, and forest products such as hardwood and softwood, are the main sources of lignocellulosic biomass that can be used for biofuel production. Cellulose and hemicellulose must be hydrolyzed first into their corresponding monomers (sugars), followed by a fermentation step using special microorganisms to convert the sugars to fuels such as

ethanol [5–7]. Hemicellulose can be easily hydrolyzed using dilute acids while cellulose needs more extreme conditions [8]. Cellulose and hemicellulose can both be hydrolyzed using cellulase enzyme mixtures consisting of at least three major types: endo-glucanase, exo-glucanase, and ß-glucosidase. These enzymes are working in a synergistic fashion to achieve hydrolysis [9]. Low corrosion problem, energy consumption, and toxicity are the main advantages of enzymatic hydrolysis process over acid hydrolysis [10].

The presence of lignin and the cellulose crystallinity present a protective barrier that prevents plant cells from being attacked by many microorganisms such as fungi and bacteria. Therefore, the structure of lignin and of crystalline cellulose must first be altered or broken down so that enzymes can easily access and hydrolyze cellulose and hemi-cellulose. This can be achieved using chemical, physical, biological, or mixed pretreatment processes [11–13].

Ozonolysis pretreatment has shown its efficiency in the degradation of lignin in lignocellulosic biomass [14–20]. Ozone is highly reactive toward compounds incorporating conjugated double bonds and functional groups with high electron densities, such as lignin. It preferably attacks lignin rather than cellulose or hemicellulose. Ozonolysis of lignin releases soluble compounds of smaller molecular weight, mainly organic acids such as carboxylic and acetic acids, which can result in a drop in pH from 6.5 to 2. The range of ozonolysis products is influenced by the structure of the lignocellulosic biomass as well as its moisture content (MC) [21].

Water content in nonsubmerged humid lignocellulosic biomass has a major effect in the ozonolysis process. Water induces cell wall swelling and consequently provides an access for ozone to functional groups of lignin. It also acts as a solvent of ozone and of some delignification products (lignin fragments). At low MC, lignin mostly reacts with gaseous ozone. Despite large amount of supplied ozone, lignin degradation is almost negligible due to the rather small contact surface area between ozone and lignin. On the other hand, when the MC is very high, similar consequences occur because ozone is now absorbed and decomposed in the bulk of the water [22].

The objective of this research was to improve the ozonolysis delignification of humid, nonsubmerged, wheat straw. It was surmised that performing the ozonolysis process in two stages with an intermediate washing step would remove many of the delignification products (lignin fragments, such as carboxylic acids), enabling ozone gas to oxidize newly exposed lignin in wheat straw rather than further degrading the lignin fragments. To the best of the authors' knowledge, no work on this matter has been reported in the literature. Acid insoluble lignin (AIL) content for both untreated and ozonated wheat straw was used as a measure of the effectiveness of the delignification process. Three parameters were studied: the Initial Water Content (IWC) of the humid straw entering first stage, the Washing Starting Time (WST; contact time during the first ozonolysis stage), and the Washing Contact Time (WCT; immersion time of wheat straw in distilled water during the intermediate washing step).

Material and Methods

Materials

Wheat (*Triticum sativum*, Soft White Superior) was harvested from a farm in Ontario, Canada in 2010. Dry bales of straw were milled using a Retsch Cutting Miller type SM 100 (Comeau Technique Ltee/Ltd., Vaudreuil-Dorion, Quebec, Canada) with a 2-mm outlet sieve. The milled wheat straw was stored in sealed plastic bags at room temperature, for

a maximum of 3 months. Working with oak sawdust, Neely [16] stated that the optimum range for IWC should be 25–35 wt. %, while Vidal and Molinier [17] working with poplar sawdust obtained an optimum water content of 70 wt. %. Therefore, the IWC was studied in the range 30–70 wt. % in this research. It was adjusted by mixing the required amount of distilled water to 5 g (oven dry weight) wheat straw humid consistency. The humid straw was transferred right away to the ozonolysis reactor.

A cellulase mixture (NS22086) consisting mainly of endo-glucanase, exo-glucanase and β-glucosidase enzymes, and β-glucosidase (NS22118) were kindly donated by Novozymes Bioenergy [23]. The activity of the cellulase mixture (NS22086) was measured using LAP 009 procedure of NREL [24] and was found equal to 106 FPU/mL.

Ozonolysis

Figure 1 shows a schematic diagram of the ozonolysis reactor set-up for this project. Compressed oxygen gas from a cylinder went through an ozone gas generator (model GL-1; WEDECO, Xylem Water Solutions, Toronto, Ontario, Canada). Ozone concentration in the outlet stream from the ozone generator was measured using OZOCAN analyzer (Ozocan Corporation, Scarborough, Ontario, Canada). A total gas flow rate was set at 1 L/min containing 3 wt. % of ozone. It entered the reactor at the bottom where humid straw had been preloaded. Unreacted ozone in the outlet gas from the top of the reactor and in the bypass streams was destructed by passing the gas through a manganese dioxide catalyst (Ozocat Corporation). The Polytetrafluoroethylene (PTFE) reactor, with a diameter of 3.5 cm and height of 20 cm, was fitted at its bottom with a stainless steel mesh (sieve number 80) and a mesh holder, acting as a holder of the humid straw and as distributor of the ozone/oxygen gas. The top part of the reactor contained a similar arrangement to prevent the straw from moving to the ozone destruction zone.

When ozonolysis was done in two stages, the ozonated straw from the first stage was removed from the reactor after a set WST. It was mixed with 100 mL of distilled water for complete immersion during a set WCT. This step was called the intermediate washing step. The aqueous suspension was then filtered through a glass microfiber filter under vacuum, and the straw was dried at 318 K. It was then either stored in a freezer at 253 K until subsequent enzymatic hydrolysis and/or analyses, or adjusted for IWC before being used in a second ozonolysis stage at the same conditions as the first one. At the end of the second stage, the straw was also filtered, dried, stored, hydrolyzed and/or analyzed.

Total solids (TS), MC, and AIL content and acid soluble lignin (ASL) content in raw and ozonated straw were

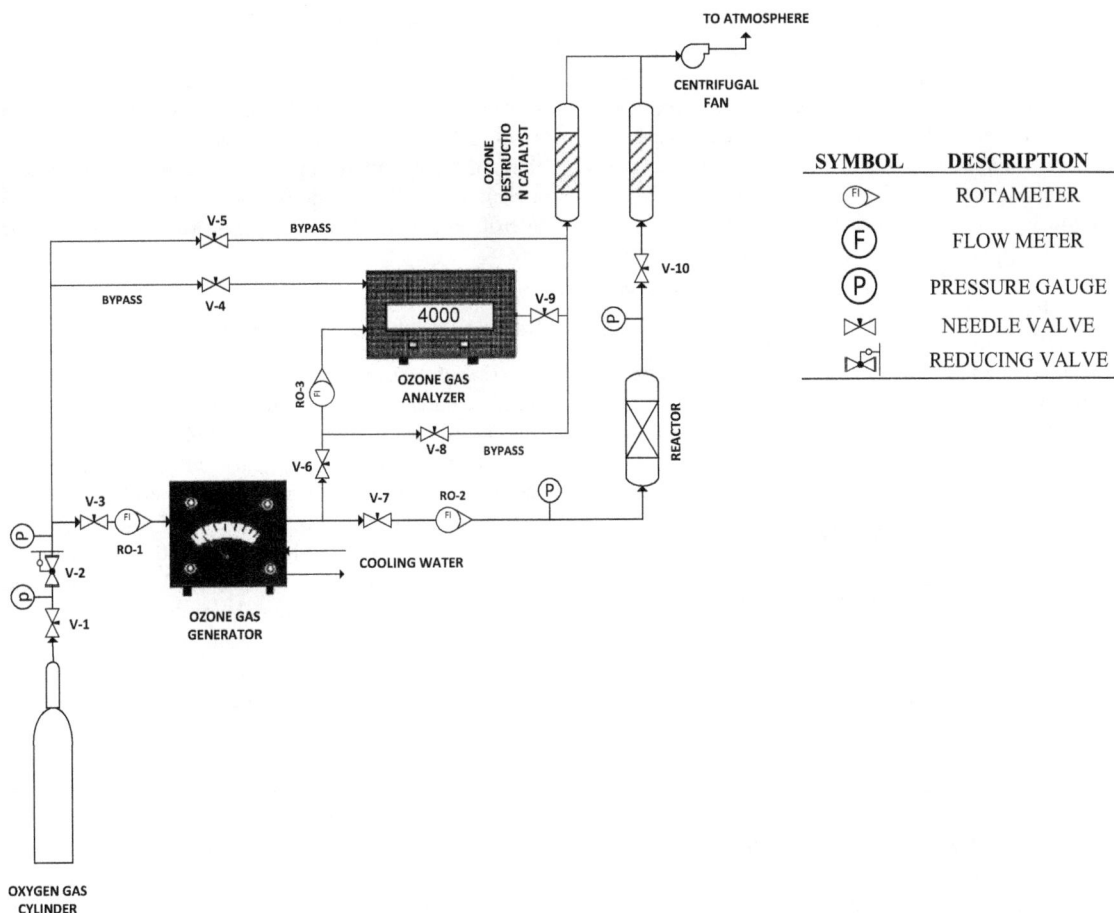

Figure 1. Schematic illustration of the ozone generation and reactor system used in this study. Operating conditions: temperature = 298 K, pressure = 1 atm, reactor volume = 192 cm^3, ozone/oxygen flow rate = 1 L/min and ozone concentration = 3 wt. %.

measured using NREL laboratory procedures LAP 001, 003, and 004, respectively [24]. The values for untreated wheat straw were: TS = 92.31 wt. %, WC = 7.69 wt. %, AIL = 20.50 wt. %, and ASL = 2.30 wt. %.

Design of experiments

Wheat straw with an IWC of 50 wt. % and fiber size <2 mm was ozonated in one stage with a total contact time of 5, 15, 30, 60, 120, or 180 min, respectively. The ozonolysis time that caused significant delignification (low AIL content) without using an excessive amount of ozone gas was chosen as the total reaction time for a one-stage or a two-stage process.

Three parameters were studied on their influence the AIL content of straw. The IWC parameter was evaluated at 30, 50, and 70 wt. %, the WCT was 1, 3, and 5 min, and the WST was 1/3 or 2/3 of the total ozonolysis time

determined above. A mixed-level factorial design (3×2^2) with two center points was used, and experiments were done in a random sequence. STATGRAPHICS® Centurion XV software (Statpoint Technologies, Inc., Warrenton, Virginia, USA) [25] generated an equation predicting the effect from each of the three parameters, and from their interactions, on the AIL content of the ozonated wheat straw. Confidence functions for experimental AIL were calculated at 97.5% probability.

Enzymatic hydrolysis

Enzymatic hydrolysis was performed on untreated and ozone-treated wheat straw at both fiber sizes of <2 mm, according to LAP 009 procedure of NREL [24], using a cellulase mixture (NS22086; 5% wt./wt. dry straw) plus β-glucosidase (NS22118; 0.6% wt./wt. dry straw). Two grams of oven-dried straw were suspended in 250 mL Erlenmeyer flasks in 50 mL acetate buffer 0.1 mol/L (pH 5.25), 1 mL aqueous sodium azide (2 wt. %) and 47 mL distilled water. Flasks were placed in an air incubator at 320 K and 68 rpm. Samples of 1.5 mL of the suspension were taken after 2, 4, 16, 40, 64, 88, 112, 136, and 160 h. They were centrifuged for 5 min at 2000 g, and the supernatant was tested for total reducing sugars (glucose equivalents) using the Dinitrosalicylic acid (DNS) method [26].

The hydrolysis yield compared the amount of reducing sugars experimentally released by the enzymatic hydrolysis of the cellulose and hemicellulose in wheat straw to the theoretical amount of reducing sugars expected to be released from the complete degradation of cellulose and hemicellulose (calculated from values reported by Mckean and Jacobs [27]) of 34 wt. % cellulose and 25 wt. % hemicellulose).

Results and Discussion

Determination of total ozonolysis (contact) time

Figure 2 shows that the AIL content of the ozonated wheat straw dropped rapidly from 20.5 wt. % to 13 wt. % in the first 30 min of a one-stage ozonolysis process but decreased very slowly after 60 min of ozonolysis time. The high rate of delignification in the first 30 min could indicate that most of ozone gas reacted with lignin present on the surface of the wheat straw. As the ozonolysis process continued, ozone gas might have reacted with some lignin decomposition fragments, and it might have become harder for ozone gas to reach lignin in the deep cavities

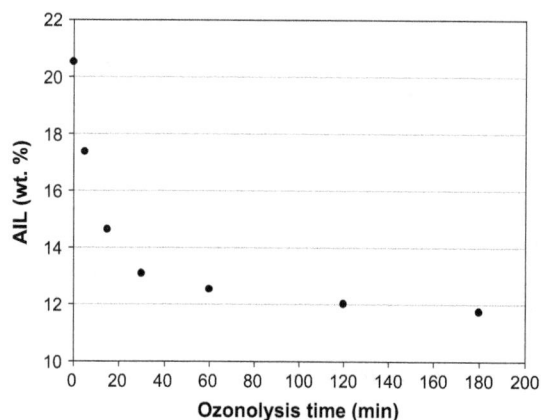

Figure 2. Acid Insoluble Lignin (AIL) content of ozonated wheat straw as a function of ozonolysis time. Five grams of wheat straw (dry weight), with fiber size <2 mm and Initial Water Content (IWC) = 50 wt. %, ozonated in a single stage at O3/O2 flow rate = 1 L/min and ozone concentration = 3 wt. %.

of the wheat straw fibers. The delignification process could also have been slowed down due to the effect of demoisturizing of wheat straw due to the ozone/oxygen gas stream flow. An ozonolysis time of 30 min was chosen as the reference one-stage process and as the total reaction time for any subsequent two-stage process because a significant delignification of wheat straw was achieved at that time without using an excessive amount of ozone gas.

Regression model analysis

Experiments were then performed to determine when to start the intermediate washing step (WST, or length of the first ozonolysis stage) from the above total two-stage contact time of 30 min, and how long the washing step should last (WCT). Instead of a full 3-level factorial design, the STATGRAPHICS® Centurion XV software allowed us to use a 3×2^2 mixed-level factorial design, evaluating WST and WCT in a 2^2 factorial, IWC at 3 levels, and two center points (increasing the number of degrees of freedom of the error to 6) in a minimum of runs. Table 1 shows the result of the 14 runs. Although the lowest average experimental AIL content of ozonated wheat straw after the second stage of the process is shown to occur at runs #8 and 11, when IWC was set at 50 wt. % and WST at 20 min, WCT did not seem to have a substantial effect on delignification because the higher values of the AIL confidence functions were close to 12 wt. %, similar to other experimental AIL values. Increasing the IWC to 70 wt. % drastically reduced the average total delignification (AIL increased to 16 wt. %). Decreasing the IWC to 30 wt. % also reduced total delignification when compared with results of 50 wt. % IWC. Decreasing ozonolysis time of the first stage (WST) from 20 min to 10 or even 15 min, at IWC of 50 wt. %, also reduced total delignification (AIL increased to around 11–12 wt. %). A similar reduction occurred when IWC was set at 70 wt. % but the effect seemed much less profound than when IWC was set at 30 wt. %.

The STATGRAPHICS® Centurion XV software calculated the regression equation which fitted the data for this mixed-level factorial design. It calculated a value of 1.0 for the variance inflation factor for each of the single effects (IWC, WST, WCT) and for the interaction effects (IWC², IWC × WST, IWC × WCT, WST × WCT), indicating a lack of confounding among these effects.

Validation of the DOE results

The analysis of variance (ANOVA) for AIL is shown in Table 2, where the variability in AIL is partitioned

Table 1. Comparison between experimental and calculated (eq. 1) AIL content.

Run #	IWC wt. %	WST min	WCT min	$AIL_{exp} \pm CF^1$ wt. %	AIL_{calc}^2 wt. %	$STDEV^3$ wt. %
1	50	15	3	10.53 ± 1.61	11.97	1.02
2	70	10	5	16.17 ± 2.60	18.49	1.64
3	50	15	3	10.66 ± 1.47	11.97	0.93
4	70	10	1	16.45 ± 2.78	18.94	1.76
5	70	20	5	15.47 ± 2.07	17.32	1.31
6	30	10	5	12.24 ± 0.95	13.09	0.60
7	70	20	1	15.49 ± 1.93	17.21	1.22
8	50	20	1	9.66 ± 2.03	11.47	1.28
9	50	10	1	11.97 ± 0.93	12.80	0.59
10	30	20	1	12.74 ± 0.22	12.93	0.14
11	50	20	5	9.59 ± 2.05	11.42	1.29
12	30	20	5	12.85 ± 0.15	12.72	0.09
13	30	10	1	13.43 ± 0.49	13.86	0.31
14	50	10	5	11.81 ± 0.42	12.19	0.27

[1] Experimental AIL ± Confidence Function at 97.5% probability.
[2] Calculated AIL using Eq. 1.
[3] Standard deviation of AIL_{exp}.

Table 2. Analysis of variance for AIL.

Source	Sum of squares	Df	Mean square	F-ratio	P-value
A:IWC	18.9728	1	18.9728	45.06	0.0005
B:WST	3.27608	1	3.27608	7.78	0.0316
C:WCT	0.216008	1	0.216008	0.51	0.5007
AA	45.7189	1	45.7189	108.59	0.0000
AB	0.31205	1	0.31205	0.74	0.4223
AC	0.07605	1	0.07605	0.18	0.6856
BC	0.226875	1	0.226875	0.54	0.4906
Total error	2.52607	6	0.421012		
Total (corr.)	71.3248	13			

into separate pieces for each of the effects. Three effects had P-values less than 0.05: IWC, the interaction term IWC^2, and WST. This indicated that these terms were significantly different from zero at the 95.0% confidence level. Although the other terms showed to be nonsignificant from the standardized Pareto chart, they were kept in the model because they had a strong interaction on AIL. The correlation matrix for estimated effects showed an almost perfectly orthogonal design, indicating that clear estimates could be obtained for these effects. The coefficient of determination (R^2) statistic indicated that the fitted model explained 96.4% of the variability in AIL. The adjusted R^2 statistic, which is more suitable for comparing models with different numbers of independent variables, was 92.3%. No indication of serial autocorrelation in the residuals occurred at the 5.0% significance level, as was revealed by the Durbin–Watson statistic. Consequently, the fitted

model with values of the variables specified in their original units was:

$$
\begin{aligned}
AIL_{calc} = {} & 29.677 - 0.755 \times IWC - 0.047 \times WST \\
& - 0.394 \times WCT + 0.009 \times IWC^2 \\
& - 0.002 \times IWC \times WST + 0.002 \times IWC \times WCT \\
& + 0.014 \times WST \times WCT.
\end{aligned} \tag{1}
$$

The accuracy of Equation 1 was also validated by comparing the experimental AIL values used to generate this equation to those predicted from it. Table 1 shows that the maximum standard deviation between those two sets of data was 1.76 wt. %.

Figure 3 shows the model predictions of the response surface for the AIL content of ozonated wheat straw when two of the studied parameters were varied from their lowest experimental value to their highest one while the third parameter was fixed at its middle value. When WCT

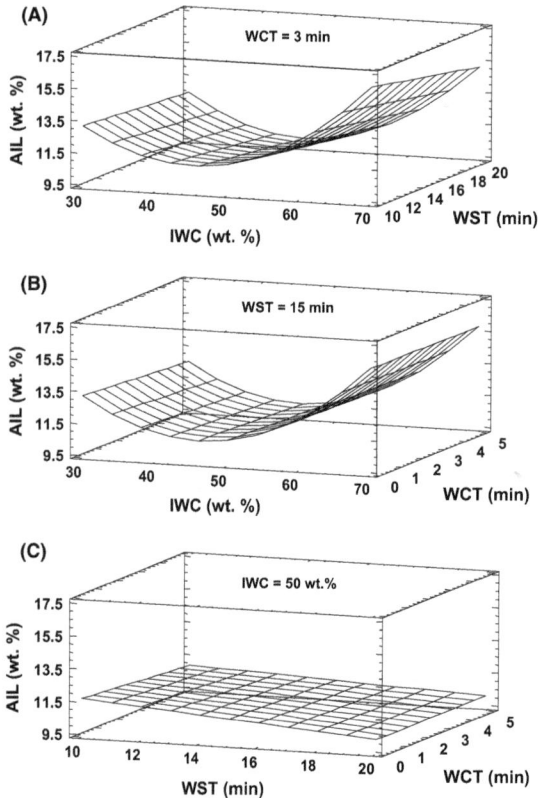

Figure 3. Predicted response surface plot for the variation in AIL content of ozonated wheat straw: (A) IWC and WST while WCT is fixed at 3.0 min, (B) IWC and WCT while WST is fixed at 15 min, (C) WST and WCT while IWC is fixed at 50 wt. %. Five grams of wheat straw (dry weight) with fiber size <2 mm were ozonated in two stages at O3/O2 flow rate = 1 L/min and ozone concentration = 3 wt. %. A 100 mL of washing water was used in the intermediate washing step.

shown). To verify whether increasing WST past 20 min could further decrease the AIL content, an experiment done at the optimum conditions but with WST of 25 min did not result in a substantial AIL decrease (data not shown). These results show significant improvement over those obtained by García-Cubero et al. [18], who reported an AIL content of 11.2 wt. % reached after 2.5 h of a single ozonolysis stage with very similar operating conditions, that is, water content equal to 40 wt. %, ozone/air flow rate equal to 1.5 L/min, ozone concentration of 3 wt. % and wheat straw fiber size of 3–5 mm. The same conclusion was achieved when comparing to results of Bule et al. [19], who achieved an AIL content of 13.0 wt. % after 120 min of a single ozonolysis process using 3 g wheat straw with particle size of 0.25 mm, and with 5.3 wt. % ozone concentration at a flow rate of 2 L/min.

Although this drop in AIL content was considered a significant improvement in the delignification of wheat straw, it was believed that the process could be further improved since the water content of the wheat straw entering the second stage was 72 wt. %, much higher than the optimal value predicted for wheat straw entering the first stage, due to the intermediate washing and filtration step. After adjusting to 45 wt. %, the AIL content of treated wheat straw after the second ozonolysis stage further dropped to 7.36 wt. %. This proves how influential the water content of the straw is on delignification by ozone.

Enzymatic hydrolysis of wheat straw

Enzymatic hydrolysis of the delignified wheat straw after ozonolysis was performed to verify the veracity of the improved process. Figure 4 shows how much sugars were released from untreated and ozonated wheat straw during their hydrolysis by cellulases. Untreated wheat straw (control) showed a steep increase to about 15% of the sugars theoretically present in the straw, in the first 5 h of hydrolysis. This might represent a period during which cellulases access any exposed cellulose surfaces produced during milling of the straw. After that period, cellulases seem to find it harder to reach the cellulose and hemicellulose inside the wheat straw, possibly due to the presence of lignin. A plateau of around 23% theoretical was reached in about 50 h.

When hydrolysis occurred on wheat straw that was delignified at the optimum values of IWC (45 wt. %), WST (20 min), and WCT (80 sec), (called "optimum" on Fig. 4), the sugar yield was about 1.3 times higher than for straw ozonated in one step (called "one step") and four times more than for untreated wheat straw. It demonstrates that a two-stage delignification process with an intermediate washing step was more effective than a

was fixed at 3 min of contact washing (Fig. 3A), maximum delignification (lowest AIL content of 10 wt. %) occurred when IWC was 45 wt. % and WST was 10 min. Figure 3B shows a similar trend in the response surface when WST was fixed at 15 min, proving that the greatest effect on AIL content came from IWC. Although a longer first ozonolysis first step (WST) led to some improvement in delignification, WCT had the smallest effect on the process (Fig. 3C). This might mean that lignin fragments produced by reaction with ozone almost instantaneously diffused from the treated wheat straw to the bulk of the washing water.

The software-calculated values of the parameters for maximum delignification were found to be for IWC at 45 wt. %, WST at 20 min, and WCT at 80 sec. Working at these values, the AIL content dropped from 20.5 wt. % for untreated wheat straw to 9.34 wt. % (data not

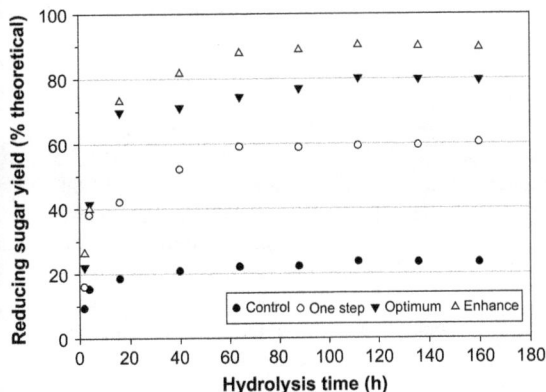

Figure 4. Enzymatic hydrolysis of untreated and ozonated wheat straw.

one-stage process for the same total treatment time because many of the lignin acid fragments are washed away, allowing ozone to attack more of the lignin present in the straw during the second contact stage. The highest sugar yield (around 90% theoretical) occurred when the water content of straw entering the second stage was further adjusted to 45 wt. % (called "enhance"). The fact that a slight drop in AIL content from 9.35 wt. % (for "optimum" straw) to 7.36 wt. % (for "enhance" straw) can result in a 10% increase in sugar yield proves that removing lignin provides better access to the straw cellulose and hemicelluloses.

Conclusion

The ozonolysis pretreatment of wheat straw in two stages, coupled with an intermediate washing step, improved the extent of delignification compared to the commonly used one-stage delignification by ozone. Maximum delignification of wheat straw was achieved with an IWC in the straw of 45 wt. %, a 20-min first stage of ozonolysis (WST), and a washing time (WCT) of 80 sec. When these conditions for a two-stage approach was used, the straw AIL content was reduced from 13.05 wt. % (for 30 min in a one-stage process) to 9.34 wt. %, and the sugar yield of the wheat straw increased from 60% theoretical to 80% theoretical.

The IWC of wheat straw was shown to have the most significant effect on the ozonolysis delignification process compared to the other two parameters, that is, WST and WCT. Adjusting the water content of straw, prior to entering the second stage, to the optimal predicted IWC value of 45 wt. % further improved the delignification process (AIL content decreased to 7.36 wt. %) and the hydrolysis into fermentable sugars (sugar yield) increased to 90% theoretical.

We have successfully proven that 30-min of a two-stage delignification process by ozone, coupled with a quick intermediate washing step and a readjustment of the straw water content to 45 wt. %, significantly reduced the lignin (AIL) content of humid nonsubmerged wheat straw from 13 wt.% to 7 wt. %. Because lignin was no longer present to preferentially bind with the cellulases, the cost of supplying enzymes for the hydrolysis will drastically be reduced. In this improved delignification process, approximately one quarter of the ozone used by a one-stage process would actually be required, representing another huge cost reduction.

Conflict of Interest

None declared.

References

1. Wyman, C. E.. 1999. Biomass ethanol: technical progress, opportunities, and commercial challenges. Annual Review Energy Environment 24:189–226.
2. Wang, M., M. Wu, and H. Huo. 2007. Life-cycle energy and greenhouse gas emission impacts of different corn ethanol plant types. Environ. Res. Lett. 2:1–13.
3. Yat, S. C., A. Berger, and D. R. Shonnard. 2008. Kinetic characterization of dilute surface acid hydrolysis of timber varieties and switchgrass. Bioresour. Technol. 99:3855–3863.
4. Hendriks, A. T. W. M., and G. Zeeman. 2009. Pretreatments to enhance the digestibility of lignocellulosic biomass. Bioresour. Technol. 100:10–18.
5. Broder, J. D., J. W. Barrier, K. P. Lee, and M. M. Bulls. 1995. Biofuels system economics. World Resource Review 7:560–569.
6. Alvira, P., E. Tomás-Pejó, M. Ballesteros, and M. J. Negro. 2010. Pretreatment technologies for an efficient bioethanol production process based on enzymatic hydrolysis: a review. Bioresour. Technol. 101:4851–4861.
7. Sousa, L. D., S. P. S. Chundawat, V. Balan, and B. E. Dale. 2009. Cradle-to-grave assessment of existing lignocellulose pretreatment technologies. Curr. Opin. Biotechnol. 20:339–347.
8. Zheng, Y., Z. Pan, and R. Zhang. 2009. Overview of biomass pretreatment for cellulosic ethanol production. Int. J. Agric. Biol. 2:51–68.
9. Wyman, C. E.. 1996. Handbook on Bioethanol: production and Utilization. Taylor & Francis, Washington, DC.
10. Conde-Mejía, C., A. Jiménez-Gutiérreza, and M. El-Halwagi. 2012. A comparison of pretreatment methods for bioethanol production from lignocellulosic materials. Process Saf. Environ. 90:189–202.

11. Hsu, T. A., M. R. Ladisch, and G. T. Tsao. 1980. Alcohol from cellulose. J. Chem. Technol. Biotechnol. 10:315–319.

12. Kumar, P., D. M. Barrett, M. J. Delwiche, and P. Stroeve. 2009. Methods for pretreatment of lignocellulosic biomass for efficient hydrolysis and biofuel production. Ind. Eng. Chem. Res. 48:3713–3729.

13. Talebnia, F., D. Karakashev, and I. Angelidaki. 2010. Production of bioethanol from wheat straw: an overview on pretreatment, hydrolysis and fermentation. Bioresour. Technol. 101:4744–4753.

14. Ben-Ghedalia, D., and J. Miron. 1981. The effect of combined chemical and enzyme treatment on the saccharification and in vitro digestion rate of wheat straw. Biotechnol. Bioeng. 23:823–831.

15. Ben-Ghedalia, D., and G. Shefet. 1983. Chemical treatments for increasing the digestibility of cotton straw. J. Agr. Sci. 100:393–400.

16. Neely, W. C.. 1984. Factors affecting the pretreatment of biomass with gaseous ozone. Biotechnol. Bioeng. 26:59–65.

17. Vidal, P. F., and J. Molinier. 1988. Ozonolysis of lignin - Improvement of in vitro digestibility of poplar sawdust. Biomass 16:1–17.

18. García-Cubero, M. T., G. González-Benito, I. Indacoechea, M. Coca, and S. Bolado. 2009. Effect of ozonolysis pretreatment on enzymatic digestibility of wheat and rye straw. Bioresour. Technol. 100:1608–1613.

19. Bule, M. V., A. H. Gao, B. Hiscox, and S. Chen. 2013. Structural modification of lignin and characterization of pretreated wheat straw by ozonation. J. Agr. Food Chem. 61:3916–3925.

20. Wu, J., S. Upreti, and F. Ein-Mozaffari. 2013. Ozone pretreatment of wheat straw for enhanced biohydrogen production. Int. J. Hydrogen Energy 38:10270–10276.

21. Contreras, S.. 2002. Degradation and biodegradability enhancement of nitrobenzene and 2,4-dichlorophenol by means of advanced oxidation processes based on ozone. PhD Thesis, University of Barcelona, Spain.

22. Mamleeva, N. A., S. A. Autlov, N. G. Bazarnova, and V. V. Lunin. 2009. Delignification of softwood by ozonation. Pure Appl. Chem. 81:2081–2091.

23. Novozymes Bioenergy. 2012. Enzymes for the hydrolysis of lignocellulose materials 2012. Available at http://www.bioenergy.novozymes.com/en/cellulosic-ethanol/samples/Documents/Application_Sheet_Cellulosic%20ethanol_enzyme_kit_Final.pdf. (Accessed 01 May 2012).

24. National Renewable Energy Laboratory (NREL). 1995. Chemical Analysis and Testing Laboratory Analytical Procedures 1995. Available at http://infohouse.p2ric.org/ref/40/39182.pdf. (Accessed 10 September 2011).

25. Stat Point Technologies Inc. 2011. Statgraphics Centurion XV, 2011. Available at http://www.statgraphics.com/downloads.htm. (Accessed 02 July 2011).

26. Ghose, T. K.. 1987. Measurement of cellulase activities. Pure Appl. Chem. 59:257–268.

27. MckeMckean, W. T., and R. S. Jacobs. 1997. Wheat Straw as a Paper Fiber Source. The Clean Washington Center, Washington.

Permissions

List of Contributors

Abdul Hai Alami
Center for Advanced Materials Research, University of Sharjah, PO Box 27272 Sharjah, United Arab Emirates
Sustainable and Renewable Energy Engineering Department, University of Sharjah, PO Box 27272 Sharjah, United Arab Emirates

Jehad Abed, Meera Almheiri, Afra Alketbi and Camilia Aokal
Sustainable and Renewable Energy Engineering Department, University of Sharjah, PO Box 27272 Sharjah, United Arab Emirates

Michael D. Kempe, David C. Miller, John H. Wohlgemuth, Sarah R. Kurtz and John M. Moseley
National Renewable Energy Laboratory, 15013 Denver West Parkway, Golden, Colorado 80401

Qurat A. Shah and Govindasamy Tamizhmani
Arizona State University Photovoltaic Reliability Laboratory, 7349 East Unity Avenue, Mesa, Arizona

Keiichiro Sakurai
National Renewable Energy Laboratory, 15013 Denver West Parkway, Golden, Colorado 80401
National Institute of Advanced Industrial Science and Technology, 1-1-1 Umezono, Tsukuba, Ibaraki 305-8568, Japan

Masanao Inoue, Takuya Doi and Atsushi Masuda
National Institute of Advanced Industrial Science and Technology, 807-1, Shuku-Machi, Tosu, Saga 841-0052, Japan

Sam L. Samuels
DuPont Company, 200 Powder Mill Road, Wilmington, Delaware 19803

Crystal E. Vanderpan
Underwriters Laboratories, 455 East Trimble Road, San Jose, California

Anton Eberhard
Graduate School of Business, University of Cape Town, Private Bag X3, Rondebosch 7701, South Africa

Tomas Kåberger
Energy and Environment, Chalmers University of Technology, SE-412 96 Gothenburg, Sweden

Arne Gladis, Pia-Maria Bondesson, Mats Galbe and Guido Zacchi
Department of Chemical Engineering, Lund University, P.O. Box 124, SE-221 00 Lund, Sweden

Bongghi Hong
Department of Ecology and Evolutionary Biology, Cornell University, 103 Little Rice Hall, Ithaca, New York 14853

Robert W. Howarth
Department of Ecology and Evolutionary Biology, Cornell University, E309 Corson Hall, Ithaca, New York 14853

Georgia Apostolou and Martin Verwaal
Design for Sustainability, Faculty of Industrial Design Engineering, Delft University of Technology, Landbergstraat 15, 2628CE Delft, The Netherlands

Angèle Reinders
Design for Sustainability, Faculty of Industrial Design Engineering, Delft University of Technology, Landbergstraat 15, 2628CE Delft, The Netherlands
Department of Design, Production and Management, Faculty of Engineering Technology, University of Twente, P.O. Box 217, 7500AE Enschede, The Netherlands

Michael D. Kempe, David C. Miller, John H. Wohlgemuth, Sarah R. Kurtz, John M. Moseley and Dylan L. Nobles
National Renewable Energy Laboratory, 1617 Cole Boulevard, Golden, Colorado 80401

Katherine M. Stika, Yefim Brun and Sam L. Samuels
DuPont Company, 200 Powder Mill Road, Wilmington, Delaware 19803

Qurat (Annie) Shah and Govindasamy Tamizhmani
Polytechnic Campus, Arizona State University, 7349 East Unity Avenue, Mesa, Arizona

Keiichiro Sakurai, Masanao Inoue, Takuya Doi and Atsushi Masuda
National Institute of Advanced Industrial Science and Technology, 1-1-1 Umezono, Tsukuba, Ibaraki 305-8568, Japan

Crystal E. Vanderpan
Underwriters Laboratories, 455 East Trimble Road, San Jose, California

Johannes Hofer, Prageeth Jayathissa, Zoltan Nagy and Arno Schlueter
Architecture and Building Systems, Institute of Technology in Architecture, ETH Zurich, John-von-Neumann Weg 9, 8093 Zürich, Switzerland

Abel Groenewolt
Institute for Computational Design, University of Stuttgart, Keplerstrasse 11, 70174 Stuttgart, Germany

Yoshihiro Nishikawa, Kiyoshi Imai and Keiji Miyao
KONICA MINOLTA, INC., 3-91 Daisen Nishi-machi, Sakai-ku, Sakai City, Osaka 590-8551, Japan

Satoshi Uchida
The University of Tokyo, 4-6-1 Komaba, Meguro-ku, Tokyo 153-8904, Japan

Daisuke Aoki, Hidenori Saito, Shinichi Magaino and Katsuhiko Takagi
Kanagawa Academy of Science and Technology, 3-2-1 Sakado, Takatsu-ku, Kawasaki-shi, Kanagawa 213-0012, Japan

Peter Ahlström and Tobias Richards
Swedish Centre for Resource Recovery, University of Borås, 501 90 Borås, Sweden

Francis Chinweuba Eboh
Swedish Centre for Resource Recovery, University of Borås, 501 90 Borås, Sweden
Department of Mechanical Engineering, Michael Okpara University of Agriculture, Umudike, Abia, Nigeria

Jinto M. Anthonykutty, Juha Linnekoski and Ali Harlin
VTT Technical Research Centre of Finland, Biologinkuja 7, Espoo, FI-02044 VTT, Finland

Juha Lehtonen
Department of Biotechnology and Chemical Technology, School of Chemical Technology, Aalto University, PO Box 16100, Aalto FI-00076, Finland

Paul Brack, Sandie E. Dann and K. G. Upul Wijayantha
Energy Research Laboratory, Department of Chemistry, Loughborough University, Loughborough, Leicestershire LE11 3TU, United Kingdom

Paul Adcock and Simon Foster
Intelligent Energy Ltd, Charnwood Building, Holywell Park, Ashby Road, Loughborough, Leicestershire LE11 3GB, United Kingdom

Kozo Fujiwara
Institute for Materials Research (IMR), Tohoku University, Katahira 2-1-1, Aoba-ku, Sendai 980-8577, Japan

Yukichi Horioka
Frontier Technology Business Research Institute Co. Ltd., Yamazaki Umenodai 11-9, Noda 278-0024, Japan

Shiro Sakuragi
Union Materials Inc., Oshido 1640, Tone-machi, Kitasoma-gun, Ibaraki 300-1602, Japan

Venu Babu Borugadda and Vaibhav V. Goud
Department of Chemical Engineering, Indian Institute of Technology Guwahati, Guwahati, Assam 781039, India

Touché Howard
Indaco Air Quality Services, Inc., Durham, North Carolina

Da Deng
Department of Chemical Engineering and Materials Science, Wayne State University, Detroit, Michigan 48202

Ahmadreza Ghahremani and Aly E. Fathy
Electrical Engineering Department, University of Tennessee Knoxville, Knoxville, Tennessee

Ali Kamel H. Al jibouri, Ginette Turcotte, Jiangning Wu and Chil-Hung Cheng
Chemical Engineering Department, Ryerson University, 350 Victoria Street, Toronto, Ontario M5B 2K3, Canada

Index

www.ingramcontent.com/pod-product-compliance
Lightning Source LLC
Chambersburg PA
CBHW061937190326
41458CB00009B/2763

9 781632 398963